U0275583

寰宇文献 Universal Library | SINOLOGY 系列

SELECTED WORKS OF BERTHOLD LAUFER

劳费尔著作集

第六卷

[美] 劳费尔 著

黄曙辉 编

 中西书局 ZHONGXI BOOK COMPANY

图书在版编目（CIP）数据

劳费尔著作集 /(美) 劳费尔著；黄曙辉编. —上
海：中西书局，2022

(寰宇文献)

ISBN 978-7-5475-2015-4

Ⅰ. ①劳… Ⅱ. ①劳… ②黄… Ⅲ. ①劳费尔 – 人类
学 – 文集 Ⅳ. ①Q98-53

中国版本图书馆CIP数据核字（2022）第207067号

第 6 卷

084

中国可以照料自己

The Oriental Review

VOLUME II   AUGUST, 1912   NUMBER 10

Published monthly by the Oriental Information Agency, No. 35 Nassau St., New York, N. Y.
Yearly subscription One Dollar and Fifty Cents. Single Copy, Fifteen Cents
MASUJIRO HONDA, Editor; TSUNEGO BABA, Associate Editor;
MOTOSADA ZUMOTO, Director and Contributing Editor (Tokio, Japan).
"Entered as second-class matter February 14, 1912, at the post office at New York, New York, under the Act of March 3, 1879."

CONTENTS

Postage is prepaid on subscriptions in the United States and its Possessions, Canada and Mexico. Two Dollars is the regular subscription price to other countries.

Remittances should be made by Draft on New York, Express Order, or Postal Money Order, payable to The Oriental Review. Currency, unless mailed in a registered letter, is at the sender's risk. Notice of change in address should be sent Two Weeks before the date of issue on which the change is to take effect. The change cannot be made unless the subscriber's Old Address is clearly indicated in addition to the New Address. All manuscripts, drawings, and photographs are received with the understanding that the Editors are not responsible for their loss or injury while in their possession or in transit. Return postage should be inclosed with each manuscript submitted, and a copy should be retained by its author.

The Oriental Information Agency was established in August, 1909, with the support of the leading financiers and merchants of Tokio and Yokohama, for the purpose of disseminating in the United States and elsewhere, reliable information concerning affairs in Japan and the Far East in general, with particular reference to the financial and economic situation in that part of the world. It is hoped and believed that the spreading of such knowledge will materially contribute to the promotion of intimate relations and incidentally also of friendship and good-will between the nations of the East and West.

CHINA CAN TAKE CARE OF HERSELF
By Berthold Laufer, Ph.D.

Dr. Laufer, prominent as an anthropologist, was born in 1874 in Cologne, Germany, and came to the United States in 1898. He has been a member of expeditions to China, Russia, and Saghalien. He is the author of Contributions to Popular Religion of Tibet, 1898; The Decorative Art of the Amur Tribes, 1902; Boas Anniversary Volume (edited) 1906; Chinese Pottery of the Han Dynasty, 1909; Romance of a Tibetan Queen; Chinese Grave Sculptures; Jade, a Study in Chinese Archæology and Religion, 1911. Also numerous papers and reviews relating to Buddhism, Tibetan literature, philology, ethnology and archæology in English, German, and French.

As a believer in Chinese ideals and as a champion of China's rights, my opinion briefly is—The Chinese were able to govern themselves before the Six Powers came into existence. China has tremendous intellectual resources and can perfectly take care of herself. Most Powers could learn a great deal from her in matters of practical ethics, economic principles and state wisdom; not to speak of art, art-industrial skill, and the great art of living. The Powers should mind their own business at home and mend the holes in their own stockings. China for the Chinese, and leave her alone. I hope to live to the day when the white peril will be wiped out from Asia and Africa. There is no shadow of a doubt that the Chinese republic will establish itself with success and will rank as the second greatest republic of the world. The Chinese are born democrats gifted with a keen sense for decency, justice and equality, and with a marvelous talent for organization. The present period may certainly be one of transition, but we must give the new government a chance and fair play, and suspend definite judgment for several years.

In case the republic should fail, China will always remain China. For over twenty years I have had unbounded confidence in China's future and shall never lose this confidence. I should, rather, be tempted to the belief that any of the Six Powers might crumble to pieces than China. China will always exist, either in this or some other form of governmental constitution, and will take a lead in the history of the world as an influential, civilizing agency. China is the land of the future, and the future is hers. It may be that the future will even impart to the world at large a higher and broader conception of life emanating from a rejuvenated China.

The best solution of the Chinese situation will come from China herself, as long as the Powers keep their greedy hands off. Japan should not be hostile to the new republic, but welcome it heartily and enter with it into a close alliance and amity. The aggressive anti-Chinese policy followed by Japan in Manchuria for the last years was a gross blunder. Japan, thus, forgot her own historical role and incurred the animosity of all Asiatic peoples. Japan must stand up for the integrity of China and join hands with her against the white man's depredatory aggressiveness. Learn from the Panislamic Movement and the Arabs! The Powers rule the world not by means of the intellectual and moral superiority which they by no means possess, but because of the discord among the nations of Asia and Africa. The storm-centre for the peace of the world is not China, but Europe. The best means for the preservation of universal peace must be sought in a counterpoise against the white peril, in a common understanding and federation of all native races of Asia and Africa against the insane encroachments of the infidels. Peoples of Asia and Africa, guard your most sacred ideals!

A SITUATION FULL OF DANGER
By Chester Lloyd Jones
Associate Professor of Political Science in the University of Wisconsin.

1. Are the Chinese able to govern themselves? At present the Chinese are in no position to undertake self government in the way this term is understood in the Western world. The introduction of Western education has as yet only touched the surface and has by no means fitted the average Chinese for the duties of citizenship under a Republican form of government. When we consider that single provinces of China have a population greater than that of the United States at the time of the Civil War, it is evident that in any case even after the Chinese have introduced a good system of education and have experimented for a long time with Republican institutions, there must still be great handicaps under which the government must work. Electoral machinery would be put through a strain heretofore unapproached if so large a country were to attempt the erection of a government to be controlled by party organization.

2. Will the Chinese Republic be successful? Because of the limitations I have already noted, it seems clear that no Chinese Republic can be successful at present if by a Republic a government such as the United States or Switzerland is understood. It might be possible to continue the government under a group of officials who would not be in a position strongly contrasted to that held by the former dynasty. If the new government attempts to disturb the long established régime of local government, I feel sure that trouble will be inevitable. China will advance by a slow adoption of Western methods and ideas.

3. In case the Republic fails, what will become of China? This no man can tell. The future of China rests on the decision of European Powers. It is now independent only because they are in disagreement. If they should come to be of one mind, or if any one group of European Powers became in-

085

婆罗多的诗颂

OCTOBER,   1912.

THE

JOURNAL

OF THE

ROYAL ASIATIC SOCIETY

OF

GREAT BRITAIN AND IRELAND.

OCTOBER 15.

October   **1912.**

PUBLISHED BY THE SOCIETY,

22 ALBEMARLE STREET, LONDON, W.

Price Twelve Shillings.

often infinitely greater than that between the *Ṛtusaṃhāra* and the other three Kāvyas ascribed to Kālidāsa. The differences between the *Meghadūta* and the *Ṛtusaṃhāra* are legitimately interesting as traces of poetic development, but they have no value as evidence for difference of authorship.

A. BERRIEDALE KEITH.

THE STANZAS OF BHARATA

In the *Mélanges d'Indianisme offerts par ses élèves à M. Sylvain Lévi,* Professor Edward Huber presents a brief paper under the title " Sur le texte tibétain de quelques stances morales de Bharata ", in which he makes an interesting attempt at shedding light on some obscure passages in this difficult text by consulting the Chinese translation of Yi-tsʻing. In criticizing Schiefner's rendering of this work, M. Huber exclusively refers to his translation which appeared in the *Memoirs of the Petersburg Academy* (vol. xxii, No. 7, 1875), but unfortunately overlooked the fact (though it is expressly indicated in the preface to this memoir, p. vii) that Schiefner has edited also the Tibetan text of this work with a Latin translation and a valuable glossary (*Bharatae Responsa tibetice cum versione latina ab Antonio Schiefner edita,* Petropoli, 1875). If M. Huber will look up this edition, he will no doubt recognize that this is a piece of thorough and creditable work which commands respect. The text is critically and carefully edited from a collation of the Kanjur prints of Narthang and Peking, and the Arabic text of Kalila and Dimna has also been utilized. M. Huber on his part availed himself of a copy of the Tibetan text made for him by a Mongol Lama in Peking after the Peking edition of the Kanjur, a copy which in all likelihood is bound to be less reliable than the edition of Schiefner.

He who is intent on furthering the understanding of

this work must take regard of a good many other things. It is known that the story of Bharata has become part and parcel of Tibetan folk-lore, and that several entirely different versions of it are in existence. Thus far three of these popular versions have been published. One under the title "The Ulūkasūtra" has been translated from a manuscript of the India Office Library by A. Schiefner in the *Mélanges asiatiques*, vol. viii, pp. 635–640 (St. Petersburg, 1879); the relations of this text to the Replies of Bharata are pointed out by him on p. 624. Another more vulgar version entitled "Hā-shang-rgyal-po and Ug-ṭad (i.e. ཨུག་སྒྲོང་), a Dialogue", translated from the Tibetan by Karl Marx, was published in JASB., vol. lx, pt. i, No. 2, pp. 37–46, 1891. Thirdly, a Tibetan text under the name རྒྱལ་བློན་གྱི་བསྟན་བཅོས་ "Çāstra of the King and the Minister" is printed in the *Tibetan Reader*, No. v, edited by Lama T. Ph. Wangdan (Darjeeling, 1898); here the Indian king, an incarnation of Māra, is called Ha-shaṅ-deva, and the minister who effects his conversion is Buddha himself transformed into an owl. Substantially, this version differs from those of Schiefner and Marx, and quite naturally, as the comical answers of the minister allow of an almost endless variation. In WZKM., vol. xiii, p. 223, I briefly alluded to a possible connexion of the Bharata series with our stories of *Eulenspiegel*; indeed, Bharata or the minister Owl (Ulūka) is in his very jokes the prototype of our *Eulen* (Owl)-*spiegel*. The three versions here mentioned have not yet been compared; of the text translated by Marx, I possess four manuscripts. But one important conclusion can be reached that, in view of the numerous variations and deviations of these texts, there is a high degree of probability that also a plurality of original Sanskrit versions of this story has existed. If this, however, was the case, it is not necessary to assume that the Tibetan and Chinese translations were made from exactly the same Sanskrit text, which seems improbable also for the reason that the two translations are separated by a long space of time. M. Huber takes it for granted that both versions have emanated from the same original, and therefore seeks the meaning conveyed by the Chinese stanzas also in the corresponding Tibetan verses. This procedure may certainly prove correct in many cases, but it must not be so in all cases. It cannot be made a general principle, as it is always possible that the Tibetan translator had a different Sanskrit wording before his eyes or interpreted the passage at variance with the Chinese translator. Under no circumstances, however, must the meaning, yielded by the Chinese phrases, be forced into the Tibetan, if it cannot naturally be deduced from the Tibetan sentence. While I gladly admit that M. Huber has largely improved on the translation of the two last stanzas quoted by him on pp. 309[1] and 310 and readily accept his result, I fail to see that his new translation on p. 307 can be deduced from the Tibetan text: nothing is there to justify the translations: " A l'improviste châtient les rois, . . . à l'improviste surviennent les bonnes aubaines." Schiefner's translation certainly is here capable of improvement; the last verse should be: "The monk ought not to think of gain." It is quite manifest that in this case the Chinese and Tibetan translations do not follow the same Sanskrit model.

M. Huber (p. 309) is quite right in attacking Schiefner's translation of *rtsa mjiṅ* རྩ་མཛིང་ by "meadow", but he is not very fortunate in the explanation of the term. "*Rtsa* signifie 'ami, parent' (bandhu) et *mjiṅ* 'cercle' (*varga, maṇḍala*). Il y a donc: 'le riche qui a peu d'amis.'" There is no word *rtsa* in the Tibetan language with the meaning of friend; *rtsa* means root, and there

---

[1] A different reading of the same stanza is quoted by Sarat Chandra Das, Tibetan-English Dictionary, p. 50*a*, which may serve as additional evidence for the existence of various versions of the text.

is a compound *rtsa lag* (*lit.* root and hands, i.e. root
and branches) which assumes the meaning of relations,
friend, usually in a Buddhist sense (= *upāsaka*). The
Tibetan–Mongol Dictionaries render it by Mongol *üri-
sadu* or *orok-sadu* (*sadu* from Skr. *sādhu*). The compound
rtsa mjiñ is simply a synonym of *rtsa lag*, and is
explained in the Dictionary *Zla-bai Od-snañ*, "the
Moonlight" (printed 1838 at Peking, fol. 95*b*), as *ünär
sadu*, "a true friend"; the literal translation of the
phrase is "the pith of the root". For the rest, the
word "rich" suggested to M. Huber by the Chinese text
only is not contained in the Tibetan; the phrase *rtsa mjiñ
c'uñ* simply means "one who has few friends".

Finally, I should like to express the wish that M. Huber
would give us a complete translation of Yi-ts'ing's text.
The work has a certain importance for the history of
folk-lore; in my opinion the jokes of Bharata must be
interpreted as riddles, the solution of which is unfortunately
placed first. If his sentences are put as queries, we obtain
veritable riddles, and it is this very feature which has
been so pleasing to the Tibetans and accounts for the great
popularity of the book in Tibet.

BERTHOLD LAUFER.

VISISTADVAITAM

The word *viśiṣṭādvaitam* is strangely mistranslated
"qualified monism". This phrase is scarcely intelligible,
and in any case does not express the fundamental teaching
of Rāmānuja. *Viśiṣṭādvaitam* is *viśiṣṭayor advaitam*,
"the identity of the two *viśiṣṭas*." *Viśiṣṭa* means
"substantive" as opposed to *viśeṣaṇa*, "adjective." Brahma
is *viśiṣṭa*; and Cit (individual souls) and Acit (matter)
are as *viśeṣaṇa* to him. Now Brahma exists in two
states, viz. in the *kāraṇāvasthā* during the periods of
dissolution, when Cit and Acit exist in a subtle (*sūkṣma*)
condition as his body, and in the *kāryāvasthā* during the

086

中国石棺

OSTASIATISCHE ZEITSCHRIFT

THE FAR EAST   L'EXTRÊME ORIENT

BEITRÄGE ZUR KENNTNIS DER KULTUR UND KUNST DES FERNEN OSTENS

HERAUSGEGEBEN

VON

OTTO KÜMMEL UND WILLIAM COHN

1912/1913
ERSTER JAHRGANG

OESTERHELD & CO. · VERLAG · BERLIN W. 15

CHINESE SARCOPHAGI. BY BERTHOLD LAUFER.

The first European author to make a communication on the occurrence of stone sarcophagi in China, as far as I know, was a surgeon of the British Army, J. LAMPREY who, in 1867, addressed a meeting of the Royal Institute of British Architects on the subject of Chinese architecture. In accordance with the nature of a lecture, the theme is not treated exhaustively nor sounded to a great depth, but intelligently presented in a large variety of topics illustrated by good sketches. "On removing some of the earthen mounds in the vicinity of the French Concession at Shanghai," LAMPREY[1] relates, "several stone coffins, or sarcophagi, were discovered. They were sufficiently large to contain a massive wooden coffin. They were hewn out of a large block of reddish sandstone grit, and had a granite slab for a lid which was made to fit closely by means of a groove in the lid, which received a corresponding ridge on the inner edge of the opening of the sarcophagus. There was no inscription by which the date could be discovered, or to show to whom the sarcophagus belonged." The measurements are not given, but from the illustration to which the figure of an erect man is added it appears that the sarcophagus placed on the ground reaches in height the waist of a man and in length a good man's size. It is, properly speaking, a mere stone chest, of rectangular shape, with straight and even walls entirely undecorated, the lid being flat in correspondence with the box-like character of the object. There are no indications pointing to a particular period, and as a type, it stands by itself, unless it is identical with the sarcophagi seen by Torrance in the caves of Sze-ch'uan to be mentioned later.

The next, in point of time, with a similar report, is E. COLBORNE BABER[2] who in a mound in the vicinity of Ch'ung-k'ing, Sze-ch'uan Province, raised a sandstone slab, about seven feet by two and a half. It proved to be the lid of a rude sarcophagus containing nothing but wet mould which may have drained in through ill-closed chinks, or have been deposited by previous desecrators. "But any case", the author adds, "the sarcophagus lies too near the surface to warrant the inference that it has ever housed a corpse; it is more probably *a blind* to divert curiosity from the situation of the true coffin, which may be expected to repose in some more recondite part of the tumulus." This explanation is, of course, the outcome of a subjective rationalism, and like all rational explanations, wrong, because the life of peoples

---

[1] *Transactions of the Royal Institute of British Architects*, 1867, p. 172.

[2] Travels and Researches in Western China (*Royal Geographical Society, Supplementary Papers*, p. 129, London, 1886).

outside of our culture-sphere is not rationalistic, or is even entirely beyond the pale of our mode of reasoning.

The first illustrations of artistic sarcophagi came to us from the Russian Orkhon Expedition, the remarkable results of which are published in W. Radloff's "Atlas der Altertümer der Mongolei" (St. Petersburg, 1892—99). A number of sarcophagi of Chinese workmanship were discovered in southern Mongolia which had been utilized by the ancient Tu-küe (Turks) in the seventh and eighth centuries. We know from the contents of the inscription that Chinese artisans sent by the Emperor to the court of the Turkish Khans were engaged in carving their inscription stones and executing sculpture-work in their tombs.[1]

A clear description of the sarcophagi is unfortunately not furnished, and nothing is said regarding their dimensions, which is a feature of great importance, nor their contents[2]. From the drawings reproduced in the "Atlas" on Plates XII, Fig. 6, XIII, Fig. 1, and CV, Fig. 1 and 2 it appears that these coffins were composed of single stone slabs, rectangular on the bottom and the two main sides, square on the smaller lateral sides. On Plate XII, Fig. 4, another sarcophagus with flat lid is outlined, decorated with cloud ornaments of Chinese style. Despite this manifest Chinese influence in design, and despite the historical fact that these coffins were, in all probability, worked by Chinese lapidaries, it is a debatable question whether they present, taken in their function as a type of sarcophagus, a really Chinese sarcophagus. In their outward form, they have the appearance of a rectangular chest, while the contemporaneous sarcophagi of the T'ang period, as far as observed in China, are imitations in stone of the well-known form of the Chinese wooden coffin, as will be seen below; further, the latter are always carved out of one mass of stone and never joined together from stone slabs. As early as in the Siberian bronze age we find graves lined with rudely hewn stone slabs, i. e. cists (Kistengräber), which in the case of children's bodies sometimes resemble veritable coffins.[3] It seems to me that the stone coffins of the Khans and princes of the Tu-küe are a direct offspring of these ancient indigenous cists and present a sort of missing link between the cist and the sarcophagus proper; they were more adapted to the latter form doubtless under Chinese influence to which the solid execution and the decorative by-play are due, but in their shapes and in their composition from single slabs they have unmistakably retained a reminiscence of their former origin.

---

[1] W. RADLOFF, Die alttürkischen Inschriften der Mongolei, III, pp. 447, 459 (St. Petersburg 1895); W. BARTHOLD, Die historische Bedeutung der alttürkischen Inschriften, pp. 12 bis 13 (St. Petersburg, 1897).

[2] The Tu-küe seem to have had various methods of burial, perhaps varying at different times. The *Wei shu* ascribes to them burial of the corpse in a pit, the *Sui shu* states that they placed the body on a horse, cremated it and gathered the ashes for burial (JULIEN, Documents historiques sur les Tou-kioue, pp. 10, 28. Paris, 1877).

[3] W. RADLOFF, Aus Sibirien, Vol. II, pp. 78, 79 (Leipzig, 1884).

22

Fig. 1. Marble Sarcophagus, T'ang Period. Dated 673 A. D. Collection of Field Museum, Chicago.

Fig. 1 represents a sarcophagus obtained by me at Si-ngan fu in April 1910 and now in the Field Museum, Chicago. It is carved from a kind of marble (crystalline limestone forming on the unpolished surface of the interior layers of finely brillant crystals) in two separate pieces, — the receptacle with rectangular, slightly projecting base hewn out of one bowlder, and the slanting, rounded lid. It measures only 71,2 cm in length, and in width 39 cm at the upper and 33 cm at the lower end; its height amounts to 31 cm at the upper, and to 24 cm at the lower end. The bottom or base is of unequal thickness, being 15,5 cm at the upper, and 8,5 cm at the lower end, while the four upright walls are on an average from 5 to 6,5 cm thick. The hollow space left in the interior, therefore, is only 62,5 cm in length, 25,5 cm in width and 15,5 cm in height at the upper end. It is self-evident that this sarcophagus could not have been used to shelter the corpse of an adult, while it could possibly lodge a child's body; but let it be stated right here that on discovery it harbored no remains whatever, neither a bone nor ashes nor any object or traces thereof, nor is any effect of such objects on the surface of the stone in the interior visible, which might be

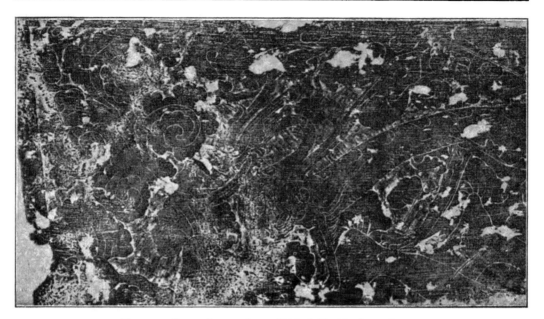

Fig. 2a. Engraving on Long Side of Marble Sarcophagus.

expected, if anything of that kind had ever been the case. The conclusion is therefore warranted that it has ever been empty since the day of its burial, and this state of affairs, as will be seen presently, agrees with the historical evidence furnished by the inscription carved into the lid. Only the four outer walls, the surface of the lid, and the edges of the receptacle and the lid have been smoothed and polished; the lid must formerly have been more tight-fitting than is true now after the wear and tear of time. The exterior of the bottom is very crudely and unevenly hewn out, and on the inner side of the lid and in the interior, the tool-marks of the lapidary running in parallel grooves are plainly visible. Instead of being polished, these portions have been smeared over with a white plaster which at the time when the work was done no doubt produced a plain surface, but which has gradually given way.

The four outer walls are decorated with designs not incised with a chisel but finely traced by means of a burin or graving tool. The lines just touch the surface without any depth and lay open the brillant white color of the marble. This process of engraving in stone is very frequently practised in the T'ang period and may be seen on numerous tombstones and other sculptures in our collection. But this work is so delicate that it defies all efforts of the camera. For this reason, the long side shown in Fig. 1 is once more reproduced from a paper rubbing in Fig. 2, and one of the small sides, that on the upper end, in Fig. 3. On each of the long sides, the main figure is a winged dragon with body long stretched out, a cluster of spirals surrounded

22*

Fig. 2b.   Engraving on Long Side of Marble Sarcophagus.

by cloud patterns emerging from its jaws. It is interesting to note that the heads of both these dragons are on the upper and higher side of the coffin, *i. e.* directed toward the head of the corpse, if the corpse were buried in it. The two front feet of the dragon show the form of mammal feet, the two hind feet distinct bird-claws. The designs are partially obscured by hardened layers of carbonate of lime which strongly adhere to the surface, but they are very clear and secure in composition and of great beauty. The subject on the small side (Fig. 3) is a phenix soaring in the clouds. Both themes are emblematic of death and resurrection.

The principal question now arises, — what was the purpose of this sarcophagus? The lid is fortunately covered with a brief inscription of six lines from which a satisfactory answer may be read.[1] The inscription is dated " on the fifteenth day of the twelfth month of the fourth year of the period *Hien hêng*" yielding the date 673 A. D. (in the T'ang period). It was composed by, or at the instigation of a certain merchant named *Fu Pao* 伏 保, a native of *K'ien-fêng* 乾 封 *hien* (modern K'ien chou) in Shensi Province. He styles himself a filial son 孝 子 who gratefully remembered the benefits received from his parents and kept in mind how well they had deserved

---

[1] The inscription is not reproduced here in its entirety, because its facsimile would require a transcript in modern characters and a lengthy palaeographical and philological discussion which would be out of place in this connection. It will be published in time together with several hundred other inscriptions on stone sculpture in our collection.

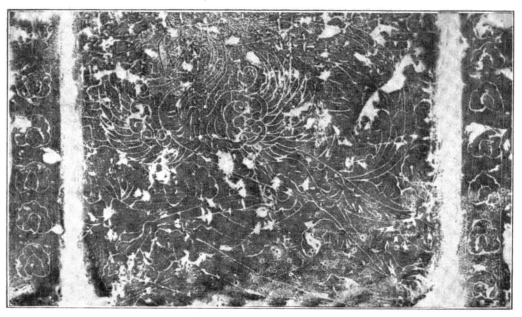

Fig. 3.   Engraving of Small Side of Marble Sarcophagus.

because of his education 憶念乳哺之恩思其掬養之功. Actuated by this motive, he had respectfully made, on behalf of his dead parents, a sarcophagus carved with ornaments 遂爲亡父母敬造銘塔一所. The word *t'a* used twice in this inscription to denote the sarcophagus in question corresponds to the Sanskrit word *stūpa* and is the general designation for a pagoda; but there is nothing strange about this phraseology if we remember that the *stūpa* and *caitya* were originally sepulchres, and then cenotaphs erected in honor of a deceased saint or in commemoration of a miracle or other remarkable incident in Buddhist life. The use of this expression, however, may lead us to think that the peculiar practise of supplying sarcophagi, as in our case, had some bearing on Buddhism, or might have even been a custom originating within its domain, a supposition which may be corroborated by the next sarcophagus to be considered. Fu Pao and his people ascertained by divination an auspicious time for the souls of his parents to change their former abode to find peace in the sarcophagus, in order to ensure, for a thousand years, a place of eternal permanency 伏保等卜得吉時於神和原遷奉安塔一所庶使千載永定方. It thus appears that the offering of this sarcophagus was an act of filial piety. The parents of the donor were dead and, according to custom, buried in a wooden coffin. The sarcophagus imitating in its shape a coffin, though on a largely reduced scale, was placed in the grave-mound, apparently above or very near to the wooden coffin in which the parents were buried

Fig. 4.   Sarcophagus from Shensi Province, T'ang Period.

and supposed to receive their souls. It was, accordingly, a soul sarcophagus into which the soul supposed to reside in the wooden coffin shifted its abode, because it was a more durable structure safeguarding an eternal existence of the soul. We do not know what the shape of wooden coffins was during the T'ang period; the present specimen allows us a reconstruction by concluding that they were built, at least in the region of Shensi, in the same form as this model, on a larger scale.

Figure 4 represents another stone sarcophagus found on the soil of Shensi Province. It had been disposed of to a Japanese dealer shortly before I returned to Si-ngan fu in the beginning of February 1910 on my way out of Tibet. A photograph of it had been taken at Si-ngan fu, a copy of which was presented to me by a Chinese friend. I did not see the object itself and can merely judge from this reproduction. Its size is about the same as that of the preceding specimen, so it must have served also the purpose of a spirit coffin. The base is higher and more projecting, so that the coffin is well set off from it. The decorations are highly Buddhistic in character, as shown above all by the plastic high-relief figure of a shaven Buddhist monk and an attendant who belongs to the type of tribute-bearer. He is holding on his hands a stone chest supposed to contain sacred books. Such quadrangular chests of large dimensions are said to have been buried with Buddhist monks in the T'ang period, and an excellent specimen covered with fine engravings is in the collection of the Field Museum. The surface of the wall from which the two figures stand out is adorned with engraved floral designs of Persian character as frequently occur in that age, and the graceful outlines of a Kinnari with folded hands and bird-feet appear in the centre. The long side of the pedestal is divided into three countersunk medallion panels, and the figure in the middle one seems to be an Apsara (Chinese *t'ien nü* 天 女). The lid seems to be covered with a combination of a cloud and floral pattern, but the photograph is not distinct enough, and the engraving too delicate to allow of a positive conclusion. The combined feature of sculpture and pictorial flat design here exhibited (of which there are numerous other examples in the contemporaneous Buddhist sculpture) is of great interest in that it exactly corresponds to a peculiar style of wall-painting which seems to have been in vogue during the T'ang epoch, and to have still been cultivated in the time of the Ming dynasty, as may be seen in many ancient temples of Shansi and Shensi. The entire wall of a temple-hall is covered with a fresco, but prominent figures moulded in clay stand out independently from the painting, but form a part of the pictorial composition.[1] Especially caves and rocks filled with hermits are favorite themes for sculptural designing of this kind. The sarcophagus in question has no inscription except four large characters of an ornamental style chiseled into the front at the head of the coffin, — reading, according to my Chinese informants, *Ta Sui ch'ao chih*, "Made (at the time of) the Great Sui dynasty". This "inscription" at once aroused my suspicion, because there is no precedent at that period for such a style which sets in as late as the Ming dynasty, and because it seemed so inadequate and out of place on this coffin and not in harmony with the whole affair. Under a volley of questions, my Chinese friends finally admitted that the "inscription" was a forgery recently perpetrated within the very city-walls of Si-ngan fu, that the sarcophagus had arrived there without the inscription, that it doubtless was a production of the T'ang period, that a clever craftsman notorious for his epigraphical counterfeits had voluntarily invented the Sui dynasty, at the instigation of a dealer from motives of greediness, and that the said Japanese had believed in the authenticity of the inscription. On hearing this revelation, I naturally expressed the wish to have the privilege of meeting this artist; but the ready promise of a special arrangement for such an interview proved quite unnecessary, for a few days later the hero of Si-ngan fu attracted by the prospects of a good bargain spontaneously called on me to invite me to an inspection of his atelier. I gladly responded and had the pleasure of meeting

---

[1] Many examples of such representations are noted and described in my diaries, but any such extracts would lead us away from the subject under consideration.

one of the cleverest and boldest forgers of our time and one of the strangest and most mysterious characters I ever knew. This man was extremely bright and intelligent, but silent, gloomy, hated by anybody, keeping aloof from his fellow-mates, and entirely living to his black art for which he had an inborn passion. He had carried on, for years, the most arduous studies in epigraphy, and there was no style of writing and no period which he could not imitate to perfection. Chinese forgers also have specialized in business; my friend of the Sui dynasty, as I styled him, made in inscription stones, in seals, and in Buddhist clay votive tablets which then were the craze of collecting China where the mania for antiquities is dominated also by fashion. Two big halls on the large compound where he lived and schemed were crowded with a bewildering mass of inscription tablets ranging from the T'ang to the Ming, of such superb execution in all details as would have outwitted the smartness of the most experienced scholar of China. He had studied the different kinds of stones employed in all periods and knew how to supply, from the vicinity of Si-ngan fu, a T'ang dynasty as well as a Yüan or a Sung stone; only the fresh looks of a recently executed piece of work were here and there treacherous, as the last touch with the cheerful feeling of age had not yet been added. In this laboratory I penetrated for the first time into the mysteries of the art of how to distinguish genuine from counterfeit seals. He certainly had a splendid collection of genuine antiquities, and one of them, a clay votive tablet of the T'ang period, was then the general talk of the whole city, on account of its real intrinsic beauty and the abnormal price which he demanded for it. Of course, he was right in this, for he showed me his imitations of the same piece which at first sight, despite an eye somewhat sharpened by long training, I could not distinguish from the original, and his productions easily brought him fifty Taels net a piece.[1] Despite his lack of popularity, this man did a tremendous business extending all over China, and had amassed a considerable fortune. Quite naturally, for the label "coming from Si-ngan fu" has an hypnotic fascination on the minds of the Chinese. The official or scholar passing through or temporarily serving in the famous city had promised his folks and friends at home to send them a souvenir of the ancient times, and so the supply must be kept going. The rule may be fairly established that the antiquities given away as presents by the Chinese to one another (of those given by Chinese to their foreign friends I do not dare to speak) are almost all forgeries, for the good reason that nobody would ever screw up his courage to utter a syllable about this fact, even if he knew it. Gifts form a part of the Chinese *Lebenslüge*, and the good friend is obliged to accept the

---

[1] It goes without saying that in China, as elsewhere, the price paid for an article constitutes no evidence whatever for its genuineness. It has, however, occured that prices paid for Chinese antiquities announced publicly to influence opinion in favor of them are, on the contrary, excellent evidence for these articles not being genuine, because the real thing could not have been secured for this price.

Sui or T'ang writing on stone with a smiling and genuine face. Politeness triumphs, and the scheme works well from beginning to end.[1]

My curiosity in this sarcophagus once being aroused, I determined to follow up its trail. When reaching Hankow in June 1910, I was told by the American Consul here that a while ago a Japanese coming from Si-ngan fu' had made a shipment to the Fine Arts Museum of Boston. It therefore was natural for me to suppose that the sarcophagus might have found its way to Boston. An inquiry made through Mr. Edward S. Morse of Salem, however, resulted in an emphatic denial on the part of the Boston Museum that this or any other sarcophagus from China had ever landed at its door. The theory now remains that the Japonese gentleman whose name is unknown to me had singled out this piece as a superior treasure and brought it to Japan[2].

The fact of a former use of sarcophagi in Sze-ch'uan has been established by Mr. Thomas Torrance,[3] who has carefully investigated the caves cut in the cliffs of the Min River, and has arrived, as I believe correctly, at the result that these were not, as formerly supposed, ancient dwelling-places of the aboriginal inhabitants of Sze-ch'uan, but Chinese burial places dating from the Han down to the Sung period. He found there two kinds of coffins, those made of clay and of stone, the former resembling in shape the present day style with convex lids, averaging from $6^1/_2$ to 8 feet in length. The sarcophagi are also of different sizes, most of them being quite large , and as Mr. Torrance is inclined to think, presumably being the outer shells of coffins; they measure roughly 8 feet in length, 3 feet in width, and from $2^1/_2$ to 4 feet in height. "In many cases the backs of these stone coffins are one with the wall of the cave. Hewn when the cave was made, the workmen thought it an advantage to leave them intact." This observation is intensely interesting in that it clearly reveals one of the lines of development which sarcophagi have taken in China. They have

[1] For this reason, whereas it is easy to see where the impostor begins, it is difficult, nay, impossible to ascertain where the impostor ends and the duped one begins. Do not always blame it on to the dealer, he is very often the dupe of the party himself. There are perhaps half a dozen of dealers in antiquities throughout China who really know what they buy and sell; the rest unanimously assert that they know not. What else could be expected where imitation created by a heavy demand of the public have been in operation in unchecked progress for over a millenium? The dealer, as a rule, is not an expert, his experience is limited, and his statements are not worth a farthing. Experts, even among the scholars, are few in China as elsewhere.

[2] These details are here merely given to assist others in tracing the whereabouts of this interesting specimen, as it would be valuable to receive a more accurate description of it. I am far from throwing any reflection on the Japanese concerned who may have acted perfectly *bona fide*, and who, after all, may be another person than the one who seems to have had the function of an agent of the Boston Museum. In fact I do not know, and my detective theory claims no other merit than to point out the probability that the object in question is in a collection of Japan and may thus eventually be discovered there by a foreign inquirer.

[3] In his interesting paper Burial Customs in Sze-ch'uan (*Journal China Branch Royal Asiatic Society*, Vol. XVI, 1910, p. 66).

grown out, or are a continuation of, the ancient stone grave-vault or cist. The fact that the live rock in the caves of the Min River was utilized to form a part of the sarcophagus plainly indicates its former stationary character and inseparable connection with the grave. The mobile and portable sarcophagus has gradually been evolved from this form, as we see in Sze-ch'uan and in the specimen discovered by Lamprey near Shanghai referred to above. Its purport was, as the circumstantial evidence allows us to infer, to lend safety and permanent protection to the wooden coffin encased in it, and therewith simultaneously also to the body.

It thus appears that essentially two main types of sarcophagi exist in China, — the sepulchral chest developed from the cist for the preservation of the wooden coffin, and the spirit or soul sarcophagus executed on a smaller scale in imitation of a wooden coffin. The former development, it will be noticed, is identical with the one which we tried to make out for the ancient Tu-küe or Turks. In spite of their technical, typological and inward differences, the two types rest on a common mental basis in the popular beliefs. Both follow the tendency to guaranty an eternal repose and durability of the mortuary abode, and a long-enduring preservation of the body. It may not be amiss to call attention to the fact that this idea is the opposite of the notions entertained by the Romans in regard to sarcophagi. It is well known that the Romans who had always practised cremation adopted stone coffins only as late as in the Christian era, on account of the then ruling belief in the caustic qualities of stone which, according to Pliny (*Historia Naturalis XXXVI*, 27) consumed the body in forty days; hence the name sarcophagus, "the flesh-eater" (from σάρξ "flesh" and φαγεῖν "to eat"). While the introduction of sarcophagi was materially an innovation in Roman burial practice, it was, psychologically, a continuation, another form of the previous disposal of the dead by the flame. The developments in China and in the West move therefore on totally different and strongly contrasting lines.

In Fig. 5, a pottery coffin of the T'ang period, likewise obtained at Si-ngan fu, is illustrated. Aside from the circumstantial evidence, the character of the red clay and the green glaze applied to the lid point decidedly to that epoch. The small dimensions indicate the use of this coffin for a child, and the height of the upper part seems to hint at the fact that it was placed in an upright sitting posture. The specimen is 51 cm long, 26 cm wide at the upper and end 24,5 cm at the lower end, with a height of 39 and 29 cm at both ends respectively, and an average thickness of 2 cm for the walls. The four walls and the lid are coated with a green glaze which, however, is entirely decomposed on the surfaces of the walls and has partially come off, while on the lid and the inner edges it is perfectly preserved and in many portions shines in brilliant silver oxidation. The receptacle is evidently shaped in a wooden mould, as is also the lid; the appliance of wooden moulds may nowadays be observed at any Chinese kiln in the making of bricks, roofing-tiles, and any angular vessels

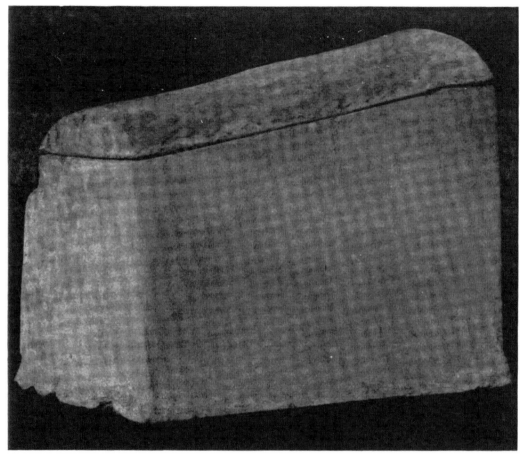

Fig. 5.   Pottery Coffin, T'ang period.

like square, hexagonal, or octagonal flower-pots.[1] During the Han period also bags of woven material, filled with sand or earth, were employed in moulding large architectural pieces of clay as used in the construction of grave-vaults and on the roofs of palaces, as may be seen from the impressions of the textile patterns left on the inner sides of the clay walls.[2] The interior of this pottery coffin, especially the four

[1] Mr. TORRANCE (l. c., p. 66) expressed his admiration for the Sze-ch'uan earthenware coffins in the words: "How they managed to make and fire such a huge thing, and to produce it so perfectly is a mystery". There is no mystery about this, and the technical process is simple and merits no particular admiration. It is only what we should expect from the notorious skill of the Chinese potter. They are the common output of workmen who can easily turn out, by means of wooden moulds, a piece of this kind in a few minutes.

[2] CHAVANNES (T'oung Pao, 1908, p. 247) has observed the same process on the tiles used for the tomb of the "Marshal" of the fifth century in T'ung-kou, Korea, north of the Yalu River,

corners, is smeared over with a thick loam, apparently to keep off more efficiently any supposed outward influences. Also this coffin seems to imitate in its shape the common large wooden coffin.

Earthenware coffins, if we can depend in this case on the Chinese traditions concerned, seem to go back into the time of a greater antiquity than sarcophagi. The *Li ki*[1] ascribes the former to the mythical age of the Emperor Shun, and states that during the Chou dynasty children below the age of eleven, for whom no mourning was worn, were buried in earthenware enclosures which appeared as an inheritance of that time. Not too excessive a stress should be laid on this passage, as has been done by some authors.[2] The *Li ki* is not an historical work but a codified ritual in which the facts and traditions of the past are frequently subjected, quite naturally, to constructions emanating from the peculiar standpoint of the Confucian school. No doubt, also these reflect ancient conditions to a certain degree, but they must never be accepted as plain fact, or as a gospel truth. The schematic sequences made out for the development of inventions and objects are, all without exception, later days' inventions to afford a pleasing interpretation for existing phenomena, but have no direct relation to the objective historical process. The chapter in the *Li ki* alluded to renders it plain that in the Chou period three modes of burial were in vogue, — wooden coffins, brick enclosures and earthenware coffins, used according to the age of the dead, for adults, youths, and children, respectively. This is a plain and credible fact; but in order to explain this state of affairs, recourse is taken to an artificial evolutionary theory to the effect that Shun made earthenware coffins, that the sovereigns of the Hia dynasty surrounded these with brick enclosures, that then the people of the Yin dynasty used wooden coffins, and that those of the Chou added the surrounding curtains and the feather ornaments. This exposition is an interesting and suggestive opinion but one of no historical value. All that the passage in the *Li ki* allows us safely to infer is that earthenware coffins were used in the Chou period for the interment of children, and that the Confucian scholars had the impression that this practice was traceable to a greater antiquity when a more extensive use of them was made also for adults.

The late Dr. BUSHELL[3] has found an account of the discovery of an ancient earthenware coffin on the south of Tan-yang-shan, which is recorded in the Annals in the fifth year (506 A. D.) of the reign of Wu Ti, the founder of the Liang dynasty; it is described as five feet high, over four feet in circumference, wide below and flat-

---

and mentions that the tiles still manufactured in that region present on their inner sides the same net due to their being fashioned over a coarse piece of hemp-cloth.

[1] LEGGE's translation, *Sacred Books of the East*, Vol. XXVII, p. 125.

[2] J. J. M. De GROOT, The Religious System of China, Vol. I, p. 282, and W. PERCEVAL YETTS, Notes on the Disposal of Buddhist Dead in China (*Journal Royal Asiatic Society*, 1911, p. 708).

[3] Chinese Art, Vol. II, p. 6.

bottomed, and pointed above, opening in the middle like a round box with a cover[1]; while the corpse was found buried inside in a sitting posture. The fact that also adults were occasionally buried in pottery coffins is revealed by the last will of the Emperor T'ai-tsu (951 A. D.) who insisted on strict simplicity in his funeral and ordered his corpse to be placed in a coffin of baked clay.[2]

Also Buddhist monks were occasionally buried in pottery coffins. The *P'ei wên yün fu* (Ch. 14, p. 177) quotes the following story from the "Records of Nanking": "The *Shêng yüan ko* of the Liang dynasty is the new name assigned to the *Wa-kuan sze* ("Temple of the Pottery Coffin"[3]). At the time of the Si Tsin dynasty (265—313 A. D.), the soil produced two clusters of dark-colored lotuses. When they were dug out, a pottery coffin was found, in which an old Buddhist monk was visible; the flowers had grown out of the lower side of his tongue. On holding an inquiry, the elders said: "There was formerly a monk reciting the Sūtra of the Lotus of the Good Law[4], who, when he died, was buried at this spot'."[5]

In regard to the age of sarcophagi, no positive assurances are given us in Chinese sources. They are not made mention of in the *Chou li* or the *Li ki*; that is to say, they were beyond the precepts of the funerary ritual and had no place in it, for the natural reason that sarcophagi are costly affairs and offer a difficult transportation problem. They occur therefore only sparsely and sporadically in the records, which is an echo of the fact that their service was but rarely enlisted, and only by the most prominent men who could afford the heavy expenditure. In an isolated passage, the *Li ki*[6] refers to a story connected with Confucius who, when living in the principality of Sung, noticed Huan, the minister of war, engaged in the preparation of a stone case for the wooden coffin (*shi kuo* 石椁), the work of which was not completed within three years. Confucius upbraided him on this occasion for his extravagance and thought a quick decay of the body to be preferable. This certainly is a mere anecdote inserted in a philosophie dialogue, and from the expression used

---

[1] Nobody could say that this description excels in clearness or would allow one to form a definite idea of the appearance of this coffin.

[2] DE GROOT, The Religious System of China, Vol. II, p. 815.

[3] The hall or villa *Shêng yüan* was situated in the south of the southern wall of Nanking. The alteration of the ancient name *Wa-kuan sze* which seems to be due to the above story into the new name *Shêng yüan* took place in the tenth century (compare Father GAILLARD, Nankin d'alors et d'aujourd'hui, pp. 130, 265, Shanghai, 1903). Gaillard writes the name of the temple 瓦官寺.

[4] The *Fa hua king* of this text apparently is an abbreviation of the title *Miao fa lien hua king* 妙法蓮華經, the Saddharmapuṇḍarīkasūtra. The pious monk had studied and recited the Sūtra of the Lotus so frequently that a pair of lotuses grew out of his tongue in the grave.

[5] 金陵志·梁昇元閣改名瓦棺寺·西晉時地產青蓮二朶·掘之得瓦棺·內見一老僧·花從舌底出·詢及父老曰·昔有僧誦法華經卒葬此地·

[6] COUVREUR, Li-ki, Vol. I, p. 165; LEGGE's translation, Vol. I, p. 149, whose rendering "stone coffin" is insufficient.

it does not become clear whether a stone grave-vault (for *shi kuo* has also this meaning) or an outer shell for the coffin, *i. e.* a pseudo-sarcophagus is here intended.

The emperors of the Chou dynasty were buried in wooden coffins. There is a strange and obscure legend related by Se-ma Ts'ien[1] in the history of the house of Ts'in, alluding to a sarcophagus found by a certain Fei-lien and inscribed with a decree of Shang Ti, the God of Heaven. If this legend is capable of teaching us something on the subject of sarcophagi in ancient China, it may be this that they were of exceedingly rare occurence, believed to be of supernatural origin and bestowed by Heaven only upon a worthy for an extraordinary act of loyalty. It is quite in keeping with the iron personality of the First Emperor Ts'in Shi when we read in Se-ma Ts'ien that he had a sarcophagus made for himself during his lifetime at the time when he constructed his sepulchre in the mountain *Li*[2]. But in the description of the Emperor's funeral, no reference is made to that sarcophagus by Se-ma Ts'ien, but only a coffin is mentioned.[3] In view of the scarcity of Chinese records relative to this subject, the archæological material before our eyes will correspondingly assume a still greater importance. Especially the contemporaneous record inscribed on the sarcophagus in Fig. 1 is a religious document of the first order. It is manifest that archæology cannot be neglected in a study of the development of religious thought in China and will impart to it new and fertile ideas.

Sarcophagi are known also from Korea, and there is a specimen in the collection of Mr. Charles L. Freer in Detroit. It is well known that sarcophagi and pottery coffins played an extensive rôle in ancient Japan, much more so than in China.[4]

[1] CHAVANNES, Les mémoires historiques de Se-ma Ts'ien, Vol. II, p. 4. I side with the translation and interpretation of the legend given by Chavannes, which seems to me correct, and cannot agree with the opinions expressed by DE GROOT (The Religious System of China, Vol. I, p. 283).

[2] CHAVANNES, ibid., p. 176.

[3] *Ibid.*, p. 195. — The two other cases relating to stone coffins quoted by DE GROOT (l. c., p. 284) afford no strong historical evidence. In the one extract which is of purely legendary tenor, Heaven again is made responsible for the appearance of a jade coffin. The *P'ei wên yün fu* (Ch. 14, p. 176b) quotes the same passage, derived by DE GROOT from the *Hou Han shu*, after the *Fêng su t'ung* and adds an allusion to the jade coffin from a poem of Li T'ai-po. The other case, I believe, does not relate at all to stone coffins, though also the *P'ei wên yün fu* quotes it under this heading. It is here the question of a natural phenomenon seen in two peculiar rock formations rising from the water, which for their resemblance with the appearance of coffins are styled "stone coffins" 石棺. Such fancy names are often applied everywhere in China to striking rock-forms. The *Ko chi king yüan* (Ch. 7) has a special section devoted to stones of peculiar shapes 象形諸石, containing several other notes of a similar character from the same work quoted by De Groot. No other quotations concerning sarcophagi are contained in the *P'ei wên yün fu* or in the *Yün fu shi i.*

[4] Compare O. NACHOD, Geschichte von Japan, Vol. I, p. 136, and the bibliographical references there given; ASTON, Nihongi, Vol. II, pp. 285, 389; YAGI SHŌZABURŌ, Nihon Kōkogaku, 日本考古學, II, pp. 108, 111 (Tōkyō, 1898); N. G. MUNRO, Prehistoric Japan pp. 344—350 (Yokohama, 1908; the statements of this author, I regret to say, are not always reliable, many of his conclusions and theories are debatable, and there are even some beyond discussion).

In view of all that has been written about this subject, it is unnecessary to canvass this ground here again, but it should be strongly emphasized in this connection that the sarcophagi and pottery coffins of Japan are entirely independent from those of China and in no historical and archæological relation with the latter;[1] they represent distinct archæological types and have sprung from religious ideas widely differing from those of the Chinese. They range, together with the total practice of funerary rites, among that group of indigenous ideas which constitute the ancient culture type of Japan in the times of pre-Chinese influence. This is easily ascertained from the outward characteristics of these objects; they are obviously imitations of houses, and the pottery coffins, in particular, are reproductions of pile-dwellings, so faithfully moulded that they could serve as an object-lesson and as a means of reconstructing the ancient house-types. The pile-dwelling, however, is the original type of Japanese domestic architecture (and occurs as such also in the culture-sphere of South-Eastern Asia, the Austronesian group of P. W. Schmidt), but is conspicuously absent in ancient Chinese culture.[2] This observation is sufficient evidence for the non-Chinese origin of these Japanese burial caskets. Their development is not difficult to grasp. When a person died in ancient Japan, he was left in his hut which was abandoned by his relatives and exchanged for another habitation.[3] This custom indicates that the dead man was and continued to be the owner of the house, and so he was buried in a likeness of this house. The house-shaped coffin was symbolic of his pursuing another form of existence in his previous dwelling-place.

A general conclusion may present itself from the subject under consideration. As soon as we try to get at the root and idea of things, as soon as we subject them

---

[1] The tradition of the *Kojiki* (B. H. CHAMBERLAIN, Ko-ji-ki, or Records of Ancient Matters, p. XLI) that the Emperor Sui-nin was the first to introduce stone tombs is not worthy of consideration; this is the favorite method dear to the hearts of all chroniclers of the East to connect origins and beginnings of things with the name of some distant monarch.

[2] It certainly does not mean much that pile-dwellings occasionally occur in China *e. g.* in the form of pavilions erected in ponds or lakes to allow of a pleasant sojourn over the water in the summer, or built by the poorer classes for economy in space and as a money-saving device owing to the cheapness of the water-plots, or finally necessitated by the natural character of a steep river-bank, as may be seen in houses on the Yangtse in Hankow and Hanyang and elsewhere. Such incidental pile-structures are found scattered all over the globe, and I even saw them in the Himalaya among the Lepcha of Sikkim wherever required by natural conditions. But there is a marked difference in principle between occasional pile-dwellings prompted by chance and circumstance, and the permanent and habitual pile-dwellings of a whole nation or a large stock of peoples where this feature enters as a prominent characteristic of ethnic life and culture, as was the case among the prehistoric pile-dwellers of Switzerland, in ancient Japan, and still among all Malayans. This point of view is overlooked in a doctor-thesis of Leipzig from Ratzel's school by JOH. LEHMANN (Die Pfahlbauten der Gegenwart, Wien, 1904). It is not merely the erection on piles which contributes to make the Japanese and Malayan house a peculiar house-type in itself, but also the roof, the gables, the curious ridge-poles (E. S. MORSE, Japanese Homes, p. 329), doors, windows, rooms, and the atmosphere of the interior. These are all things antipodes to China.

[3] CHAMBERLAIN, *l. c.*, p. XL.

not only to an historical but also to a psychological analysis, we recognize the vast diversity and differentiation of all ideas prevailing in the various culture-groups. The alleged sameness and uniformity of human culture established by an antiquated school of ethnology is a fable of those unable to see or shunning serious research. Cultural objects have not emanated from presupposed "elementary ideas" (*Elementargedanken*) lingering like germs in the human psyche, but man's soul was everywhere a wonderfully kaleidoscopic organism mirroring the gay butterfly play of imagination in all products of thought and work. The sarcophagus was a thing of the Egyptians, the Greeks, the Romans, the Etruskans, but it was another thing to the Chinese, another thing to the Japanese. The only point in common between the East and the West is that the sarcophagus presents a stone receptacle for burial, and China even offers the unique spectacle of a sarcophagus not for the burial of a body but of a soul. But how superficial and external are these coincidences, and how meaningless when compared with the contrast of religious significance! Another ingenious method dear to the minds of certain ethnologists — to form a general classificatory notion which usually is quite arbitrary, and to trace the "evolution" of this notion through the universe regardless of space and time — is also well illustrated in the present case. There is no evolution of the sarcophagus as such; in the most varied cultures, at very different periods, sarcophagi have been made as the results of different religious points of view and purposes. Each human product, whether industrial or artistic, and each idea must be studied within the space and time by which they are bounded, in connection with the tradition of the peoples by whom they are created, and in relation to the total history of that culture-group from which they have originated. *Nulla salus* outside of this principle!

087

中国的鱼型符号

$1.00 per Year   NOVEMBER, 1912   Price, 10 Cents

The Open Court

A MONTHLY MAGAZINE

Devoted to the Science of Religion, the Religion of Science, and the Extension of the Religious Parliament Idea

Founded by Edward C. Hegeler

AN OPEN JADE RING; CHINESE SYMBOL OF SEPARATION.
(See pages 673-675.)

The Open Court Publishing Company

CHICAGO

Per copy, 10 cents (sixpence). Yearly, $1.00 (in the U.P.U., 5s. 6d.).

FISH SYMBOLS IN CHINA.

BY BERTHOLD LAUFER.

[The Field Museum of Chicago contains a great number of valuable jade ornaments which, together with many archeological objects, were collected by Dr. Laufer. We take pleasure in here furnishing our readers with illustrations of some of them together with Dr. Laufer's explanations,[1] and we begin by reproducing a peculiar Chinese girdle ornament called *küeh,* which consists of a ring open in one part and symbolizing separation. Wu Ta-Ch'eng, Dr. Laufer's authority, published the picture of one of them which he considers the oldest type of *küeh.*]

THE symbolism relative to the incomplete rings called *küeh* is peculiar. Wu Ta-ch'êng alludes to it in figuring a specimen in his collection (see accompanying illustration) in which I believe

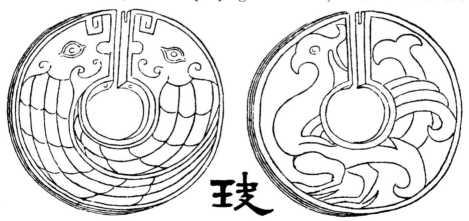

KÜEH. AN OPEN JADE RING, SYMBOL OF SEPARATION.
The Chinese archeologist Wu Ta-Ch'eng regards the figures on the obverse as two dragons, though they look more like fishes.

the oldest type of these rings may be found. It is carved from green jade with a black zone and has a double dragon (*shuang lung*) engraved on the one face and "the scarlet bird" (*chu kio* or

[1] Berthold Laufer, *Jade; a Study in Chinese Archæology and Religion.* Chicago, Field Museum, 1912.

chu niao), the bird of the southern quarter, on the other face. The form as here outlined exactly agrees with that on a tile disk of the Han period (*Chinese Pottery of the Han Dynasty*, Plate LXVII), Fig. 4). It is not known what its proper significance is on the tile nor in this connection on the ring. The break in the ring is effected by a narrow strip sawn away between the two dragon-heads which cannot touch each other; it symbolically indicates the rupture or the breaking-off of cordial relations between two people.

The gloomy half-ring *küeh* originally meant separation, banishment, nay even capital punishment; or, what could not appeal either

JADE GIRDLE PENDANTS. PAIRS OF FISHES.

to the people at large, the decision in literary disputes. But this entire symbolism must have died out during the Han period; for then a new style of girdle-ornament gradually seems to have come into general use, carved into graceful designs not pointing to any serious disaster for the wearer. It is useless to raise here a question of terminology, and to argue that these ornaments differ from the ancient half-rings and may have developed from another type which may have even existed in the Chou period under a different name. This may be, but the fact remains that the long series of these objects is designated *küeh* by the native archeologists, and that in some of them the type, and above all, the designs of the *küeh,*—and these are presumably the oldest in the group of the new *küeh,*—have been faithfully preserved.

The two illustrations of double fishes, here reproduced, are carved from green jade. In the first their fins are connected, and they are holding in their mouths the leaved branch of a willow (*liu*), according to the Chinese explanation. It should be added that, during the Han period, it was customary to pluck a willow-branch (*chê liu,* see Giles No. 550), and to offer it to a parting friend who was escorted as far as the bridge *Pa* east of Ch'ang-ngan where the branch of separation (*küeh!*) was handed to the departing friend.[2] The significance of this ornament is therefore simple enough: we must part, but we shall remain friends as these two fishes are inseparable. It reveals to us at the same time how the *küeh,* so formidable in the beginning with its message of absolute divorce, was mitigated into a more kind-hearted attitude which made it acceptable

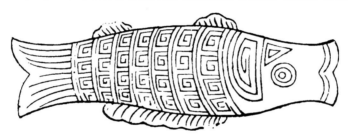

JADE GIRDLE PENDANTS. SINGLE FISH.

to all people—it became a parting-gift, a farewell trinket. The date of this piece is set at a period covering the Wei and Tsin dynasties, i. e., roughly the third and fourth centuries A. D., but I have no doubt that the pattern goes back to the creative period of the Han.

The second figure displays a similar design of a pair of fishes, the same carving being brought out on both faces. Also here, the editors explain the plant design as that of a willow. The leaves are represented here on the bodies behind the gills, and a leaf-shaped wreath (with the perforation of the ancient *küeh*) appears between the lower fins. Another difference is that the tips of the tails here touch each other which seems to hint at a more intimate union of the party concerned, while there is a gap in the previous piece in correspondence with the break in the ancient half-ring.

[It is noteworthy that the fishes frequently appear in pairs in the Christian catacombs where the idea of a parting suggests itself very obviously. Here the two fishes are usually separated by an anchor, the common symbol of

[2] Pétillon, *Allusions littéraires,* p. 172.

hope, so as to suggest very plainly the idea of a parting with the hope of meeting again. We may add that the pair of fishes as they appear in the zodiac are very different in nature and presumably in meaning, and should not be confounded with either the Chinese fish, with the *küeh,* or the Christian fishes in the catacombs; and further the figure of the single fish has again a significance of its own. In Chinese it means loneliness, independence and uniqueness. We here reproduce such a single fish.]

The scales are conceived of as meander fretwork; but I do not know whether, for this reason, this fish is associated with thunder. The peculiar feature is, at all events, its single-blessedness in distinction from the common fish couples. There is a huge fish in the Yellow River, called *kuan* (Giles, No. 6371, Pétillon, *loc. cit.,* p. 500)[3]

JADE GIRDLE PENDANT. CICADA.

supposed to be a kind of spike, noted for its solitary habits of life, and therefore an emblematic expression for anybody deprived of company like an orphan, a widower, a bachelor, or a lonely fellow without kith or kin.[4] A girdle-ornament of this design was perhaps a gift for a man in this condition.

Among the jade amulets placed on the corpse to prevent its decay the fish occurs on the eye and lip-amulets. But there are also

[3] The Chinese theory that this species is not able to close its eyes is certainly mere fancy, as in all fishes the accessory organs of the eye like the lids and lachrymal glands are poorly developed.

[4] In this sense, it is mentioned as early as in the Shu king. In one poem of the Shi king, No. 9 of the songs of the country of Ts'i, Wên Kiang, the widow of Prince Huan of Lu, is censured for returning several times into her native country of Ts'i where she entertained an incestuous intercourse with her own brother, the prince Siang. The poet compares her to the fish *kuan* who is restless and sleepless at night for lack of a bed-fellow (see Legge, *Shi king,* Vol. I, p. 159, and Vol. II, p. 293).

instances of large separate carvings representing fishes which have no relation to the body, but have been placed in the coffin for other reasons.

The Field Museum of Chicago contains two mortuary jade fishes unearthed from graves of the Han period. One of them[5] is a marvelous carving of exceedingly fine workmanship, all details having been brought out with patient care. It represents the full figure of a fish, both sides being carved alike, 20 cm. long, 11 cm. wide, and 2 cm. thick, of a dark spinach-green jade. A small piece has been chipped off from the tail-fin. There is a small eye in the dorsal fin and a larger one below in the tail-fin. It is therefore likely that the object was suspended somewhere in the coffin; it is too large and too heavy (it weighs 1¼ pounds) to have served for a girdle-ornament. In this way,—with comparatively large bearded head and short body,—the Chinese represent a huge sea-fish called *ngao* (Giles No. 100).

Such large and fine jade carvings are likely to have had a religious significance, and the following passage may throw some light on this subject:

"In the Han Palace Kun ming ch'ih a piece of jade was carved into the figure of a fish. Whenever a thunderstorm with rain took place, the fish constantly roared, its dorsal fin and its tail being in motion. At the time of the Han, they offered sacrifices to this fish in their prayers for rain which were always fulfilled."[6]

The middle figure on the same plate, a fragment, perhaps only the half of the original figure, is represented carved in the shape of a fish of leaf-green jade clouded with white specks, on the lower face covered with a thick layer of hardened loess. It is 11.5 cm. long, 4.2 cm. wide, and 9 mm. thick.

In the July number of the *Journal of the Anthropological Society of Tokyo* (Vol. XXVII, 1911), there is an article by Prof. S. Tsuboi describing some interesting figures of animals of chipped flint, one of them representing a well-formed fish (p. 132).

While the religious symbolism formerly connected with the fish has almost disappeared it continues as a favorite ornament, and jade girdle pendants in the shape of fishes are still much in use. The third figure of the same plate represents such a modern carving of white jade showing a fish surrounded by lotus-flowers (9.8 cm. long, 4 cm. wide). The contrast between this modern and the two ancient pieces in design and technique is evident.

[5] The upper figure in the adjoined plate.

[6] *Si king tsa ki,* quoted in *P'ei wên yün fu,* Ch. 100 A, p. 6 a.

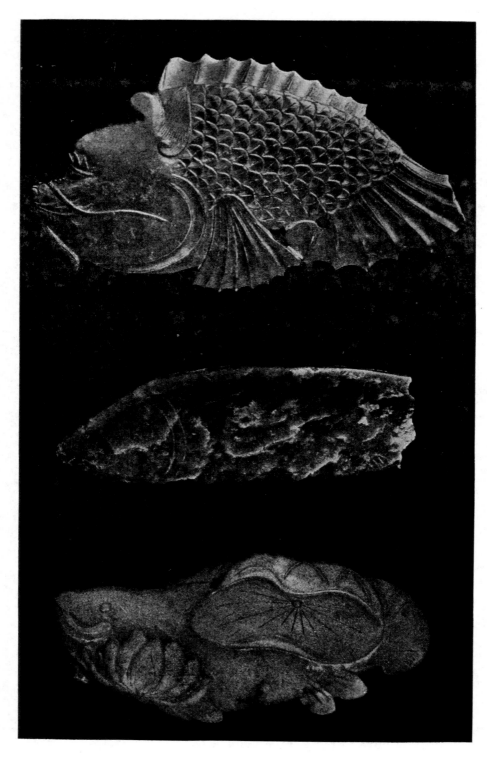

JADE CARVINGS. FISHES.

The butterfly carved from white and brownish - yellow jade is a unique specimen among mortuary offerings. It is alleged by those who found it that it was taken from the grave-mound of the famous Emperor Ts'in Shih (B. C. 246-211) near the town of Lin-tung which is 50 *li* to the east of Si-ngan fu. I am not fully convinced that this is really the case, though any positive evidence *pro* or *contra* this assertion is lacking; but there is no doubt that, judging from its appearance and technique, this is a burial object of considerable age and unusual workmanship, such as is likely to

MORTUARY JADE CARVING. BUTTERFLY.

have been buried with a personage of high standing only. It is a flat carving (12.6 × 7.6 cm., 0.5 cm. thick) both in open work and engraved on both faces, the two designs, even in number of strokes, being perfectly identical. The work of engraving is executed with great care, the lines being equally deep and regular. We notice that a plum-blossom pattern is brought out between the antennæ of the butterfly; it is the diagram of a flower revealing a certain tendency to naturalism, which seems to bring out the idea that the butterfly is hovering over the flower. We further observe four designs of plum-blossoms, of the more conventional character, carved *à jour* in the wings. The case is therefore analogous to that illustrated on a Han bronze vase (*Chinese Pottery of the Han Dynasty,* p. 283).

It is known that in modern times the combination of butterfly and plum-blossom is used to express a rebus (*mei tieh*) with the meaning "Always great age" (W. Grube, *Zur Pckinger Volkskunde,* p. 139).[7] It is difficult to say whether, in that period to which this specimen must be referred, this notion was already valid, though the possibility must be admitted in view of the early rebuses traced by A. Conrady (preface to Stentz, *Beiträge zur Volkskunde Süd-Schantungs*). It would, however, be erroneous to believe that the rebus in all cases presented the prius from which the ornament was deduced, for most of these ornamental components are much older and may even go beyond an age where the formation of rebuses was possible. The rebus was read into the ornaments, in well-nigh all cases; while other single ornaments were combined into complex compositions with the intention of bringing out a rebus. It is not the rebus which has created the ornaments, but it is the ornament which has elicited and developed the rebus; the rebus has merely shaped, influenced and furthered the decorative compositions as, e. g., occurring in the modern Peking embroideries figured by Grube. In the present case, it is quite obvious that the association of the butterfly with a floral design rests on natural grounds, and was originally not provoked by a mere desire of punning, which is the product of a subsequent development.

A very curious feature of this specimen is that the two upper large plum-blossoms in the wings are carved out in loose movable rings turning in a deeply hollowed groove but in such a way that they cannot be taken out, a clever trick such as the later authors designate as "devil's work" (*kuei kung*). This peculiarity certainly had also a significance with reference to the mortuary character of the object. Such movable pieces are designated by the Chinese as "living" (*huo*); so we have here two "living" plum-blossoms in distinction from the two "dead" plum-blossoms below, and the two former might have possibly conveyed some allusion to a future life.

---

[7] There is also the interpretation *hu-tieh nao mei,* "the butterfly playfully fluttering around plum-blossoms," alluding to long life and beauty (*Ibid.,* p. 138, No. 15).

088

《菲律宾的中国陶器》续篇

FIELD MUSEUM OF NATURAL HISTORY.

PUBLICATION 162.

ANTHROPOLOGICAL SERIES.   VOL. XII, No. 1.

CHINESE POTTERY
IN THE PHILIPPINES

BY

FAY-COOPER COLE

WITH POSTSCRIPT BY

BERTHOLD LAUFER

The Robert F. Cummings Philippine Expedition

GEORGE A. DORSEY
Curator, Department of Anthropology

CHICAGO, U. S. A.
July, 1912.

In the spring of 1906 Mr. Robert F. Cummings of this city expressed his intention of providing the Field Museum of Natural History with funds to defray the expenses of an extended series of Ethnological investigations in the Philippine Islands.

Working under this liberal endowment the following expeditions have been in the field:

In 1906 Mr. S. C. Simms visited the Igorot of Benguet, Lepanto and Bontoc, and the Ifugao of Nueva Viscaya. During 1907–8 Mr. F. C. Cole worked among the Tinguian, Apayao and Kalinga tribes of Northern Luzon, and the Batak of Palawan.

The late Dr. William Jones reached the Philippines in the fall of 1907 and proceeded to the Ilongot of the Upper Cagayan river, Luzon. After residing a year in that district he was murdered by members of a hostile village. Following Dr. Jones' death Mr. Simms returned to the Philippines, secured the material gathered by Dr. Jones and completed the Igorot and Ifugao collections, visiting for this purpose the Mayayao and Amburayan Igorot, in addition to certain points touched on the first expedition.

In the fall of 1909 Mr. Cole returned to the Islands and devoted nearly two years to the study of the pigmy blacks of Bataan province, the Bukidnon of North Central Mindanao, and the several tribes residing about the Gulf of Davao in Southern Mindanao.

While the primary object of these expeditions was to gather museum collections, much time was given to the study of the mental and material culture, as well as of the language, folklore and anthropometry of the tribes visited. The results of these studies will appear from time to time in the Anthropological Series of this Museum. The present paper forms the first issue of Mr. Cole's researches.

GEORGE A. DORSEY.

I

CHINESE POTTERY IN THE PHILIPPINES

When the Spaniards first set foot in the Philippines, they found evidences of trade with an advanced nation. When near Leyte, Magellan stopped for a time at a small island whose chief "embraced the captain-general to whom he gave three porcelain jars covered with leaves and full of rice wine." [1] Later when Pigafetta and his companions went ashore, they were treated to wine taken from a large jar, and when the meal was served, "two large porcelain dishes were brought in, one full of rice, and the other of pork with its gravy." [2] When they reached Cebu (April 7, 1521), they were informed by the king that they were welcome "but that it was their custom for all ships which entered their ports to pay tribute, and that it was but four days since a junk from Ciama (*i. e.* Siam) laden with gold and slaves had paid tribute." The tribute was refused but friendly relations were established, whereupon the king "had refreshments of many dishes, all made of meat and contained in porcelain platters, besides many jars of wine brought in." [3]

When Pigafetta visited the king of Zubu (Cebu), he found him "seated on a palm mat on the ground, with only a cotton cloth before his privies. . . From another mat on the ground he was eating turtle eggs which were in two porcelain dishes, and he had four jars of palm wine in front of him covered with sweet smelling herbs and arranged with four small reeds in each jar by which means he drank." [4]

Later they were conducted to the house of the prince "where four young girls were playing, one on a drum like ours, but resting on the ground; the second was striking two suspended gongs alternately with a stick wrapped somewhat thickly at the end with palm cloth; the third, one large gong in the same manner; and the last, two small gongs held in her hand, by striking one against the other, which gave forth a sweet sound. . . These gongs are made of brass and are manufactured in the regions about the Signio Magno which is called China." [5] After the death of Magellan, the fleet sailed to the south

[1] BLAIR and ROBERTSON, The Philippine Islands, Vol. XXXIII, p. 15.

[2] *Ibid.*, p. 119.

[3] *Ibid.*, p. 139.

[4] *Ibid.*, p. 149. This is still the method of drinking in Mindanao (compare Pl. I).

[5] BLAIR and ROBERTSON, (PIGAFETTA) Vol. XXXIII, pp. 149-151.

3

until they reached Mindanao. There they made peace with the king, and Pigafetta went ashore with the ruler, in order to see the island. He describes the country, people, their customs and foods, and did not fail to note that "in the house were hanging a number of porcelain jars and four metal gongs."[1] Here they also learned more of the large island of "Lozon" (Luzon) lying to the northwest, "where six or eight junks belonging to the Lequian (Liukiu) people go yearly."[2] Proceeding further to the south, they encountered the island of Borneo where they found many evidences of an advanced civilization and an active trade with neighboring countries. Here they saw beautiful porcelain jars, cups and dishes, silks and carpets.[3]

The chronicles of succeeding expeditions left many references to Chinese articles and trade.[4] In the account of Loaisa's Expedition, we are told of the Island of Bendenao (Mindanao) where two junks from China come each year for purposes of trade. "North of Bendanao is Cebu, and according to the natives it also contains gold, for which the Chinese come to trade each year."[5] Again in 1543, Alvarado says of Mindanao: "Upon capturing this island we found a quantity of porcelain and some bells. They are well supplied with perfumes from the Chinese who come to Mindanao and the Philippinas."[5]

The first (recorded) encounter of the Spaniards with the Chinese seems to have been during a trip from Panay (May 8th, 1570) to Luzon and Manila. When off the Island of Mindoro they learned that "two vessels from China, the inhabitants of which the natives call Sangleys (i. e. merchants), were in a river near by." Salcedo was dispatched to reconnoiter the ships, and to request friendship with them, but the Chinese made a warlike display, whereupon they were attacked by the Spaniards who after a short fight took possession of the junks. "The soldiers searched the cabins in which the Chinese kept their most valuable goods, and there they found silk, both woven and in skeins, gold thread, musk, gilded porcelain bowls, pieces of cotton cloth, gilded water jugs, and other curious articles, although not in a large quantity considering the size of the ships. The decks of the vessels were full of earthen jars and crockery, large porcelain vases, plates and bowls, and some fine porcelain jars which they call *sinoratas*."[6] They also found iron, copper, steel and a small quantity of wax which the Chinese had

[1] *Ibid.*, p. 205.
[2] *Ibid.*, p. 207.
[3] *Ibid.*, p. 215.
[4] Blair and Robertson, Vol. III, p. 42; Vol. II, p. 72; Vol. III, p. 57.
[5] Blair and Robertson, Vol. II, pp. 35, 69.
[6] Blair and Robertson, Vol. III, p. 74.

purchased.   From their captives they learned that three more Chinese boats were trading only three leagues away.   Later, on crossing to Luzon, at a point near the town of Balayan, they found that two Chinese ships had just been trading there, and that in a quarrel two Chinamen had been made captives and others had been killed.   Proceeding to Manila bay, the Spaniards found four Chinese vessels, with earthenware jars and porcelains, trading.   In the city they learned that forty Chinese and twenty Japanese were regular residents there.   Friendly relations appeared to have been established when the Moro raja treacherously attacked the Spaniards.   In return the Spaniards burned a part of the city, in the ruins of which they found many objects of porcelain.

After the Spaniards had become established in Manila, the trade with China steadily increased, [1] not only in that city but in other ports of the Islands.   At first the articles dealt in were of little value to the Spaniards, for "they brought some trifle, although but a small quantity, as the natives with whom they come principally to trade commonly use, and for them are brought only large earthen jars, common crockery, iron, copper, tin and other things of that kind.   For the chiefs, they brought a few pieces of silk and fine porcelain." [2]   Of such little use were these articles to the newcomers that it was proposed, in 1574, to stop the trade. [3]   However, the Chinese were quick to accommodate themselves to the new conditions, and we soon find them supplying many articles, such as "sugar, barley, wheat, and barley flour, nuts, raisins, pears, and oranges; silks, choice porcelains and iron; and other small things which we lacked in this land before their arrival." [4]   Each year this trade increased until the number of the traders was in the thousands, and the Spaniards became dependent upon them for their sustenance.   Even the natives relied on this trade to such an extent that the old industries languished and the colony became each day less able to support itself.   However, in addition to the foodstuffs which the colony needed they brought silks and other articles which entered into direct competition with the products of the mother country, and this resulted in the royal decree of 1586, which prohibited all such trade. [5]   This edict failed of its purpose, and in hopes of devising a plan whereby the competition would be eliminated, the outflow of gold to China be stopped, and the return of the natives to their old pursuits be accomplished, a meeting was called, and leading Filipino were summoned

[1] *Ibid.*, pp. 167, 172, 181, 225.
[2] BLAIR and ROBERTSON, Vol. II, p. 238; Vol. III, pp. 243-5.
[3] *Ibid.*, Vol. III, p. 226, note.
[4] Letters of Lavezaris, *Ibid.*, Vol. III, p. 276.
[5] BLAIR and ROBERTSON, Vol. VI, pp. 28, 29, 90, 150, 283, 286.

to give evidence under oath concerning the extent and nature of Chinese trade. It was believed that if trade in Chinese cloth and the like could be stopped, the natives and Chinese would continue to trade without using money; "for if they should wish to barter in the Islands — which is not forbidden them — they can and will obtain goods as they formerly did, in exchange for such articles as *siguey* (a small white snail), dye wood, and carabao horns; to this mode of trading the Chinese will adapt themselves and the outflow of money will cease." [1] The nine Filipino chiefs, from villages near Manila, agreed that before the Spaniards came to the Islands the people raised cotton, which they made into cloth for their own garments and did not depend on the Chinese, "for although one or two ships came from China each year at that time, these brought no cloths or silks, but only iron and earthenware and *camanguian*, [2] while since the arrival of the Spaniards, often twenty or thirty ships come each year." [1]

The inquiry was without result, and the Chinese increased in numbers and power until 1596, when about twelve thousand were expelled from the Islands.[3] Despite hostile laws and massacres, they continued to increase and spread out over the Islands throughout the time of Spanish rule, and to-day they dominate the trade with the natives of the Archipelago. The commerce with the Spaniards, whom the civilized natives imitated, was so much more lucrative than that previously carried on with the various villages that the old trade in pottery and the like seems practically to have ceased. Despite the constant references of the early writers to the Chinese and their trade the importation of earthenware and common glazed pottery seems not to have been mentioned after about the year 1600.

While the greater part of the Chinese wares doubtless entered the Islands through direct trade, a considerable amount came in through trade with "Borneo, Maluco, Malacca, Sian, Camboja, Japan and other districts." [4] "A few years before the Spaniards subdued the Island of Luzon, certain natives of Borneo began to go thither to trade, especially to the settlements of Manila and Tondo; and the inhabitants of one island intermarried with those of the other." [5] "The cargoes of these traders consisted of fine and well made palm mats, a few slaves for the natives, sago, and *tibors;* large and small jars, glazed black and very fine, which are of great service and use." [5] Legaspi tells of captur-

[1] BLAIR and ROBERTSON, Vol. VIII, pp. 82–84.

[2] Incense.

[3] BLAIR and ROBERTSON, (MORGA). Vol. IX, p. 266.

[4] BLAIR and ROBERTSON, Vol. III, p. 298; Vol. V, pp. 73, 105; Vol. XVI, p. 176; BECCARI, Wandering in the Great Forests of Borneo.

[5] MORGA, *Ibid.*, Vol. XVI, pp. 134, 185.

ing, near Butuan, a junk whose crew were Borneo Moors. They had with them silk, cotton, porcelain and the like. They also traded in bells, copper and other Chinese goods.[1]

Inter-island trade among the Filipino seems to have reached considerable proportions prior to the arrival of the white man. Some of their trips carried them to the ports of Borneo, and one account credits the Tagalog and Pampango with sailing "for purposes of trade to Maluco, Malaca, Hanzian (Achen?), Parani, Brunei, and other kingdoms." [2] Pigafetta tells of their party seizing a junk in the port of Borneo in which "was a son of the king of Luzon, a very large island." [3] In 1565, Legaspi learned that two Moro junks from Luzon were in Butuan trading gold, wax, and slaves.

These Moro from Luzon also came to Cebu to arrange with Legaspi for the right to trade, and when they met with success, two junks from Mindoro were induced to go there also. "They carried iron, tin, porcelain, shawls, light woolen cloth and the like from China." [4]

It will thus be seen that pottery and other articles of Chinese origin might have had a rapid spread along the coasts of the Archipelago, from whence they slowly penetrated into the interior by means of trade.[5] It seems, however, that even upon the arrival of the Spaniards, some of this ware had assumed great value in the eyes of the natives, and in 1574 we find the native chiefs sending "jewels, gold, silks, porcelains, rich and large earthen jars, and other very excellent things" in token of their allegiance to the King of Spain.[6] It was also the custom at that time for the family of the deceased to bury with the body "their finest clothes, porcelain ware, and gold jewels," [7] and when this became known to the Spaniards they began to rifle the graves in order to secure these valuable objects. This continued until it became necessary for Legaspi to order that "henceforth no grave or burial place be opened without the permission of his Excellency." [8]

There is some evidence that burial in jars was early practiced in the Philippines. ADUARTE, writing in 1640, describes the finding, by a crew shipwrecked on the Batannes islands, of "some jars of moderate size covered with others of similar size. Inside they found some dead

[1] BLAIR and ROBERTSON, Vol. II, p. 207; Vol. III, p. 57, note; Vol. XXXIV, p. 224; BARROWS, History of the Philippines, pp. 99–101.

[2] BLAIR and ROBERTSON, Vol. XXXIV, p. 377.

[3] Ibid., Vol. I, p. 265.

[4] Ibid., Vol. II, pp. 117, 142.

[5] Ibid., Vol. V, 121; BARROWS, History of the Philippines, p. 182.

[6] BLAIR and ROBERTSON, Vol. III, p. 249; Vol. IV, p. 290.

[7] Ibid., Vol. II, p. 139.

[8] Ibid., Vol. II, p. 173.

bodies dried, and nothing else."[1] Dr. MERTON MILLER of the Philippine Bureau of Science recently opened a number of mounds found on the Island of Camiguin lying north of Luzon. In them he found jars placed one over the other, in the manner just described, and containing some human bones as well as a few beads.[2] Mr. EMERSON CHRISTY, also of the Philippine Bureau of Science, while exploring ancient burial caves in the Subuanan district of Mindanao, found a number of large Chinese jars, some containing human bones and accompanied by agate beads. Fragments of large jars were also found in the burial cave of Pokanin in Southern Mindoro [3] (compare Pl. II). Dr. FLETCHER GARDNER, who first visited the place, described the cave as follows: [4]

"It is situated about half way between the towns of Bulalacao and Mansalay in Southern Mindoro. It is on the seaward face of a cliff about 500 feet high and 200 yards wide and is about 200 feet above high water mark. In the summer of 1904, while hunting for guano, I accidentally discovered this cave and procured the skulls and other bones which I am sending you. The nearest inhabitants, who live within half a mile of the cave at the little sitio of Hampangan Mangyans, have known that these remains were there but deny that the bones are those of their ancestors. As two or three members of the sitio assisted me in procuring and carrying away the bones I am satisfied that they believe the statement to be true, but as will be seen from the remains of basketry and fabrics enclosed with the bones these products are practically the same as those of the inhabitants of the sitio above mentioned. I believe that during the great Moro raid of 1754 when seventy-five slaves were taken from Manol and Mansalay the Mangyan at that time inhabiting the neighborhood were driven into the interior and abandoned this cave for burial purposes. . . The bones were covered with about three inches of dust and nitrous earth, which argues a very long time without disturbance."

From this evidence it seems not at all unlikely that jar burial may have been practiced by the Filipino, especially those in direct trade relations with Borneo, in which country such burials are common.[5] In this connection it is interesting to note that Dr. Hirth believes jar burial to have been introduced into Borneo by the Chinese traders from Fukien, and its introduction was probably later than the lifetime of Chao

[1] Aduarte, *Ibid.*, Vol. XXXI, p. 115. JAGOR, Travels in the Philippines, pp. 259–61.

[2] *Philippine Journal of Science*, Feb., 1911, pp. 1–4.

[3] The contents of this cave are now in the Field Museum of Natural History.

[4] Extract from letter to Field Museum.

[5] LING ROTH, The Natives of Sarawak and British North Borneo, Vol. I, pp. 150–154; FURNESS, Home Life of Borneo Head-hunters, p. 139.

Ju-kua, in the early part of the thirteenth century.[1] Ancient remains other than those just cited are of rare occurrence in the Philippines; so I shall quote somewhat at length the very interesting account, given by Jagor, of excavations in Ambos Camarines, Luzon.

"In 1851, during the construction of a road a little beyond Libmánan, at a place called Poro, a bed of shells was dug up under four feet of mould, one hundred feet distant from the river. It consisted of Cyrenae (*C. suborbicularis, Busch.*), a species of bivalve belonging to the family of Cyclades which occurs only in warm waters, and is extraordinarily abundant in the brackish waters of the Philippines. On the same occasion, at the depth of from one and a half to three and a half feet, were found numerous remains of the early inhabitants, skulls, ribs, bones of men and animals, a child's thigh-bone inserted in a spiral of brass wire, several stags' horns, beautifully formed dishes and vessels, some of them painted, probably of Chinese origin; striped bracelets, of a soft, gypseous, copper-red rock, glancing as if they were varnished; small copper knives, but no iron utensils; and several broad flat stones bored through the middle; besides a wedge of petrified wood, embedded in a cleft branch of a tree. The place, which to this day may be easily recognized in a hollow, might, by excavation systematically carried on, yield many more interesting results. What was not immediately useful was then and there destroyed, and the remainder dispersed. In spite of every endeavor, I could obtain, through the kindness of Señor Fociños in Nága, only one small vessel. Similar remains of more primitive inhabitants have been found at the mouth of the Bígajo, not far from Libmánan, in a shell-bed of the same kind; and an urn, with a human skeleton, was found at the mouth of the Pérlos, west of Sitio de Poro, in 1840.

"Mr. W. A. Franks, who had the kindness to examine the vessel, inclines to the opinion that it is Chinese, and pronounces it to be of very great antiquity, without, however, being able to determine its age more exactly; and a learned Chinese of the Burlingame Embassy expressed himself to the same effect. He knew only of one article, now in the British Museum, which was brought from Japan by Kaempfer, the color, glazing and cracks in the glazing of which (*craquelés*) correspond precisely with mine.[2] According to Kaempfer, the Japanese

---

[1] F. Hirth, Ancient Chinese Porcelain (*Journal of the China Branch of the Royal Asiatic Society*, Vol. XXII, N. S., pp. 181–3, 1888).

[2] Referring to this paragraph Dr. C. H. Read of the British Museum says: "There must be some mistake in Jagor's book. No such jar given by Kaempfer is in the Museum, and I cannot understand my predecessor, Sir. A. W. Franks, making such a statement. I may mention that I knew Dr. Jagor intimately and regard him as more than usually accurate."

found similar vessels in the sea;[1] and they value them very highly for the purpose of preserving their tea in them."

MORGA writes: "On this island, Luzon, particularly in the provinces of Manilla, Pampánga, Pangasinán, and Ylócos, very ancient clay vessels of a dark brown colour are found by the natives, of a sorry appearance; some of a middling size, and others smaller; marked with characters and stamps. They are unable to say either when or where they obtained them; but they are no longer to be acquired, nor are they manufactured in the islands. The Japanese prize them highly, for they have found that the root of a herb which they call *Tscha* (tea), and which when drunk hot is considered as a great delicacy and of medicinal efficacy by the kings and lords in Japan, cannot be effectively preserved except in these vessels; which are so highly esteemed all over Japan that they form the most costly articles of their showrooms and cabinets. Indeed, so highly do they value them that they overlay them externally with fine gold embossed with great skill, and enclose them in cases of brocade; and some of these vessels are valued at and fetch from 2,000 tael to 11 reals. The natives of these islands purchase them from the Japanese at very high rates, and take much pains in the search for them on account of their value, though but few are now found on account of the eagerness with which they have been sought for.

"When Carletti, in 1597, went from the Philippines to Japan, all the passengers on board were examined carefully, by order of the governor, and threatened with capital punishment if they endeavoured to conceal 'certain earthen vessels which were wont to be brought from the Philippines and other islands of that sea,' as the king wished to

[1] This is not a fact but a legend. ENGELBERT KAEMPFER (The History of Japan, Glasgow reprint, Vol. III, p. 237) relates a story, told him by Chinese, regarding an island Maurigasima near Formosa famous in former ages for its fine porcelain clay. "The inhabitants very much inrich'd themselves by this manufacture, but their increasing wealth gave birth to luxury, and contempt of religion, which incensed the Gods to that degree, that by an irrevocable decree they determin'd to sink the whole island." Then follows the long story of the virtuous king who managed to escape the disaster miraculously, and to flee into the province of Fukien. The island sank, and with it all its ceramic treasures. They were subsequently taken up by divers and sold to Chinese merchants of Fukien who traded them to Japan at immense sums. There is consequently a double error in the above statement of Franks: it is not the Japanese who found jars in the sea, nor does Kaempfer say that they were celadons or similar to them; on the contrary, he describes them as "transparent, exceeding thin, of a whitish color, inclining to green," which is almost the opposite to a celadon. That legend, as far as I know, has not yet been traced to a Chinese source. BRINKLEY (Japan, Vol. VIII, p. 267) shows little understanding of folklore, if he calls it a foolish fable; it doubtless ranks among the category of familiar stories of sunken isles and towns in Europe. Brinkley's explanation that the story was probably invented by some Japanese Swift to satirise the irrational value attached to rusty old specimens of pottery is decidedly untenable, if for no other reason, because, according to Kaempfer's statement, the legend is Chinese in origin. The pottery in question is, in my opinion, Chinese ware of Fukien, and the legend emanates from the potters of Fukien. [B. L.]

buy them all. . . 'These vessels were worth as much as 5, 6, and even 10,000 scudi each; but they were not permitted to demand for them more than one Giulio (about a half Paolo).' In 1615 Carletti met with a Franciscan who was sent as ambassador from Japan to Rome, who assured him that he had seen 130,000 scudi paid by the king of Japan for such a vessel; and his companions confirmed the statement. Carletti also alleges, as the reason for the high price, 'that the leaf cia or tea, the quality of which improves with age, is preserved better in those vessels than in all others. The Japanese besides know these vessels by certain characters and stamps. They are of great age and very rare, and come only from Cambodia, Siam, Cochin China, the Philippines, and other neighbouring islands. From their external appearance they would be estimated at three or four quatrini (two dreier) . . It is perfectly true that the king and the princes of that kingdom possess a very large number of these vessels, and prize them as their most valuable treasure and above all other rarities . . . and that they boast of their acquisitions, and from motives of vanity strive to outvie one another in the multitude of pretty vessels which they possess.'

"Many travellers mention vessels found likewise amongst the Dyaks and the Malays in Borneo, which, from superstitious motives, were estimated at most exaggerated figures, amounting sometimes to many thousand dollars.

"St. John relates that the Datu of Tamparuli (Borneo) gave rice to the value of almost £700 for a jar, and that he possessed a second jar of almost fabulous value, which was about two feet high, and of a dark olive green. The Datu fills both jars with water, which, after adding plants and flowers to it, he dispenses to all the sick persons in the country. But the most famous jar in Borneo is that of the Sultan of Brunei, which not only possesses all the valuable properties of the other jars but can also speak. St. John did not see it, as it is always kept in the women's apartment; but the sultan, a credible man, related to him that the jar howled dolefully the night before the death of his first wife, and that it emitted similar tones in the event of impending misfortunes. St. John is inclined to explain the mysterious phenomenon by a probably peculiar form of the mouth of the vessel, in passing over which the air-draught is thrown into resonant verberations, like the Aeolian harp. The vessel is generally enveloped in gold brocade, and is uncovered only when it is to be consulted; and hence, of course, it happens that it speaks only on solemn occasions. St. John states further that the Bisayans used formerly to bring presents to the sultan; in recognition of which they received some water from the sacred jar to sprinkle over their fields and thereby ensure plentiful harvests. When the sultan was asked whether he would sell his jar for £20,000, he answered that no offer in the world could tempt him to part with it." [1]

This desire for old jars was by no means confined to the traders and Japanese, for the tribes of the interior had secured a great number of them at a very early period, and later when the supply from the coast had ceased, they began to mount in value until a man's wealth was, and still is, largely reckoned by the number of old jars in his possession (compare Pl. III). As they were handed down from one generation to another, they began to gather to themselves stories of wondrous origin and deeds, until to-day certain jars have reputations which extend far beyond the limits of the tribes by which they may be owned. While among the Tinguian of Abra, the writer continually heard tales of a wonderful jar called *Magsawi* (Pl. IV). It was credited with the ability to talk; sometimes went on long journeys by itself; and was married to a female jar owned by the Tinguian of Ilocos Norte. A small jar at San Quintin, Abra, was said to be the child of this union and partook of many qualities of its parents. [2] The history of this jar as related by its owner, Cabildo of Domayco, is as follows: "Magsawi, my jar, when it was not yet broken talked softly, but now its lines are broken, and the low tones are insufficient for us to understand. The jar was not made where the Chinese are, but belongs to the spirits or Kabonían, because my father and grandfather, from whom I inherited it, said that in the first times they (the Tinguian) hunted Magsawi on the mountains and in the wooded hills. My ancestors thought that their dog had brought a deer to bay (which he was catching), and they hurried to assist it. They saw the jar and tried to catch it but were unable; sometimes it disappeared, sometimes it appeared again, and, because they could not catch it they went again to the wooded hill on their way to their town. Then they heard a voice speaking words which they understood, but they could see no man. The words it spoke were: 'You secure a pig, a sow without young, and take its blood, so that you may catch the jar which your dog pursued.' They obeyed and went to secure the blood. The dog again brought to bay the jar which belonged to Kabonían (a spirit). They plainly saw the jar go through a hole in the rock

---

[1] JAGOR, Travels in the Philippines, p. 162. In *Zeitschrift für Ethnologie*, Vol. I, 1869, pp. 80–82, JAGOR describes an ancient burial cave in Southern Samar. In it were found broken pieces of crudely decorated pottery associated with human remains.

[2] Other jars credited with the ability to talk were seen by the writer, and similar jars are described by travelers in Borneo. See LING ROTH, Natives of Sarawak and British N. Borneo, Vol. II, p. 286; HEIN, Die bildenden Künste bei den Dayaks auf Borneo, p. 139; also ST. JOHN, Life in the Forests of the Far East.— The idea of sex in jars is widespread throughout the Archipelago.

which is a cave, and there it was cornered so that they captured the pretty jar which is Magsawí, which I inherited." [1]

Other jars of equal fame "were found in caves in which the spirits dwelt," or were called into being by supernatural agencies. References to these wonderful jars abound in the folktales, the following quotations from which will serve to show the character of all. [2]

"Not long after he started, and when he arrived in the pasture, all the jars went to him, and all the jars stuck out their tongues; for they were very hungry and had not been fed for a long time. The jars were *somadag, ginlasan, malayo,* and *tadogan,* and other kinds also. [3] When Aponîtolau thought that all the jars had arrived, he fed them all with betel-nut covered with *lawed* leaves. As soon as he fed them, he gave them some salt. Not long after this they went to the pasture, and they rode on the back of a carabao. As soon as they arrived, all the jars rolled around them and stuck out their tongues, and Aponîbolinayen was afraid, for she feared that the jars would eat them. The wide field was full of jars. Aponîtolau gave them betel-nut and *lawed* wine and salt. As soon as they fed them, they went back home." (Extract from the tale about Gimbangonan.)

"And they took many things to be used at the wedding. So they agreed on the marriage price, and Bangan and his wife said, the price must be the *balaua* [4] nine times full of different kinds of jars. As soon as the *balaua* was filled nine times, Daluagan raised her eyebrows, and immediately half of the jars vanished, and Aponîbolinayen used her (magical) power, and the *balaua* was filled again, so that it was truly filled. When they had danced, all the guests took some jars, before they went home." (From the Kanag tale.)

"'Now we are going to pay the marriage price according to the custom,' said Aponîbolinayen, 'our custom is to fill the *balaua* nine times with different kinds of jars.' So Aponîbolinayen said 'Ala, you *Alan* [5] who live in the different springs, and *Bananayo* [5] of Kadanan

---

[1] Similar stories of jars turning to animals and *vice versa* are encountered in the Southern Philippines and in Borneo. See Ling Roth, Natives of Sarawak and British North Borneo, Vol. II, p. CLXXVI; Hein, Die bildenden Künste bei den Dayaks auf Borneo, pp. 132–134.

[2] The following are extracts from Tinguian folktales. During the dry season bonfires are built in various parts of the village and around them the men and women gather, the former to make fishnets, the latter to spin. Meanwhile some good story-teller chants these tales.

[3] Each type of jar has its particular name.

[4] A small spirit house built during a certain ceremony.

[5] Lesser spirits.

and you *Liblibayan*,[1] go and get the jars which Kanag must pay as the price for Dapilísan.'   As soon as she commanded them, they went and filled the *balaua* nine times."   (Tale of Dumalawí.)

"So they danced and the big jars which she had hung about her neck made a noise, and the earth shook when she moved her body. The people did not agree, and they said: 'Five times full, if you do not have that many (jars) he may not marry Aponíbolinayen.'   He was so anxious to marry her that he told his parents to agree to what they said.   As soon as they agreed, Langaan used magic so that all the jars which the people wanted were already in the *balaua*.   The day came when they agreed to take Linggíwan to Aponíbolinayen, and he carried one jar.   As soon as they arrived there, they made the rice ceremony."[2]   (Extracts from tale of Ginambo and Gonígonau.)

"Soon after they started, they met the *doldoli* (a jar) in the way. 'Where are you going, young men,' it said.   'Where are you going,' you ask; we are going to secure the perfume of Balewán, for though we are still far from it we can smell it now.'   The jar replied: 'Ala, young men, you cannot go there, for when anyone goes there, only his name goes back to his town,' (*i. e.* he dies), but the boy replied: 'We are going anyway.   That is the reason we are already far from home, and it is the thing which the pretty maiden desires.'   'If you say that you are going anyway, you will repent when you reach there.'   So they left the jar and walked on."   (From Balewán tale.)

"The food was of thirty different kinds, and they were ashamed to be in the house of Ilwísan which had in it many valuable jars, for the *Alan* (spirit) had given them to him."   (Aponíbolinayen tale.)

Great prices are offered and sometimes paid for the more renowned jars, and successful war parties are accustomed to return home with numbers of such trophies.

Every wild tribe, encountered by the writer, in the interior of Luzon, Palawan and Mindanao, possesses these jars which enter intimately into the life of the people (Pl. V–VIII).   Among many the price paid by the bridegroom for his bride is wholly or in part in jars (Pl. IX–X).   When a Tinguian youth is to take his bride, he goes to her house at night, carrying with him a Chinese jar which he presents to his father-in-law, and thereafter he may never address his parents-in-law by name.   The liquor served at ceremonies and festivals is sometimes contained in these jars (Pl. XI–XVI), while small porcelain dishes

[1] Lesser Spirits.
[2] This is still the custom when the groom finally claims the bride.

contain the food offered to the spirits. Porcelain plates are used by the mediums when summoning the spirits, and having served in such a capacity are highly prized; so much so that they are never sold during the lifetime of the medium, and after her death only to an aspirant for mediumship honors (Pl. XVII). When about to call a spirit into her body, the medium sets herself in front of the spirit mat, and covering her face with her hands, she trembles violently, meanwhile chanting or wailing songs in which she bids the spirits to come and possess her (Pl. XVIII). From time to time she pauses, and holding a plate on the finger tips of her left hand, she strikes it with a string of sea shells or a bit of lead, in order that the bell-like sound may attract the attention of the spirits. Suddenly a spirit takes possession of her body and then as a human the superior being talks with mortals (Pl. XIX).

In districts where head-hunting is still in vogue, a Chinese jar is readily accepted as payment in full for a head, and many feuds are settled on this basis. In 1907 the writer accompanied a war party from Apayao to a hostile village several days' march distant. The two villages agreed to make peace on the terms of one jar for each head the one town held in excess of the other, and on this basis the Apayao paid eleven jars to their erstwhile enemies.

Most tribes of the interior have pottery of their own manufacture. These generally bear distinctive names according to the uses to which they are put. Thus among the Tinguian a jar used for greens or vegetables has a definite name, while another in which meat is cooked has its own designation.

In Northern Luzon the women of certain towns have acquired such fame as potters that their wares have a wide distribution, and the industry has almost vanished from neighboring villages.

The general method employed by the potters (Pl. XX–XXI) is as follows: The clay after being dampened is carefully kneaded with the hands, in order to remove stones and bits of gravel. A handful of the mass is taken up and the bottom of the bowl roughly shaped with the fingers. This is placed on a wooden dish, which in turn rests on a bamboo rice winnower — forming a crude potter's wheel. The dish is turned with the right hand while the woman shapes the clay with the fingers of the left or with a piece of dampened bark cloth. From time to time a coil of fresh clay is laid along the top of the vessels and is worked in as the wheel turns. Further shaping is done with a wooden paddle, after which the jar is allowed to dry. In a day or two it is hard enough to be handled, and the operator then rubs it, inside and out, with stone or seed disks, in order to make it perfectly smooth. The jars are placed in dung or other material which will make a slow fire and are burned for a night, after which they are ready for service.[1] Some tribes understand the art of glazing with pitch, but this is not generally practiced throughout the Islands. These jars are generally red in color, and in form quite distinct from those of Chinese manufacture. They are in daily use and have a value of only a few centavos.

[1] The writer found this process both in Luzon and Mindanao. Dr. JENKS found a slightly different method of production at Bontoc (see JENKS, The Bontoc Igorot, pp. 117–121). This process is illustrated by a life sized group in the Field Museum of Natural History. Pl. XXII.

POSTSCRIPT

By Berthold Laufer

At the request of Mr. Cole I take the liberty to append a few notes on the subject of Chinese pottery in the Philippine Islands. From the very interesting information furnished by Mr. Cole on the subject, it becomes evident that two well-defined periods in the trade of Chinese pottery to the Islands must be distinguished. The one is constituted by the burial pottery discovered in caves, the other is marked by the numerous specimens still found in the possession of families and, according to tradition, transmitted as heirlooms through many generations. Let us state at the outset that from the viewpoint of the Chinese field of research a plausible guess may be hazarded as to what these two periods are,— the mortuary finds roughly corresponding to the period of the Chinese Sung dynasty (960–1278 A. D.), and the surface finds to that of the Ming dynasty (1368–1643).[1] By this division in time I do not mean to draw a hard and fast line for the classification of this pottery, but merely to lay down a working hypothesis as the basis from which to attack the problem that will remain for future investigation. There is the possibility also that early Ming pieces are to be found in the graves or caves and, on the other hand, the existence of Sung and After-Ming specimens, say of the seventeenth and eighteenth centuries, in the hands of the natives will no doubt be established with the advance of search and research. But these two cases, if they will prove, will surely remain the exceptions, while the formula as expressed above carries the calculation of the greatest probability.

It is well known that during the middle ages a lively export trade in pottery took place from China into the regions of the Malayan Archipelago, India, Persia, Egypt, the east coast of Africa, and Morocco. Quite a number of ancient specimens of China ware have been discovered in all those countries and wandered into collections of Europe. The curiosity of investigators was early stimulated in this subject, and to A. B. Meyer, Karabacek, Hirth, A. R. Hein, F. Brinkley and others, we owe contributions to this question from the ceramic and trade-

---

[1] Certainly I have here in mind only those specimens prized by the natives as heirlooms and looked upon by them as old. There is assuredly any quantity of modern Chinese crockery and porcelain spread over the Philippines, which, however, is of no account and not the object of legends and worship on the part of the natives.

17

historical standpoint, while active explorers, particularly on Borneo, have brought to light considerable material in the way of specimens. For the Philippines, little had been done in this direction, and it is the merit of Mr. Cole to render accessible to students a representative collection of that pottery which may be designated as "second period," and which is of the highest interest as palpable evidence of the intercourse between China and the Philippines during the Ming period.

The establishment of the two periods is reflected also in the traditions of the Malayan tribes. Mr. Cole (p. 12) relates that the *Magsawi* jar was not made where the Chinese are, but belongs to the spirits or Kabonīan. There are other jars clearly recognized as Chinese by the natives. In regard to the latter, the tradition is still alive; the former are of a more considerable age or were made in a period, the wares of which could no more be supplied by the Chinese, so that the belief could gain ground that they had never been made by the Chinese, but by the spirits. Among the Dayak of Borneo, this state of affairs is still more conspicuous. There, the oldest jars have been connected with solar and lunar mythology. Mahatara, the supreme god, piled up on Java seven mountains from the loam which was left after the creation of sun, moon and earth. Ratu Tjampu, of divine origin, used the clay of these mountains to make a great number of *djawet* (sacred jars) which he kept and carefully guarded in a cave. One day when his watch was interrupted, the jars transformed themselves into animals (compare Cole, pp. 12, 13) and escaped. When a fortunate hunter kills such game it changes again into a jar, which becomes the trophy of the hunter favored by the gods. According to another tradition, the god of the moon, Kadjanka, taught the son of a Javanese ruler, Rāja Pahit, to form jars out of the clay with which Mahatara had made sun and moon; all these jars fled to Borneo, where they still are.[1] I do not believe that these traditions point to Java as a place from which pottery found its way to Borneo; Java has merely become a symbol for the mysterious unknown. This mythical pottery attributed to the action of gods, it seems to me, is to be identified with Chinese pottery of the Sung period, while that accompanied by mere narrative traditions seems to correspond to that of the Ming period. This sequence of myth and plain story has its foundation in long intervals of time and in many changes as to the kinds and grades of pottery introduced from

---

[1] A. R. HEIN, Die bildenden Künste bei den Dayaks auf Borneo, p. 134 (Wien, 1894), and F. S. GRABOWSKY, *Zeitschrift für Ethnologie*, Vol. XVII, 1885, pp. 121–123. Grabowsky is of the opinion that Perelaer, to whom the second tradition is due, can never have heard it from the lips of a Dayak, but simply ascribed to them this tradition originating from Java.

China. This does not mean that a piece ascribed to the spirits will necessarily be a *Sung*, and one credited with a tale always a *Ming*, for interchanges, adjustments and confusions of traditions are constantly at operation.

As no material regarding the earlier period of burial pottery (except a small fragment) exists in the Field Museum, I must be content with a few suggestive remarks regarding the latter. Chinese-Philippine trade must have existed early in the thirteenth, and very likely in the latter part of the twelfth century, as I tried to establish on a former occasion,[1] chiefly guided by the accounts of a Chinese author, Chao Ju-kua, who around 1220 wrote a most valuable record of the foreign nations then trading with China. His work has been translated and profusely commented on by Prof. Hirth.[2] Chao Ju-kua mentions three times the export of porcelain, by which also pottery not being porcelain must be understood, in the barter with the Philippine tribes. Unfortunately he does not tell us of what kind, or from which locality this pottery was, but one interesting fact may be gleaned now from a comparison of the Philippine place-names known to him with those reported by Mr. Cole as having yielded finds of burial jars. Dr. Miller, Mr. Cole informs us, discovered jars containing human bones and beads in mounds opened by him on the Island of Camiguin, lying north of Luzon. This name is doubtless identical with Ka-ma-yen mentioned by Chao Ju-kua as forming the "Three Islands" with Pa-lao-yu (Palawan?) and Pa-ki-nung,[3] and he gives a lively description of the barter with the Hai-tan (Aëta) living there, with the express mention of porcelain. Fragments of large jars, says Mr. Cole, were also found in the burial cave of Pokanin in southern Mindoro; now Chao Ju-kua describes a country in the north of Borneo which he calls *Ma-yi(t)* and identified by me with Mindoro, the ancient name of which was *Mait*. Mindoro, where Spaniards and Chinese met for the first time in 1570, was an old stronghold of the latter, and probably at an earlier date than Luzon. These coincidences cannot be accidental, and must further be taken in connection with the fact to which Mr. Cole justly calls attention, that jar burial may have been practised, especially by those Filipino in direct trade relations with Borneo. It seems to me that we are bound to assume an historical connection between the two and an influencing

[1] The Relations of the Chinese to the Philippine Islands, p. 252 (*Smithsonian Miscellaneous Contributions*, Vol. L, Part 2, 1907).

[2] A complete translation of the work jointly edited by Hirth and W. W. Rockhill has been printed by the Academy of St. Petersburg and is soon expected to be out.

[3] See Hirth, Chinesische Studien, p. 41.

of the Filipino by the Borneo custom.[1]   On both sides, we encounter almost the same kinds of Chinese ceramic wares, the same veneration for them, and a similar basis of folklore and mythology associated with them, so that the belief in an interdependence seems justifiable.   The one fact stands out clearly: Chao Ju-kua, a reliable author of the Sung period, himself a member of the imperial house, relates the export of pottery to Borneo and the Philippines (in the case of Borneo also that of celadons) at his time, the beginning of the thirteenth century, a trade which may have set in at a much earlier date.   This pottery can but have been the contemporaneous pottery of the Sung period, and we are, for this reason, entitled to look to the Philippines for Sung pottery. As the pottery found in the caves is, in all probability, older than that now possessed by the natives, there is the greatest likelihood of identifying this burial pottery with the productions of the Sung period.   The investigations of the antiquities of the Philippines are in their beginnings, and further results and more tangible material must be awaited before definite verdicts can be arrived at.   The pottery fragments must be carefully gathered and examined; it is obvious that they will be of immense value in helping to make out the periods of these burial places. The *terminus a quo* is given by the eleventh century.   The small vessel

---

[1] The subject of jar-burial remains one to be investigated.   It is still practised in China among the Buddhist priesthood and, according to the observations of W. Perceval Yetts (Notes on the Disposal of Buddhist Dead in China, *Journal R. Asiatic Society*, 1911, p. 705), occurs throughout the region of the Middle and Lower Yangtse.   The same author informs us (p. 707) that the earthenware tubs required for this purpose resemble those commonly used for holding water or for storage of manure.   "Occasionally two ordinary domestic tubs (*kang*) joined mouth to mouth are made to act as a coffin, though usually tubs specially manufactured for funeral purposes are obtained.   These are made in pairs, and are so designed that the rim of the lid of the uppermost tub fits closely over the rim of the other, producing a joint easily rendered airtight by the aid of cement.   A pair thus joined together forms a chamber resembling a barrel in shape."   Most of these vessels are said to come from the kilns of Wu-si in Kiangsu Province.   The ancient earthenware coffins, however, considered by Mr. Yetts in this connection, must be separated from these burial jars, as they are pre-buddhistic in origin; such a pottery coffin with green-glazed lid attributed to the T'ang period, is in the Chinese collection of the Field Museum.   E. Boerschmann (Die Baukunst und religiöse Kultur der Chinesen, Vol. I, P'u t'o shan, p. 175) states that the cremation and preservation of Buddhist priests in large urns of glazed pottery is generally practised; that in the pottery kilns of all provinces such jars are made up to 1.50 m in height and shipped far away, and that a district on the Siang River in Hunan, a little north of the provincial capital Ch'ang-sha, is a well-known place for their production.   The jars are mostly glazed brown, concludes Boerschmann, and adorned with reliefs alluding to death, *e. g.* two dragons surrounding a dragon-gate and a pearl in the entrance, indicating that the priest has passed the gate of perception and reached the state of perfection. This information sheds light on the fact that it was dragon-jars which were utilized on Borneo for purposes of burial.
   An interesting practice of jar-burial is revealed by Paul Pelliot (Le Fou-nan, *Bulletin de l'Ecole française d' Extrême-Orient*, Vol. III, 1903, p. 279) from a passage in the *Fu-nan ki*, written by Chu Chi in the fifth century A. D.   It relates to the kingdom of Tun-sün, a dependance of Fu-nan (Cambodja), which seems to have been largely under the influence of Brahmanic India.   Over a thousand Brahmans

mentioned by Jagor is most probably a piece of celadon pottery. Prof. Eduard Seler has been good enough to inform us that it is not preserved in the Berlin Museum, but he describes a similar piece extant there, a fragment of a plate or a flat bowl found by Dr. Schetelig in a cave of Caramuan, Luzon, on the Philippines. "The material," Prof. Seler says, "is a red-burnt hard clay including small white bits of what is apparently calcareous matter. The well-known salad-green glaze exhibiting numerous fine crackles covers the entire surface except the circular foot. On the lower face, the marks of the potter's wheel are visible. On the glazed surface shallow grooves are radially arranged." This description, beyond any doubt, refers to a specimen of celadon pottery of the Sung period, and I am especially interested in the fact that it is hard, red-burnt stoneware, and not porcelain. The former authors always spoke of celadon porcelains exclusively, an error first refuted by Captain F. Brinkley,[1] who justly says that all the choice celadons of the Sung, Yüan, and even the Ming dynasties were stoneware, showing considerable variation in respect to fineness of *pâte* and thinness of biscuit, but never becoming true translucid porcelain. The majority of celadon pieces in the Sung period seem to have been stoneware, while the porcelain specimens increase during the

from India were settled there, married to native women and engaged in reading their sacred books. When they are sick, says the Chinese report, they make a vow to be buried by the birds; under chants and dances, they are conducted outside of the town, and there are birds who devour them. The remaining bones are calcined and enclosed in a jar which is flung into the sea. When they are not eaten by the birds, they are placed in a basket. As regards burial by fire, it consists in leaping into a fire. The ashes are gathered in a vase which is interred, and to which sacrifices are offered without limit of time. The inference could be drawn from this passage that the practice of burial in jars is derived from India. "Among the tribes of the Hindukush," reports W. Crooke (Things Indian, p. 128), "cremation used to be the common form of burial, the ashes being collected in rude wooden boxes or in earthen jars and buried." This was the case also in the funerary rites of ancient India (W. Caland, Die altindischen Todten- und Bestattungsgebräuche, pp. 104, 107, 108) when the bones after cremation were gathered in an urn; according to one rite, the bones collected in an earthenware bowl were sprinkled with water, the bowl was wrapped up in a dress made from Kuça grass and inserted in another pottery vessel which was interred in a forest, or near the root of a tree or in a clean place in a durable relic-shrine. Among the Nāyars or Nairs of Malabar, the pieces of unburnt bones are placed in an earthen pot which has been sun-dried (not burnt by fire in the usual way); the pot is covered up with a piece of new cloth, and all following the eldest, who carries it, proceed to the nearest river (it must be running water), which receives the remains of the dead (E. Thurston, Ethnographic Notes in Southern India, p. 215, Madras, 1906). The latter practice offers a parallel to the burying of the jar in the sea, as related above in regard to Tun-sün. Nowadays, the bones after cremation are gathered on a gold, silver, or copper plate in Cambodja (A. Leclère, Cambodge: La crémation et les rites funéraires, pp. 76, 82, Hanoi, 1906). On jar-burial on the Liu-kiu Islands compare the interesting article of M. Haber- landt, Über eine Graburne von den Liukiu-Inseln (*Mitteilungen der Anthropol. Gesellschaft in Wien*, Vol. XXIII, 1893, pp. 39–42); the specimen figured is doubtless a Chinese production as used for the burial of the ashes of a Buddhist monk.

[1] China, Keramic Art, p. 34 (London, 1904).

Ming epoch.   To this conclusion, at least, I am prompted by a series of celadons gathered by me in China and including specimens of the Sung, Ming, and K'ien-lung periods.   It is somewhat a matter of surprise that a larger number of celadons has not been discovered on the Philippines.   Judging from the account of a Japanese writer on ceramics, translated farther below, there must have been a large quantity of this fine and curious pottery on the Islands in former times, and the search of the Japanese for ceramic treasures there in the sixteenth and seventeenth centuries was chiefly prompted by their craving for celadons.   Maybe the Japanese have taken hold of the best specimens, maybe these are still hidden away in solitary caves or untouched burial mounds.   We hope that these remarks will instigate present and future explorers on the Islands to keep a vigilant watch on celadons, and to pick up even small fragments, always with exact statements of locality, site, nature of the find (underground, surface, cave, mound, etc.) and traditions of the natives, if there are any, because they may be of great significance.   Everything relating to celadons is of utmost historical importance; in almost every case, in my opinion at least, it is possible to define the age or period of a piece of celadon, and also the place of its production,— China, Japan, Korea, or Siam.   The Sung celadons are inimitable and could never be imitated, and the varying character of this pottery through all ages affords a most fortunate clue to chronological diagnosis.

In glancing over the collection of pottery brought home by Mr. Cole, we are struck, first of all, by a certain uniform character of all these pieces, if we leave aside the three small dishes reproduced on Plate XVII, which in correspondence with their different ceramic character enter also a different phase of religious notions.   Only in the latter lot a single piece of porcelain is found (Pl. XVII, Fig. 3).   All other specimens are characterized as stoneware of an exceedingly hard, consistent and durable clayish substance; most of them are high and spacious jars of large capacity; all of them are glazed, and well glazed, and betray in the manner and color of glazing as well as in their shapes and decorative designs a decidedly Chinese origin; all of them have a concave unglazed bottom, most of them are provided with ears on the shoulders for the passage of a cord to secure convenient handling and carrying; none of them is impressed with a seal, date-mark, or inscription of any other kind.   All of them are the products of solid workmanship executed with care and deliberation, apparently with a side-glance at a customer who knew.   On the whole, two principal types are discernible,— dragon-jars and plain jars.   Both groups are distinguished

at the same time by different glazes, and it may be surmised at the outset that they originate from different kilns.

The three jars on Plates VI–VIII exactly agree with one another
in shape and glaze (evidently an iron glaze) the color of which moves
from a light-yellow to a dark-brown.   In the form of rim, neck and
shoulders, the identity is perfect.   The shoulders are decorated with
five massive lion-heads [1] formed in separate moulds and stuck on to the
body of the vessel, perforations running horizontally through the jaws.
The designs, wave-bands and a couple of dragons with the usual cloud-
ornaments, are incised in the body of the clay and in the two specimens
on Plates VI and VII not covered by the glaze, while in the case of the
specimen in Plate VIII the outlines and scales of the dragon have been
overlaid with a glaze of darker tinge, resulting in a flat-relief design.
The dragon-jar in Plate V differs from those three in form and technique,
and is an extraordinary specimen.   The clay walls are of much thinner
build and covered with a fine dark-greenish slip.   Six ears (two of which
are broken off) rest on the shoulders; they are shaped into the very
frequent conventional form of elephant heads ending in curved trunks.
The two dragons are turned out in moulds and playing with the pearl
(not represented in the illustration) designed as a spiral with flame.

In this connection, attention should be drawn to the dragon-jars
of a similar type discovered in large numbers on Borneo.   The *Tung
si yang k'ao*, an interesting Chinese work describing the far-eastern
sea trade of the sixteenth century and published in 1618 (Ming period)
relates that the people of Bandjermasin on Borneo at first used banana
leaves in the place of dishes, but that, since trade had been carried on
with China, they had gradually adopted the use of porcelain; that they
liked to bargain for porcelain jars decorated with dragons on the surface; and that they would keep the bodies of the dead in such jars instead of burying them.[2]   Despite everything that has been written on
the subject of these jars, their descriptions, from a ceramic and historical
point of view, are still rather unsatisfactory.   The illustrations referred
to below are made from sketches, not from photographs.   A. B. Meyer
and Grabowsky describe the glazes as brown or mottled brown, one

[1] A. B. Meyer (Altertümer aus dem Ostindischen Archipel, p. 7, Leipzig, 1884)
describing similar jars from Borneo speaks of five Rākshasa or lion-heads.   They are,
according to Chinese notion, nothing but lion-heads.   The Rākshasa heads are quite
different in style, are always characterized by long protruding tusks, and never occur
as decorations on Chinese pottery.

[2] Hirth, Ancient Chinese Porcelain, p. 182.— The Dayak designation *rangkang*
for these jars seems to me to be suggested by the Chinese name *lung kang* ("dragon-
jar").— For illustrations of Borneo dragon-jars see F. S. Grabowsky, *Zeitschrift
für Ethnologie*, Vol. XVII, 1885, Pl. VII, or A. R. Hein, Die bildenden Künste bei den
Dayaks auf Borneo, p. 133 (Wien, 1890).

glazed white being the only exception. Not having had occasion to see any of them, I think I should not be too positive in my judgment, but can merely give it as my impression that the Borneo dragon-jars are very similar in shape, glaze and design to those from the Philippines, and that both seem to have originated from the same Chinese kiln.

Unfortunately, our knowledge of Chinese pottery is far from being complete, and anything like a scientific history of it does not yet exist. Our collectors have been more interested in porcelains, and the subject of common pottery has been almost wholly neglected. Porcelain is nothing but a variety of pottery and can be properly understood only from a consideration of the subject in its widest range. Porcelain and stoneware appear in China as parallel phenomena, that is to say, the same processes of glazing and decorating have been applied to both categories alike, and certain porcelain glazes have their precedents in corresponding glazes on non-porcellanous clays. The study of this ware is therefore of importance for the history of porcelain, and it has besides so many qualities and merits of its own that it is deserving of close investigation for its own sake. If we had at our disposal such complete collections of pottery from China as we have from Japan, it would presumably be easy to point out the Chinese specimens corresponding to those of the Philippines, and to settle satisfactorily the question as to the furnace where they were produced. Such a collection, whose ideal object it would be to embrace representative specimens, ancient and modern, of the many hundreds of Chinese kilns, will probably never exist, as it would require for itself a large museum to be housed. From my personal experience, restricted to the more prominent kilns of the provinces of Shantung, Chili, Honan, Shansi, Shensi and Kansu, I may say that dragon-jars of the Philippine type are not turned out there at the present day, nor can ancient specimens of this kind be obtained there. Both facts are conclusive evidence, for if once made, some vestiges of them would have survived in modern forms, in view of the stupendous persistency of traditions among the potters. A priori it may be inferred that the Philippine pottery came from those localities which were in closest commercial touch with the Islands, i. e. the provinces of Fukien and Kuangtung in southern China. The fictile productions of the latter province are included under the general term Kuang yao, Kuang being an abbreviation of the name of the province, yao meaning "pottery." The city of Yang-kiang in the prefecture of Chao-k'ing, not far from the coast, may be credited, in all likelihood, with the manufacture of the dragon-jars.   Dr. BUSHELL [1] thus describes the productions of this locality: "A peculiarly dense, hard, and refractory stoneware is fabricated here, the body of which ranges from reddish, brown, and dark gray shades to black.   All kinds of things are made at this place, including architectural ornaments, cisterns, fish bowls and flower pots for gardens, tubs and jars for storage, domestic utensils, religious images, sacred figures and grotesque animals, besides an infinity of smaller ornamental and fantastic curiosities.   These potteries are distinguished for the qualities of the glazes with which the dark brown body is invested.   One of them, a *soufflé* blue, was copied in the imperial porcelain manufactory by T'ang Ying [in the eighteenth century], from a specimen specially sent from the Palace at Peking for the purpose." Nothing accurate is known about the history of this factory, and additional proof is required to show that dragon-jars were once manufactured there.   It is not very likely that jars strictly identical with those found on Borneo and the Philippines will ever turn up in China, unless by excavations on the ancient sites of the kilns.   Chinese collectors of exquisite ceramic treasures were not interested in this common household ware which the religious spirit of the Malayan tribes has faithfully preserved.   The age of these dragon-jars is illustrated by the fact stated by several observers that the Dayak refused to buy any later imitations made in China which speculative dealers tried to palm off on them, and that any remembrance of their Chinese origin is lost. The same is the case, according to the statement of Mr. Cole, on the Philippines.   This fact is singular, as the natives there have been in constant relations with the Chinese, as a Chinese colony has been settled at Manila for centuries, and it can be accounted for only by the explanation that at one remote period dragon-jars of a superior quality, at least in the eyes of the natives, were fabricated which were not rivaled by the later productions.   This assumption will be quite plausible to one familiar with ceramic developments in China exhibiting different aspects and ever-varying processes and qualities through all periods.   For this reason, I feel inclined to set these dragon-jars in the epoch of the early intercourse of the Chinese with the Philippines, the end of the Sung or the early Ming period, say roughly the time of the thirteenth to the fifteenth century. [2]

[1] Chinese Art, Vol. II, p. 13 (London, 1906).

[2] In China, large vessels of the shape of these dragon-jars, usually of much larger size, are still used everywhere for the storage of the water-supply needed in the household.   They find their place in a corner of the courtyard and are filled, according to want, with the water drawn from wells, which is brought in by carriers or on wheel-barrows.   They are called *kang* or *wêng*, and no doubt represent an an-

The other group of pottery in the Cole collection is characterized by well-made thick and oily glazes ranging in color from a peculiar light-blue to shades of grass-green, dark-green, olive-green, and lilac, sometimes combined on one surface.   There can be no doubt that all these pieces represent *Kuang yao*, either made at Yang-ch'un, or at Yang-kiang, in Kuangtung Province.   None of them is a real celadon, though some of the glazes, in particular the jar on Plate XII, come near to it, to a certain degree.[1]   Similar glazes are still turned out at Yi-hsing on the Great Lake (*T'ai hu*) near Shanghai, but they are inferior in quality to these specimens.   They owe their attractions entirely to the glaze brilliant with its varying colors blue speckled, flecked with green, or green being the prevailing tint, the blue looking out from beneath it in spots or streaks; in one example (Pl. IX, Fig. 2), fine purplish lines like bundles of rays are brought out around the shoulders under the glaze.   The only exception is represented by the jar in Plate XI, which is covered by a dark olive-green glaze, (also in its interior) interspersed with yellowish and brownish spots.   It is possibly a *Sung* production, while the others may belong to the Ming period.   The only decorated jar is that in Plate XIV which is adorned with a flat-relief band of floral designs. The jar in Plate XII has the four ears worked into animal-heads which differ in style from the lion-heads on the dragon-jars.   The larger jars are used in China for holding water, the smaller specimens are wine-vessels.

In regard to the three small pieces grouped on Plate XVII, I have no positive judgment, for lack of material that could be adequately compared with them.   The most interesting of these specimens is that in Fig. 1.   The ornaments of this stoneware dish are laid out in a cinnabar-red paint over a buff-colored glaze; this paint is produced either by means of vermilion or silicate of copper.   A ring on the inner side of the dish is left unglazed; the lower side is completely glazed with exception

cient type of pottery.   During the middle ages, the province of Chêkiang enjoyed a certain fame for their manufacture (see S. W. BUSHELL, Description of Chinese Pottery and Porcelain, p. 130).   At the present time, the best are made in the kilns of Yi-hsing in the province of Kiangsu.— Porcelain jars decorated with dragons are mentioned as having been made in the imperial factory established under the Ming (ST. JULIEN, Histoire et fabrication de la porcelaine chinoise, p. 100).   The extensive rôle which the dragon played during that period is too well known to be discussed here anew.   But as early as the Sung period (and possibly still earlier) the dragon appears as a decorative motive on pottery.   In our Chinese collection in the Field Museum, *e. g.*, there is a large *Sung* celadon plate the centre of which is decorated with the relief figure of a dragon.   Dragons and many other motives were doubtless applied to common pottery centuries before they made their début on porcelain.

   [1] I am inclined to think that such pseudo-celadons have caused travellers in the Archipelago unfamiliar with the ceramics of China or having merely a book knowledge of the subject to see celadons in many cases where there are none, and am seconded in this opinion by Dr. BUSHELL (*l. c.*, p. 13).

of the raised rim on which the dish stands. Nothing like this dish is known to me from China, and I should rather suspect a Japanese origin for it. However, he who will take the trouble to peruse the Japanese account on Luzon pottery, translated below, will receive the impression that it may belong to that still mysterious class styled "Luzon ware" by the Japanese author.

The tiny cup in Fig. 2 is covered with a grayish glaze with an impure yellowish tinge and has a floral design in black-blue overglaze painting; three ornaments along the outward rim resemble fishes. Fig. 3 represents a blue and white porcelain dish, as said before, the only porcelain in this collection; scenery of mountains and water, a rock and a building in the foreground, are painted under the glaze in a darkened blue of poor quality. This piece is of crude and coarse workmanship, and I do not remember having seen anything similar in China. I believe I do not go far amiss in assigning it to the early attempts of the Japanese to imitate the Chinese cobalt-blue, which was first studied by Shonzui on his visit to King-tê-chên in 1510. Also the mark on the bottom (Fig. 3b) betrays a decidedly Japanese trait, and the dish is probably connected with the great export era of Japanese porcelain in the seventeenth century. BRINKLEY (Japan, Vol. VIII: Keramic Art, p. 87) remarks: "With regard to the possibility of Japan's porcelain having found its way to Eastern countries in the early years of its manufacture, it appears from the evidence of a terrestrial globe in 1670 and preserved in the Tōkyō Museum, that Japan had commercial relations with the Philippines, Cambodja, Tonkin, Annam, Siam, and various parts of China, in the beginning of the seventeenth century."

The exaggerated valuation affixed to these pieces of pottery by the Malayan tribes is not by any means justified by their merits, but seems to be largely the consequence of the wondrous stories associated with them. It is accordingly a mere ideal estimation resulting from social and religious customs. Hardly any of these pieces can lay claim to unusual ceramic or artistic qualities, and from a Chinese ceramic viewpoint they are average common household productions, which would not be very costly affairs when made in China at the present time. While the natives have apparently linked their own ideas and beliefs with this pottery, the question is justified as to whether the impetus for the formation of this ceramic lore was possibly received from Chinese traders. It would be plausible to assume that these were clever enough to trade off on the innocents not only the jar, but also a bit of a marvelous story about its supernatural qualities, which was capable of increasing the price by not a few per cent. It was not even necessary for them to strain their imagination to an extraordinary degree, while on the lookout for such stories, as they abound in the domain of their own folklore, so that an optimist might feel inclined to think of them as honest rogues who themselves believed what they told their customers in a mere good-natured attempt to be entertaining.

In the *T'ao shuo* "Discourse on Pottery" written by Chu Yen in 1774 and translated by Dr. BUSHELL, [1] we find the following tradition on record:

"Chou Yi-kung (a celebrated military commander during the Sung dynasty) sent a teacup as a present to a poor friend, who after his return home prepared tea and poured it into the cup, whereupon there immediately appeared a pair of cranes, which flew out of the cup and circled round it, and only disappeared when the tea was drunk."

"Such wonderful stories," continues the Chinese author of the treatise, "may not be impossible like the transformations which happen spontaneously in the furnace. Porcelain is created out of the element 'earth,' and combines in itself also the essential powers of the elements 'water' and 'fire.' It is related in the *Wu ch'uan lu*, that when the military store-house at Mei-chün, in the province of Sze-ch'uan, was being repaired, a large water-jar was found inside full of small stones. After the religious worship on the first day of each moon, another lot of water and stone used to be added, and this was done for an unknown number of years, and yet even then it was not quite full. We read again in the *Yu ya chih*, that while Ts'ao Chu was a small official at Ch'ien-k'ang, Lu was officiating as Prefect, and there stood in front of his Yamen a large jar of the capacity of five hundred piculs, from the interior of which used to come out both wind and clouds. These are similar stories, and are quoted here on that account."

In the same work (p. 47) a story referred to the year 1100 is told to the effect that at a wine banquet of friends the sounds of a pipe and flute were suddenly heard, faintly echoing as if from above the clouds, rising and falling so that the musical notes could almost be distinguished, and how upon investigation it was discovered that they came out of a pair of vases, and how they stopped when the meal was over. Here we meet an interesting analogy with the Philippine talking jars discussed by Mr. Cole. Another magic legend is related regarding a scholar who bought an earthenware basin to wash his hands in. The water remaining on the bottom froze on a cold winter day, and he saw there a spray of peach blossom. Next morning there appeared a branch of peony crowned with two flowers. On the following day a winter landscape was formed, filling the basin, with water and villages of bamboo houses, wild geese flying, and herons standing upon one leg, all as complete as

[1] Description of Chinese Pottery and Porcelain, p. 127 (Oxford, 1910).

a finished picture. The scholar had the basin mounted and enclosed in a silk-lined case; and in the winter, he invited guests to enjoy the sight. The logic of this story is intelligible: designs and scenery as painted on pottery here appear on a plain, coarse basin by a magical process which is suggested to imagination by the flowers formed in an ice-crust.

While these stories seem to have emanated from the literary circle of society and savour of bookish estheticism, there are also others into which more popular elements enter, and which characterize themselves as originating from Taoism. There is a saying in regard to the mysterious ways of the Taoists capable of concentrating Heaven and Earth in a vase. The legend goes that a certain Fei once noticed a stranger jumping into a vase and completely disappearing in it. Fei, in utmost surprise, hurried to the scene and respectfully greeted the old man who invited him to enter also the marvelous vase. He gladly accepted the offer and found a palace with a table covered with exquisite dishes and wines which he heartily enjoyed. The old man possessed the faculty of placing the finest sights of nature in this jar and called himself Vase-Heaven (*Hu T'ien*), subsequently changed into *Hu kung*, "Mr. Vase." [1] Based on this legend, a potter at the end of the Ming period gave himself the sobriquet "the Taoist hidden in a Vase" (*Hu yin tao jên*).[2]

Taoist priests are generally called in by the people to expel evil spirits. They are able to capture the demons and sometimes put them in an earthenware vessel closed with a cover containing some magic character, and the devils are thus safely carried away by the priests. These and other spirits are sometimes sold to the people as imbued with the power of conferring prosperity on their owners, at prices ranging from twenty to forty Mexican dollars.[3]

If the Chinese were lovers of fine porcelains and celebrated them in verses, the Japanese may be called maniacs and worshippers of pottery. In view of their relations with the Philippines and the interchange of pottery between the two, a subject discussed farther below, it may not be amiss to allude briefly to the ceramic folklore of Japan, which, after all, may have stimulated to a certain degree the imagination of the Philippine tribes. It is well known that tea was the chief agency in the refinement of pottery, in Japan as in China, and also in a refinement of life and social manners. The tea-plant was intro-

---

[1] PETILLON, Allusions littéraires, p. 70.

[2] HIRTH, Ancient Chinese Porcelain, p. 200.

[3] Compare E. Box, Shanghai Folklore (*Journal China Branch Royal Asiatic Society*, Vol. XXXIV, p. 125).

duced into Japan from China in the thirteenth century, and at the close of the fifteenth, tea-clubs were formed which practised an elaborate tea-ceremonial growing into a sort of esthetic and religious cult. Needless to say that these tea-tasting competitions were derived also from China and in full swing there as early as the Sung period.[1]

The Japanese devotees of the tea-cult were intent on supplying their cherished pieces of pottery with a history and with poetical names; they were animated with a soul, and wrapped up in precious brocades, treated as gems and relics. They were eagerly bought and sold at prices far out of proportion with their real value. It is recorded, says BRINKLEY (Japan, Vol. VIII, p. 270), that the Abbot Nensei, in exchange for a little tea-jar of Chinese faience, known as "First Flower," obtained in 1584 a vermilion rescript excusing himself and his descendants from the payment of all taxes forever; and it is further a fact that amateurs of the present time disburse hundreds of dollars for specimens of *Soto-yaki* that scarcely seem worth the boxes containing them. Kuroda, the feudal chief of Chikuzen, had a triple case made for a Chinese tea-jar presented to him, and appointed fifteen officials who were all held responsible for its safety (*Ibid.*, p. 319). Of wonderful tales of Japan connected with pottery, the story of the dancing tea-jar which assumed the shape of a badger (*tanuki*) [2] may be called to mind as an analogy to the personification and zoomorphy of Malayan jars.

In 1854 Tanaka Yōnisaburō wrote a book under the title *Tōkikō* "Investigations of Pottery," which was published in 1883 at Tōkyō in two volumes of moderate size. This author has devoted a noticeable study to the pottery introduced into Japan from foreign countries, and shows that many pieces taken for Japanese are in fact of foreign origin. He dwells at length on the pottery of Luzon, which was highly appreciated in Japan, and which seems to have acted as a stimulus to the productions of her kilns. Owing to the importance and novelty of this subject, a complete translation of two chapters of the Tōkikō is here added. In the first chapter, foreign pottery, inclusive of that of Luzon, is considered in general; in the second chapter, Luzon pottery is dealt with more specifically. The general designation of this pottery is *Namban*. The latter is a Chinese word composed of *nan* "south" and *Man*, originally a generic term for all non-Chinese aboriginal tribes inhabiting the mountain-fastnesses of Southern China. It is usually translated "the southern Barbarians," but it is very doubtful

---

[1] BUSHELL, *l. c.*, p. 124. The Japanese tea-ceremonies have been described in many books. Of monographs, W. HARDING SMITH, The Cha-No-Yu, or Tea Ceremony (*Transactions of the Japan Society London*, Vol. V, pp. 42–72) and IDA TROTZIG, Cha-No-Yu Japanernas Teceremoni (Stockholm, 1911) may be mentioned.

[2] First told in English garb by A. B. MITFORD in his Tales of Old Japan.

to me whether any such sting adhered to the name in the beginning. In the ancient Chinese texts, the Man tribes are frequently spoken of with dignity and respect, and Chinese authors do not shun to admit many cultural elements which the Chinese owed to them.   The term Man may occasionally be used contemptuously,— and in what community would an extratribal name not be turned to such an occasional use?— but this certainly does not mean that a stigma is implied in each and every case.   In the Chinese accounts of the conflicts with the Spaniards on the Philippines, the Spaniards are sometimes entitled *Man* instead of their usual name, because the chronicler gives vent to his exasperation at their outrages, and there, it is doubtless intended for savages.[1]   The Japanese adopted from the Chinese the term Nan-Man or Namban and applied it first to all foreign regions south of their home (with the exception of China), its meaning being simply "foreign tribes of the south" or "southern foreigners" including Formosa, the Philippines, the Malayan Archipelago, Malacca, and the two Indias.   Subsequently, it was transferred also to the Portuguese, Spaniards and Dutch who made their first appearance in the southern waters, and it finally assumed the general meaning "foreign," especially in connection with foreign products, like *namban kiwi* "foreign millet," *i. e.* maize, *namban tetsu*, "foreign iron."   The church built by the Jesuits at Kyōto in 1568 and destroyed in 1588 after Hideyoshi's edict of proscription was called *Namban-ji*, "Temple of the Foreigners."

[1] LAUFER, The Relations of the Chinese to the Philippine Islands, pp. 262, 271, 276.   (*Smithsonian Miscellaneous Collections*, Vol. L, Part 2, 1907.)

TWO CHAPTERS FROM THE TŌKIKO

I. Objects of the Namban

The pottery of the Namban Islands which are Amakawa,[1] Luzon,[2] Mo-u-ru,[3] Eastern India,[4] Cochin,[5] Annam, Nekoro,[6] and Taiwan (Formosa) is usually named according to the locality where it is manufactured. In case that its place of origin is not obvious, the people simply speak of Namban objects, as Namban is a general designation for all these places. While the best productions of the Namban are tea-canisters (*cha-ire*), we have no reason to doubt that they produce also utensils of other character. When I investigated a pitcher (*mizusashi*) shaped like ⌘, experts took it for the ware called Enshu Mikirigata Takatori.[7] It was made from a black-purple clay covered with a silvery lustre and brilliant with black marks. I had it exposed

---

[1] The name is transcribed in the text only in Katakana signs, not given in Chinese characters, which would facilitate its identification. Judging from its phonetic composition, it sounds Japanese, and *amakawa* is indeed a Japanese word (meaning "the inner bark of a tree"). No such geographical name, as far as I know, occurs in Japan, the Luchu Islands, or the Philippines. It is mentioned farther below in this text that it forms with Luzon and Formosa the group of Three Islands (*Mishima*) and produces pottery of white clay and grayish glaze.

[2] In Japanese pronunciation: *Rusun* (Chinese: *Lü-sung*).

[3] Presumably the Moluccas; written only in Katakana.

[4] In Japanese: *Tō Indu*. In other passages the word *Tenji* (Chinese: *T'ien-chu*) is used for India.

[5] The Chinese designation *Kiao-chih* is used.

[6] Possibly the Nikobars.

[7] The designation of the famous master of tea-ceremonies (*chanoyu*) Kobori Masakazu (1576–1645) and a group of pottery manufactured according to his instructions in Takatori in the province of Chikuzen (see F. BRINKLEY, Japan, Vol. VIII, Keramic Art, p. 318; OUÉDA TOKOUNOSOUKÉ, La céramique japonaise, pp. 89, 93). This name is given in distinction from the Ko-Takatori (Old Takatori) started by Korean settlers in that district. It is not very likely that the above mentioned pitcher is of real Takatori make, as a glaze of that description does not occur among Takatori productions known to us, which generally are of white, light-blue or ash-colored glazes, or take the Chinese "transmutation glaze" (*yao pien*) as model. Our author evidently means to express the same opinion which leads him to class the piece in question among foreign or Namban wares.— In this connection, it is interesting to note that in the ancient pottery kilns at Sawankalok, Siam, small vases and bottles have been discovered by Mr. T. H. LYLE, described by Mr. C. H. REID as being "of a fine pottery covered with mottled glaze, the shapes often elegant, and sometimes highly finished, recalling the fine tea-jars made at Takatori in the province of Chikuzen in Japan" (*Journal Anthropological Institute*, Vol. XXXIII, 1903, p. 244).

32

to a fire, and the glaze assumed a golden hue.   The clay was a mixture of yellow and red earths and changed into a brown.   It proved to be a Namban production.

Further among Yashiro Karatsu-hakeme [1] wares, there was a specimen of black-purplish clay emitting, when struck, a metallic sound.   I had a piece broken out, and clay and glaze on examination under a lense attested to its being Namban.   Among old Hakeme, that kind known as *kōdai* [2] with black-purplish clay and dark-brown [3] and silvery lustre is Namban Hakeme.   When investigating some pieces without marks among Bizen, [4] Imbe, [5] Karatsu, [6] and Tamba, [7] they proved to be Namban.

Mishima ("Three Islands") pottery is that made on the three islands of Amakawa, Luzon, and Formosa.   Among this so-called class of Mishima, the large pieces with purple-black clay and green glaze (*sei-yaku*) are Luzon pottery; [8] those of white clay and grayish [9] glaze are Amakawa.   As to Formosa, I have as yet no proofs, but pieces popularly called Hagi Mishima[10] with a light lustre and decorated with

[1] Karatsu or Nagoya on the north-west coast of Hizen has been the harbour of entry and exit for the greater part of the traffic between Japan, China and Korea; the name Karatsu means "port for China."   BRINKLEY (*l. c.*, pp. 307 *et seq.*) and EDWARD S. MORSE (Catalogue of the Morse Collection of Japanese Pottery, pp. 37 *et seq.*) have devoted full discussions to the pottery productions of Karatsu.   Those with a broad brush-mark of white are termed *hakeme, i. e.* brush-marked.   BRINKLEY maintains that the potters of Karatsu were chiefly imitators, and that, their best efforts being intended for the tea-clubs, they took as models the rusty wares of Korea, Annam, *Luzon*, etc., or the choicer but still sombre products of the Seto kilns. If this statement be correct, the specimen alluded to above might be also a Karatsu imitation of a Namban pottery.

[2] *Lit.* high terrace.

[3] Jap. *shibu*, the juice expressed from unripe persimmons (*kaki*), from which a dark-brown pigment for underglaze decoration was obtained in Korea (BRINKLEY, p. 49).

[4] The province of Bizen is celebrated for its hard reddish-brown stoneware described by BRINKLEY (pp. 328 *et seq.*) and MORSE (pp. 49 *et seq.*).

[5] Imbe is a district in the province of Bizen.   Under the name *Imbe-yaki*, "pottery of Imbe," or *Ko-Bizen*, "Old Bizen," the ware made there at the end of the sixteenth century is understood (BRINKLEY, p. 329).   Nearly every piece of Imbe ware bears a mark of some kind, usually impressed (MORSE, p. 49) so that the pieces without marks seem to be the exceptions justifying to some extent the suspicion of a foreign origin.

[6] It is difficult to understand what is meant by unmarked pieces of Karatsu, as the Karatsu potters were not in the habit of marking their productions, and have left no personal records (BRINKLEY, p. 311).   See also the last paragraph of this chapter where the presence of marks on Karatsu is utilized as evidence of its foreign origin.

[7] On the pottery of the province of Tamba see BRINKLEY, p. 398, and MORSE, pp. 178, 347, 360.

[8] Apparently celadons.

[9] Jap. *shiro-nezumi*, "white rat."

[10] Manufactured in the province of Nagato, with a pearl gray glaze (BRINKLEY, p. 343; MORSE, p. 81).

a row of round knobs, or water pitchers with black marks on the bottom appear to be Formosan.   I shall deal with this subject in a subsequent book.   Among Gohon [1] Mishima, there are Korean and Mishima. Specimens called Kumo-tsuru Mishima [2] with good lustre and fine writings are Amakawa.   Mishima is merely a general designation.   It should be specified as Higaki [3] Mishima, Rei-pin [4] Mishima, Hana [5] Mishima, Hakeme [6] Mishima, Muji [7] Mishima. [8]

Among the Irapo [9] I tested the clay of Old Irapo with the brush-mark (*hakeme*) Kukihori Genyetsu Irapo, [10] and found it to be Namban clay.   Its make-up is crooked (*yugami*), and it is hard like Korean.   As regards the name of the potter Genyetsu, he was usually called Kuki-hori. [11]   Writing the latter name with the Chinese characters for *kugi* ("nail") and *hori* ("to carve") is of recent origin.   Kukihori is the name of a locality.   His style is not limited to the Irapo, but some of the Gohon [12] are like it.   Considering a rice-bowl, [13] a confusion with Korean ware is possible; in regard to tea-canisters (*cha-ire*), how-ever, they are obviously Namban.   The Genyetsu Irapo very seldom go by the mark "made by Genyetsu" (*Genyetsu-saku*).   It is the same case as with the Ki-Seto of Hakuan [14] under whose name originals and

[1] Gohon is the name of a pottery made in Korea at the instigation of Iyemitsu, the third Shōgun of the Tokugawa family (1623–49) which was imitated in the kilns of Asahi in Yamashiro Province (T. OUÉDA, La céramique japonaise, p. 89).   BRINK-LEY (p. 356) remarks that in the Asahi ware imitations are occasionally found of the so-called Cochinchinese faience, but that they are rare and defective.   This fact may account for the above definition of Mishima.

[2] *I. e.* Mishima with clouds (*kumo*) and cranes (*tsuru*); also to be read *Un-kwaku* in Sinico-Japanese pronunciation.   According to BRINKLEY (p. 48), this design was a favorite in the Korean celadons manufactured at Song-do.   In all probability, celadons are involved also in this case.

[3] *Higaki* means a hedge or fence (*kaki*) formed by the tree *hi* or *hinoki*, *Thuya obtusa.*

[4] Evidently a transcription of the name Philippines, the first syllable being dropped.   Japanese lacks the sound *l* and substitutes *r* for it.

[5] *I. e.* flowery or decorated.

[6] Decorated with brush-work.

[7] Plain or undecorated.

[8] Here the term Hana Mishima is repeated, though occurring only in the pre-ceding line.   The book is somewhat carelessly written.

[9] The *irapo* were low-priced bowls serving in Korea for making offerings to the dead on the cemeteries (T. OUÉDA, *l. c.*, p. LVIII).

[10] *I. e.* an *irapo* bowl made by the potter Genyetsu from Kukihori.

[11] Written with Katakana signs.

[12] See above note 1.

[13] Jap. *chawan, lit.* a tea-bowl, by which a large bowl to eat rice from is understood at present, while a tea-cup is called *cha-nomi-wan*, "bowl for tea-drinking."

[14] The name of a potter in the latter part of the fifteenth century about whom very little is known.   BRINKLEY (p. 274) and MORSE (p. 200) place him in the latter part of the fifteenth century, TOKOUNOSOUKÉ OUÉDA (p. 8) in the first half of the seventeenth.   His name is connected with the production of a yellow faience, the

imitations are included. It is a mistake to designate all Gohon as Korean.[1]

Toyotomi Hideyoshi (1536–1598) despatched a ship from Sakai to Luzon and had a genuine jar (*tsubo*) made there. At that time, not only jars were brought home from there, but it is also probable that he sent to Luzon samples of Furuori Enshu.[2] Among Namban ware we find a cup to wash writing-brushes in (*fude-arai*) called Hana Tachibana, [3] copied from Raku ware, [4] and also a plain bowl in the style of Shigaraki Enshu Kirigata.[5] The fact that the lord of Enshu allowed the seal of this ware to be placed only on the pottery for his own royal household and on that of Ido [6] is of deep significance.

The Mishima bowls (*domburi*) now in general use are made of pur-

yellow ware of Seto (*Ki-Seto*), some of which are attributed by tradition directly to his hand. The later copies of his work were, as in so many other cases, named for him, and this makes the point of coincidence with the Irapo of Genyetsu.

[1] This is a repetition of what was stated above in regard to the Gohon.

[2] The Takatori pottery named for the lord of Enshu, Kobori Masakazu (see above, p. 32). The term *furuori* (or according to Sinico-Japanese reading *ko-shoku*) means "ancient weavings," and possibly refers to a group of pottery decorated with textile patterns. If the above statement should really prove to be an historical fact, it would shed light on the piece of alleged Enshu pottery discussed by our author in the beginning of this chapter and explained by him as Namban. We could then establish the fact of an interchange of pottery between Takatori and Luzon which would have resulted in mutual influences and imitations.

[3] *I. e.* decorated pottery with an orange glaze. This ware was produced toward the middle of the eighteenth century at Agano, Buzen Province; its glaze was granulated so as to resemble the skin of an orange, hence known as *tachibana* (BRINKLEY, p. 403). The process is of Chinese origin (ST. JULIEN, Histoire et fabrication de la porcelaine chinoise, p. 195; S. W. BUSHELL, Description of Chinese Pottery and Porcelain, p. 58).

[4] Raku is the designation of a hand-made pottery originating from a Korean potter Ameya Yeisei who settled at Kyōto in 1525. His son Chōjiro was protected by Hideyoshi and presented by him with a gold seal bearing the character *Raku* ("Joy") derived from the name of his palace Jūraku erected at Kyōto in 1586; hence the mark and name of this pottery.

[5] Shigaraki is a place in the Nagano district, Ōmi province, where pottery furnaces were at work as long ago as the fourteenth century. Large tea-jars for the preservation of tea-leaves were the dominant feature of its manufacture. A tea-jar of this kind, of extraordinary size, glazed a light-reddish tinge with splashes of pale-green overglaze on the shoulders, is in the collections of the Field Museum. The variety of Shigaraki mentioned in the text is usually called *Enshu-Shigaraki*, named after Kobori Masakazu, the lord of Enshu, to whom reference was made above (p. 32). According to BRINKLEY (p. 369), the productions with this label offer no distinctive features, but are valued by the tea-clubs for the sake of their orthodox shapes and sober glazes.

[6] Ido is a keramic district in Korea from which Shinkuro and Hachizo hailed, two Korean captives who after Hideyoshi's expedition to Korea settled at Takatori in Chikuzen and started a kiln there. During the early years of their work they used only materials imported from their native country, and these productions were therefore designated as Ido. Kobori Masakazu, the feudal chief of Enshu, interested himself in the Korean potters and became influential in the perfection of their work. The *Ido-yaki* seems to have served also the Korean potter of Hagi as a model, for the chief characteristic of his productions was grayish *craquelé* glaze with clouds of salmon tint (BRINKLEY, p. 344).

plish clay, and the glaze is decorated in the style of nipples (*chichimi moyō*). They are much neater than Korean ware. As they are fired under an intense heat, their shapes are well curved, and their sound is metallic. They are all of Kukihori style.

Old Ido (*Furu-Ido* or *Ko-Ido*), Green Ido (*Ao-Ido*), and Ido-waki which are green and hard are manufactured in Eastern India. The book *Wa-Kan-cha-shi* ("Records regarding Tea in Japan and China") says that this pottery comes from India, and that even those pieces said to be produced in Korea have come over from India; the assertion of some that it is called Ido as being made in the style of a certain potter Ido is erroneous; Furu-Ido and Ao-Ido are entirely different from other Ido both in clay and glaze. This explanation of the Book on Tea is correct: the Ido mentioned above are of Indian make, and the other Ido are Korean.[1] There are also Shiūsan[2] Ido and Sowa[3] Ido which appear to be kinds of pottery of India Ido. Their glaze is blistered and of low grade. Ao-Ido is the celadon[4] of India. Among the objects left in the temple Kin-chi-in by Tōdō Takatora[5] (1556–1630), there are also Ido which seem to be celadons (*seiji*).

Namban Totoya[6] pottery has a blue-black glaze uncrackled. Its clay is black and purplish, and its sound is metallic. Some have three or four apertures[7] in the body, and others more. The old ones are called Kaki-no-heta ("Persimmon-calyx"). Among this class, also incense-boxes (*kōgo*) and pitchers (*mizusashi*) are found.

As regards Namban celadon (*seiji*), it has a black-purple clay and green glaze (*sei-yaku*) running in white streams (*tamari-yaku*) here and there. It has a metallic sound and is popularly called Muji Kumo-tsuru or Un-kwaku,[8] or Hagi make. As regards the production of the green, it is called Karatsu Kumo-tsuru. What the ancients called Muji Kumo-tsuru is this.

The pottery designated as Old Kumo-tsuru and Kumo-tsuru is a production of the Namban. Its style of painting is fine, and the mark

---

[1] Regarding these Korean Ido see BRINKLEY, pp. 51–52.

[2] The name is composed of the two characters for "ship" (*shiu*, Jap. *fune*) and "mountain" (*san*, Jap. *yama*). The name is derived from Chou-shan, a place in the province of Fukien, China, where a hard white porcelain was made.

[3] Transcribed in Katakana.

[4] Jap. *seiji*, "green porcelain," identical with the Chinese name for celadons. Probably the celadons of Siam are meant here (T. H. LYLE, Notes on the Ancient Pottery Kilns at Sawankalok, Siam. *Journal Anthropological Institute*, Vol. XXXIII, 1903, pp. 238–245).

[5] A daimyō who served Nobunaga and then Hideyoshi and retired on his master's death into the monastery Kōyasan.

[6] Transcribed in Katakana.

[7] *Lit.* eyes.

[8] *I. e.* plain, with clouds and cranes, a favorite design in celadons.

of a tripod vessel (*gotoku*) with which it is provided is also a tripod vessel of the Namban.

On the preceding pages the difference between Korean and Namban pottery has been explained. Further details will follow. Namban are the various countries as described in the previous notes (*koguchi-gaki*).

Namban pottery provided with the seal of the oven from which it originates is usually not recognized as such by our contemporaries, though clay and glaze point to its being Namban. The mark ⊤ [1] finely made on Imbe ware, the mark ✕ on a jar (*tsubo*) of Bizen, the mark ⊤ three times on a tea-canister (*cha-ire*) of the same ware, and the mark Roku-zō on a pitcher (*mizusashi*), and marks on several other potteries represent the national writing of Luzon (*Luzon-no kokuji*).[2] Also a deep-brown glazed tea-canister (*shibu yaku-no cha-ire*) on which the character ⊤ [3] is written consists of Namban clay. Some of these marks may have been produced by Japanese who crossed over; but others may have been made by the natives (*Man-jin*), for it cannot be ruled that Namban has no marks of the furnace. There is, *e. g.*, on a fire-pan (*hi-ire*) of Annam the mark *Ta-kang* [4] impressed by means of a seal, which is the name of the maker. The tea-canisters called Chōsen-Garatsu (Korean Karatsu) [5] which have a plant-green (*moyegi*) glaze and purplish clay, or also dark-brown (*shibu*) glaze with purplish clay are taken by our contemporaries for real Karatsu-make on account of their seals of the furnace, but I consider them as foreign manufac-

---

[1] This is a Chinese character (*ting*, Jap. *tei*). Imbe pottery is characterized by a great variety of peculiar marks the significance of most of which is unknown (see MORSE, pp. 49 *et seq.*).

[2] The following characters, found in Philippine alphabets, resemble somewhat the markings on these vessels: ⊤ Pampanga; ξ Tagalog; ξ Ilocano, equivalents for *la;* } Visayan; ⊤ Pampanga and Tagbanua for *na;* ✕ Tagbanua for *ka*. [F. C. C.]

[3] Denoting the numeral 7.

[4] The two characters are transcribed according to the Annamite pronunciation.

[5] BRINKLEY, p. 310.

tures in view of the presence of the seals. Such canisters were the models of the Oribe ware.[1] Karatsu pots are not made with a view to durability and therefore not in need of affixing a seal of the furnace. According to clay and glaze, they are objects of the Man. It may be that Japanese who went abroad imported this ware, or some may have imported the clay and glaze and baked the vessels at home. At any rate, it is not the clay and glaze of Karatsu.

II. LUZON

Of pottery vessels of Luzon, there is a large variety. As a rule, people call only jars (*tsubo*) and tea-canisters (*cha-ire*) Luzons. Owing to the fact that all other articles of Luzon bear out a similarity to those of Hagi, Karatsu, Seto, Bizen, Tamba, Takatori, Higo, Oribe, and Shino, [2] Luzons are erroneously believed to be restricted to the above two articles. Comparing the specimens discovered by me with those imported at present by Chinese junks, I may give the following descriptions of the various wares.

1. Tamba looks very much like Luzon. Luzon is of hard clay and lustrous glaze. Greenish-yellow glaze is splashed (*fukidasu*) over the bottom. Our home-made ware (*i. e.* Tamba), however, is soft, and greenish-yellow glaze is painted on the bottom. It frequently

[1] BRINKLEY, p. 275.

[2] These are, with the exception of Shino, names of pottery-producing localities in Japan; the wares themselves are simply named for the places of production. Most of them have been referred to in the preceding chapter. Hagi is the chief town in the province of Nagato where pottery kilns were started in the sixteenth century by a Korean whose descendants have continued the manufacture down to the present time. Higo is the principal province on the island of Kiūshū where pottery-making, also under Korean influence, commenced in 1598. The Shino pottery, a rude stoneware of thick, white crackled glaze, decorated with primitive designs in dark-brown (*shibu*) pigments, was originated in 1480 by Shino Ienobu, a celebrated master of the tea-ceremonies (BRINKLEY, p. 276); MORSE (p. 191) gives 1700 as the earliest date to which pieces recognized under the name of Shino go back, but the type of this pottery must have been made long before this date, as the gray, white-inlaid Shino is accorded an age of three hundred and fifty years.

Our author Tanaka has a different story to tell regarding the origin of Shino. In his second volume (p. 9) he relates that Shino Munenobu utilized a white-glazed water-basin from Luzon and turned it into a rice-bowl, which gave rise to the name "bowl of Shino" (*Shino chawan*); later on, this bowl was handed down to Imai Mune-hisa, but the book *Mei-butsu-ki* ("Records of Famous Objects") says that it is Chinese; imitations of this bowl made in Owari are called *Shino-yaki;* there are many wares from Luzon and Annam which are like Shinoyaki, and which should be carefully distinguished according to clay and glaze. This account plainly shows how hazy and uncertain Japanese traditions regarding their potters and pottery are. The man Shino Munenobu is called by BRINKLEY Shino Ienobu, by MORSE Shino Saburo or Shino Oribe (pseudonym Shino So-on), by T. OUÉDA Shino Soshin. Has he really lived, and when? If he lived in the latter part of the fifteenth century, as maintained by a weak tradition, he is not very likely to have obtained any pottery from Luzon, as there is no evidence of Japan having had any intercourse with the Philippines at such an early date.

happens that Luzon is mistaken for a Tamba.[1] The distinction must be made by examining the particular features, as they closely resemble each other in their general make-up.

2. Matsumoto Hagi [2] is of soft (*yawaraka*) clay, its glaze is not transparent (*sukitōru*), and its sound is mellow (*yawaraka*). Luzon has a white clay and lustrous glaze, its lustre being more vigorous than the green of a snake (*jakatsu*);[3] it has a clear sound. There are Tamba which are alike Matsumoto; they are of yellow clay.[4]

3. Takatori is of red clay and crackled glaze. Luzon is of white and yellow clay, with uncrackled glaze, and has the design of a whirl (*uzu*) on the handle.

4. Among Seto there are Luzons.[5] Among these there are pitchers, bowls and tea-canisters with gold glaze and black streaks running over it. They are found scattered among those called "certain wares" (*naniyaki*).

5. Oribe and Luzon resemble each other. Luzon is hard and lustrous; Oribe is soft and of poor lustre.

6. There are also Shino which are identical with Luzons. Luzons have a transparent glaze, and on the bottoms and handles of the bowls

---

[1] The notice of BRINKLEY (p. 399), presumably derived also from a Japanese source, that the early productions of Tamba,— a peculiar faïence having reddish paste and blisters on its surface,— are supposed to resemble an imported ware attributed to Siam, is remarkable in this connection. BRINKLEY, further, alludes to splashed glazes on Tamba which occasionally occur and are not without attractions, and Mr. MORSE (p. 179) describes a Tamba jar of rich brown Seto glaze with splashes of lustrous brown, mottled with greenish-yellow; but neither mentions splashes or paints on the bottom. OUÉDA TOKOUNOSOUKÉ (La céramique japonaise, p. 90) says regarding the ancient Tamba pieces that their surface is uneven or rough like the Korean vases or those of the *Namban*,— the only previous instance in our literature where this term has been used with reference to pottery.

[2] Matsumoto is a place in the Abu district, province of Nagato. The Korean Rikei who opened pottery work in Hagi, on his search for suitable clay, first discovered it at Matsumoto, and there he settled (BRINKLEY, p. 344; MORSE, p. 82).

[3] *Jakatsu* is the name of a peculiar glaze invented in China, imitated in a ware of Satsuma; its dark gray and green glaze is run in large, distinct globules, supposed to resemble the skin of a snake (but not the scales on a dragon's back, as BRINKLEY, p. 137, says). In China, this glaze (called "snake-skin green," *shê p'i lü*) first appears in the era of K'ang-hi (1662–1722) and is still imitated at King-tê-chên (ST. JULIEN, Histoire et fabrication de la porcelaine chinoise, pp. 107, 195).

[4] Mr. MORSE (p. 178) speaking of the earliest Tamba made in Onohara evidently alludes to this passage when he says: "These are probably the ones mentioned in Tōkikō as resembling old Hagi." But Morse maintains that these pieces have reddish clay.

[5] Seto is a small village in the province of Owari. The Seto ware (*setomono* or *setoyaki*) which has become the generic term for all ceramic manufactures of Japan was originated by Tōshiro, the so-called Father of Pottery (regarding his life see MORSE, pp. 183–184). In Vol. II, p. 11 of his work, Tanaka remarks that among a kind of yellow Seto (*Ki-Seto*), to which we referred above (p. 34), with lustrous glaze and metallic sound. Luzon, Annam, and Fukien wares are mixed, that the latter has fine white clay, while Seto clay is coarse.

there are designs of *tomoye* and three apertures. Shino has nothing of the kind.

7.   The tea-canisters. of Luzon are of the best quality. Those which might be confounded with Tamba are first and second grades. Those looking like Seto are coarse and low grades of Luzons. They are frequently found together with those ranging as second qualities among the Namban.[1] The water-dishes (*mizu-ire*)[2] and oil-dishes (*abura-ire*) among the Namban with yellow clay and splashes of dark-brown (*shibu fukidasu*) are Luzons.

8.   I obtained a tea-canister of the shape ⌴ similar to Tamba.

On its shoulders, four plum-blossoms and two seals are impressed by means of a stamp. The writing was first illegible, but when I rubbed

it, it appeared as follows:[3] . The symbol in the latter

seal may be the character ☐ *Lü* in the national writing of Luzon.[4]

This vessel was of yellow clay and tea-colored glaze with splashes of dark-brown (*shibu*).

9.   Pearl-gray celadon (*shukō seiji*) is the celadon (*seiji*) of Luzon.

10.   Sun-koroku should be written Rusun- (*i. e.* Luzon) kōroku.[5]

[1] Namban is here expressed in the text by "insular objects."

[2] Dishes containing the water to be poured on the ink-pallets and used in rubbing a cake of ink.

[3] In the Japanese text, the two seals are placed the one below the other; for the sake of convenience, they are here arranged side by side.

[4] This supposition is probably correct. The case is as follows. The second portion of the seal plainly contains two Chinese characters reading *sung ch'i;* this character *sung* is used in writing the second syllable in the name of Luzon, Chinese *Lü-sung*, Japanese *Ru-sun*. It is therefore logical to conjecture the character for *Lü* preceding that for *sung*. The sign in the first seal, however, is not obviously identical with the latter, but apparently a variation of it in ornamental style, which, as suggested by our author, may have developed on Luzon itself. If we adopt this reading, we obtain the legend: *Lü-sung ch'i* (Chinese) or *Ru-sun tsukuru* (Japanese), which means: "Luzon make," or "made on Luzon." I see no reason to doubt the credibility of our informant, and take it for granted that a vessel with such a seal really was in existence. This fact, then, is of great historical importance, for it demonstrates that pottery may have actually been manufactured on the Philippines either by Chinese or Japanese, or by both.

[5] Mr. MORSE (p. 321) alludes to this passage in the following notice: "The work *Tōkikō* says that the word Sunkoroku ought to be written Rosokoroku. It further adds that *Sun* stands for the Chinese dynasty, and *Koroku* the name of a pottery." But it will be seen from the above text that our author means to express a different sense. He is far from identifying the word *Sun* with the *Sung* dynasty, but proposes to interpret it as *Lü-sung*, *Rusun*, Luzon (the reading *Roso* is certainly possible, but the Tōkikō, in the first passage where the word occurs, transcribes the characters in Kana as *Rusun*). The pottery called Sunkoroku is, according to Morse, a hard stoneware with dull yellowish or grayish clay (that having the former

On Luzon it is the designation for a dyed article.   On a flowerpot of Mitani Riōboku Fukushiu, the legend *Karamono Kōroku* (*i. e.* Chinese Kōroku) is inscribed.   Kōroku is an article of pottery.   It is so called by combining the names of the utensil and the locality.   It is soft because it is not thoroughly baked.   Among later imports some with black designs and pale-yellow glaze are encountered.   Its sound is solid.

11.   Luzon is compact and dense both in clay and glaze.   After years, when washed, it appears like new, and its age may be doubted. This is due to the intense heat of the tropical regions.

12.   The genuine jars and tea-canisters have their bottoms concave.[1] The "Book on Tea" (*Cha-kei*) says: "When placed on the bottom and on the sides of the body, tea keeps well in these jars."   Luzon, therefore, is serviceable for tea.

13.   The best qualities are of white clay; the middle grades are of yellow clay mixed with white clay and sand; and the lowest grades are of purplish-black clay.

14.   All Luzon pieces have the wheel-mark (*rokuro*) .   On the incense-boxes (*kōgō*) it is always found inside of the body and on the lid.   On the basins (*hachi*), censers (*kōro*) and bowls (*chawan*) it is outside on the bottom.   On the pitchers (*mizusashi*) it is on the handle. Among the so-called Koshido of Iga Shigaraki, Luzons are numerous. They should carefully be distinguished.   Those of stronger lustre and free from any defilements are Luzons.   One will surely find two vertical spatular marks on the right.

The following varieties are encountered among Luzons: Tea-canisters with plum-blossoms impressed by means of a stamp, and a

color being the oldest) with a peculiar archaic decoration of scrolls and diapers, rarely landscapes, carefully drawn in dark brown; whatever the origin of the style of decoration, it forms a most unique type.   There are fifteen pieces of this pottery in the Morse collection at Boston, and one of these is dated 1845.   It may hence be inferred that the first part of the nineteenth century is the period when the Sunkoroku was in vogue.   The Japanese concerned seem to agree in assigning to it a foreign origin.   T. Ouéda (La céramique japonaise, p. 69) explains the word as the name of a centre of foreign manufacture the products of which were imitated. Brinkley (p. 171) holds a more elaborate theory.   He makes Sunkoroku a variety of Satsuma copied from a faience of archaic character manufactured near Aden, and valued by the Japanese for the sake of its curiosity and foreign origin.   "The *pâte* is stone-gray, tolerably hard, but designedly less fine than that of choice Satsuma wares.   The glaze is translucid, and the decoration consists of zigzags, scrolls, diapers, and tessellations in dark brown obtained from the juice of the *Kaki*.   The Indian affinities of this type are unmistakable.   It is not without interest, but a somewhat coarse gray faience with purely conventional designs in dark brown certainly cannot boast many attractions.   The original ware of Aden is, in some cases, redeemed from utter homeliness by a curious purplish tinge which the glaze assumes in places."   It is evident that this pottery is different from that of our Japanese author, which is stated to be soft.

[1] This is the case in all specimens of Philippine jars in Mr. Cole's collection

thin yellow-green glaze. The same with a combination of black and gold glaze. The same with gold glaze. The same with black glaze. The same with tea-colored (brownish) glaze and provided with ears. The same with green-yellow glaze. The same with yellow glaze. The same of the shape of a rice-kettle. The same with four nipples.[1] The same with projecting bottom. The same called *Usu-ito-giri*.[2] The same called *Hi-tasuki*.[3] The same with candy-brown[4] glaze. Monrin.[5] Tegami.[6] Oil-pitchers. The same with ears. The same, Utsumi and Daikei.[7] The same called Nasubi.[8] The same called Warifuta.[9] Various shapes of Bizen. Shapes of Iga, and other kinds.

Of Mishima there are the following: Undecorated common ones (*muji-hira*). The same, of the black variety of the country Go.[10] With painting of a trout (*ayu*). Various kinds with brush-marks (*hakeme*). Old Mishima. Deep bowls (*domburi*). Various Mishima. Gourd-shaped fire-holders with brush-mark.

Of white porcelain, there are the following: Pitchers (*katakuchi*). Hand-jugs (*te-bachi*). Boat-shaped jugs (*fune-bachi*).[11] Various bowls. Tachimizu. Gourd-shaped fire-holders. Plain basins. Fire-holders

---

[1] *I. e.* knobs. The Chinese archæologists avail themselves of the same expression in describing the knobs on certain ancient bronze bells and metal mirrors. Compare p. 36.

[2] *I. e.* cut with a thin thread. The thread was used to cut off the superfluous clay at the bottom of the piece before removing the latter from the wheel, a contrivance first applied by the famous potter Tōshiro of the thirteenth century (BRINKLEY, p. 266). The term is here simply used in opposition to the pieces with projecting bottom.

[3] *I. e.* vermilion cord; *tasuki* is a cord used for girding up the sleeves while working. These vessels doubtless had a cord brought out in relief around the neck, as may be seen, *e. g.*, also in Chinese terra cotta of Yi-hsing.

[4] *Ame* or *takane* is a kind of jelly made from wheat or barley flour.

[5] Or *Bun-rin*. BRINKLEY (p. 319) mentions a tea-jar named *Fun-rin cha-tsubo*, without explaining this designation.

[6] Jars with ears.

[7] *Utsumi* (Chinese: *nei hai*) means inland sea, and *daikei* (Chinese: *ta hai*) great sea; expressions to denote certain varieties of pottery.

[8] *I. e.* egg-plant.

[9] *I. e.* with divided lids.

[10] Chinese: *Wu*. Wu was the name of an ancient kingdom in China inhabited by a non-Chinese stock of peoples and comprising the territory of the present province of Kiangsu, the south of Anhui, and the north of Chêkiang and Kiangsi. An ancient tradition has it that the Japanese called themselves descendants of the ancestor of the kings of Wu (CHAVANNES, Les mémoires historiques de Se-ma Ts'ien, Vol. IV, p. 1), and the oldest cultural relations of Japan with China refer to this region. The Japanese understand the name *Wu* (or *Go* according to their pronunciation) in the sense of middle China, also as China in general, sometimes more specifically as the region of the Yangtse Delta, or as Nanking.

[11] A jardinière in the shape of a boat, of Shino pottery, is figured in *Collection Ch. Gillot*, p. 104.

with ears. Kaya-tsubo.[1] Wine-cups [2] with ears, on stands. Kōroku water-pitcher. Kōroku deep bowls (*domburi*). The same, hexagonal incense-box. The same, incense-box in straight lines.

Of Oribe shapes, the following are known: Three incense-boxes. Water-pitcher (*mizutsugi*). Flower-vase. Bowl. Basin.

Of Shino shapes, a bowl, incense-box, water-pitcher, wine-cup, saucer, basin, jug (*katakuchi*), and others, are known.

Of black-glazed ware, flower-vases with ears, various water-pitchers, and tachimizu with ears are known.

Of Iga shapes, water-pitchers with ears, one made by Kōson,[3] and various pieces similar to Iga and Shigaraki are known.

Various pieces resembling Seto, Tamba, Takatori, Yasshiro, Karatsu, etc.

Of plainly burnt ware:[4] bottle with vermilion cord (*hi-tasuki tokuri*); fire-holder; gourd-shaped water-pitcher; large and small kayatsubo.

Of gourd-shaped pieces there are: jugs (*katakuchi*); hexagonal ones; tachimizu; flower-vases with horizontal rope and ears; rippled bowls with sea-slug glaze;[5] basin in the shape of a fish; water-pitcher with dark-brown (*shibu*) glaze and potato-head (*imo-gashira*).

Of Shibu ware there are water-pitchers with indented rim; green-glazed *katakuchi*, and the same of black glaze and gold glaze.

This account is exceedingly interesting, but must certainly not be accepted on its face value. The author apparently suffers from a certain degree of Luzonitis by seeing Luzon ware in every possible case, and without rendering himself a clear account of what this Luzon pottery is. Judging from the extensive trade carried on between China and the Philippines, the large bulk of foreign pottery brought to the Islands must have been of Chinese origin, and the descriptions given by our Japanese author, however succinct they may be, hardly allow of any other inference than that the pieces referred to are Chinese. If we adopt this point of view, an embarrassing difficulty arises at once. If it is here the question of plainly Chinese pottery, why does the Japanese scholar not make any statement to this effect? Is it believable that a Japanese expert in ceramics who is bound to know Chinese pottery thoroughly, and who writes about it with authority in the same

---

[1] *Lit.* mosquito-net jars.

[2] *Choku, lit.* pig's mouth.

[3] Apparently provided with this mark.

[4] *Suyaki-mono, i. e.* unglazed pottery.

[5] *Namako-gusuri,* so called from the likeness which the *flambé* glaze bears to the greenish-blue mottled tints of the sea-slug (*namako*), a Chinese glaze imitated in Satsuma ware (Brinkley, p. 137).

book, should have failed to recognize the Chinese character of pieces brought from Luzon over to Japan? If he does not allude to any Chinese relationship, but classifies this ware as a distinct group of Luzons,— what is, or could then be, the specific character of these productions to differentiate them from Chinese or any other? One point is obvious at the outset,— that this Luzon ware cannot be due to any native tribes of the Philippines. The descriptions refer to highly glazed pieces of an advanced workmanship, such as have never been turned out by the aborigines, whose primitive unglazed or polished earthenware could hardly have tempted the Japanese, not to speak of having elicited their admiration, as we read on the preceding pages. In order to understand, on the part of the Japanese, the assumption of an individual, artistic Philippine pottery coveted by them and deemed worthy of imitation, we have three possibilities to take into consideration: the trade of Siam and Cambodja with the Islands by which pottery of these countries has doubtless reached them, particularly the celadon made in Siam; a special manufacture of pottery in China for the needs of the Philippine market; and possibly, to a certain extent, a home production on the Islands through Chinese or Japanese settlers (or both).[1] By availing themselves of local clays and glazing materials, these may have accomplished a ware of fairly peculiar qualities and yet not much removed from what they had learned in the lands of their birth. Such an hypothesis would indeed meet the requirements of the situation advantageously and satisfactorily. The only objection to be made to it,— and it is certainly a strong one,— is that no record of any Sino-Japanese pottery-making on the Islands exists, either in Spanish accounts, or in native traditions, or in Chinese and Japanese literature. On the other hand, no valid reason could be advanced against the possibility of its existence, and in the same manner as the ruins of the celadon kilns of Siam, for a long time disowned, have finally been discovered, we may expectantly look forward to a future similar

---

[1] In the *Seiyō-ki-bun*, an old Japanese manuscript by Arai Hakuseki, translated by S. R. Brown (*Journal North China Branch Royal Asiatic Society*, N. S., Vols. II and III, Shanghai, 1865 and 66), there is the following passage relative to the Japanese settlement on Luzon (Vol. II, p. 80): "In the southwest part of Luzon, there is a mountain which produces a large amount of silver. More than three thousand descendants of Japanese emigrants live there together and do not depart from the customs of their fatherland. When their officers make their appearance abroad, they wear two swords and are accompanied by spear-bearers. The rest of these Japanese wear one sword. The Spaniards have laws for the government of this colony of Japanese, and do not let them wander about in the country indiscriminately. Four years ago twelve Japanese who had been driven off from our coast by a storm, arrived at Rusun [Luzon], and the Spaniards assigned them a place with the rest of their countrymen." The political platform of these Japanese colonizers, who seem to have been settled before 1598, was an *entente cordiale* with the Spaniards and hostile attitude toward the Chinese, in their own interest (see *Relations of the Chinese*, etc., p. 269).

discovery on the Philippines. One palpable piece of evidence pointing in this direction is furnished by our author in the description of a tea-canister bearing the Chinese seal "Luzon make" (p. 40). The only plausible explanation for this, if the report is correct,— and I see no reason to take umbrage at it,— is that a jar with such a special mark could but have been produced on the very soil of Luzon.

Conspicuous among the pottery recorded in the Tōkikō are the celadons. They are attributed to the Namban in general, to India and to Luzon in particular. The black-purple clay, the green glaze, the metallic sound, the designs of clouds and cranes, all pronounced characteristics of celadons, are insisted on by our author. The search of the Japanese for celadons in the Philippines is the more remarkable, as they received these vessels from China and Korea and subsequently manufactured them in their own country. Celadon was imitated at Okawachi in the province of Hizen, though the time of its beginnings seems not to be known. According to Brinkley (p. 99) the color of the glaze in some of the best specimens is indescribably beautiful; only a practiced eye can perceive that, in point of delicacy and lustre, the advantage is with the Chinese ware. In the first part of the seventeenth century, celadon was produced at Himeji in the province of Harima on the Inland Sea (*Ibid.*, p. 372), in the eighteenth century by the potter Eisen at Kyōto (p. 210), later on at Meppō (p. 378), from 1801 at Inugahara (p. 380), quite recently by Seifu at Kyōto (p. 417), and at Otokoyama in the province of Kishiu (p. 377).

Hideyoshi, the Taikō, a liberal patron of the ceramic industry which was revived and promoted under his untiring activity, had a genuine jar made for himself on Luzon, as stated by our author. This is in accord with contemporaneous Jesuit relations. The Jesuit Ludwig Froez (Frois) wrote in 1595: "In the Philippines jars called *boioni* are found which are estimated low there but highly priced in Japan, for the delicious beverage *Cie* (tea) is well preserved in them; hence what is counted as two crowns by the Filipino, is much higher valued in Japan and looked upon as the greatest wealth like a gem."[1] Hideyoshi monopolized the trade in this pottery and is said to have confiscated similar jars on their arrival from Japanese Christians who had purchased them at Manila, and to have prohibited any further trade in them under penalty of death.[2] But the same Hideyoshi was visited in his castle at Ōsaka by Chinese merchants who brought him the choicest ceramic productions of their country. Many a noble pair of celadon vases

[1] Quoted by O. Münsterberg, Chinesische Kunstgeschichte, Vol. II, p. 247.

[2] O. Nachod, Die Beziehungen der Niederländischen Ostindischen Kompagnie zu Japan, p. 57 (Leipzig, 1897). Compare also Cole, above, p. 10.

thus came into his possession, and were presented by him to temples throughout the country where several of them are still carefully preserved.[1]   For this reason, we are bound to presume, either that the celadons hunted by the Japanese on the Philippines were different from those imported from China, or that the Chinese imports did not suffice to fill the demand, and that the commercial opportunities afforded on the Philippines must have had a special attraction for them.   This may indeed be inferred from the political events of the time.   Hideyoshi's military expedition to Korea in 1597 was a blow directed against China. During the rule of the Ming dynasty (1368–1643), commercial relations between China and Japan were crippled; Japanese corsairs pillaged the coasts of southern China, and fear of them led to the exclusion of Japanese trading-vessels except admission on special passports, and but few Chinese junks stealthily made for Japan.   Only the advance of the Manchu dynasty brought about a change in these conditions, and after the Dutch had lost the possession of Formosa (1662), China's trade with Japan began to flourish.   While Hideyoshi, owing to the high ambitions of his politics, observed a hostile attitude toward China, he cast his eyes Philippineward.   In 1592, he despatched a message to the Spanish Governor, demanding the recognition of his supremacy; otherwise he would enforce it by an invasion and devastation of the Islands.   The frightened Governor, not prepared for such an attack nor willing to lose the profitable trade relations with Japan, sent an embassy under the leadership of a Dominican to the Taikō to whom he offered a treaty of amity.   Hideyoshi promised to desist from military action, on payment of a yearly tribute.   In 1593, the conditions of this treaty were stipulated, according to which the Japanese promised to despatch annually to Manila ships freighted with provisions, to stop piracy, and to grant passports to Spanish captains for the safety of their ships.[2]

In many cases where our Japanese author believes to recognize Namban or Luzon types among well-known Japanese wares, I am under the impression that such coincidences, partially, may be due to the common ancestorship of these pieces being in China.   The traditions of Japanese potters rest on those of China, and even in comparatively modern productions of Japanese furnaces, many ancient Chinese forms are rather faithfully preserved.   Mr. MORSE (p. 320), in speaking of Satsuma, has the following interesting remark: "One of the types of Ninagawa [3] resembles very closely in form a jar found among ancient Chinese pieces discovered in caves in Borneo, an example of which is

[1] BRINKLEY, l. c., p. 31.
[2] NACHOD, l. c., pp. 58, 60.
[3] A Japanese writer on pottery.

in the Trocadero Museum in Paris." This can only mean to say that the piece in question is derived from a Chinese type, which was also the parent of the Borneo jar.

But whatever our criticism of this Japanese record may be, it reveals a good many interesting facts hitherto unknown to us. It unrolls a picture of a former intimate contact between the two cultures, and undeniably shows that at a time the Philippines must have been a rich storehouse of fine pottery of various descriptions coveted and imitated by the Japanese. We are thus confronted with the fact that historical problems worthy of investigation are connected with the Philippines, and that the question of foreign pottery in existence on the Islands is much more complicated than it appears on the surface. Inquiries should be made in Japan as to any surviving examples of this so-called Luzon pottery and its possible influences on indigenous manufactures. Further research conducted in the Philippines may bring to light additional material toward the solution of this problem.

PLATE 1.

Chinese jars filled with the liquor provided for a funeral. Central Mindanao.
Compare p. 3 note.

FIELD MUSEUM OF NATURAL HISTORY.

DRINKING RICE WINE THROUGH REEDS.

PLATE II.

Pokanin burial cave. Southern Mindoro.
Compare pp. 8, 19.

BURIAL CAVE.

PLATE III.

Collection of jars in the possession of a wealthy Tinguian in the Abra Sub-province, Northern Luzon. The famous talking jar stands in the center.
Compare p. 12.

CHINESE POTTERY JARS.

Compare p. 12.

PLATE IV.

Two renowned jars from the Sub-province of Abra, Northern Luzon. The jar on the left is Magsawî, the famous talking jar.

TWO FAMOUS CHINESE POTTERY JARS.

Chinese Dragon-Jar.
Catalogue No. 109159.   Height 53.7 cm., diameter of opening 12.6 cm.
From Abra Sub-province, Northern Luzon.   A man's wealth is largely reckoned in these jars; they also are used as part payment for a bride.
Compare pp. 14, 23.

CHINESE DRAGON-JAR.

CHINESE DRAGON-JAR.

CHINESE DRAGON-JAR.

CHINESE DRAGON-JAR.

PLATE IX.

Chinese Blue Glazed Wine-Jars.

No. 1. Catalogue No. 109163. Height 22.2 cm., diameter of opening 9 cm.
No. 2. Catalogue No. 109161. Height 21.7 cm., diameter of opening 9 cm.
From Abra Sub-province, Northern Luzon. These jars are always a part of the
price paid for a bride. They are sometimes used as liquor receptacles.
Compare pp. 14, 26.

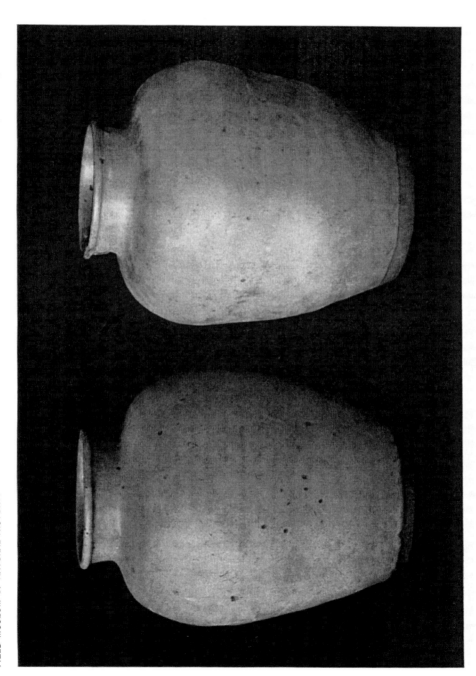

1   CHINESE LIGHT-BLUE GLAZED WINE-JARS.   2

CHINESE JAR WITH BLUE AND GREEN MOTTLED, CRACKLED GLAZE.

CHINESE DARK OLIVE-GREEN GLAZED WINE-JAR.

PLATE XII.

Chinese Light-Blue Glazed Jar.
Catalogue No. 109160. Height 39.5 cm., diameter of opening 11.7 cm.
From Abra Sub-province, Northern Luzon. Such jars are not in daily use, but frequently contain the liquor served at ceremonies and festivals.
Compare pp. 14, 26.

CHINESE LIGHT-BLUE GLAZED JAR.

PLATE XIII.

Dancing about the jars of liquor at a ceremony. Abra Sub-province, Northern Luzon.
Compare p. 14.

CHINESE JARS ENTERING A DANCING-CEREMONY

CHINESE BLUE AND GREEN GLAZED LIQUOR-JAR.

Chinese Green-Glazed Jar.
Catalogue No. 128645.   Height 61 cm., diameter of opening 15.3 cm.
From North Central Mindanao.   The possession of such a jar is a sign of wealth.
The lashings are attached in order that the jar may be more easily carried when
filled with liquor.
Compare pp. 14, 22, 26.

CHINESE GREEN-GLAZED JAR.

CHINESE GREEN-GLAZED WINE-JAR.

1   2   3a   3b

EARTHENWARE AND PORCELAIN DISHES EMPLOYED BY THE MEDIUMS WHEN SUMMONING SPIRITS.

PLATE XVIII.

Mediums summoning the spirits to partake of the food offered on the table. The ancient Chinese jar in the foreground contains rice wine.
Abra Sub-province, Northern Luzon.
Compare p. 15.

MEDIUMS SUMMONING THE SPIRITS.

PLATE XIX.

Mediums directing a ceremony, Abra Sub-province, Northern Luzon.
The wine offered to the participants is contained in the Chinese jar seen beside
one of the mediums.
Compare p. 15.

MEDIUMS DIRECTING A CEREMONY.

PLATE XX.

Tinguian potters at work, Abra Sub-province, Northern Luzon.
Compare p. 15.

TINGUIAN POTTERS.

TINGUIAN POTTERY VENDERS.

IGOROT POTTERY MAKERS.

089

中国民间传说中的祈福螳螂

$1.00 per Year   JANUARY, 1913   Price, 10 Cents

The Open Court

A MONTHLY MAGAZINE

Devoted to the Science of Religion, the Religion of Science, and the Extension of the Religious Parliament Idea

Founded by EDWARD C. HEGELER

BEL MERODACH AND THE DRAGON.
From a Babylonian monument. The god's hands are reversed. (See pages 17 and 19.)

The Open Court Publishing Company

CHICAGO

Per copy, 10 cents (sixpence). Yearly, $1.00 (in the U.P.U., 5s. 6d.).

THE PRAYING MANTIS IN CHINESE FOLK-LORE.*

BY BERTHOLD LAUFER.

TS'AI YUNG (133-192 A. D.)[1] a scholar and statesman of the Han dynasty, was once invited to a party, and on reaching the house, heard the sound of a lute played inside. It was a tune to a war-song expressing a desire for murder. Ts'ai, for fear of being killed, at once returned. The host and his guests pursued him, and when questioned, Ts'ai gave the reason for his retreat. The guests said: "When you approached, we seized the lute, as we noticed on a tree in the courtyard a mantis trying to catch a cicada; three times the mantis had reached it, and three times it failed in its attack. We feared that the mantis might miss the cicada (and therefore played the warlike tune)." Ts'ai was thus set at ease.

This story is the outcome of popular notions regarding the mantis which is looked upon as a formidable warrior endowed with great courage. The habits of the mantis are well known: the so-called flower-mantis in tropical regions resembles the flowers of certain plants, and in these flowers it lurks awaiting smaller insects upon which it feeds. What we term the "praying" attitude of the mantis in which its knees are bent and the front-legs supported on a stem, is nothing but this lying in ambush for other insects. Good observers of nature, the ancient Chinese were very familiar with its peculiar traits; they called it "the insect-killer" (sha ch'ung) or "the heavenly horse" (t'ien ma) from its speed, and greatly admired its bravery.[2] Its eagerness to catch cicadas is repeatedly emphasized, and above all, immortalized by the famous story of the philosopher Chuang-tse.

---

* See the author's book, *Jade, A Study in Chinese Archæology and Religion.*

[1] Giles, *Biographical Dictionary*, p. 753.

[2] Compare the Chinese drawing of the mantis.

"When Chuang-tse was wandering in the park at Tiao-ling, he saw a strange bird which came from the south. Its wings were seven feet across. Its eyes were an inch in circumference. And it flew close past Chuang-tse's head to alight in a chestnut grove. 'What manner of bird is this?' cried Chuang-tse. 'With strong wings it does not fly away. With large eyes it does not see.' So he picked up his skirts and strode towards it with his crossbow, anxious to get a shot. Just then he saw a cicada enjoying itself in the shade, forgetful of all else. And he saw a mantis spring and seize it, forgetting in the act its own body, which the strange bird immediately pounced upon and made its prey. And this it

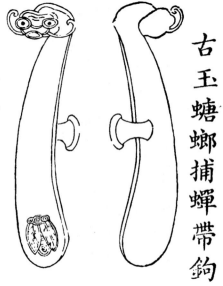

MANTIS CATCHING THE CICADA.
On jade buckle.

was which had caused the bird to forget its own nature. 'Alas!' cried Chuang-tse with a sigh, 'how creatures injure one another. Loss follows the pursuit of gain.'"

Surely, this pretty allegorical story has impressed the minds of the Chinese people deeper than the insipid account regarding Ts'ai Yung; and the Han artists, it is more credible, drew on Chuang-tse as the source for the motive of the mantis struggling with the cicada. Also Giles comments in his translation: "This episode has been widely popularized in Chinese every-day life. Its details have been expressed pictorially in a roughly-executed wood-cut, with the addition of a tiger about to spring upon the man, and a well into which both will eventually tumble. A legend at the side

蚟　　　螳

THE PRAYING MANTIS.

reads,—All is Destiny!" And in this thought, I believe, we should seek also the explanation of the motive on the Han jade buckle. Certainly, it does not mean such a banality as that frigid "kill!" intimated by the philistine scribbler of the *Ku yü t'u p'u,* but it was a *memento mori* to admonish its wearer: "Be as brave as the mantis, fear not your enemy, but remember your end, as also the undaunted mantis will end!"

In another passage Chuang-tse exclaims: "Don't you know the story of the praying-mantis? In its rage it stretched out its arms to prevent a chariot from passing, unaware that this was beyond its strength, so admirable was its energy!"

JADE GIRDLE-PENDANT, CICADA.
Showing upper and lower surfaces.

This is an allusion to another famous story contained in the *Han shih wai chuan,* a work by Han Ying who flourished between B. C. 178-156. It is there narrated: "When Duke Chuang of Ts'i (B. C. 794-731) once went ahunting, there was a mantis raising its feet and seizing the wheel of his chariot. He questioned his charioteer as to this insect who said in reply: 'This is a mantis; it is an insect who knows how to advance, but will never know how to retreat; without measuring its strength, it easily offers resistance.' The Duke answered: 'Truly, if it were a man, it would be the champion-hero of the empire.' Then he turned his chariot to dodge it, and this act won him all heroes to go over to his side."

090

中国的鲤鱼比赛

$1.00 per Year   JUNE, 1913   Price, 10 Cents

The Open Court

A MONTHLY MAGAZINE

Devoted to the Science of Religion, the Religion of Science, and the Extension of the Religious Parliament Idea

Founded by EDWARD C. HEGELER

The Open Court Publishing Company

CHICAGO

Per copy, 10 cents (sixpence).   Yearly, $1.00 (in the U.P.U., 5s. 6d.)

THE CHINESE BATTLE OF THE FISHES.

BY BERTHOLD LAUFER.

IN his article "The Fish as a Mystic Symbol in China and Japan," Dr. Carus reproduced among other illustrations a stone bas-relief of the Han period representing a battle of the fishes,[1] and aptly described it as "an army of fishes going to war, thus presupposing the existence of a Chinese fish-epic which may have been a battle of the fishes corresponding to the Homeric Battle of the Frogs and Mice." Neither Dr. Carus nor I were able at that time to point to a source of ancient Chinese lore from which this representation of a fish-epic might have been derived. I believe I am now able to supply this want, and to trace the tradition which may have given the impetus to this curious artistic conception.

It is well known that under the reign of the first Emperor Ts'in Shi (B. C. 221-210) the belief prevailed in the existence of three Isles of the Blest, P'êng-lai, Fang-chang and Ying-chou, supposed to be far off in the eastern ocean, and to contain a drug capable of preventing death and securing immortality. The desire of the emperor to possess this drug prompted him to send an expedition out in search of these islands. The party consisted of several thousands of young boys and girls headed by the magician Sü Shi.

"Several years elapsed," Se-ma Ts'ien,[2] the father of history, tells us, "and they were not able to find the drug. Because they had incurred great expense and feared a reprimand, they made this false report: 'The drug of P'êng-lai can be found, but we were always prevented from so doing by the large *kiao*[3] fish and therefore could not reach the place. We wish to propose that an excellent

---

[1] *The Open Court,* July, 1911, p. 402.

[2] E. Chavannes, *Les mémoires historiques de Se-ma Ts'ien,* Vol. II, p. 190.

[3] A species of shark.

archer be sent with us so that when the fish appears, he can shoot it with arrows from the repeating crossbow.'[4]

"Emperor Ts'in Shi dreamed that he was fighting with the God of the Ocean who had the appearance of a man. He applied to a scholar of profound knowledge, an interpreter of dreams, who said to him: 'The God of the Ocean cannot be seen, because he is guarded by the large fishes and dragons. If your Majesty will offer prayers and sacrifices, and be ready and attentive, the good gods may be invoked.'

"The emperor, accordingly, ordered those going to sea to take along implements for catching the large fish, whereas he himself, armed with a repeating crossbow, and waiting for the large fish to come forth, kept in readiness to aim at it. He went from Lang-ya[5] to the mountain Yung-ch'eng,[6] without seeing anything; arriving at Chi-fu,[7] he perceived a large fish which he aimed at and killed." Shortly afterwards the emperor died.

In another chapter of his "Historical Memoirs" Se-ma Ts'ien gives a different version of the story:[8]

"Emperor Ts'in Shi dispatched Sü Fu to sea in search of the marvelous beings. On his return Sü Fu forged an excuse and said: 'I saw a great god in the ocean who thus addressed me: Are you the envoy of the Emperor of the West?—I replied in the affirmative.—What are you looking for?—I replied: I wish to ask you for the drug prolonging the years and increasing longevity.—The god said: The offering of your king of Ts'in is trifling; you may see this drug but must not take it.—Thereupon the god conducted me toward the south-east, and we arrived on the island of P'eng-lai. I saw the gate of the palace Chi-ch'eng, where stood an emissary of copper color and having the body of a dragon; his splendor illuminated the sky above. Then greeting him twice I said: What offering can I make to you?—The God of the Ocean said: Give me sons of good family with virgin daughters, as well as workmen of all trades. Then you will obtain the drug.'

"Emperor Ts'in Shi was very well satisfied and sent three thousand young boys and young girls; he gave Sü Fu seeds of the five kinds of grain and workmen of all trades. Sü Fu set out on

[4] Such crossbows with a magazine from which six to eight darts can be shot off in rapid succession are still manufactured and utilized in China.

[5] On the south coast of Shantung Province.

[6] In the prefecture of Lai-chou on the north coast of Shantung Province.

[7] On the north coast of Shantung.

[8] Chavannes, loc. cit., p. 152.

THE ENVOY OF TS'IN SHI AND THE OCEAN-GOD.
(Repeated from *The Open Court*, July, 1911.)

his route; he found a calm and fertile place where he stopped and made himself king, and never returned."

From both these versions of the tradition, an understanding of the bas-relief in question may be derived. Indeed the submarine kingdom of the God of the Ocean is there displayed before our eyes. It is the sea, not a river, which is intended, as above all evidenced by the representation of sea-mammals. A seal is manifestly outlined on the upper left margin just above the canopy of the chariot, and there are reasons to believe that the Chinese first became acquainted with seals and other marine mammals through these very sea expeditions under Emperor Ts'in Shi. The oil obtained from seals was utilized for burning in the lamps placed in the emperor's tomb, and it was believed that they could not burn out for a long time.[9]

On the right-hand side of the slab is represented a four-footed mammal (slightly damaged) holding a spear in its forepaws. The center of the composition is occupied by the dignified personage driving the chariot drawn by three huge sea-fishes. The powerful God of the Ocean, "of the appearance of a man," guarding the Fortunate Isles and their treasure, the drug of immortality, may now be recognized in him; he is holding a jade emblem of rank in his hands. The man kneeling in front of his chariot, likewise provided with such an emblem, is apparently the magician, the envoy of the emperor, requesting the aquatic ruler for the drug. The armed warriors astride the fishes, and the fishes and frogs armed with bucklers and swords surrounding their lord on all sides, are his valiant body-guard ready to fight the unwelcome intruders, or perhaps on the warpath toward the shores of Shantung to punish the audacious emperor for his high-minded ambitions.

The subject of this bas-relief may therefore be defined as the struggle of Emperor Ts'in Shi with the God of the Ocean and his fish-creatures.

[9] Chavannes, loc. cit., p. 195.

091

中国祖先图像的发展

JOURNAL OF

RELIGIOUS PSYCHOLOGY

Including its

Anthropological and Sociological
Aspects

EDITED BY

G. STANLEY HALL

AND

ALEXANDER F. CHAMBERLAIN

VOL. 6

1913

Louis N. Wilson, Publisher
Clark University, Worcester, Mass.

JOURNAL OF
RELIGIOUS PSYCHOLOGY

| VOL. 6 | APRIL, 1913 | No. 2 |

THE DEVELOPMENT OF ANCESTRAL IMAGES IN CHINA[1]

BY BERTHOLD LAUFER, PH. D.,

Associate Curator of Asiatic Ethnology, Field Museum of Natural History, Chicago.

The field of Chinese research has undergone remarkable changes during the last years,—changes, based first of all on an extension of material, a widening of our horizon, and hence naturally resulting in a modification of method. The former students of China made the abundant records of Chinese literature the almost exclusive foundation of their study and occasionally had recourse to the observation of modern life as a convenient vehicle of interpretation. The actual monuments of the past were but rarely consulted, as they were difficult of access, and at first sight hardly promised a successful point of attack. Chinese objects of art remained for a long time the sport of collectors and amateurs only, and whatever the artistic merits of their spoils may have been, it was not material imbued with qualities to stimulate scientific curiosity and enterprise. Meanwhile, we are gradually preparing to enter upon a new era, and, if Chinese archeology cannot yet lay claim to existence as a well-established science, the beginnings which it has made bid fair for the future. The utilization of the remains of the past has acted as an efficient ferment on the elucidation of nearly all periods of Chinese culture and revealed phases in the development of religious thought entirely unsuspected ten years ago. For the first time we were allowed to obtain an

---

[1] Read at the meeting of the American Anthropological Association at Cleveland, Ohio, January 2, 1913.

insight into the psychic qualities of the ancient Chinese and into the religious concepts and sentiments of the great masses of which the one-sided and colored books of the Confucian school have hardly preserved any record. The monuments in clay, metal and stone which have risen from the graves thus manifest their fruitfulness in many respects; they are living witnesses of a bygone age and helpful building-stones in the reconstruction of the ancient culture; they fill a serious gap left by the literary documents, and are bound to shed new light on the correct understanding of the ancient texts which, as we now recognize more and more, have in many cases been gravely misunderstood or misinterpreted by the scholars of China. The brilliant school of French Sinologues who, it must be openly admitted, are the ingenious leaders in this field, has paved the way for a more critical attitude toward Chinese texts and traditions, and applied to them a vigorous and rigorous method parallel to that of classical philology. The former generations of scholars, as can hardly be expected otherwise, too easily accepted any Chinese statements and views on their face value, and too often laid down as facts, where solely opinion, reflection, or reinterpretation, was involved. In short, our own method of interpretation of data is thoroughly reformed, and the results obtained from a combination of the two factors of archeology and philology seem to assume a more solid and satisfactory shape.

The subject which at present seems in most urgent need of a revision is that of the ancient religion of the Chinese in the primeval period before the rise of Confucianism and Buddhism. For reasons of state policy the authors of the sacred books of Confucianism have vehemently suppressed the greater part of ancient popular lore and beliefs, retaining only what seemed to them reasonable and in keeping with their standard of morality, or relating myth and legend in forms mutilated or newly digested, as best suited their purpose. An attempt at reconstructing the ancient mythology of China, for this reason, is fraught with great obstacles. The gradual development into conventionality, formalism and dogmatism has, moreover, obscured the clear conception of the ancient ideas and converted the oldest history into the dream of a golden age which is purely the outcome of fabulous speculation. One of the favorite dogmas of later official Confucianism is the sweeping statement that in times of antiquity there was no image of any shape nor idol-worship of any kind, and it is no wonder that this axiom has crept as a gospel-truth into all foreign books on the religion of the Chinese and still holds sway in the minds of students.

The tendency of this tenet is obvious: it was directed as a blow against Buddhism, whose idolatry is so odious to the literati, and should serve, positively, as a glorification of the purity and sublimity of the dearly cherished antiquity. But the existence of images representative of deities in the earliest periods of Chinese culture can no longer be disclaimed, and a close study of this subject, thus far, has brought to light four large classes of images:

1. A group of images of geometric forms, usually carved from jade of required color, used in the worship of the cosmic deities, Heaven, Earth, and the Four Quarters.
2. A group of zoomorphic images in the shape of tigers, fishes, dragons, and numerous other monsters.
3. Human clay figures of shamanistic origin serving the purpose of healing disease and protecting the corpse in the grave, and representations of the shaman himself.
4. Statues of heroes and ancestors.

Under the title of Ancestral Images I take the liberty to present a few notes on the last subject. The principal idea by which I am guided is that the so-called ancestral tablets, such as are still found and worshiped in every family of China, have developed from real anthropomorphic images representing the ancestor. The subject, simultaneously, has an interesting bearing on the question as to how ancestral worship arose, was formed, and grew. It will be most advantageous to follow up the problem in a retrospective manner.

What is called at present the ancestral tablet is merely a conventional mode of translating a much more significant Chinese term which means "the wooden lord or master (*mu chu*)." It consists of a rectangular slip of planed wood, rounded on the top and inserted in a pedestal, and inscribed in black ink on both sides according to a stereotyped formula: Seat of the soul of such and such a one, giving the posthumous name of the deceased, followed by rank and title, if any, and by the date of his birth and death, with the addition of the name of the oldest son who has erected the tablet. Everybody on his death receives a special posthumous name, and the name of his lifetime is forever tabooed. During the first hundred days after burial, this tablet receives daily offerings of food and drink, and is, in every respect, treated as the representative of the dead man. In other words, it is apparently conceived as his animate image. At the lapse of that period, an auspicious day is chosen to convey the tablet into the ancestral temple, such as is owned only by families of high standing, or into a shrine kept in the house, as is done by the majority of people; the ancestral shrine is always arranged in close proximity to the one sheltering the domestic gods, so that the deified ancestors, in the estimation of the living ones, are placed on a par with the latter. From this moment the tablet has ceased to act as a live image and sinks to the function of a symbol. It is to the son a reminder of his departed father, an object on which, to speak with the Chinese, the bereaved one will fix his heart. The tablet, however, becomes animate periodically, i. e., on the occasion of a sacrifice to the ancestor, when his soul is supposed to descend into, and to be present in, the tablet, in order to take part in the offerings. To an ethnological mind, an observation here presents itself spontaneously, and this is, that the symbol of ancestral worship becomes effective through the medium of writing, that it is the function of writing which has resulted in the use of a conventional tablet, and, consequently, that, in times when writing was not yet in existence or not yet fully developed, the object of ancestral worship cannot have been a tablet, but must have been something of a different character. A survival of the older form is still preserved in the most important custom connected with the making of a tablet. This is the ceremony of punctuating the tablet, which in northern China has shriveled into a much abridged, conventionalized form. As already mentioned, the tablet is inscribed with the posthumous name of the dead followed by the words "seat of the soul of," *shên chu*. The last word, meaning "seat," is first written incompletely, the dot on the top being omitted. In the punctuation ceremony, this dot is added by a high official, for the alleged reason of rendering a particular honor to the dead. This practice becomes quite intelligible through another more complete series of rites still observed in the southern parts of China. There the tablet, wrapped up with red cloth, is carried in the funeral procession in an open palanquin preceding the bier. When the coffin is lowered into the grave, the eldest son kneels down at the end of the pit where the feet of the dead are, being surrounded by his brothers in the same posture. Then he turns his face toward the sun and receives from the hands of a relative or friend the tablet, which he holds with both his hands on his back, bending his head. A member of the family who is highest in rank, or an official, who is on friendly terms with the family, now removes the red cover from the tablet and touches it six times with dots in vermilion, two pairs of dots in the middle denoting eyes and ears, one above and another below representing head and feet and interpreted from the thoughts of nature philosophy as symbolizing heaven and earth, man being considered the product of the union of the two. This explanation is recited in a verse by the officiating writer during the ceremony, and this verse, doubtless, is very old:

I am painting heaven, may heaven always be bright over his grave.
I am painting earth, may the site of the grave be of powerful effect.
I am painting the ears, may the ears well listen (i. e. to the prayers of the descendants).
I am painting the eyes, may his eyes penetrate through all (i. e. be open to the needs of the descendants).

In this punctuation rite, accordingly, a mystic and magic shorthand figure of a man is drawn, and it is this very act which renders the tablet alive and useful, capable of serving as the domicile of a soul. A most interesting parallel of this practice is offered by the process of the Taoist priests of enlivening the images destined for worship in their temples. When a new religious image has left the workshop of the artist, it is brought in solemn procession to a temple where an older image of the same deity exists. With offerings of incense, the participants invoke this deity to allow part of its inherent soul substance to migrate into the newly completed image. The notion that images are not merely representative but animated with a soul and living realities is most ancient in China and by no means due to Buddhism; on the contrary, the images of Buddhism have originally had quite another significance, and the Chinese people have merely transferred their own long inherited ideas regarding images to those instituted by Buddhism. As soon as the new image is installed in its temple, the ceremony of consecration, called the "opening of the eyes," is performed by Taoist priests, who celebrate a mass (sometimes lasting for several days) and then, accompanied by prayers and music, daub with a dot the eyes, mouth, nose, ears, and even hands and feet of the image, by availing themselves of blood or vermilion, vermilion being a substitute for blood as in the case of the ancestral tablet. By this procedure, the organs of sense of the god are opened. This very act enables him to become conscious of his senses, and to commune with the external world; now he is capable of listening to the prayers of his votaries, acting upon them and enjoying their offerings. This usage apparently rests on the same psychical foundation as the one mentioned before in regard to the ancestral tablet, and the parallel becomes still more striking, as the Taoist images are also accompanied by wooden tablets inscribed with the name of the god, of the same form and scope as the ancestral tablets. It hence follows, and the fact is, that the ancestors are conceived as gods, and that the same designation *shên* 'spirit, ghost, soul' is applied to both. The employment of blood in the consecrating of images goes back to times of great antiquity when the ancestral temple and all paraphernalia of a sacred character were sanctified with the blood of sacrificial animals. The single dot added in the north of China to the word *chu* denotes the eye of the ancestor. It is an echo of the ancient idea that the drawing of the eye is a magic rite rendering the image alive (even a sacred act to the great artist). In ancient astronomy, the four stars composing Cancer, one of the signs of the Zodiac, were conceived as the sojourn of the manes of the ancestors and designated the Eye of Heaven, because the eye of the ancestors looked down upon the descendants.

While a survival of the image character is thus manifest in the ancestral tablets and their association with the established cult of the national gods becomes apparent, there are also real ancestral images still in existence of a decidedly realistic shape. In many parts of southern China a paper doll representing the dead is made for the occasion of the funeral ceremony; it is designated "the body for the soul" (*hun shên*) and often dressed in the costume of an official, the supposition being that this image should serve as a resting-place for the roaming soul who may leave it at will to partake of the offerings. On the last day of the ceremony, the paper image is cremated with all the sacrificial paraphernalia, in order to be the body of the soul in the other world. Cremation, in funeral ceremonies, does not mean destruction but a transposition of an object from the visible into the spirit world. The soul-body impersonated by the paper image, accordingly, has the same function as the tablet, except that the rôle it plays on earth is limited in time and is then continued in the beyond. Universal, at the present time, is the custom of having a portrait of a deceased father and mother painted. These painted portraits are allowed to take the place of the tablets during the sacrifices, but altogether have more of a social and sentimental significance. The time when they came into vogue is not exactly known, but surely goes back into the first centuries of our era; it is clear that this practice could arise only at a period when the art of portrait-painting was fully developed. In other words, the painted ancestral portrait is a thought-manifestation *sui generis,* which was not by any means evolved from a primitive conception of the ancestral image. It was merely an adaptation of pictorial art to the specific needs of ancestral worship. It is also doubtful whether the paper images alluded to can be considered in the light of a survival, and historical investigation seems to prompt us to the belief that they should rather be regarded as a comparatively recent innovation actuated by the more sensual needs of the masses. The present co-existence of ancestral tablets, images, and portraits must not lead us astray into a preposterous evolutionary construction of these three phenomena, but we must consult the data of history, in order to arrive at a proper understanding of the development.

In studying ancestral worship in its historical aspect, we are, first of all, confronted by the surprising fact that the present condition of affairs is only about 800-900 years old (dating from the time of the Sung dynasty), *i. e.,* considering Chinese developments, quite a recent affair. In this modern period, every individual is entitled to render a cult to his ancestor, and obliged to have the ancestral tablet; ancestral worship is universal and individual, while in earlier times it was neither the one nor the other. Among a number of Chinese scholars the opinion prevails that, in times of antiquity, the officials and the literary class had not the privilege of a tablet; the former were honored by a sort of figure or puppet, made of a textile in silk supported in the interior by a wooden frame, and the literati were distinguished by a similar contrivance of plaited straw.   This subject, like many others, is one of great controversy among the scholars of China, and no unanimity of opinion has been reached.   The one important fact remains that the present regulations were set up only as late as in the twelfth century, and that prior to that date neither all classes of the population were covered by this practice nor were tablets in general vogue.   The whole question is closely interwoven with that of the development of social organization in China, on which (it is sad but true) no sensible investigation as yet exists.   In the beginning, the social unit was not the natural family but the territorial clan group, each comprising from five to several thousands of families.   In the ancient law only these groups play a conspicuous rôle, while the family has no independent existence in the political organism.   There were, correspondingly, at first only clan or tribal ancestors, whereas the family ancestor is a much later product.   Briefly stated, the development was this, that the family gradually grew in importance within the boundaries of the clan.   This movement was furthered by the *débâcle* of the feudal organization of the Chou period in the third century B. C. and the progress of Confucian teachings which placed the family in the foreground. The strengthening of the family resulted in an extending and deepening of ancestral worship.   The number of tribal ancestors was comparatively small, and the object and meaning of clan ancestral worship differed to a marked degree from the later family ancestral worship.   Confucius inculcated the worship of ancestors as an action of filial piety, as a continuation of that respect and devotion which the good son owed to his parents during their life; while adopting and endorsing this ancient feature of the national religion, he impressed on it the seal of moral duty and obligation.   This is certainly the final conclusion but not the beginning nor the real cause of ancestral worship.   The chief agency instrumental in the formation of ancestral worship in ancient China seems to me to be the idea of protection.   The ancestors supposed to reside in Heaven had full charge of their descendants on earth and could avert from them disease, drought, or other calamities, bestow blessings upon them, and punish them for misdemeanor. The clan ancestor, as a rule, was a powerful hero of superior bodily and mental strength, who had impressed his fellow mates by an extraordinary deed or well deserved of them by a new invention, a new law or institution. In all likelihood, they were not in all cases blood ancestors at all but culture-heroes, chosen as ancestors by the clan or tribe, and then designated as "spiritual ancestors." Often enough it happened that a clan adopted the ancestor of another more powerful clan as its own or affiliated itself with other clans under the guidance of the same ancestor god. The ancestor was the chief protector of the clan in all conditions of life and responsible for its welfare; events of importance were communicated to him before his shrine, and many state-affairs were transacted in his temple; he was invoked for success in war and consulted by means of divination in difficult predicaments; vows were made to him to ensure the realization of a wish, for the fulfilment of which he was rewarded by a sacrifice. These ancestors were worshiped under images of stone or wood, and as we have now recognized the essential features in the development of ancestor worship, we are prepared to grasp the difference between ancestral image and ancestral tablet. The clan ancestor, whether real or alleged, whether real or mythical, was a personage of individual traits imbued with well defined characteristics and equiped with certain attributes in popular imagination; he was capable of having a statue erected to him. No doubt, and there are records to this effect, also family ancestors were revered in the shape of a simulacrum in ancient times, but the rapidly increasing number of this class of ancestors necessarily led to a generalization of the entire ancestral subject and resulted in the mechanical tablet, generalized and conventionalized as the ancestor himself. The clan ancestors then disappeared entirely, and a host of colorless souls or spirits, one exactly like the other, remained, corresponding to the same instrument of expression.

This is merely a brief abstract of a detailed investigation; it would certainly lead too far to lay all references of documentary evidence before you, but I may be allowed to cite one brief story which is characteristic of the ancient cult of ancestral images:

Ting Lan, a native of Ho-nei in Honan (first century A.D.), on the death of his mother, carved a figure of her in wood and continued to wait upon it as though it were his mother in the flesh. One day a neighbor came in to borrow something, and his wife consulted the figure, which shook its head; whereupon the neighbor in a great rage struck it over the face. When Ting Lan came in, he noticed an expression of grief on the figure's features, and on hearing what had happened, at once went off and gave the neighbor a thrashing. This led to a charge of assault, but when the constables came to arrest him, tears were seen trickling down the face of the figure. Ting's filial piety being thus recognized by the gods, he was not only acquitted, but the Emperor even sent an order for his portrait. (Giles, *Biographical Dictionary*, p. 736.)

One of the most curious customs observed in ancient China was the representation of the ancestor by a living substitute. This was a privilege due only to the sovereigns and the high dignitaries. The representative of the ancestor was chosen among the grandsons or greatgrandsons of the dead man; his son could never take this place, as it was his duty to perform the sacrificial rites. This substitute was called the "corpse" (*shi*). He was clad in the upper garment of his grandfather (all clothing of the dead was conscientiously preserved), and when still an infant, was carried on the arms of a servant to the place of the ceremony. His face was turned toward the south, and the tablet was set up at his right hand side. All participants in the sacrifice, i. e., those descending in the same lineage as the defunct, had to salute the substitute, all, even old men, prostrating twice before him and offering him meats and beverages. Only when he had tasted of each of the offerings were those present allowed to partake of them. The substitute of a deceased woman was selected from among the spouses of one of the grandsons. This function was so highly esteemed that every one who met a substitute on his road was obliged to alight from his carriage in order to salute him,— a homage which even a sovereign could not dodge. This practice prevailed only during the Chou dynasty and was completely abolished at the end of the third century B. C., when the great revolutionary movement of the Ts'in dynasty smashed the entire feudal system and the venerable culture of the Chou. Unfortunately, the contemporaneous traditions regarding this unique custom are very fragmentary, and the explanations afforded by Chinese authors all come down from later ages when the custom itself had died out. I will not intrude upon you by reproducing all these reflections, however interesting they may be, but quote only the two most weighty opinions. Pan Ku, an ingenious writer and historian of the Han period, remarks:

"The substitute figures in the ceremony of the sacrifice to the ancestors, because the soul does not emit sounds which could be perceived, nor has a shape which could be seen, and because the affection and yearning of a pious son has no object on which to concentrate his sentiments; for this reason he chooses a substitute to whom he offers viands, whereupon he breaks the bowls, as if his own father were satiated. The substitute drinking to satisfaction conveys to him the illusion that it is the soul of the deceased who has quenched his thirst."

It is noteworthy that according to this point of view the substitute is not the seat, the dwelling-place of the soul of the ancestor, but merely its likeness, the conception of which is inspired in the son by emotional reasons. Chu Hi (1130-1200), however, the great expounder of Confucian doctrines in the Sung period, states:

"In times of old, the substitute acted in the sacrifices; since the descendants are, as it were, the continuation of the life of the ancestors, the substitute has the same life (lit., breath) as the dead one, and the soul of the ancestors will doubtless rest in the persons of their descendants, it inhabits them, and clothes them as if with a garment."

This view is interesting but somewhat subjective, for nothing can be found in the ancient texts that would lend support to a belief in the migration of the soul into the bodies of the descendants. Other Chinese authors rebuke this custom as repellant and credit its abolition to the progress of Confucian wisdom. The interpretation of Chu Hi, however, may switch us on another track. It was an ancient custom described at great length in the Rituals to call back the soul of a person immediately after death. The essential part of the ceremony was that somebody with the best garment of the dead ascended the roof of the house, and moving the garment to and fro, exhorted the soul to return. It was hoped that the soul might recognize the garment and slip into it. This conception indicates the intimate relation of the soul to the clothes, and it is possible that in the case of substitution the soul of the grandfather was conceived as temporarily returning to his garment which had been donned by his grandson. The ancient culture of China does not present a unit, but an assemblage of the most varied tribes, speaking different languages, and diverse in the degree and formation of culture. We, therefore, meet varying practices in ancestral worship in early times, and it seems to me the living representative of the ancestor need not be set in direct causal connection with the ancestral image; this means, the living substitute was not necessarily a substitute for the image nor can the image be explained as a portraiture of the living substitute, for the employment of ancestral images is well authenticated for the same period.

Fortunately, we are now also in a position to point out ancestral images in archeological monuments, and many curious facts of archeology are now rendered intelligible by this investigation. I must abstain from dwelling here upon this subject which would require a good deal of illustrative material, but must refer to a subsequent publication where it will be treated in detail.

The development of ancestral images may be finally summed up as follows: First, the great dead of past generations magnified in the eyes of subsequent men were worshiped as guardian-spirits and adopted as ancestor-gods by the clan-groups; crude images were erected in their memory and honored with regular sacrifices to insure their blessings for the community, or as a recompense for favors received. The notions of deified heroes, of divine qualities, and ancestral qualification were still blended in these tribal ancestor spirits, and remained for a long period latent also in the family ancestors, particularly in those of imperial, royal and other noble houses. The clan images were therefore adapted to the individual ancestors, whose souls were believed to dwell in their images while simultaneously a living substitute for the ancestor during the sacrifices was a prerogative of aristocracy. The ancestral wooden tablet, on the whole, represents a secondary development, as it could spring up only at an advanced period of civilization when writing had reached a perfect state. The magic of writing takes the place of the visual elements of the likeness, and the form of the tablet is suggested by two factors,—the wooden or bamboo slips, which were the ordinary writing material before the invention of rag-paper in 105 A. D., and an assimilation to the external shape of the images, which were simple rectangular stone slabs or pillars with a rude face carved on the top. The tablet is throughout a counterpart of the image and inhabited by the soul or spirit of the ancestor. As the growth of ancestral worship advanced, the tablet became more and more formal and conventional; ancestral worship developed into a thoroughly democratic institution, and the effect was that ancestors sank in value and became at a discount. While in the beginning on a par with the rest of the gods, they could no longer withstand their keen competition, for gods multiplied in China at a tremendous rate of speed, and a clear and net distinction between gods and ancestors was the consequence. Images, therefore, fell into disuse for ancestors, and were reserved exclusively for the gods. This movement was fostered by the coming into power of Buddhism and Taoism, which set up a systematic code of elaborate iconography. The immense host of Buddhist and Taoist images has swallowed the ancient national gods and finally brought about the purely formal and ritualistic character of the ancestral tablets. In the consecration ceremony, their relation to the primeval cult is still manifestly retained, and what they may have lost in artistic expression, at the expense of their greater rivals, is largely compensated by an increase in spirituality, by their magnificent simplicity, by their social power and ethical influence, by their equalizing democratic tendency, which is surely one of the most imposing emanations of Chinese genius.

092

《甘珠尔》，根据藏文原版编译

DOKUMENTE DER INDISCHEN KUNST

DOKUMENTE DER INDISCHEN KUNST

ERSTES HEFT

MALEREI

DAS CITRALAKSHAŅA

NACH DEM TIBETISCHEN·TANJUR HERAUSGEGEBEN UND ÜBERSETZT

VON

BERTHOLD LAUFER

MIT EINER SUBVENTION DER KÖNIGLICH BAYERISCHEN
AKADEMIE DER WISSENSCHAFTEN AUS DER HARDY-STIFTUNG

LEIPZIG

OTTO HARRASSOWITZ

1913

Druck von W. Drugulin, Leipzig

VORWORT

In der Erforschung der indischen Kunstgeschichte stehen wir am Vorabend einer neuen Zeit. Großmütig zahlen die Länder Ostasiens die Schuld der Zeiten heim und geben uns dankbar zurück, was sie vor Jahrhunderten von Indien empfangen haben. Alle Gebiete des Ostens haben während der letzten Jahre eine erstaunliche Fülle neuer Schätze beigesteuert, die als lebendige Zeugen buddhistischen Geisteslebens den Ruhm Indiens in alle Welt verkündet haben. Die großartigen Funde, die Grünwedel, Le Coq, Pelliot und Stein aus Turkistan mitgebracht haben, die fortgesetzte unermüdliche Arbeit des Archæological Survey in Indien, die hingebende und gründliche Forschung der Franzosen an den Monumenten von Indochina, die hervorragende und feinsinnige literarische Tätigkeit der Japaner, die Veröffentlichung der buddhistischen Skulpturen Chinas durch Chavannes haben uns in der buddhistischen Kunst eine Kulturgröße ersten Ranges erkennen lassen, die an räumlichem und zeitlichem Umfang, an schöpferischer Erfindung, an Tiefe des Gedankens und der Empfindung der christlichen Kunst als ebenbürtige Schwester zur Seite tritt. Jahre werden vielleicht dahingehen, bis die jetzt in unseren Museen aufgespeicherten Sammlungen verarbeitet, bis der gewaltige Stoff für die Wissenschaft eingeheimst sein wird, mit dessen

Hilfe das noch offene Buch der Geschichte der buddhistischen Kunst geschrieben werden kann.

Auf den folgenden Blättern ist der Versuch gemacht, ein bescheidenes Scherflein zur Lösung der Probleme beizutragen, die uns dieses Kunstgebiet aufgibt. Der Mangel an dokumentarischem Material, das einer Betrachtung der indischen Kunst als sichere Grundlage dienen kann, ist oft und mit Recht beklagt worden. Der tibetische Tanjur ist vielleicht die einzige buddhistische Quelle, die authentische Texte dieser Art enthält, und diese zu erschließen ist die Aufgabe dieser kleinen Sammlung von Dokumenten[1]. Seit dem Jahre 1895 habe ich mich mit diesen Texten wie mit der buddhistischen Kunst beschäftigt und hätte wahrscheinlich dieses Schmerzenskind noch länger im Schreibpulte ruhen lassen, wenn mir nicht äußere Umstände eine baldige Veröffentlichung hätten ratsam erscheinen lassen. Die neue Phase, in die das Studium der buddhistischen Kunst gegenwärtig eingetreten ist, wie oben kurz gekennzeichnet, weckte den Wunsch, dem Kreise der Fachgenossen dieses Material nicht länger vorzuenthalten. Für mich persönlich hat eine reichhaltige Sammlung tibetischer Malereien und altchinesischer buddhistischer Skulpturen und Bronzen, die der Bearbeitung harrt, die Notwendigkeit ergeben, aufs neue vertieftere Rechenschaft über die geistigen Grundlagen zu suchen, auf denen sich diese Kunst Ostasiens aufbaut, und zu diesem Zwecke müssen wir eben nach Indien wandern.

In der Einleitung sind die hauptsächlichen Ergebnisse, die sich aus dem Studium dieses Textes aufdrängen, kurz zu-

---

[1] Den Grund, warum sie Dokumente der indischen Kunst (und nicht der buddhistischen Kunst, wie ich vorgezogen hätte) betitelt sind, wird man aus der Einleitung ersehen.

sammengefaßt. Der starken Versuchung, von diesem Ziele aus Streifzüge in das Gebiet der indischen Kunstgeschichte zu unternehmen, habe ich, von einigen Andeutungen abgesehen, widerstanden, weil es zunächst nur darauf ankommt, den nicht immer ganz leichten Text, wie er uns vorliegt, richtig zu erfassen und ihm seine tatsächlichen Werte abzugewinnen. Daß vieles aus demselben zu lernen ist, und weit mehr als ich vorläufig anzudeuten für gut befand, werden die Kunsthistoriker am besten zu beurteilen wissen. Für eine Reihe von Erscheinungen, die wir bisher in der tibetischen und chinesischen Malerei nicht zu deuten vermochten, wird er in Zukunft von prinzipieller Bedeutung werden.

Zu meiner eigenen Überraschung habe ich beim Abschluß dieser Studie wahrgenommen, daß die Überzeugungen, zu denen mich dieses Dokument der indischen Malerei geführt hat, sich den von Havell mehr intuitiv als exakt gewonnenen Anschauungen stark nähern. Wie Havell bin ich zu der Ansicht gelangt, daß die indische Kunst auf einer alten indischen Grundlage, auf einheimischen Überlieferungen, auf nationalen Inspirationen beruht, zunächst was die Malerei betrifft. Auch in seiner Schätzung der Gandhārakunst dürfte der neue, ebenso begeisterte als temperamentvolle Prophet indischer Art und Kunst wohl Recht behalten[1]. Der klassische Einfluß auf

---

[1] Man schließe aus diesen Bemerkungen nicht, daß ich mit allen Anschauungen Havell's übereinstimme; doch ist hier nicht der Ort für eine Kritik derselben. Die Archäologen sind natürlich nicht so schlimm als sie von Havell gemacht werden; niemand von uns ist so schlecht als die andern ihn machen. Auch der Gegensatz zwischen Archäologen und Kunstbetrachtern, den er künstlich konstruiert, ist nicht haltbar. Daraus, daß die Archäologen nicht beständig mit Schönheitsphrasen operieren, ist doch nicht zu folgern, daß sie kein ästhetisches Gefühl haben. So wird Havell ungerecht gegen Grünwedel, Foucher und andere verdiente Forscher, indem er völlig verkennt, daß wir doch zunächst einmal verstehen müssen, was

die Gandhārakunst ist wesentlich formaler und technischer Natur und berührt kaum die innere Seite des indischen Lebens; die indische Kunst ist darum nicht griechisch, ebensowenig als es die christliche Kunst ist[1]. Sie ist von indischen Gedanken geschaffen und beseelt und das Spiegelbild einer Ästhetik und Weltanschauung, die aus indischem Boden hervorgesprossen sind. Die indische Kunst hebt nicht bei Gandhāra an, sondern hat viele Jahrhunderte früher gelebt, bevor ein Grieche den Boden Indiens betrat, mit ihren eigenen Anschauungen und Theorien. Gandhāra trat als ein lokal begrenztes Intermezzo auf, als eine Episode, die kam und verschwand; sie war keine Offenbarung für die Inder; denn die Hast, mit der sie Gandhāra wieder abgeschüttelt haben, ist Beweises genug, daß sie hinreichende Kraft besaßen, ihre eigenen Wege zu wandeln. Aus dem vorliegenden Texte erkennen wir klar, daß die Inder eine in alte Zeit zurückgehende Tradition der Malerei besessen und eine theoretische Konstruktion des

die Kunstwerke darstellen, bevor wir zu kunstgeschichtlichen Schlüssen und Urteilen kommen. Hätte Havell Grünwedels treffende Charakteristik der indischen Malerei in Baumanns „Allgemeiner Geschichte der bildenden Künste" gelesen, so hätte er auch sicher erkannt, daß Grünwedel der ästhetischen Bedeutung der indischen Kunst durchaus gerecht geworden ist. Aber bei all seinen wunderlichen Ausfällen muß man offen anerkennen, daß Havell ein in ganz unenglischer Art origonaler Kopf ist, der unbeirrt von Schulmeinungen selbständig sehen und denken kann und mit seiner künstlerischen Intuition der Dinge ungeheuer anregend wirkt.

[1] Es ist leicht zu verstehen, daß die Bedeutung der Gandhārakunst, da sie durch ihre bequeme Anlehnung an die klassische Archäologie allgemeines Interesse erregt hat, etwas überschätzt wurde, und daß diese Überschätzung die Gesamtperspektive der indischen Kunst einigermaßen verschoben hat. Im übrigen wird man gut tun, die abwartende Methode zu befolgen, denn wir wissen schließlich über Gandhāra herzlich wenig. Systematische Arbeiten in dem großen Ruinenfelde sind noch gar nicht ausgeführt worden, und bis dahin bleibt die Gräko-Gandhāra-Frage vorläufig noch Hypothese oder Indizienbeweis.

Malens geschaffen haben, die sie nie und nimmermehr von den Griechen hätten erlernen können, weil es bei diesen nichts Ähnliches gegeben hat. Jeder Atemzug in diesem Lehrbuch der Malerei, vom ersten bis zum letzten, ist echt indisch.

Für das Verständnis des Textes glaube ich getan zu haben, was sich zurzeit tun ließ. Daß in vielen Punkten weitere Aufklärung erforderlich sein wird, davon bin ich vollkommen überzeugt. Mein Bestreben war zunächst, eine philologisch getreue Übersetzung herzustellen und durch möglichst zahlreiche Verweisungen auf Sanskritäquivalente auch den des Tibetischen unkundigen Sanskritisten in den Stand zu setzen, die helfende und bessernde Hand anzulegen. Eine für die Zwecke des allgemeinen Kunstwissenschaftlers berechnete Übersetzung, aus der sich das neue Material als eine gegebene Tatsache leicht abstrahieren ließe, mochte und wollte ich nicht liefern. Damit soll nicht gesagt sein, daß sie für den denkenden Kunstwissenschaftler, der auch indisch denken kann, nicht benutzbar sei. Die kleine Auflage von zweihundert Exemplaren beweist, daß sie *pro domo*, für den engeren Kreis der Fachgenossen, bestimmt ist, deren Kritik abzuwarten bleibt, und deren Urteil für andere wie für mich erst entscheiden muß, in welcher Richtung sich die praktische Nutzanwendung aus diesem Texte ziehen läßt. Ich darf nicht den geringsten Anspruch darauf erheben, als Sanskritphilologe zu gelten, noch eine fachwissenschaftlich exakte Kenntnis des indischen Altertums zu besitzen, und erwarte daher Belehrung und Aufklärung von den Indologen. Wenn mir das eine oder andere entschlüpft sein sollte, was das Gefühl des strengen Indologen verletzen könnte, so bitte ich im voraus um seine Vergebung. Auch von dem Jainismus, den ich in der Einleitung zu streifen gezwungen war, fehlt mir alle Kenntnis, und ich hoffe, daß die Jainaforscher hier ein kräftig Wörtchen sagen und die Frage eines möglichen Jainaeinflusses in das rechte Licht setzen werden. An dem isolierten Punkte, in dem ich hier wirke, fehlt es mir leider gänzlich an indologischen wie anderen Fachgenossen, deren Rat und Meinung mir wertvoll wäre, und so muß es denn bei dieser etwas mühsamen Fernmethode des Gedankenaustausches verbleiben.

Der Königlich Bayerischen Akademie in München spreche ich für die liberale Unterstützung des Druckes ehrerbietigen Dank aus, insbesondere den Herren Mitgliedern der Hardy-Kommission, die auf die Initiative des Herrn Geheimrat Kuhn die Veröffentlichung dieser Arbeit ermöglicht haben.

FIELD MUSEUM
CHICAGO ILL.
AUGUST, 1912.   BERTHOLD LAUFER.

INHALT

EINLEITUNG

> Wie der Sumeru der erste unter den Bergen, wie
> unter den Strömen die Gaṅgā die erste, und die
> Sonne der erste unter den Himmelskörpern ist, wie
> der Garuḍa der König der Vögel, wie unter den
> Göttern Indra der erste ist, so ist die Malerei die
> erste der Fertigkeiten.
>
> Citralakshaṇa V. 366—370.

I.

In seiner kurzen, auf den Angaben des Index-Bandes
fußenden Inhaltsangabe des tibetischen Tanjur hat ALEXANDER
CSOMA darauf hingewiesen, daß die dritte Abteilung, die Sūtra
(tib. *mdo*), die sich in einer bestimmten Reihenfolge mit Wissenschaft und Literatur befassen, unter anderem auch „einige Andeutungen auf die mechanischen Künste" enthalte[1]. Unter
den von ihm aufgeführten Titeln begegnen wir einem Werke
sku-gzugs-kyi mts'an-ñid mit der Übersetzung „Beschreibung
eines Bildes Buddhas hinsichtlich der Proportion der verschiedenen Glieder seines Körpers"[2]. Tatsächlich finden sich
aber vier auf die Kunst bezügliche Werke im Tanjur, und
zwar in Band 123 der Sūtra, deren genaue Anzeige das Ver-
dienst von GEORG HUTH[3] ist. Hier finden wir unter dem Titel
„Darstellende Künste" die folgenden vier Werke aufgeführt:

---

[1] *Asiatic Researches*, Vol. XX, First Part, p. 569 (Calcutta, 1836).

[2] *Ibid.*, p. 583 (oder *Annales du Musée Guimet*, Vol. II, p. 375).

[3] Verzeichnis der im tibetischen Tanjur, Abteilung mDo (Sūtra),
Band 117—124, enthaltenen Werke (*Sitzungsberichte der Berliner Akademie,*

1

1. *Daçatalanyagrodhaparimaṇḍalabuddhapratimālakshaṇa-nāma.* „Über das Wesen der plastischen Darstellung Buddhas, dem zehn Spannen breiten Nyagrodha-Baum an Umfang gleich."

2. *Saṃbuddhabhāshitapratimālakshaṇavivaraṇanāma.* „Von dem Saṃbuddha verkündeter Kommentar über die Größenverhältnisse der (Buddha-)Statue".

3. *Citralakshaṇam.* „Theorie der Malerei."

4. *Pratimāmānalakshaṇanāma.* „Theorie der Größenverhältnisse der Statuen."

Diese vier Schriften sind aus dem Sanskrit ins Tibetische übersetzt worden. Die Sanskrit-Originale sind aller Wahrscheinlichkeit nach verloren gegangen. Über die Zeit ihrer Abfassung wird nichts vermeldet. Aus dem Inhalt der Werke selbst lassen sich keine bestimmten chronologischen Anhaltspunkte eruieren; aus inneren Gründen bin ich geneigt, dem Citralakshaṇa ein höheres Alter als den drei übrigen zuzuschreiben. Ein Verfassername ist nur zu Nr. 4 angesetzt, nämlich *Etei-bu* oder nach anderer Schreibung *Aṭei-bu*[1]. Dieser Name entspricht, wie wir aus dem Texte dieser Schrift feststellen können, Skr. *Ātreya*, der als ein Ṛishi der Vorzeit und Fortsetzer des Geschlechts des Viçvakarman gilt[2]. Da nun das zweite und dritte Kapitel des Citralakshaṇa dem Gotte Viçvakarman in den Mund gelegt werden, so geht schon hieraus hervor, daß dieses Werk der Abhandlung Nr. 4 zeit-

---

1895, S. 271). Die Gesamtzahl der Çāstra in der Abteilung Sūtra wird im *dPag bsam ljon bzaṅ* (ed. SARAT CHANDRA DAS, p. 424, letzte Zeile) auf 519 oder 522 angegeben.

[1] Im Tanjur-Index von Co-ne (s. S. 50): *Ātrei-bu*; im *dPag bsam ljon bzaṅ*: *A-trai-bu.*

[2] Es wird nämlich am Schlusse der einzelnen Kapitel angegeben, daß sie aus dem Werke *E-tre-yai* (oder *E-ṭe-yai*) *t'ig-le*, d. i. wohl *Ātreyatilaka* stammen. Der Name Ātreya ergibt sich daraus mit voller Gewißheit.

lich vorangehen und aus einer früheren Kunstschule stammen muß.

In diesem Hefte ist der tibetische Text des Citralakshaṇa kritisch ediert und übersetzt; die drei anderen Werke werden im zweiten Hefte folgen[1]. Das Citralakshaṇa nimmt eine gesonderte Stellung in dieser kunsttechnischen Literatur ein. Nicht nur, daß sich das Werk ausschließlich mit der Malerei befaßt, es steht auch zu Nr. 1, das der Errichtung der Buddha-Statuen gewidmet ist, in diametralem Gegensatz. Jeder würde erwarten, daß, wie uns Nr. 1 die buddhistische Tradition der Plastik vermittelt, Nr. 3 dieselbe Überlieferung in bezug auf die Malerei erschließen würde. Dem ist aber zu unserer großen Überraschung nicht so: das Citralakshaṇa ist kein buddhistisches Werk und enthält keine Spur buddhistischer Tradition. Nicht einmal der Name Buddhas ist darin erwähnt, und es fehlen alle spezifisch buddhistischen Termini. Nur brahmanische Götter werden in der Einleitung angerufen. Ein brahmanischer König, ein Brahmane, der Todesgott Yama und Brahma sind die handelnden Personen des ersten Kapitels, in dem die Ursprungslegende von der Malerei berichtet wird. Dann übernimmt Viçvakarman die Hauptrolle und erzählt als Parallele

---

[1] A. GRÜNWEDEL hat in den Noten zu seiner „Mythologie des Buddhismus in Tibet und der Mongolei" (S. 203), ebenso in „Die Kunst des alten Indiens" (in BAUMANN, „Allgemeine Geschichte der bildenden Künste" I, 2, S. 611) nachdrücklich auf die Bedeutung dieser Abhandlungen im Tanjur aufmerksam gemacht. In anbetracht der Tatsache, daß ihre Existenz bereits seit 1895 bekannt war, ist der Satz von VINCENT A. SMITH (A History of Fine Art in India and Ceylon, p. 304, Oxford, 1911) "The Hindus have never taken sufficient interest in art for its own sake to write treatises, practical, historical, or critical, on the subject" doch zum mindesten etwas voreilig. Und Smith selbst (pp. 304—306) kommentiert Tāranātha's Angaben; wie hätte Tāranātha über indische Kunstschulen schreiben können, wenn ihm die Inder nicht das Material dazu geliefert hätten?

1*

zu der Entstehung der Malerei auf Erden ihren früheren
mythischen Ursprung unter den Göttern im Himmel, wobei
auf rein vedische Traditionen zurückgegriffen wird. Im dritten
Kapitel, das wesentlich technischen Inhalts ist, und in dem die
Maße der Körperteile und die wesentlichen Merkmale der zu
malenden Figuren dargelegt werden, ist keine einzige Anspielung auf spezifisch buddhistische Dinge enthalten. Von
Anfang bis zu Ende befinden wir uns in einer rein brahmanischen, oder vorläufig besser gesagt, in einer nicht-bud-
dhistisch indischen Atmosphäre.

Wenn wir den Charakter dieses Textes als außerhalb der
Grenzpfähle des Buddhismus liegend erkennen, so wäre es
hingegen eine voreilige Schlußfolgerung, demselben nur aus
diesem Grunde einen vorbuddhistischen Ursprung zuzuschreiben,
oder ihn gar als die allgemein brahmanische oder gemein-
indische Tradition über Malerei zu betrachten; er stellt vielmehr nur die Überlieferung einer bestimmten Sekte oder Malschule dar[1]. Es wird ausdrücklich hervorgehoben, daß es der
Arten der Malereien verschiedene gibt, und daß in dem vorliegenden Werke drei Kunstlehren, die des Viçvakarman, des
Prahlāda und des Nagnajit zusammengefaßt werden; das erste

---

[1] Von vornherein müssen wir uns auf den Standpunkt stellen, daß das
Citralakshaṇa nur das geistige Spiegelbild einer bestimmten Periode der
Malerei sowie einer bestimmten Malschule (und Sekte) sein *kann*. Dafür
sind nicht nur die angeführten literarischen Deduktionen maßgebend, sondern
auch die Befunde der indischen Malerei selbst. Der Inhalt des Citralakshaṇa
umspannt nur ein gewisses Gebiet der Malerei und bezieht sich deutlich auf
eine Frühperiode derselben, während andere weite Gebiete indischer Malkunst von diesem Werke *nicht* bedeckt werden. Wir haben also hier *ein*
Lehrbuch eines gewissen Zweiges der älteren indischen Malerei vor uns,
aber nicht *das* Lehrbuch *der* indischen Malerei. Über die Bedeutung von
citralakshaṇa vergl. Abschnitt V der Einleitung, S. 28.

Kapitel[1] bezeichnet sich am Schlusse als das Citralakshana des *Nagnajit*[2], d. i. Bezwingers der Nackten[3]. Dieser eigentümliche Terminus wird im Gespräche des Brahma mit dem Könige so erklärt, daß er auf die Preta, die Geister der Abgeschiedenen, bezogen wird, und weil der König im Kampfe mit Yama die von diesem herbeigerufenen Preta besiegt hat, empfängt er den Titel *Nagnajit*, d. h. also in diesem Falle, Besieger der nackten Preta. Diese Erklärung ist indessen rein äußerlich, denn wir sehen des weiteren, daß Viçvakarman, der Künstlergott, im Munde des Brahma als ein Nagnajit bezeichnet, daß am Schlusse des ersten Kapitels ein „Citralakshana des Nagnajit" als literarische Quelle zitiert wird, die auch im Eingang des Kapitels erwähnt ist, und daß der Verfasser in der Einleitung den Viçvakarman und den Nagnajit anruft. Es folgt daraus, daß Nagnajit zunächst der Gattungsbegriff für solche Maler ist, die nackte Figuren darstellen, also einen Nacktkünstler bezeichnet. Und der König erhält diesen Titel, nicht nur, weil er die Preta überwunden, sondern vor allem, weil er der erste der Menschen ist, der ein gemaltes Porträt hergestellt hat. Dieses Porträt ist das Bild eines toten Knaben, eines Preta, der nackt dargestellt ist, und wenn Brahma dem Könige ans Herz legt, er möge auch in Zukunft

---

[1] Das ganze Werk besteht aus drei Kapiteln.

[2] Das Petersburger Wörterbuch in kürzerer Fassung erklärt *Nagnajit* als Eigennamen 1. eines Fürsten der Gandhāra, 2. eines Autors über Bildhauerei und eines Dichters (nach MONIER WILLIAMS: Autor eines Werkes über Architektur. Muß wohl in beiden Fällen „Malerei" heißen). Der Gandhārafürst wird im ʾAitareyabrāhmana VII, 34 erwähnt (LASSEN, Indische Altertumskunde, 1. Aufl., Band I, S. 622); er wurde ein Jainamönch und ist einer der vier Pratyekabuddha der Jaina (H. JACOBI, Jaina Sūtras, Part II, p. 87, *Sacred Books of the East*, Vol. XLV).

[3] So auf Grund der tibetischen Übersetzung des Namens: *gcer-bu t'ul-byas, gcer t'ul*.

ein „Nacktbezwinger" *(nagnajit)* der Preta bleiben, so bedeutet das nichts anderes als daß er sich auch fürderhin der Malerei widmen soll. Die von dem Gotte Viçvakarman vorgetragene Lehre von den Maßen bezieht sich auf den nackten Menschen und zeigt, daß die altindischen Maler in erster Linie den menschlichen Körper in seiner Nacktheit studiert und die Götter, Könige und Helden nackt dargestellt haben[1]. Diese Beobachtungen mögen uns auf die rechte Fährte leiten auf der Suche nach den indischen Kreisen, aus denen diese Gedanken über die Malerei hervorgegangen sind. Es kann sich dabei wohl nur um die beiden Sekten der Ājīvaka oder Ājīvika (tib. *kun-tu atsʻo-ba-pa*) und Jaina oder Nirgrantha (tib. *gcer-bu-pa*)[2] handeln, brahmanische Asketen, die als ernste Rivalen

---

[1] Die Bemerkung E. B. HAVELL's (Indian Sculpture and Painting, p. 217, London, 1908) "painting nudity for its own sake, or as a test of skill, was never the aim of an Indian artist" bezieht sich natürlich nur auf die späte mittelalterliche Phase der indischen Malerei, wenn sie überhaupt in dieser Allgemeinheit Gültigkeit hat. Es braucht kaum hervorgehoben zu werden, daß die nackten Darstellungen der alten Inder, wie im primitiven Kulturleben der Völker im allgemeinen, auf einer ganz anderen psychischen Grundlage beruhen als unsere Nacktmalerei, die nur ästhetische Zwecke verfolgt.

[2] Diese Identifikation wird von JÄSCHKE und CHANDRA DAS in ihren Wörterbüchern gegeben. Die tibetische Übersetzung *gcer-bu-pa* bedeutet indessen „Sekte der Nackten" und ist daher nicht auf Skr. *nirgrantha* „die Ungefesselten", vielmehr auf Skr. *acelaka* „die Unbekleideten" oder Skr. *nagnaka, nagnāṭa* gegründet. *Nagna acela* vereinigt findet sich, von dem Guru Goçāliputra (Gosāla) gesagt, in Çikshāsamuccaya (ed. BENDALL, p. 331); Skr. *digambara* wird tib. *nam-mkʻai gos-can* wiedergegeben. Im Chinesischen liegt die Sache ähnlich; die chinesischen Buddhisten erklären *Nirgrantha* sowohl in dem Sinne von „ungefesselt" als „nackt" (EITEL, Handbook of Chinese Buddhism, p. 108), und ST. JULIEN (Mémoires sur les contrées occidentales, Vol. II, p. 42) faßt infolgedessen das Wort als „unbekleidete, nackte". *Nirgrantha,* sagt J. EDKINS (Chinese Buddhism, p. 73), *means a devotee who has cut the ties of food and clothing, and can live without feeling hungry or cold.* Wir sehen also, daß die Tibeter und Chinesen, die ja, wie

der ersten Buddhisten auftraten und in den ältesten Büchern
des buddhistischen Kanons beständig erwähnt werden. Wahrscheinlich sind die beiden Sekten gleichzeitig mit dem Buddhismus oder etwas früher gestiftet worden, und König Açoka
hat ihnen seinen Schutz angedeihen lassen[1]. Die Ājīvika sind
identisch mit den alten Vaishṇava, Anhängern des Nārāyaṇa
(Vishṇu), und besonders den Paramahaṁsa, die nach dem
Vaikhānasa Dharmasūtra[2] nackt gingen und Kuhmist schlangen[3].
Im Anfang unseres Werkes wird Nārāyaṇa angerufen. Die
Buddhisten waren der Nacktheit wie der Askese gram und
spotteten der nackten Brahmanen[4]. Mahāvīra, der Stifter der
Jaina, schaffte alle Kleidung ab und folgte darin dem Beispiele
Gosālas, des Begründers der Ājīvika Sekte[5]. So ist es wahr-

wir an dem praktischen Beispiel von Hüan Tsang beobachten, ihre Nachrichten über die Jaina aus tendenziösen buddhistischen Quellen schöpften,
dem Worte Nirgrantha eine durch die indischen Buddhisten vermittelte, miß-
verständliche Interpretation zugrunde gelegt haben. Denn daß die Tibeter
nirgrantha wirklich in dem Sinne von *gcer-bu* gefaßt haben, sehen wir an
ihrer Übertetzung des Namens des Stifters der Jaina, *gÑen-bu gcer-bu* (CHANDRA
DAS, Dictionary, p. 967b), was wörtlich Skr. *Jñātriputra Nirgrantha* entspricht (s. auch Mahāvyutpatti ed. CSOMA und ROSS, p. 26 Nr. 6); ebenso
wird in demselben Wörterbuch *nirgrantha* durch *gcer-bu-pa*, und *acelaka*
durch *gos-med* übersetzt (*ibid.*, p. 25 Nr. 16 und 17).

[1] H. KERN, Manual of Indian Buddhism, pp. 74, 112; RHYS DAVIDS,
Buddhist India, pp. 143, 290; A. GUÉRINOT, Répertoire d'épigraphie Jaina,
pp. 23, 69.

[2] Dieses Werk ist nach A. A. MACDONELL (A History of Sanskrit Literature, p. 262) im dritten Jahrhundert n. Chr. entstanden.

[3] G. BÜHLER, Indian Studies III, p. 21. Die beste Auskunft über die
Ājīvika gibt der treffliche Artikel von HOERNLE in Hastings' Encyclopædia
of Religion and Ethics, Vol. I, pp. 259—268.

[4] E. HUBER, Sūtrālaṁkāra, p. 103. Aus der Phrase *celui qui peut
rompre les Liens* (p. 104) geht wohl hervor, daß es sich um Nirgrantha
handelt.

[5] HOERNLE, *l. c.*, p. 265b. — Die Vorliebe der Asketen für die Nacktheit ist in erster Linie in religiösen, teilweise auch in sexuellen Motiven

scheinlich, daß das künstlerische Verständnis für die Darstellung des nackten Körpers in den Kreisen der nackt gehenden Büßer der Ājīvika und Nirgrantha entstanden ist, die in der Beschaulichkeit ihres Büßerlebens täglich Gelegenheit hatten, die natürlichen Körperformen zu studieren. Der Gedanke hat sich in Indien selbst auf dieser Grundlage des religiösen Lebens entwickelt, und wir brauchen nicht den Umweg über Babylonien und Griechenland zu machen, um das Nackte in der indischen Kunst zu verstehen.

In merkwürdiger Übereinstimmung mit diesen Beobachtungen befinden sich die ältesten, von dem leider zu früh verstorbenen Dr. T. BLOCH[1] entdeckten indischen Freskomalereien in der Jogīmārāhöhle, die dem dritten Jahrhundert v. Chr. angehören. Die Bilder sind leider noch nicht reproduziert

---

(wahrscheinlich so bei Gosāla) zu suchen. Die Nacktheit verlieh übernatürliche Macht (V. HENRY, La magie dans l'Inde antique, p. 109, Paris, 1909); EITEL (Handbook of Chinese Buddhism, p. 104) zitiert *nagna* und *mahānagna* als Bezeichnungen kriegerischer Geister oder Barden von übernatürlicher Kraft, die nackt erscheinen; die chinesische Erklärung des Sanskritwortes ist „nackten Leibes" (*lu shên*) oder „Geister von großer Kraft" (*ta li shên*). Wahrscheinlich liegt hier eine Beziehung zwischen Nacktheit und Kraft vor. Die Anschauung, daß sich in den Kleidern Läuse aufhalten, die beim Gebrauche Schaden leiden könnten, daß beim Betteln um Kleidung unreine Leidenschaften erregt werden, daß Kleidertragen und andere Genüsse nur Laien zukommen (H. JACOBI, ZDMG, Band XXXVIII, 1884, S. 13) beruht natürlich auf sekundärer Entwicklung. Weitere Materialien (aber mit ungenügenden Erklärungen) findet man bei R. SCHMIDT, Liebe und Ehe in Indien, S. 16 u. f. (ein Buch, mit dem ich mich übrigens wenig befreunden kann).

[1] Caves and Inscriptions in Rāmgarh Hill (*Archæological Survey of India.* Annual Report, 1903—4, Calcutta, 1906, pp. 123—131). Ein besonderes Interesse knüpft sich insofern an die Inschrift von Jogīmārā, als sie den Namen eines Künstlers Devadinna berichtet, der nach BOYER ein Bildhauer, nach BLOCH, was mir wahrscheinlicher ist, ein Maler war (*lupadakhe = rūpadakshaḥ*).

worden; aus den kurzen Schilderungen von Bloch geht hervor, daß nackte Figuren eine hervorragende Stelle einnehmen. Um einen Baum, in dessen Zweigen ein Vogel und eine menschliche Figur, wahrscheinlich ein Kind, dargestellt sind, stehen eine Anzahl menschlicher Figuren, ähnlich der auf dem Baume, alle unbekleidet, das Haar in einen Knoten auf der linken Seite des Kopfes gebunden. Ferner ist eine männliche Figur mit untergeschlagenen Beinen sitzend dargestellt, offenbar nackt, mit drei bekleideten Begleitern, die stehen; neben dieser Gruppe zwei ähnliche sitzende Figuren mit drei Begleitern. *The subjects cannot be interpreted at present*, bemerkt VINCENT A. SMITH[1], *but the nudity of the principal figures suggests a connexion with the Jain rather than the Buddhist religion, if the cave and paintings had any religious significance, which is doubtful[2].* Sicher ist jedenfalls, daß uns hier Szenen aus dem indischen Wald- und Büßerleben vorliegen.

Über eine künstlerische Betätigung von seiten der Ājīvika wissen wir nichts. Dagegen haben die Jaina in der Kunst Indiens eine große Rolle gespielt und besonders in der Architektur Hervorragendes geleistet. Die Annahme ist daher nicht überraschend, daß auch Handbücher der Kunst aus ihrer Mitte hervorgegangen sein sollten, und wir werden uns wohl kaum in der Vermutung irren, wenn wir die Verfasserschaft des Citralakshaṇa einem Maler zuschreiben, der entweder selbst ein Anhänger der Jaina war oder in enger Beziehung zu ihnen und ihrer Kunst stand. Damit ist nicht gesagt, daß das

---

[1] A History of Fine Art in India and Ceylon, p. 274 (Oxford, 1911).

[2] Mr. Smith scheint den Bericht Blochs nicht sehr genau gelesen zu haben. An der Bedeutung der Grotte kann nach der Inschrift doch kein Zweifel sein. Sie diente einer Gesellschaft von Schauspielerinnen als Ruhestätte und war von Sutanukā, einer Devadāsī, für ihre Kolleginnen zu diesem Zweck gestiftet worden.

ganze Werk, so wie es hier vorliegt, jainistischen Ursprungs sein muß. Es besteht aus drei ungleichen Teilen, die aus verschiedenen Quellen fließende Traditionen zu enthalten scheinen. Die im ersten Kapitel berichtete Sage vom Ursprung der Malerei macht den Eindruck einer gewissen Altertümlichkeit und mag sehr wohl auf eine alte brahmanische Quelle zurückgehen; die Jaina schöpfen ja wie die Buddhisten aus der älteren indischen Literatur[1]. Das erste Kapitel führt den Titel *Nagnavrata*[2] und wird als aus dem Werke „Merkmale des Gemäldes[3] des Nagnajit" (also *Nagnajitcitralakshaṇa*) entlehnt bezeichnet, worunter wir uns demnach ein altes Werk vorzustellen haben, das von der Malerei handelte und einen gewissen Nagnajit zum Verfasser hatte. In diesem Namen haben wir bereits eine deutliche Anspielung auf die Jaina erkannt. Sollte der hier in Rede stehende Nagnajit mit dem Naggati (= Nagnajit) der Jainatexte als identisch, und dieser letztere als ein ehemaliger König von Gandhāra zu betrachten sein, so wäre darin vielleicht eine Anknüpfung der Jainakunst an die von Gandhāra zu erblicken. Die Kunstlehre des Nagnajit ist aber wohl identisch mit dem in unserem Texte erschlossenen Vortrag des Gottes Viçvakarman, der in seiner Eigenschaft als Maler der Götter auch ein „Nagnajit" genannt wird. Der

---

[1] E. LEUMANN hat diesen Beziehungen eingehende Untersuchungen gewidmet, vor allem in seiner interessanten Abhandlung „Beziehungen der Jaina-Literatur zu andern Literaturkreisen Indiens" (*Actes du sixième Congrès international des Orientalistes à Leide*, Vol. III, pp. 469—564), in der die Beziehungen der Jaina zur buddhistischen, zur epischen und astronomischen Literatur der Brahmanen und zur Tantraliteratur dargelegt werden.

[2] Falls meine auf Grund des Tibetischen (*gcer-bu brtul-ba*) erschlossene Zurückübersetzung ins Sanskrit korrekt ist (s. S. 144). Eine freie Wiedergabe des Titels wäre: Übung der Nacktmalerei.

[3] Über diese Übersetzung vergl. weiter unten Abschnitt V, S. 28.

individuelle Nagnajit, dem sowohl in der Einleitung wie am Schlusse dieses Kapitels ein Werk Citralakshaṇa zugeschrieben wird, muß also nach der indischen Vorstellung seine Inspirationen von Viçvakarman empfangen und vielleicht als Schüler oder Inkarnation desselben gegolten haben. Er gehörte jedenfalls zu einer Schule, die sich nach Viçvakarman benannte und ihn wohl als Schutzpatron verehrte. Er ist indessen nicht der eigentliche Verfasser dieses Werkes noch des ersten Kapitels, das vielmehr von einem ungenannten Verfasser herstammt, der den Stoff für dieses Kapitel aus dem ihm vorliegenden Werke des Nagnajit geschöpft hat. Daß dem so ist, läßt sich klar aus vier Tatsachen ersehen:

1. Aus der Fassung des Kolophons des ersten Kapitels, dem zufolge dieses *Nagnavrata* betitelte Kapitel aus dem Citralakshaṇa des Nagnajit entlehnt, aber nicht von ihm verfaßt ist.

2. Eine auf Nagnajit bezügliche Angabe findet sich in den Endschriften des zweiten und dritten Kapitels nicht, die also einer anderen Quelle entlehnt sein müssen.

3. In der Einleitung ruft der Verfasser den Viçvakarman an und verneigt sich dann dem Nagnajit zu Füssen, dann vor den übrigen Meistern. Da sich kaum annehmen läßt, daß Nagnajit sich selbst das Weihrauchfaß geschwungen hätte, so müssen wir wohl einen unbekannten Verfasser supponieren, der den Nagnajit als seinen Meister verehrte.

4. Das Werk ist in der Einleitung als ein zusammenfassender Abriß dreier Schriften über Malerei bezeichnet, die durch die Namen Viçvakarman, Prahlāda und Nagnajit charakterisiert werden, in denen ich drei verschiedene Praktiker und Theoretiker erblicke (s. S. 129). Wir sehen also deutlich, daß „Nagnajit" nur eine der von unserem Autor kompilierten

Quellen bedeutet und nur für das erste Kapitel benutzt ist, und sind zu der weiteren Schlußfolgerung berechtigt, daß die Schriften der beiden anderen Autoren Viçvakarman und Prahläda die Vorlagen für die Darstellung des zweiten und dritten Kapitels abgegeben haben. Das vorliegende Werk ist demnach nicht der erste indische Versuch in dieser Richtung, sondern hat zum mindesten schon drei Vorgänger gehabt, woraus wir schließen dürfen, daß Spekulation über Gegenstände der Malerei bereits ein alter Bestandteil indischen Denkens gewesen ist, und daß die praktische Ausübung der Malerei, die naturgemäß der theoretischen Betrachtung vorausgeeilt sein muß, ein altes einheimisch-nationales Erbgut Indiens gebildet hat.

Das Verhältnis des Nagnajit zu dem uns unbekannten Verfasser des Werkes ist schwierig zu beurteilen. War er selbst ein Jaina, und der nach ihm arbeitende Verfasser desgleichen? Oder war Nagnajit ein in vorjainistischer Zeit lebender Künstler, der später von den Jaina als einer der ihren beansprucht und in ihren Kunstkanon aufgenommen wurde? Vorläufig möchte ich mich für die letztere Annahme entscheiden, da wir mit den Tatsachen der Kunstentwicklung selbst rechnen müssen und gezwungen sind, die Existenz einer künstlerisch entwickelten Malerei bereits für die alte brahmanische Periode anzunehmen. „Nagnajit" wäre somit als das Symbol der Jainakunst, als ihr geistiger Stammvater zu betrachten, als ein Maler nackter Götterfiguren, die wir eben in der Kunst der Jaina wiederfinden. Auf diese Übereinstimmung mit der im indischen Kunstschaffen uns entgegentretenden Wirklichkeit ist natürlich der größte Wert zu legen. Wie immer die historische Entwicklung der Dinge gewesen sein mag, die Jaina Anklänge in dem bloßen Namen Nagnajit können nicht fortgeleugnet werden, und die lebendigen Traditionen der Jainakunst stimmen damit aufs merkwürdigste und beste überein. Die Nacktheit ist der charakteristische Zug der Jaina, auf den sie selbst den größten Nachdruck legen, und der von den Buddhisten immer und immer wieder in den Vordergrund gerückt wird. So stellen sie denn auch ihre Palladine, vor allem die vierundzwanzig Propheten *(Tirthakara)*, unbekleidet dar[1]. Es ist bekannt, daß Hüan Tsang im Reiche Siṁhapura den durch eine Inschrift gekennzeichneten Ort gesehen hat, wo Mahāvīra oder Vardhamāna, der Stifter der Jaina, die Erleuchtung erlangt haben soll, daß er bereits die Sekte der Çvetāmbara erwähnt, und daß ihm die große Ähnlichkeit der Statue ihres göttlichen Meisters mit der des Tathāgata auffiel, die sich nur durch die Bekleidung von jener unterscheide, während die Schönheitsmerkmale des Mahāpurusha bei beiden übereinstimmend auftreten[2].

II.

Ist das Citralakshaṇa nicht-buddhistischen Ursprungs, so sind wir nunmehr vor die Frage gestellt, wie das Werk in die Hände der Buddhisten gelangt ist und seine Aufnahme in den buddhistischen Tanjur gefunden hat. Die Übersetzung ins Tibetische beweist, daß die indischen Buddhisten die Ab-

---

[1] J. FERGUSON and J. BURGESS, The Cave Temples of India, pp. 485 u. f.; J. ANDERSON, Catalogue and Handbook of the Archæological Collections in the Indian Museum, Part II, pp. 196 u. f. Ein gutes und allgemein zugängliches Beispiel bei F. KIELHORN, On a Jain Statue in the Horniman Museum (JRAS, 1898, p. 101).

[2] ST. JULIEN, Mémoires sur les contrées occidentales, Vol. I, pp. 163—164; S. BEAL, Buddhist Records of the Western World, Vol. I, pp. 144—145; TH. WATTERS, On Yuan Chwang's Travels in India, Vol. I, pp. 251—253.

handlung adoptiert und für ihre Zwecke benutzt haben müssen; sie beweist ferner, daß sie keinen eigenen Leitfaden der Malerei besessen haben, denn diese Schrift ist die einzige ihrer Art im Tanjur enthaltene, und ein rein buddhistisches Werk über Malerei hat sich bisher nicht nachweisen lassen. Die Buddhisten müssen also triftige Gründe gehabt haben, weshalb sie in Ermangelung eines Besseren zu dem Erzeugnis einer gegnerischen Sekte ihre Zuflucht genommen haben[1]. Für ihre ehrliche Gesinnung spricht die Tatsache, daß sie dasselbe unverändert gelassen haben; wäre es ihnen doch ein Leichtes gewesen, durch einige geringfügige Verschiebungen und Zusätze den Text in ihrem Sinne umzuarbeiten[2] oder zu färben.

---

[1] Es dürfte kaum berechtigt sein, mit dem Argument zu operieren, daß der Name des Jainaverfassers von den Buddhisten in absichtlicher Tendenz unterdrückt worden sei; denn auch in der ersten von buddhistischer Seite verfaßten Abhandlung des Tanjur, die sich mit der Anfertigung der Buddhastatuen beschäftigt, fehlt ein Autorname. Man muß überhaupt auf der Hut sein, unsere abendländischen Vorstellungen den Gedanken des Ostens unterzuschieben, wo das Sekten- und Religionsgetriebe einen ganz anderen Charakter aufweist und niemals die scharfen Gegensätze angenommen hat als bei uns. Das gilt für Indien wie für Tibet, China und Japan, wo die Welt doch stets friedlicher und gemütlicher war als bei uns. Unsere engherzige bureaukratische Anschauung, daß sich ein Individuum von Rechts und Staats wegen öffentlich zu einem bestimmten Glauben bekennen muß, fehlt in Ostasien vollständig, wo es jedermann stets unbenommen war, den ihm passenden Weg der Erlösung zu suchen, und wenn nötig, zu wechseln. Aber die europäische Zivilisation hat ja nun einmal das Monopol auf die Toleranz für sich beschlagnahmt!

[2] Beispiele dafür sind vorhanden. So wenn in der tibetischen Übersetzung der *Praçnottararatnamālikā* in der Anrufung der Name Çivas durch den des Mañjuçrī ersetzt wird (ED. FOUCAUX, La guirlande précieuse des demandes et des réponses, p. 7, Paris, 1867). Es ist interessant und in diesem Zusammenhang erwähnenswert, daß dieses Werk von Brahmanen, Buddhisten und Jaina (s. *ibid.*, p. 8 Note) in gleicher Weise in Anspruch genommen wird, offenbar weil alle drei in dieser ethischen Spruchliteratur auf gemeinsamem Boden stehen. Dieselbe Annahme erscheint mir für das

Man braucht im dritten Kapitel nur Buddha statt Cakravartin einzusetzen, und die Übersetzung in die Sprache der Buddhisten ist fertig. Das Werk muß also ein gewisses kanonisches Ansehen genossen haben, das den Gedanken einer Verfälschung der ursprünglichen Überlieferung von vornherein ausschloß, und darin mögen wir eine Garantie für die Güte dieser Überlieferung selbst erblicken. Die Verbindungsbrücke von den Jaina — oder in welcher Sekte auch immer das Citralakshaṇa entstanden sein mag — zu den Buddhisten ist leicht zu erkennen. Wir haben sie bereits in den Stichworten, Cakravartin — Buddha, angedeutet. Der Cakravartin steht im Mittelpunkt der gesamten Lehre von der Malerei: die Maße seiner Körperteile, sein Aussehen, seine Merkmale, sein Gang werden mit liebevollem Eingehen in die kleinsten Einzelheiten geschildert; er ist somit der vornehmste Gegenstand der Malerei, zugleich auch der schwierigste und komplizierteste, und steht daher im Vordergrund der Lehrmethode auch aus pädagogischen Gründen. Wer den Cakravartin malen kann, wird die übrigen Gestalten leicht beherrschen, da es sich bei diesen um ein Weniger, eine Reduktion des Typus des großen Mannes, handelt. In der Schilderung seiner körperlichen Eigenschaften werden zahlreiche Ausdrücke gebraucht, die uns aus den beiden Reihen der zweiunddreißig großen und der achtzig kleinen Schönheitsmerkmale des Mahāpurusha bekannt sind[1].

Gebiet der Malerei denkbar, deren wichtigster Gegenstand in der ältesten Zeit der Cakravartin war, der auch allen dreien gemeinsam angehört.

[1] Die Identität der tibetischen Bezeichnungen läßt sich durch Vergleich mit den Listen des Lalitavistara und der Mahāvyutpatti feststellen. In der Übersetzung ist in jedem einzelnen Falle darauf hingewiesen, nicht nur aus philologischen Gründen, weil die Übersetzung selbst davon gefördert wird, sondern auch weil diese Übereinstimmungen für die Kunstgeschichte von großem Wert sind.

Eine Geschichte der Entwicklung dieser aus so vielen hetero-
genen Elementen gebildeten Serien zu schreiben wäre eine
dankbare Aufgabe; sicher sind sie nicht das Werk der Buddhisten, sondern waren schon in altbrahmanischer Zeit vorhanden und repräsentieren das Ideal des nationalen Helden.
Wir wissen, daß Buddha mit dem Cakravartin identifiziert
worden[1] und in den buddhistischen Texten beständig so bezeichnet wird. Das Ideal des Cakravartin wurde infolgedessen
auf den Buddha übertragen, und was die künstlerische Betätigung betrifft, dürfen wir hinzusetzen, auf *die* Buddha, und
ferner auf die Bodhisatva. Wenn die Buddhisten für die
Malerei den Kanon einer heterodoxen Sekte angenommen
haben, so kann sie nur der Beweggrund geleitet haben, daß
sie hier die Methode vorgezeichnet fanden, wie der Cakravartin malerisch dargestellt werden soll. Daraus folgt zunächst
die Tatsache, daß sie Buddha und Bodhisatva auch wirklich
so gemalt haben[2]; freilich müssen wir uns hier sogleich vor

---

[1] Vgl. besonders A. GRÜNWEDEL, Buddhistische Kunst in Indien,
S. 136—140.

[2] Es ist zu beachten, daß Buddhatypen in der Malerei auftreten, die
kein Gegenstück in den Skulpturen haben. Man braucht nicht das Feld-
geschütz der Anthropologie aufzufahren, um zu erkennen, daß Buddhatypen
von Ajaṇṭā wie z. B. der von GRÜNWEDEL reproduzierte (BAUMANN, Allg.
Geschichte der bildenden Künste, S. 694) einen echten Inder darstellen und
mit den Skulpturen von Gandhāra in keinem Zusammenhang stehen. Mag
immerhin eine Malschule in Gandhāra bestanden haben, wie vermutet, aber
nicht bewiesen ist, die Malschulen, die in den Grotten von Ajaṇṭā vertreten
werden, sind meines Erachtens echt indisch und haben keinen Gandhāra-
geruch. Wie dem auch sein mag, unsere Texte gebieten uns Vorsicht in
der Konstruktion vorschneller Beziehungen und Wechselwirkungen zwischen
Skulptur und Malerei; was nach unseren Schultraditionen bei uns stattgefunden hat, braucht darum in Indien noch lange nicht der Fall gewesen
zu sein. Skulptur und Malerei sind in diesen Texten scharf geschieden und
haben andersgeartete Traditionen, die auf weit getrennte Epochen zurück-

Übereilungen hüten und nicht die Schlußfolgerung verallgemeinern, daß die Buddhisten im allgemeinen ausschließlich nur nach diesem und keinem anderen Rezept gearbeitet hätten. Wie in der Geschichte des Buddhismus die Zugehörigkeit aller Erscheinungen zu einer bestimmten Sekte und ihre Umgrenzung in Zeit und Raum nie außer acht gelassen werden darf, so müssen wir uns auch stets gegenwärtig halten, daß den einzelnen Sekten bestimmte Kunstschulen entsprechen, und daß es außerdem räumlich wie zeitlich weit getrennte Malschulen gegeben hat. Ließe sich die Zeit der Abfassung des Citralakshana genau definieren, so wären wir vielleicht in der Lage, uns eine fester umschriebene Meinung über diejenige buddhistische Richtung, Sekte oder Schule zu bilden, welche dieses nicht-buddhistische Lehrbuch zu ihrem Eigentum gemacht hat. Aber das Werk enthält leider kein Kolophon, keinen Verfasser- und Ortsnamen, und es läßt sich nur aus inneren Gründen vermuten, daß ihm ein verhältnismäßig hoher Grad von Altertümlichkeit zuzuschreiben ist. Vorläufig müssen wir uns mit der interessanten Tatsache begnügen, daß die Buddhisten, vielleicht schon in alter Zeit, Anregungen zur Malerei und ein Vorbild des Cakravartin-Buddha aus nicht-buddhistischen indischen Kreisen, höchstwahrscheinlich von den Jaina, empfangen haben. Es hat also vor dem Buddhismus eine einheimische Malerei in Indien gegeben, und wir werden wohl auch in der buddhistischen Kunst mit Jaina Einflüssen zu rechnen haben, die eine gründliche Untersuchung verdienen. Vieles, was uns bisher in der buddhistischen Kunst rätselhaft

gehen. Also tritt damit zunächst die Mahnung an uns heran, Skulptur und Malerei getrennt für sich zu betrachten und die Malerei fürs erste innerhalb der hier erschlossenen, wirklichen indischen Tradition anzuschauen und zu würdigen anstatt in Hypothesen, die solider Beweisunterlage ermangeln.

2

geblieben ist, wird mit großer Wahrscheinlichkeit darauf zurückzuführen sein[1].

---

[1] Vor allem vielleicht in der Ikonographie der Arhat und Sthavira,
deren Geschichte noch ganz im Dunkeln liegt. Interessant in diesem Zusammenhang ist die Tatsache, daß Albērūnī von den Arhat als "the class
called Nagna" spricht (SACHAU, Alberuni's India, Vol. I, p. 121). — Ich weiß
sehr wohl, daß bisher die der obigen entgegengesetzte Anschauung ausgesprochen worden ist, und daß es nicht an Meinungen fehlt, die alle brahmanische und Jaina-Kunst aus der buddhistischen ableiten wollen. Dies ist
eine der Übertreibungen der panbuddhistischen Idee. Niemand wird leugnen
wollen, daß die buddhistische Kunst zur Zeit ihrer höchsten Entfaltung
fruchtbare Anregungen überallhin ausgestreut und das brahmanische wie
jainistische Pantheon beeinflußt hat. Daraus folgt aber weder das höhere
Alter der buddhistischen Kunst noch die ausschließliche Beherrschung der
brahmanischen und Jaina-Kunst durch den Buddhismus. Die Zeit der gegen-
seitigen Beeinflussung ist nur eine Etappe in der späteren historischen Entwicklung. Die Periode, die hier in Rede steht und durch den Text des
Citralakshaṇa vertreten wird, liegt aber weit jenseits dieser späten Phase
und beweist die Ursprünglichkeit brahmanischer und vielleicht auch jaini-
stischer Kunstideen und Kunstgebilde gegenüber den buddhistischen. Die
späten Kunstdenkmäler, die uns in Indien erhalten sind, vermögen an den
Schlußfolgerungen, die wir aus diesem Texte ziehen müssen, nichts zu ändern.
Kunstwerke, die dem hier beschriebenen Cakravartin entsprechen, sind eben
noch nicht aufgefunden worden, aber sie werden möglicherweise noch entdeckt werden. Diese Periode, nennen wir sie die nagnajitische, muß vor
der Zeit des Buddhismus gelegen, und die Jaina scheinen ihre Traditionen
absorbiert und erhalten zu haben. Nicht die Jaina, sondern die Buddhisten
waren die Entlehner. Wenn die Buddhisten das ganze brahmanische Götter-
system adoptiert haben, dann ist auch die größte Wahrscheinlichkeit vorhanden, daß sie die Ikonographie dieser Götter von den Brahmanen empfangen haben; es ist undenkbar, daß sie die künstlerische Gestaltung derselben selbst erfunden haben sollten.

Die Nachrichten aus der indischen epischen und dramatischen, auch
aus der jainistischen und buddhistischen Literatur über Malerei, sollten ein-
mal kritisch gesichtet und zusammengestellt werden. E. B. HAVELL (Indian
Sculpture and Painting, pp. 156—163, London, 1908) hat bereits dankens-
werte Vorarbeiten dazu geliefert, die willkommene Fingerzeige geben. Im
Buddhismus wird das Gemälde öfters zu Vergleichen verwendet, z. B. in
Vasubandhu's Kommentar zum Gāthāsaṃgraha (SCHIEFNER, *Mélanges asiatiques*,
Vol. VIII, 1878, p. 580).

III.

Der Kern des zweiten Kapitels, das die Entstehung der Opferhandlungen zubenannt ist, besteht darin, daß die Götterbilder in ursächlichen Zusammenhang mit dem Opferkult gebracht werden. Anknüpfend an die vedischen Vorstellungen von der Entstehung des Alls, wird eine sich allmählich entwickelnde Anthropomorphie der Götter angenommen, die in dem Augenblick, wo ihre Schönheit die höchste Form erreicht hat, ihre eigenen Bilder malen, auf Anweisung Brahmas, mit der Bestimmung, daß sie unter diesen Bildern Verehrung und Opfer von seiten der Menschen empfangen sollen. Diese Stelle ist höchst merkwürdig und hat vielleicht kaum eine Parallele in der übrigen indischen Literatur. LASSEN[1] bemerkt: „Bilder der Götter waren den Indern der ältesten Zeit ganz unbekannt, und sie werden höchst selten in den epischen Gedichten erwähnt. Nach dem Gesetzbuche waren die Priester, welche bei den Götterbildern dienten, ausgeschlossen von den Opfern, welche den Göttern und dem Manu dargebracht wurden." Nach A. A. MACDONELL[2] werden materielle Objekte gelegentlich in der späteren vedischen Literatur als Symbole erwähnt, die Gottheiten darstellen; dergleichen (vielleicht ein Götterbild) ist sogar in einer Stelle des Rigveda zu verstehen, in welcher der Dichter frägt: „Wer will diesen meinen Indra für zehn Kühe kaufen? Wenn er seine Feinde erschlagen hat, mag er ihn mir zurückgeben." Verweisungen auf Götterbilder beginnen in den späteren Zusätzen zu den Brāhmaṇa und in den Sūtra zu erscheinen. E. B. HAVELL[3] hält es für ganz

---

[1] Indische Altertumskunde (1. Aufl.), Band I, S. 793 (Bonn, 1847).

[2] Vedic Mythology, p. 155 (Straßburg, 1897). Vgl. dazu E. W. HOPKINS, The Religions of India, p. 95 (Boston, 1895).

[3] The Ideals of Indian Art, p. 126 (London, 1911).

2*

sicher, daß anthropomorphische Götterbilder lange vor den frühesten Gandhāra-Skulpturen in Indien verehrt wurden, und beruft sich auf eine Stelle im Mahābhārata, wo als ein Zeichen künftigen Unheils „die Bilder der Kurukönige in ihren Tempeln zittern und lachen, tanzen und weinen."

Für die Lebenszeit Buddhas dürfen wir wohl brahmanische Tempel mit Götterbildern in großer Zahl annehmen. Im Lalitavistara wird erzählt, daß, als der Bodhisatva am Tage nach der Geburt in den Göttertempel *(devakula)* gebracht wurde, die Statuen aller Götter, des Çiva, Skanda, Nārāyaṇa, Kuvera, Candra, Sūrya, Vaiçravaṇa, Çakra, Brahma, der Welthüter usw., sich von ihren Standorten erhoben und dem Bodhisatva zu Füßen fielen[1].

Wie das erste Kapitel die sagenhafte Entstehung der Malerei auf Erden schildert, so behandelt das zweite, historisch zurückgreifend, ihren mythischen Ursprung in der Götterwelt des Himmels. Die hier zum Ausdruck kommende Idee von der religiösen Bedeutung des Malens ist der indischen Kunst verblieben und mit dem Buddhismus in alle Länder Ostasiens gewandert, wo immer der Pinsel geführt wurde. Das Gemälde der Gottheit wird zum Gegenstand des Opfers, es vernichtet die Sünden, befreit von Krankheit und Leidenschaft, vertreibt die Dämonen und beschert inneren Frieden. Eine segensreiche Wirkung geht von ihm aus für den Künstler wie für seinen Besitzer. Materialistisch vergröbert tritt uns dieser Gedanke im Tantra Kultus des tibetischen Lama entgegen, der die Gottheit bannt, indem er sie malt, und ihr dann im siegreichen Zauber seine Wünsche abringt[2]; idealistisch verfeinert und

---

[1] E. Windisch, Buddhas Geburt, S. 150.

[2] Für das Malen bestimmter Götter gelten genaue Vorschriften für den Maler, die sich sogar auf seinen Charakter und seine Stimmung er-

verklärt im Leben und Schaffen der großen chinesischen
Meister, denen die Kunst Sache des Herzens und tiefer reli-
giöser Begeisterung war, die in ihren Bildern das Mittel zur
Erlösung sahen und in ihrem Meisterwerke lebten und ins
Nirvāṇa eingingen[1].

IV.

Ein Lehrbuch der Malerei kann naturgemäß erst auf
Grund einer langen praktischen Ausübung und Erfahrung in
der Kunst entstehen. Der Mensch spekuliert nicht, um dann
das Ergebnis seiner Gedanken in die Tat umzusetzen, sondern
er handelt, unabhängig von seinem Willen, und sucht dann
seine Handlungsweise sich und anderen zu erklären. Die

strecken. Einen charakteristischen Fall derart hat GRÜNWEDEL (Mythologie
des Buddhismus, S. 102) aus dem Çrīmahāvajrabhairavatantra angezeigt:
„Der Maler muß ein guter Mann sein, nicht zu bedächtig, nicht zum Zorn
geneigt, heilig, gelehrt, seine Sinne hütend, gläubig und mildtätig, frei von
Begehrlichkeit, mit solchen Tugenden muß er begabt sein. Die Hand eines
solchen Malers darf auf Çūraleinwand malen." Er muß im Verborgenen
zeichnen; er darf malen, wenn außer dem Maler noch ein Sādhaka dabei
ist, aber nicht so, daß es ein anderer, ein Mann der Welt, sieht. Vor einem
Fremden zeigt man das Bild nicht, auch legt man es in eines Fremden
Nähe nicht aus. Will man lesen, anbeten oder essen, so erhält es Preis
und als seine charakteristische Speise Mahāmāṃsa. Religiöse Weihe ist
also die Grundbedingung für eine erfolgreiche Wirkung der Malerei.

[1] Es war, wie LAURENCE BINYON (The Flight of the Dragon, p. 111)
mit beredten Worten ausführt, "the home of the painter's soul". Der
Japaner, der· nach Binyon als höchsten Ausdruck der Bewunderung die
Worte gebrauchte „ein Bild, vor dem man sterben könnte", ist keineswegs
eine Ausnahme. In China habe ich wiederholt in diesem Sinne reden
hören, und es ist weit mehr als eine bloße Phrase. Der leidenschaftliche
chinesische Liebhaber alter Malereien bewahrt sein Lieblingsbild (wie seinen
Sarg) bis an seinen Lebensabend, um in seinem Anblick zu verscheiden; er
mag, der Not gehorchend, seine Sammlung veräußern, aber diesen einen
Schatz wird er getreulich behüten und mit ihm auf Kranichschwingen in
die Seligkeit eingehen. Es verlohnte sich der Mühe, der Entwicklung dieses
Erlösungsgedankens der Kunst, von Indien bis China, tiefer nachzuspüren.

wirkliche Ursache mag ihm verborgen bleiben, aber die Motivierung, die innere Rechtfertigung, wird gefunden und dem Geschehenen als Deutung untergelegt. Sehr anmutig und sinnig ist dieses Verhältnis von Theorie und Wirklichkeit in dem indischen Buche von der Malerei zum Ausdruck gebracht. Der König, dem jede Kenntnis des Malens während seines religiöser Beschauung gewidmeten Lebens fern gelegen hatte, wird unversehens, aus einem äußeren Anlaß, zum ersten Maler und schafft das erste Bild spontan unter der Inspiration Brahmas. Der Gott leitet seine Hand: es ist die unbewußte, aus sich selbst treibende schöpferische Tätigkeit des Künstlers. Ein naives Staunen ergreift ihn über das von ihm Geschaffene, und er beginnt darüber nachzudenken, wie das Werk entstanden ist. Er wendet sich andächtig an Brahma, den schöpferischen Geist der Kunst, um Belehrung über das Wesen der Malerei und wird an Viçvakarman verwiesen, den göttlichen Künstler der Tat, der von Brahma beides, die Philosophie und die Ausübung der Kunst, empfangen hat. Nicht der Urquell des schöpferischen Genius, nicht das transzendente Ideal offenbart sich dem schwachen Menschen, der die Vermittlung des allschaffenden Meisters bedarf, um einen Funken des himmlischen Feuerbrandes auf die Erde zu verpflanzen. Der König, der erste irdische Maler, erhält *nach* seiner ersten unbewußt vollbrachten Kunstleistung die theoretische Anweisung über die Malerei aus dem Munde des Viçvakarman, gleichsam als ein Symbol für die Erscheinung, daß eine Periode praktischer Betätigung den theoretischen Reflexionen über die Kunst vorausgegangen ist. Diese Erscheinung sehen wir ja im Entwicklungsgang aller großen Künstler sich wiederholen; zuerst vom inneren Drange sieghaften Schaffens getrieben, werden sie allmählich zum Nachdenken über das Wesen ihrer

Kunst geführt, um sich selbst Rechenschaft über ihr Verhältnis zu den Gesetzen des Schönen abzulegen. Wer dächte bei dem Wahrheit suchenden indischen König nicht an unseren Albrecht Dürer, als ihn die Sehnsucht nach dem schönsten Menschenbilde dazu trieb, einen Idealkörper geometrisch zu konstruieren?[1] Ebenso die Inder. Die trockene Aneinanderreihung von Maßzahlen[2], mit denen das dritte Kapitel angefüllt ist, darf uns nicht zu Vorurteilen in der Einschätzung des Geistes leiten, der aus diesem Schema entgegenweht. Es stellt eine höchst bemerkenswerte Tat in der gesamten Kunstgeschichte dar, eine Tat, von der in unserer Kultur erst Leonardo Da Vinci und Dürer geträumt haben. Dem Inder war die Malerei, wie unser Text sagt, eine Wissenschaft, ein Veda, und sie wurde daher nach wissenschaftlicher Methodik betrieben. Jener kindlichfrohe Hang zur klassifizierenden Systematik, der die Grammatik wie die Philosophie beherrschte, vererbte sich auch auf die Malerei, die ohne ein durchdachtes System von Regeln und Vorschriften nicht hätte lebensfähig bleiben können. Aus der Mitte der Typen, deren Gestaltung der malenden Kunst als vornehmste Aufgabe zufiel, trat alle andern mächtig überragend der Universalherrscher hervor, der Kaiser des Jambudvīpa, der Cakravartin. Die Lehre, wie diese Idealgestalt nach alt geheiligter Überlieferung gemalt werden soll, stellt den Inhalt des dritten Kapitels des Citralakshana dar.

Das dritte Kapitel setzt sich aus verschiedenen Bestandteilen zusammen. Daß wir es nicht im Zustand des Originals

---

[1] R. WUSTMANN, Albrecht Dürer, S. 20.

[2] Die im Texte angegebenen Maße beziehen sich auf den Cakravartin in Lebensgröße; er wird natürlich als Übermensch aufgefaßt und überschreitet das Normalmaß um ein Beträchtliches. Der Maler hat ihn gewiß nie in diesen Größenverhältnissen dargestellt, sondern auf die seinem Zwecke passenden Proportionen reduziert.

besitzen, geht aus der Bemerkung hervor, daß die Definitionen der sechsunddreißig Arten von Mienen später gelehrt werden sollen (S. 155); davon ist indessen in unserem Werke nicht mehr die Rede. Wir müssen also wohl annehmen, daß dieser Abschnitt verloren gegangen ist. Auf zwei offenbar interpolierte Stellen ist S. 161 und 164 hingewiesen; vielleicht sind noch einige Verse mehr dieser Art vorhanden, in denen Wiederholungen des bereits Gesagten vorkommen, aber solche Wiederholungen mögen auch vom Autor beabsichtigt sein.

In der Hauptsache lassen sich im dritten Kapitel zwei Hauptbestandteile unterscheiden, die wohl als älterer und jüngerer zu betrachten sein dürften. Dieser jüngere Abschnitt umfaßt die fünfundvierzig Verse 1011—1056, die schon äußerlich durch das veränderte Versmaß charakterisiert sind. Während sonst in dem ganzen Werke (mit einigen vereinzelten Ausnahmen) das siebensilbige Metrum angewendet ist, besteht dieser Teil aus neunsilbigen Versen und unterscheidet sich ferner von der vorausgehenden Partie in stilistischer Hinsicht durch seine blühende, poetische Sprache. Das Thema ist ein Preislied auf die hervorragenden körperlichen Eigenschaften des Cakravartin mit Rücksicht auf seine malerische Darstellung. Die Tatsache, daß dieses Stück aus einer anderen Quelle stammt, ist daraus zu erkennen, daß hier erstens Dinge wiederholt werden, die im Vorausgehenden schon erörtert worden sind, und daß zweitens Eigenschaften des Cakravartin hervorgehoben werden, die mit den vorhergehenden Angaben in Widerspruch stehen. Die wichtigsten dieser Widersprüche sind zwei. Im ersten Teile wird die Körperfarbe des Cakravartin als goldgelb beschrieben, in diesem Teile als weiß wie das Mondlicht und die Campablume; dort macht er den Eindruck eines Sonnenheros, hier den eines Mondheros[1]. Im ersten Abschnitt werden die Geschlechtsteile in ihren Massen ausführlich beschrieben, und wir müssen infolgedessen wohl annehmen, daß es eine Periode in der indischen Malerei gegeben hat, wo der Penis wirklich dargestellt worden ist[2]. In dem zweiten eingeschobenen Abschnitt jedoch wird gesagt, daß der Cakravartin seine Geschlechtsteile wie der Elephantenkönig eingezogen trägt, das heißt als Anweisung für den Maler, daß dieselben nicht darzustellen seien. Darin liegt ein prinzipieller Gegensatz zwischen den beiden Quellen. Daß hier wieder solare und lunare Elemente zu Tage treten, ist wohl wahrscheinlich, doch mögen sich die Mythenforscher mit dieser Frage auseinandersetzen. Hier kommt es lediglich darauf an, sich der Bedeutung dieses Gegenstandes für die Entwicklung der Kunst in Indien bewußt zu werden und in diesen beiden unvereinbaren Gegensätzen eine Scheidung zweier scharf getrennter Perioden zu konstatieren. In der einen älteren Periode wird der Cakravartin mit dem Organ der Zeugung gemalt, in der zweiten offenbar weit jüngeren Periode ohne dasselbe. Die beiden hier vorliegenden Quellen müssen also verschiedenes Alter haben, aus verschiedenen Kunsttraditionen geflossen sein, und die Verse 1011—56 als spätere Interpolation in den ursprünglichen Text betrachtet werden.

Das dritte Kapitel wird zwar am Ende als „das Kapitel von den Maßen" bezeichnet, aber die Maßangaben über den Cakravartin nehmen nur einen Teil desselben ein und sind mit

---

[1] Man beachte den eingehenden Vergleich mit dem Monde S. 175, vor allem die Erklärung des Nimbus aus dem Mondlicht.

[2] In der Angabe, daß das Gewand des Cakravartin weißfarbig und lose angeordnet ist, sehe ich keinen Widerspruch zu der sich zwingend aus dem Texte ergebenden Annahme, daß er nackt dargestellt wurde. Das Gewand war ein leichter nur die Umrisse des Körpers einhüllender Überwurf, der die Nacktdarstellung nicht beeinträchtigte.

V. 926 abgeschlossen. Hier beginnt eine Beschreibung der Zähne, der Zunge, des Haupt- und des Körperhaars, mit besonderer Berücksichtigung der Farben, sowie der Körperhaltung und des Ganges des Cakravartin. Die Schilderung geht hier über die Bedürfnisse des Malers hinaus, jedoch wohl in berechneter Absicht, um ihm tief einzuprägen, welche Wirkungen das Bild des Cakravartin hervorbringen soll. Mit anderen Worten, während vorher die substantielle Meßkunst gelehrt wurde, ist hier der geistige Inhalt, der vergeistigte Ausdruck des Bildes das Thema. So wenn die Stimme des Cakravartin mit der des Elephanten, des Pferdekönigs, des Donners verglichen wird, so wird damit nicht ein laokoonartig geöffneter Mund dem Maler zur Pflicht gemacht, sondern nur die Forderung gestellt, daß seine Gestalt so eindrucksfähig sein muß, daß man seine Stimme wirklich zu hören glaubt.

Wie der Cakravartin, so sind auch die Götter darzustellen, nur jugendlicher; ihr unbehaarter Körper gleicht dem eines sechzehnjährigen Jünglings (S. 172), und für ihr Gesicht sind nicht die Farben des Cakravartin zu verwenden (S. 170). Außer diesen beiden Haupttypen werden noch die Weisen, die gewöhnlichen Könige oder Könige der Menschen und schließlich das Volk behandelt. Wir entnehmen hier die wichtige Tatsache, daß das exakte Maßschema für den Cakravartin ausschließlich besteht, nicht aber für die anderen Figuren, die nach freiem Ermessen zu schaffen sind und nur proportionsmäßig korrekt sein sollen[1]. Wir sehen demnach, daß die Kunstlehre den Künstler in keine Zwangsjacke einschnürte und ihm Freiheit des Schaffens gewährleistete. Das ihm gegebene Mahnwort ist einfach: Schönheit, oder wie es im Texte negativ

---

[1] Vgl. auch die Anmerkungen zur Übersetzung S. 182.

ausgedrückt ist, Vermeidung häßlicher Wirkungen. Der angehende Maler soll seine Übungen bei den Weisen beginnen, wohl aus pädagogischen Gründen, weil diese am einfachsten darzustellen sind, und schreitet erst dann zu dem komplizierten Körper des Cakravartin vor. Das Malen von Dämonen soll der junge Student nicht einseitig übertreiben, da diese sonst leicht Gefahr über seine Mitmenschen heraufbeschwören können; auch hier herrscht wieder die Anschauung, daß das Bild ein Wirkliches und Lebendiges ist, daß die darin dargestellte Gestalt Form und Leben annehmen kann, wie wir so oft in der Geschichte der chinesischen Maler hören. Interessant ist die Tatsache, daß König Rāma, der Held des Rāmāyana, als Gegenstand der Malerei genannt wird; drei andere Könige bilden eine Gruppe mit ihm, deren Namen noch zu identifizieren bleiben (S. 179); sie stammen nicht aus dem genannten Epos. Die Bestimmung einer weiteren Gruppe von vier Namen, die, nach ihren relativen Höhenmaßen zu urteilen, stufen- oder rangweise auf den Cakravartin folgen, muß dem Scharfsinn der Indologen vorbehalten bleiben.

Aus der Angabe, daß es zwölftausend Çāstra gibt, die sich mit den Merkmalen (d. h. gleichzeitig Malereien) der vom Weibe Geborenen befassen und in fünf Klassen eingeteilt werden, und aus der Prophezeiung zukünftiger Grantha über das Thema dürfen wir schließen, daß es mancherlei Lehrbücher gegeben hat, welche über die malerische Darstellung von Männern und Frauen Aufschluß gaben. Sollte sich in den indischen Literaturen nichts derartiges erhalten haben? Schade, daß die Frauen so kurz behandelt werden. Aber ein Lichtblick ist doch gegeben in dem kurzen Satze, daß sie züchtig oder ehrbar erscheinen sollen. Die Frau muß also in der alten Malerei ein ernster und würdevoller Gegenstand gewesen sein.

V.

Was ist ein *Citralakshaṇa*, was bedeutet der Name? GEORG HUTH hat mehr gefühlsmäßig als richtig „die Theorie der Malerei" als Übersetzung des Titels vorgeschlagen, und zur Entschuldigung mag ihm füglich dienen, daß man eben keinen Buchtitel richtig übersetzen kann, ohne vom Inhalt des Werkes Kenntnis zu nehmen. Diese Übersetzung ist nur bedingt richtig oder höchstens als freie Übertragung zulässig. Eine eigentliche Theorie wird in dem Werke ja gar nicht vorgetragen, sondern es werden die körperlichen Merkmale *(lakshaṇa)* des Cakravartin als Anweisung für den Maler dargelegt. Aus der Summe der Erfahrungstatsachen, die sich aus einer langen Übung des Malens ergeben haben, wird die allgemeine Norm für die Darstellung des Cakravartin abgeleitet und als maßgebende Richtschnur für die Zukunft aufgestellt. So ist das Resultat eine Zusammenfassung praktischer Regeln. Wir haben bereits gesehen, daß sich viele der hier aufgezählten Eigenschaften des Cakravartin in den zweiunddreißig unter der Bezeichnung *lakshaṇa* bekannten Schönheitsmerkmalen des großen Mannes *(mahāpurusha)* wiederfinden, wie sie im Buddhismus aufs neue systematisch zusammengestellt und dem Buddha selbst zugeschrieben werden. Und unter dem Titel *citralakshaṇa* ist offenbar nichts anderes zu verstehen als „die charakteristischen Körpermerkmale oder die wesentlichen Kennzeichen des Gemäldes", *scil.* des Cakravartin[1]. Es ist mir

---

[1] In einer Beschreibung Buddhas in Açvaghoshas Sūtrālamkāra (HUBER's Übersetzung, p. 397) heißt es: „Die zweiunddreißig Lakshaṇa und achtzig Anuvyañjana schmücken ihn und lassen ihn einem Gemälde gleichen." Die Beschreibung scheint sich in der Tat an eine Malerei anzulehnen. Einige der in dieser Schilderung gebrauchten Ausdrücke wie „sein Körper gleicht einer Masse geschmolzenen Goldes" (d. h. in seiner Farbe), die dem Lotusblatt gleiche Zunge, der nach rechts gewundene Nabel, lassen sich im Citra-

zweifelhaft, ob Skr. *citra* und das entsprechende tib. *ri-mo* in dem generellen Sinne unseres Wortes Malerei als der Malkunst gebraucht werden; sie bezeichnen wohl nur das konkrete Bild. Wie dem auch sein mag, was weit wichtiger ist, die hier vorliegende Bedeutung von *citra* ist erst eine abgeleitete, sekundäre, nicht die ursprüngliche. Das Wort *citralakshaṇa* ist von Hause aus ein Terminus der Physiognomik und bedeutet „die Merkmale der Linien des Körpers", insbesondere der Hand und der Finger. Die wichtigsten Charakterzüge des Cakravartin sind aus der Physiognomik geschöpft und von da in die Malerei entlehnt worden.

Es wäre ungerecht, wollte ich verschweigen, was mich auf diesen Gedanken gebracht hat. Zunächst war es GRÜNWEDEL's treffende Bemerkung in seinem Handbuch der buddhistischen Kunst in Indien (S. 137): „Die Spezialisierung von Körpereigentümlichkeiten des ‚großen Mannes' stützte sich auf die altgeübte Kunst der Zeichendeutung und bildete . . . die Grundlage zu künstlerischen Versuchen." Sodann fiel mir auf, daß Dharmadhara, der als einer der Übersetzer einer Abhandlung über Physiognomik im Tanjur fungiert[1], zusammen mit dem Tibeter Grags-pa rgyal-mts'an aus Yar-luṅs, gleich-

---

lakshaṇa nachweisen. So wurde, wenn ein Bild Buddhas bei einem Maler bestellt wurde, der Nachdruck auf die Lakshaṇa gelegt; vgl. BEAL, Buddhist Records of the Western World, Vol. II, p. 102. Beal's Übersetzung "I wish now to get a figure of Tathāgata painted, with its beautiful points of excellence" ist wohl korrekt, denn im Texte steht *miao siang* 妙相, und *siang* 相 entspricht Skr. *lakshaṇa*. JULIEN (Mémoires, Vol. I, p. 110) übersetzt: je voudrais faire peindre la figure admirable de Jou-lai (Tathāgata), was den Sinn weniger passend wiedergibt.

[1] HUTH, Verzeichnis, *l. c.*, S. 275, Nr. 26. Der von Huth gebrauchte Titel Chiromantie ist zu eng gefaßt; die Hand spielt ja eine Hauptrolle in diesen beiden Texten, aber auch der Fuß und andere Körperteile werden in Betracht gezogen, so daß der weitere Begriff Physiognomik vorzuziehen ist.

zeitig mit der Assistenz desselben Tibeters die oben S. 2 unter 2 und 4 zitierten, auf die Kunst bezüglichen Schriften übersetzt hat. Da nun als Regel aufgestellt werden kann, daß Sanskrittexte gleichartigen oder verwandten Inhalts denselben Übersetzern zugewiesen wurden, weil sie mit der technischen Sprache des betreffenden Gebietes einmal vertraut waren, so ließ sich voraussagen oder ahnen, daß zwischen den Texten über Kunst und Physiognomik ein geistiges Band oder wenigstens ein gemeinsamer Fond technischer Ausdrücke obwalten müsse. In dieser aprioristischen Annahme habe ich mich nicht getäuscht, denn die Lektüre der beiden Abhandlungen über Physiognomik im Tanjur hat ergeben, daß sich die Mehrzahl der im Citralakshaṇa aufgezählten Lakshaṇa des Cakravartin dort mit denselben tibetischen Ausdrücken wiederfinden, woraus wir schließen dürfen, daß auch die entsprechenden Termini der Sanskritoriginale identisch waren, und es ergibt sich ferner, daß erst aus dem Bereiche der Physiognomik die Gründe zu holen und zu verstehen sind, warum gewisse Eigenschaften dem Cakravartin zugewiesen werden. Wir müssen uns also auf den Standpunkt des Übersetzers Dharmadhara stellen und von seiner Übersetzungsmethode profitieren, indem wir in allen Fällen, wo Gleichheit des Wortlauts vorliegt, auch Identität der Sache annehmen[1].

----

[1] Mehrfach ist in den Anmerkungen zur Übersetzung auf diese Übereinstimmungen hingewiesen. Eine vollständige Bearbeitung der beiden Texte über Physiognomik wäre sehr lehrreich — für die Kunstgeschichte. Die darin vorgetragene Weisheit ist gewiß nicht schwerwiegend, aber wenn wir uns auf den Standpunkt stellen, daß Männer und Frauen, wie sie dort in ihrem physischen Habitus beschrieben werden, auch gemalt worden sind, so dürfen wir die Bedeutung dieser Dinge für das Verständnis altindischer Malereien nicht unterschätzen, namentlich der zahlreichen durch Unterschriften nicht bestimmten Bilder in den Grotten von Ajaṇṭā, deren Stoffe aus dem indischen Volksleben geschöpft sind. Mit dem Leitfaden der Physiognomik

Wie die Malerei, so hat es die Physiognomik mit den Körpermerkmalen des Menschen *(lakshana* oder *vyañjana*, tib. *mts'an* oder *mts'an-ñid)* zu tun[1] und benutzt zu diesem Zwecke hauptsächlich die Linien *(citra*, tib. *ri-mo*[2]) der Hände und Füße. Dasselbe Wort, auf das Gebiet der Malerei übertragen, wird nun in dem Sinne von „Gemälde, Bild, Porträt" gebraucht. Daß dieser Bedeutungsübergang wirklich stattgefunden hat, kommt ja im ersten Kapitel des Citralakshana (S. 140) in naiver Deutlichkeit zum Ausdruck, wenn Brahma, nachdem der König das erste Bild gemalt hat, den neuen Gebrauch des Wortes *citra* in dem Sinne eines Gemäldes konstatiert. Man wäre also zu der Annahme versucht, daß, wie *citra* ursprünglich die charakteristischen Linien auf der Gesamtfläche der Haut des menschlichen Körpers bezeichnete, eine aus Linien gebildete Zeichnung entwickelt wurde, in der die Linien die charakteristischen Züge einer Persönlichkeit wiederzugeben versuchten. Die Malerei dürfte sich also in Indien aus einer zeichnerischen Linienkunst entwickelt haben, die als Projektion der von der Physiognomik aufgestellten Regeln zu gelten hat[3]. Wenn

---

in der Hand könnten wir in vielen Fällen enträtseln, was für ein Typus oder Charakter hier dargestellt sein soll.

[1] Die erste Schrift über Physiognomik im Tanjur beginnt mit den Worten: སྐུལ་བཟང་སྐྱེས་པ་བུད་མེད་ཀྱི་མཚན་ཉིད་རབ་ཏུ་བ་འདད་པར་བྱ་ „Die Körpermerkmale der unter einem glücklichen Stern geborenen (Skr. *subhagajāta)* Frauen sollen erklärt werden." Es ist klar, daß tib. *mts'an-ñid* hier nicht die Bedeutung von „Theorie" haben kann.

[2] In diesem Sinne wird tib. *ri-mo* beständig in den beiden Abhandlungen über Physiognomik gebraucht.

[3] Diese eigenartige Entwicklung der Malerei in Indien ist auch von allgemein wissenschaftlichem Interesse. Sie zeigt wieder einmal, daß Entwicklungen nicht nach dem subjektiven Klassifikationsschema der in Evolutionen verrannten Richtung der Ethnologie verlaufen, sondern, daß sich äußerlich gleichartige Erscheinungen oder Resultate von ganz verschiedenen

irgend etwas, so spricht gerade dieser Zug eine beredte Sprache
für die Originalität und Unabhängigkeit der indischen Malerei;
denn anderswo hat es eben eine solche Entwicklung nicht
gegeben. Damit soll keineswegs gesagt sein, daß sich die
Malerei ausschließlich und unmittelbar als eine Folgeerscheinung
aus der Physiognomik ergeben hat; sie war nur eine Quelle,
und vielleicht nur eine der Quellen, aus der die Malerei geschöpft hat. Die gemeinsame Grundlage ist klar, aber die
Mittel und der Zweck gehen im Verlauf der Entwicklung weit
auseinander. Die Physiognomik will den Menschen über seine
Zukunft belehren und prophezeit aus der Beschaffenheit seiner
körperlichen Merkmale Glück oder Unglück, Reichtum oder
Armut, hohe oder niedere Stellung, und insbesondere seine
Altersgrenze. Aber es ist beachtenswert, daß schon in der
Physiognomik das ästhetische Empfinden eine nicht unbedeutende
Rolle spielt. In seinen Ursprüngen mit dem Sexualtrieb eng

Grundlagen her entwickeln können. In China sehen wir die Malerei, in
engstem Zusammenhang mit der Kalligraphie, sich aus der Schrift, d. h. da
die Schrift im Anfang nichts anderes als eine Reihe symbolisch gedeuteter
Ornamente war, — aus der Ornamentik entwickeln. Im alten Indien ist
von einer Beziehung der Malerei zur Schrift nicht die geringste Spur zu entdecken, da es dort keine ornamentale Schrift gegeben hat; die Symbole
der Physiognomik werden zum treibenden Faktor der Entwicklung, was in
China und anderwärts nicht stattgefunden hat. Es kann also keine Rede
davon sein, daß sich die Malerei in China und Indien auf der gleichen
Linie entwickelt hätte. Es sind vielmehr zwei verschiedene Linien, die von
zwei weit auseinanderliegenden Punkten herlaufen, die sich in Richtung und
Verlauf scharf unterscheiden, um schließlich an einem Punkte zu enden, in
dem äußerlich eine gleichmäßig aussehende Erscheinung erzielt wird, die
wir als Malerei bezeichnen. Und so geht es mit den meisten Erscheinungen
im Kulturleben. Entwicklungen sind mannigfach und hundertfältig, und was
sich in dem einen Kulturkreis *so* entwickelt hat, mag sich in einem andern
unter ganz anderen Umständen und Begleiterscheinungen abgespielt haben.
Das banale *si duo idem faciunt, idem non est* ist in Wahrheit das Leitmotiv
der Kulturgeschichte. Vgl. auch die Addenda am Schlusse.

verbunden, bricht es in der Schilderung der Eigenschaften einer trefflichen Frau wuchtig durch; denn die Physiognomik lehrt auch, welche Frauen begehrenswert sind, und welche es nicht sind[1]. Hier liegt natürlich das Rassenideal zugrunde, und an diesem Punkte war der gemeinsame Boden für die Physiognomik und die Malerei vorbereitet. In dem Suchen nach dem Schönheitsideal des indischen Mannes und der indischen Frau ist unschwer der geistige Zusammenhang zu erkennen, der beide innig verbindet, und der indische Geist wäre nicht indisch, wenn er dieses Ideal nur als allgemeinen Begriff erfaßt und nicht methodisch in einem Kanon ausführlich formulierter Regeln analysiert hätte[2].

---

[1] Hier einige Beispiele aus der tibetischen Version. „Das dem Licht der aufgehenden Sonne vergleichbare Mädchen mit dem Antlitz des Vollmonds, mit großen Augen und roten Lippen, selig, wer ein solches Mädchen findet! Blau wie das Blatt des blauen Utpala, gazellenäugig, weiß wie der Elephant, mit allen Edelsteinen und Goldornamenten geschmückt, desgleichen schwarzäugig und von weißer Haut ist ein Mädchen schön, großäugig und rotlippig, selig, wer ein solches Mädchen findet!" Üble Eigenschaften einer Frau sind dicke Waden, struppiges Haupthaar, hängende Lippen, lange Rippen, ungleichmäßige Zähne von dunkelblauer Farbe, vertrockneter Gaumen, Lippen, Zungen und Glieder; geschwollenes Gesicht, ungleichmäßige Brüste, lange Nase und Knie. Die besten Qualitäten sind fleischige Oberschenkel und Wangen, weiße und gleichmäßige Zähne, lotusblattgleiche große Augen, Lippen wie die Vilvafrucht, hohe Nase, Gang eines Elephanten, nach rechts gewendeter und zurückgelegter Oberkörper, glatt und von schöner Farbe; die geheimen Teile anziehend und kräftig entwickelt; wohltönende Stimme und schönes Haar; wenig Schweiß, geringe Körperhaare, Mäßigkeit in Schlaf und Essen.

[2] Es ist beachtenswert, daß auch in der Medizin die charakteristischen Symptome einer Krankheit *lakshana* heißen (JOLLY, Medicin, S. 38), und daß im Tibetischen dafür gleichfalls das Wort *mts'an-ñid* verwendet wird. Im Bower Manuskript (HOERNLE, Part II, p. 164) wird es auch von den unterscheidenden Merkmalen der verschiedenen Arten von Myrobalanen gebraucht. So kommt es also bei der Bestimmung indischer (und ebenso buddhistisch-ostasiatischer) Bilder im wesentlichen darauf an, eine richtige

3

VI.

Es wäre durchaus verfehlt, der indischen Malerei (und Kunst überhaupt) einen rein religiösen Ursprung zuzuschreiben oder das religiöse Element darin einseitig zu betonen. Überall auf Erden, auch auf den primitivsten Stufen des Lebens, ist der Kunsttrieb dem Menschen eingeboren und macht sich im Schmuck des Körpers, der Kleidung und der Waffen geltend, im Ornament, das mit der Religion in keiner Beziehung steht. Wird dem Ornament eine religiöse Deutung untergeschoben, wird es mit Gedanken des Mythus in Verbindung gesetzt, oder werden mit Vorliebe Geister und Götter gemalt oder geschnitzt, so stellt der Künstler seine Kraft in den Dienst der Religion; aber es ist nicht die Religion, welche die Kunst erst schafft, denn Kunsttrieb und Ornament sind älter als alle Mythen und alle Religion. Die Religion ist nicht die Ursache der Kunst, sondern wirkt nur auf sie ein; je entwickelter sie ist, desto mehr wird sie die Gedanken der Kunst befruchten und den Künstler zu höheren Leistungen anspornen. Ist es ein Zufall, daß das erste nach der indischen Legende gemalte Bild nicht eine Gottheit, sondern das Porträt eines toten Knaben ist, dessen Namen nicht einmal überliefert wird, der nicht das geringste vollbracht hatte, um Anspruch auf Ruhm bei der Nachwelt zu erheben? Ist es ein Zufall, daß die Lehre von der Malerei in der Person des Cakravartin, des indischen Weltherrschers, gipfelt, anstatt, wie wir erwarten

---

Diagnose auf Grund der ihnen einwohnenden Lakshaṇa zu stellen. Dazu gehören nicht nur die offensichtigen in den Händen gehaltenen Attribute, sondern auch die Farben der Haut und Kleidung, Form und Stil des Haupthaars, und wenn vorhanden, des Bartes; Form und Ausdruck der Augen; Vorhandensein der Kopfbedeckung; Nimbus und Schmuck, Art des Schmuckes und der Kleidung; Stellung der Arme, Hände, Finger, Beine und Füße; Sitz, Umgebung und Hintergrund usw.

könnten, in den Gestalten eines Indra oder Brahma? Dürfen
wir in dem Umstand, daß der erste Maler ein König, und
daß der König der Könige das wichtigste Motiv der alten
Malerei war, ein Zeichen dafür erblicken, daß die Malerei ursprünglich an den Königshöfen gepflegt und unter dem Schutze
machtvoller Könige entwickelt wurde? Waren es die Kshatriya,
und nicht etwa die Brahmanen, denen der Anstoß zu der
edeln Kunst zu verdanken war?[1] Im Rāmāyana werden ja
Gemäldegalerien der Könige (*citraçālā*) beschrieben (S. 179).

Man beachte von Anfang an die Betonung des ästhetischen
Moments im Citralakshaṇa; der Gedanke, daß das Schöne die
Aufgabe und der Zweck der Kunst ist, durchweht das ganze
Werk. Man lese Brahmas überschwengliche Lobrede auf die
Malerei (S. 142) und lege sich die Frage vor, was für eine
Wirkung diese Kunst auf die Zeitgenossen hervorgebracht
haben muß, die eine solche Bewunderung und Wertschätzung
einzuflößen imstande war. Es sind nicht nur, wie Viçvakarman
erklärt, die Maße und Abzeichen, die Proportionen und Formen,
die der Maler lernen muß, sondern auch die Ornamente und
Schönheiten (S. 144), und wenn es auch der Malerei obliegt,
zur Verehrung der Götter beizutragen, so ist doch der Nachdruck auf das Schöne zu legen (S. 149). Im dritten Kapitel
lasse man nicht außer acht, wie häufig bei der Schilderung

---

[1] Über Künstler an den Höfen der Könige vgl. R. Fick, Die sociale
Gliederung im nordöstlichen Indien zu Buddhas Zeit, S. 186 u. f. Fick hat
leider die Maler nicht beachtet, aber sie spielen doch in den buddhistischen
Texten ihre Rolle, wie schon Schiefner durch seine indischen Künstler-
anekdoten gezeigt hat (bei Ralston, Tibetan Tales, pp. 360—363); vgl.
ferner E. Huber, Sūtrālaṁkāra, pp. 117, 170, und S. Lévi, Açvaghoṣa, le
Sūtrālaṁkāra et ses sources (*Journal asiatique*, 1908, p. 88). Eine Gemälde-
galerie bei den Jaina finde ich erwähnt bei W. Hüttemann, Die Jnāta-
erzählungen im sechsten Anga des Kanons der Jinisten, S. 39 (Straßburg, 1907).

3*

der einzelnen Körperteile hervorgehoben wird, daß sie hübsch
oder schön[1] gemacht werden sollen, und daß schließlich das
Prinzip aufgestellt wird: „Die anmutigen Züge aller Wesen
sollen gemalt werden" (S. 173), und: „Häßliche Wirkungen
müssen vermieden werden" (S. 180). Es kann somit keine Rede
davon sein, daß das Malen des Cakravartin nur eine auf bloßer
Routine beruhende, mechanische Konstruktionsarbeit ist, sondern daß sie einen ästhetischen Sinn, einen künstlerischen Geist
voraussetzt. Man darf sich auch nicht zu der einseitigen Anschauung verführen lassen, als wenn die beschreibende Reihe
der Körpermerkmale auf eine rein physische Komposition
hinausliefe, die des geistigen Ausdrucks ermangelt. Denn die
physischen Kennzeichen sind nach der indischen Anschauung
der Spiegel der Seele, die Objektivierung des Charakters,
gleichsam ein fleischgewordenes Karma. Im Vordergrund
stehen für den Maler die Augen, die Füße und die Hände,
deren Erörterung den breitesten Raum einnimmt und offenbar
mit besonderer Vorliebe behandelt ist. Die Klassifikation der
verschiedenen Arten von Augen entspricht verschiedenen
Typen und Charakteren, aber es ist leicht zu ersehen, daß
keine noch so mundgerechte Gebrauchsanweisung jemand be-
fähigen kann, z. B. ein Auge gleich der Cowriemuschel zu malen
und damit den Ausdruck des Zornes oder Schmerzes darzustellen, wenn er nicht das künstlerische Talent besitzt, das
geistige Element sichtbar hervorzuzaubern. Man darf sich
daher unter der alten Malerei nicht einen handwerksmäßigen
Betrieb vorstellen, den Hinz oder Kunz hätte erlernen können,
weil sie hier auf ein Schema leicht faßlicher Formeln reduziert
ist, sondern man muß den Vorhang vor dem verschleierten

---

[1] Der tibetische Text bedient sich der drei Wörter *sdug*, *bzań* und *mdses*.

Bilde lüften und die geistige Potenz erfassen, in der jene
Formeln wurzeln, indem man das Auge auf die uns hinter-
lassenen Kunstwerke richtet. Das Werk der Malerei war kein
bloßer materieller Prozeß, sondern eine geistige Tat von un-
mittelbarer Wirkung auf die Seele des Künstlers. „Das Malen
der Augen der Götter bringt Reichtum und Glück ein", sagt
das Citralakshaṇa (S. 155), und weiter: „Wer ein solches Gesicht gemalt hat, wird beständig Güter erwerben" (S. 157).
Wir sehen, daß die Idee von der belebenden und beseelenden
Kraft des Auges, von der wir so viel in den Anekdoten über
die alten chinesischen Maler hören, bereits im alten Indien
Wurzeln geschlagen hat, und daß der seelische Ausdruck zu
den unerläßlichsten Forderungen der Kunst gehörte.

VII.

Schließlich bleibt noch die praktische Verwertung und der
Einfluß des Citralakshaṇa in der Malerei des Lamaismus zu
betrachten. Aus dem Umstand, daß in dem Abschnitt über
die Malerei in dem historischen Werke *dPag bsam ljon bzaṅ*
(S. 184) eine kurze Inhaltsangabe des ersten Kapitels des Citralakshaṇa mitgeteilt wird, ist zu ersehen, daß dieses Werk die
Lektüre gelehrter Lamen gebildet hat[1]. Die in Tibet und der

---

[1] Die indische Sage vom Ursprung der Malerei scheint indessen unter
den lamaischen Völkern nie volkstümlich geworden zu sein, wohl weil sie
ihrem Gedankenkreise doch zu fern lag. Unter den Malern (*zurāčī*) der
Kalmüken hat die Tradition Wurzeln geschlagen, daß die Anfänge der
Malerei auf Çākyamuni selbst zurückgehen, der vor seinem Tode den da-
maligen Malern befohlen haben soll, sein Bild zu zeichnen; sie erschraken
aber und konnten nicht zeichnen; da ließ er sie sein Bild von seinem
Schatten abnehmen, wie auch geschah. Diese Nachricht findet sich bei
A. Žitetski, Skizzen aus dem Leben der Astrachanschen Kalmüken, p. 64
(*Nachrichten der Kais. Gesellschaft von Liebhabern der Naturw., Anthr. und
Ethn.*, russisch, Bd. 77, Heft 1, Moskau, 1893). Die kalmükische Tradition

Mongolei noch jetzt von den Lamen befolgte Methode des Malens beweist mehr als alles andere, daß sie nach den im dritten Kapitel des Citralakshaṇa niedergelegten Regeln arbeiten. Es folgt aus den hier gegebenen Vorschriften, daß der Maler zuerst ein geometrisches Schema zeichnete und darin proportionsmäßig die einzelnen Maße eintrug; auf Grund der Angaben in unserem Texte läßt sich dieses Schema leicht rekonstruieren, indem man zuerst eine Mittellinie mit der Maßeinheit 108, der Höhe des Cakravartin, entwirft. Das ist genau dasselbe Verfahren, wie es von den tibetischen Malern geübt wird, wenn sie die Umrisse der Körperformen in ein rechteckiges Liniennetz eintragen, in dem jede Körperlinie, wie bei uns jeder Ort auf dem Gradnetz einer Karte, ihren bestimmten Platz zugewiesen hat[1]. Es gibt jedenfalls auch originaltibetische auf die Technik der Kunst bezügliche Schriften. Das „Verzeichnis der Tibetischen Handschriften und Holzdrucke im Asiatischen Museum" von I. J. Schmidt und O. Böhtlingk (S. 59, Nr. 6) zitiert ein Werk unter dem Titel *Saṅs-rgyas-kyi gzugs brñan bris-pai tig tsad* mit der Übersetzung „die Proportion (der einzelnen Teile) beim Abmalen der Gestalt Buddhas" und dem Zusatz „Europäische Handschrift, Tibetisch-Mongolisch". Der Titel zeigt Anklänge an die oben S. 2 unter Nr. 1 aufgeführte Abhandlung aus dem Tanjur, auf der das Werk wahrscheinlich basiert sein dürfte. Ein des Tibetischen Unkundiger könnte sich leicht verführen lassen, in dem

---

ist aus dem Divyāvadāna geschöpft. Der Schatten ist anscheinend eine späte Erfindung, die wohl aus den Skizzenzeichnungen (*tig tsad*, s. weiter unten) hergeleitet wurde. Vgl. Burnouf (Intoduction à l'histoire du buddhisme indien, p. 304) und Grünwedel (Buddhistische Kunst in Indien, S. 68).

[1] Vgl. H. H. Godwin Austen, On the System employed in outlining the Figures of Deities and Other Religious Drawings, as practised in Ladak, Zaskar [i. e. *Zaṅs dkar*] etc. (JASB, Vol. XXXIII, 1865, pp. 151—4).

ebenda S. 45 (Nr. 427, 4) angezeigten Titel eine Abhandlung über Kunst zu wittern, auf Grund der Übersetzung „Der bestens erklärte Kern, oder: Lob der Einheit oder des Zusammenhanges der bildlichen Darstellungen (Buddhas)". Die Übersetzung ist indessen verfehlt; der tibetische Titel *rten ạbrel bstod-pa legs bšad sñiṅ-po* läßt sich also ins Sanskrit zurückübersetzen: *pratītyasamutpādastotrasubhāshitagarbha*; die Schrift enthält also einen Preis auf den Satz vom Kausalnexus, schwerlich aber etwas über Kunst.

Der berühmte Lama Sum-pa mk'an-po Ye-šes dpal-ạbyor (1702—1775), Verfasser einer Geschichte des Buddhismus, deren kunsthistorisches Kapitel hier S. 184 übersetzt ist, soll schon in seiner Jugend die religiöse Malerei studiert haben[1] und gilt als Verfasser eines ikonographischen Handbuchs. Der von CHANDRA DAS in verstümmelter Form und nur zur Hälfte mitgeteilte Titel desselben lautet nach Lama TSYBIKOV[2]: སྐུ་གསུང་ ཐུགས་རྟེན་ཐིག་རྩ་མཚན་འགྱུར་ཅན་མེ་ཏོག་འཕྲེང་བ་མཛེས་ཞེས་བྱ་བ་ „Der schöne Blumenkranz: Grundzüge der Zeichnungen von Buddhastatuen, Malereien und Caitya[3], nebst Kommentar"[4]. Das Werk

---

[1] JASB, Vol. LVIII, Part I, 1889, pp. 37, 38.

[2] *Musei Asiatici Petropolitani Notitia* VII, p. 074, Nr. 27.

[3] Über diese Bedeutung von *sku gsuṅ t'ugs* vgl. die Anmerkung auf S. 184.

[4] Unter *t'ig rtsa* ist die mit Tusche ausgeführte Umrißzeichnung zu verstehen, die Hilfslinien, welche die Grundlage der Figur bilden und vor dem Auftragen der Farben gezeichnet werden. Das oben erwähnte Werk ist wohl identisch mit dem von A. D. RUDNEV (Bemerkungen über die Technik der buddhistischen Ikonographie, p. 8; *Publications du Musle d'Anthropologie* etc. *de St.-Pét.*, Nr. V, 1895) erwähnten ཐིག་རྩ་མཚན་འགྱུར་ (nach mongolischer Aussprache *tik za čan del*); nach Rudnev heißen ikonographische Lehrbücher bei den Mongolen *tik* oder *tighei nom* (*nom* = Skr. *sūtra*), und es sollen Bücher mit Darstellungen und Erklärungen der Körpermaße jeder

ist in Versen abgefaßt und von zahlreichen Zeichnungen auf fünf Blättern begleitet. Es gibt Anweisungen zum Zeichnen der verschiedenen Gottheiten und Caitya wie zur Kalligraphie der Alphabete. Nach der Ansicht der mongolischen Maler (*zurāčï*)[1] soll es, wie Tsybikov mitteilt, das wichtigste der bekannten Werke über Ikonographie sein. Das Petersburger Exemplar ist eine handschriftliche Kopie eines sehr schlechten Druckes, in dem die Zeichnungen von Coiji (? Cʻos-rje) Bancuk Badmayev hinzugefügt sind. Tsybikov erwähnt auch eine vom Abte des rÑa-rgod Klosters (bei Lha-braṅ in Kansu), namens Nam Seṅge, verfaßte „Erklärung des Sinnes des grundlegenden Werkes über Ikonographie (*tʻig rtsa*)", das ein Kommentar des vorhergehenden Werkes zu sein scheint. Ferner verdanken wir demselben Lama willkommene Nachrichten über einige kleinere Traktate in tibetischer und mongolischer Sprache, die sich auf dasselbe Gebiet beziehen[2]. Ein mongolischer Text von 7 fol., nach der einzigen Handschrift eines unbekannten Autors kopiert, befaßt sich mit den Proportionsmaßen des Leibes des vollendeten Buddha; eine andere mongolische Schrift von 12 fol., die dasselbe Thema in bezug auf die Gottheiten behandelt, ist Übersetzung eines bekannten tibetischen Werkes über Ikonographie von Adja Gegen; der Name des Übersetzers ist nicht genannt, und es sollen außer dieser kaum noch andere mongolische Übersetzungen des Werkes vorhanden sein. Von Interesse scheint eine tibetische Schrift, mit inter-

---

Gottheit sowie mit Beschreibungen, wie man sie darstellen soll, vorhanden sein. Der interessanten Abhandlung ist eine schöne Heliogravüre nach einer Photographie von Baradin beigegeben, die einen burjatischen Lamamaler bei seiner Arbeit zeigt.

[1] Geschrieben *jirughačï* (Stamm *jiru-*).

[2] *L. c.*, pp. 072—074, Nr. 14—26.

linearer mongolischer Version, unter dem Titel སངས་རྒྱས་ཀྱི་ཆག་ ཚད་ལེགས་པར་བཤགས་སོ་ (6 fol.) zu sein, die eine kurze Darlegung der Körpermaße der Buddha, Bodhisatva, der schrecklichen Götter und Göttinnen, der Vidyādevī u. a. enthält und das Werk „eines der bekannten alten Lehrer Tibets" mit Namen བཅོམ་ལྡན་རིག་པའི་རལ་གྲི་ („Das siegreiche Schwert des Wissens")[1] sein soll. Die übrigen hier aufgezählten Schriften sind meist kurze Traktate von 1 fol. im Umfang, die das Malen bestimmter Gottheiten behandeln und sich wohl mehr an die Sādhana-literatur anlehnen.

Interessant ist es auch, daß in den Lebensbeschreibungen der tibetischen Heiligen und Kirchenfürsten, denen sämtlich ein kanonisches Schema zugrunde liegt, wunderbare Abzeichen bei ihrer Geburt erwähnt werden, die sich mit dem Schema der Malerei decken. So zeigte sich an dem Körper des im Jahre 1542 geborenen Dritten Dalai Lama bSod-nams rgya-mts'o ein weißroter Glanz, und er war mit allen Lakshaṇa und Anuvyañjana eines Mahāpurusha versehen; seine beiden Augen waren schön wie Lotusblätter[2]. Der Zusammenhang der Malerei mit der Physiognomik tritt in Tibet noch eklatant zutage. Mit Recht macht Dr. HERMANN BECKH[3] in einer Be-

---

[1] Wahrscheinlich identisch mit dem berühmten Abte von sNar-t'an, bekannt als sNar-t'ai Rig-pai Ral-gri, „das Schwert des Wissens von Nar-thang", der Druckleiter des Kanjur und Tanjur. Bei HUTH (Geschichte des Buddhismus in der Mongolei, Band II, S. 165) wird er bCom-ldan Rig-ral von sNar-t'an genannt.

[2] G. HUTH, Geschichte des Buddhismus in der Mongolei, Band II, S. 201.

[3] *Zeitschrift des Vereins für Volkskunde in Berlin*, 1912, Heft 3, S. 330. Man wird bemerken, daß die Phraseologie der Lakshaṇa bei der Beschreibung der Königstochter stark verwertet ist; als ich das Buch in Ostasien übersetzte, stand mir leider keinerlei Literatur zur Verfügung, auf die ich mich hätte berufen können.

sprechung des „Roman einer tibetischen Königin" darauf aufmerksam, daß wir in der Prüfung der Abzeichen der Prin-
zessin K'rom-pa rgyan „das Gebiet der Physiognomik betreten,
die in Tibet wie anderwärts geübte Kunst, aus Einzelheiten
der Körperbeschaffenheit auf geistige Eigentümlichkeiten zu
schließen".

Im übrigen wird jeder, der in der Lage ist, sich mit tibetischer Malerei zu beschäftigen, an der Hand praktischer Beispiele den Einfluß unseres Werkes feststellen können. Theoretische Erörterungen sind in diesem Falle nur von Wert,
wenn sie von Demonstrationen guter Bilder begleitet sind,
was natürlich nicht im Rahmen dieser mehr andeutenden als
ausführenden Einleitung liegt, sondern einer Darstellung der
lamaistischen Malerei überlassen bleiben muß. Malerei läßt
sich auch aus den trefflichsten Büchern nicht erfassen, man
muß selber sehen, studieren und denken und dieselben Origi-
nale immer und immer wieder vornehmen.

VIII.

Was die Textausgabe anbelangt, so ist die Palastausgabe
des Tanjur des Kaisers K'ien-lung (im Asiatischen Museum zu
St. Petersburg, als **A** bezeichnet)[1] zugrunde gelegt und mit

---

[1] Mit den jetzt üblich gewordenen Bezeichnungen roter und schwarzer
Tanjur kann ich mich leider nicht befreunden. Abgesehen davon, daß sie
wenig geschmackvoll gewählt sind, sind sie in sachlicher Hinsicht irreführend,
da dadurch der Eindruck hervorgerufen wird, als gäbe es nur einen bestimmten rot gedruckten und nur einen bestimmten schwarz gedruckten
Tanjur, und als wenn dieses rein äußerliche Merkmal einen wesentlichen
Gegensatz zwischen beiden Ausgaben bezeichnete. Dem ist aber nicht so.
Zunächst gibt es mindestens drei (aller Wahrscheinlichkeit nach sogar vier)
rot (d. h. mit einem hellroten Schwefelquecksilber oder Merkurisulfid, das
aus Zinnober gewonnen wird, tib. *mts'al*) gedruckte Ausgaben des Kanjur
und Tanjur, eine im Palast von Peking 1700 auf Befehl des Kaisers K'ang-hi

der Narthang(སྣར་ཐང་)-Ausgabe (von mir in Peking erworbenes, vom Dalai Lama mitgebrachtes Exemplar in der John Crerar Library, Chicago; als **B** bezeichnet) kollationiert. Dieses Verfahren ist von rein praktischen Gründen diktiert, weil der **Pekingdruck** leichter lesbar ist, aber keineswegs durch die

---

gedruckte (s. LAUFER, *Bulletin de l'Académie des Sciences de St. Pétersbourg*, 1909, pp. 567—574) und eine ebenda zur Zeit des Kaisers K'ien-lung (1736—1795) gedruckte (identisch mit dem Exemplar des roten Tanjur im Asiatischen Museum zu St. Petersburg); diese beiden Ausgaben sind nach dem in China üblichen Sprachgebrauch als „Palastausgaben von Peking" oder als „Peking-Druck" mit dem Zusatz des betreffenden Kaisernamens zu zitieren. Die Pekingausgaben werden bereits von ST. JULIEN (*Journal asiatique*, 1849, p. 360) erwähnt. Ein rot gedrucktes Tripiṭaka ist ferner im Kloster von Li-t'ang (West-Sze-ch'uan, China; gegründet 1579) vor dem Jahre 1792 erschienen, da bereits in dem 1792 verfaßten *Wei Ts'ang t'u chi* erwähnt (s. ROCKHILL, JRAS, 1891, p. 271). Ich habe einige Bände dieser rot gedruckten Ausgabe im Besitze eines Lama gesehen und schließe aus der Tatsache, daß auf dem rechten Rande jedes Blattes der Titel des betreffenden Werkes in chinesischer Sprache aufgedruckt ist (was in den tibetischen Ausgaben nicht der Fall ist), daß die Li-t'ang-Edition ein Wiederabdruck der Palastausgabe von K'ien-lung ist, wo sich derselbe Zug wiederfindet; ich habe bereits darauf hingewiesen, daß zur Zeit K'ang-hi's eine viersprachige Konkordanz des Tripiṭaka aufgestellt worden ist. Es sind mir ferner im östlichen Tibet Nachrichten zugekommen, daß im Kloster von Derge ein roter Kanjur und Tanjur gedruckt worden seien; doch weigerten sich die Lama von Derge, als ich mich dank der Zuvorkommenheit des dort residierenden chinesischen Tao-t'ai einige Tage dort aufhalten durfte, mir ihre Bücherschätze zu zeigen, so daß ich keine Angaben über diese Ausgabe machen kann. Eine große Anzahl von Büchern erhielt ich in Derge heimlich durch Vermittlung von Chinesen und habe jetzt gegründete Aussicht, die Indexbände des Kanjur und Tanjur von da zu erhalten, so daß ich vielleicht in Zukunft darüber berichten kann. Sicher ist, daß in Derge ein Schwarzdruck des Kanjur erschienen ist; ein Exemplar desselben, vorzüglich gedruckt, von W. W. Rockhill in Peking beschafft, befindet sich in der Library of Congress in Washington. Schwarz gedruckte Ausgaben des Kanjur und Tanjur sind in fast allen größeren Klöstern Tibets und der Mongolei hergestellt worden und sollten stets nach dem Druckort, wenn möglich auch nach der Sekte, von der sie stammen, zitiert werden.

bibliographische Sachlage gerechtfertigt. Denn der Narthangdruck ist älter als die Palastausgabe und verdient daher schon
wegen dieser zeitlichen Priorität den Vorrang. Aber noch
mehr: die Narthang-Ausgabe bewahrt die ältere Tradition des
Bu-ston, während sich die K'ien-lung-Ausgabe an die Revision
des Fünften Dalai Lama anlehnt und überhaupt in textkritischer
Hinsicht jeder tieferen Selbständigkeit entbehrt. Sie ist nur
eine gut gedruckte Ausgabe, aber keine unabhängige Version,
die für die Geschichte der Tanjurrezensionen eine Bedeutung
hätte. Diese Tatsache habe ich aus dem umfangreichen Index-
band des Tanjur von Co-ne festgestellt, in welchem die Literaturgeschichte des Tanjur ausführlich behandelt ist (Näheres
S. 50). Die uns jetzt vorliegende Narthang-Ausgabe ist die
dritte in Narthang veranstaltete Auflage des Tanjur und hat
bereits den Einfluß der revidierten Ausgabe des Fünften Dalai
Lama erfahren. Wie sich herausstellen wird, stammt das
Citralakshaṇa aus dieser letzteren Ausgabe, nicht aus der des
Bu-ston, in der es fehlte, und daher rührt die große Übereinstimmung des Textes in beiden Drucken. Die folgende
Liste übereinstimmender Lesarten, wo es sich um eigentümliche Schreibungen oder offenkundige Irrtümer handelt, erbringt
den Beweis dafür, daß beide Drucke den Text des Citralakshaṇa aus derselben Quelle geschöpft haben, und diese ist die
Ausgabe des Fünften Dalai Lama. Erwähnenswert in diesem
Zusammenhange ist, daß in dem vierten Tanjurwerke über
Kunst (S. 2) der Verfassername bald *Eṭeya*, bald *Etreya* geschrieben wird, und daß die Peking- und die Narthang-Ausgabe
an derselben Stelle genau dieselbe Schreibart übereinstimmend
befolgen.

Beide Drucke haben folgende eigentümliche Schreibungen
oder Irrtümer gemeinsam: V. 162 *tsom* für *rtsom*; V. 234 *sñas*

für *brñas*, dagegen V. 252 beide das regelrechte *brñas*; V. 207 *dris* für *bris*; V. 333 *mts' an-ñid bdag*, wo zweifellos *dag* zu lesen ist; V. 342 *rigs byed* statt *rig byed*; V. 366 *klu rnams* statt *c'u* (oder vielleicht *kluñ*) *rnams* (daß hier ein Wort für „Fluß" zu suchen ist, geht offenbar aus dem folgenden *Gañ-gā* hervor); V. 367 *ma ste* (bezw. *sta*) statt *ñi-ma*; V. 437 *žig* statt *bžig*; V. 472 *mts' on* statt *mts' an*; V. 487 *rigs* statt *rig*; V. 640 *mgron-bu* für *agron-bu*; in V. 729 fehlt in beiden Drucken, wahrscheinlich durch Auslassung der Partikel *ni*, eine Silbe; V. 795 *me-yi abras-bu* für *mig-gi abras-bu*; V. 797 *sku-bar* für *bsku-bar*; V. 800 *sma* für *sme*; V. 805 und 820 *t'e-boñ* für *mt'e-boñ*, an anderen Stellen übereinstimmend *mt'e-boñ*; V. 841 und 1026 *ap'ra* für *p'ra*; V. 863 *yañ* für *yañs*; V. 897 *lhag* für *ltag*; V. 962 und 978 *t'oñ-ka* für *mt'oñ-ka*, dagegen V. 1029 übereinstimmend *mt'oñ-ka*; V. 963 *ci-le* für *ci-la*; V. 1001 *zlo gar* für *zlos gar*; V. 1020 *sar-pa* für *gsar-pa*; V. 1022 *mt'on tiñ* für *mt'on t'iñ*; V. 1027 *rgyal* für *rgyas*; V. 1121 *Ri-boñ mc og* statt *Ri-bo mc'og*, dagegen V. 1088 beide *Ri-bo aion*.

In folgenden acht Fällen gewährt **B** eine korrekte Lesart, wo **A** zweifellos im Irrtum ist: V. 317 *lha* (**A**: *zla*); V. 377 *t'ad* (**A**: *ts'ad*); V. 378 *ñes* (**A**: *des*); V. 505 *de-dag* (**A**: *de des*); V. 509 *rañ-gi ni* (in **A** ist *ni* ausgelassen, so daß dem Verse eine Silbe fehlt); V. 540 *rtse-mo* (**A**: *tse-mo*); V. 547 *bstan* (**A**: *pa stan*); V. 803 *ts'igs* (**A**: *ts'egs*).

Man ersieht also, daß die Narthang-Ausgabe in manchen Fällen die richtige, in anderen Fällen die bessere Lesart enthält. Auf der anderen Seite bringt auch die Peking-Ausgabe Berichtigungen von Fehlern des Narthangdrucks, obwohl es auch da Berichtigungen gibt, die nicht immer als wirkliche Verbesserungen gelten können. Im großen und ganzen müssen wir aber zugestehen, daß der Pekingdruck „der Güter höchstes nicht ist", und daß der ältere Narthangdruck unbedingt zu Rate gezogen werden muß[1]. Daran kann die Tatsache nichts ändern, daß der Pekingdruck in technischer Hinsicht vollendet, schön, klar und leicht lesbar ist, während der Narthangdruck an technischen Mängeln leidet und oft schwer zu entwirren ist. Es beruht aber auf Verkennung der Sache, die Narthang-Ausgabe vom Standpunkt des Druckes aus als „schlecht" zu bezeichnen. Die in Narthang geschnittenen Holztafeln sind so gut und sorgfältig geschnitzt als irgendwelche Druckblöcke in Asien. Der Faktor, der die Klarheit oder Unklarheit des Druckes entscheidet, ist einzig und allein das Papier. Von denselben Holztafeln kann man auf gutem, zähem und undurchsichtigem Papier gute Abdrücke, und auf dünnem, durchsichtigem, von zahlreichen nichtmazerierten Holzfasern durchsetztem Papier miserable Abdrücke herstellen. Das Experiment läßt sich leicht demonstrieren, und ich kann Interessenten des Druckes und Druckverfahrens in meiner Bibliothek identische tibetische Drucke auf guten und schlechten Qualitäten von Papier zeigen, die von genau denselben Holztafeln abgezogen sind, die einen gut, die andern schlecht lesbar. Die rein technische Verfassung der meisten auf minderwertiges Papier gedruckten Narthang-Ausgaben darf uns aber gewiß nicht abschrecken, von diesen den rechten Gebrauch zu machen. Das

---

[1] Dieses Urteil stimmt mit dem gelehrter Lamen überein, die ich über den Wert der Ausgaben des Kanjur und Tanjur zu befragen Gelegenheit hatte. Sie alle weisen, und mit Recht, auf die Narthangausgabe als die beste hin; darunter ist natürlich der innere Wert des Textes, nicht die äußere technische Qualität zu verstehen. Seit langem bin ich mit Untersuchungen über die Geschichte der verschiedenen Kanjur- und Tanjurausgaben und der Entwicklung des Kanons in Tibet beschäftigt, und es stellt sich mehr und mehr heraus, daß die meisten der zahlreichen Drucke auf den von Narthang zurückzuführen sind.

Berliner Exemplar ist ja vielleicht mehr als andere gestraft und wird auf immer den Schrecken des Anfängers bilden. Mit Entsetzen denke ich selbst an die verzweifelten Stunden zurück, als ich an demselben im Jahre 1895 debütierte und mir schließlich das Petersburger Exemplar verschreiben ließ. Die Fähigkeit, solche Drucke zu lesen, hängt naturgemäß viel von dem Grade der Kenntnis und Beherrschung der Sprache ab. Unser Narthangdruck in Chicago besitzt ein erträgliches Papier von mittlerer Qualität; der Kanjur ist in guter Verfassung und fließend lesbar, der Tanjur ist minder gut, läßt sich aber immerhin bewältigen. Für kursorische Lektüre sind meines Erachtens die schwarzen Drucke vorzuziehen, während die Rotdrucke den meisten Augen nicht zuträglich sind und nach langem Lesen empfindliche Wirkungen hinterlassen. Es liegt mir selbstverständlich fern, die Pekinger Palastausgaben, die ich ja schon bei früherer Gelegenheit als Meisterwerke der Buchdruckerkunst gepriesen habe, hier herabsetzen zu wollen. Aber es war einmal nötig, da noch nicht die Aufmerksamkeit darauf gelenkt worden ist, das literarische Verhältnis beider Ausgaben klarzulegen, um ihren Wert für die Textkritik zu bestimmen. Man soll niemanden nach seinem Gewand beurteilen, am wenigsten ein Buch. Die Palastausgabe hat keinen höheren Wert für die Textkritik, weil sie in prächtigem Kaiserornat erscheint, sondern nur eine größere praktische Bedeutung für die leichtere und schnellere Festsetzung des Textwortlauts, und die Narthang-Ausgabe verdient keine Verachtung, weil sie im durchlöcherten, unscheinbaren Mönchskleide auftritt, sondern ernstliche Beachtung als die treue Behüterin der orthodoxen Tradition. Für die Herstellung eines kritisch gesicherten Textes müssen selbstverständlich beide Ausgaben, und wo die Möglichkeit geboten ist, auch andere Ausgaben benutzt werden.

Der Text des Citralakshaṇa ist uns im Tanjur in guter
Verfassung überliefert[1] und wird wohl das verlorene Sanskrit-

[1] Wenn Dr. HERMANN BECKH in seiner sehr verdienstlichen und gewissen-
haften Ausgabe des Udānavarga (S. IV) von der „im allgemeinen schlechten
Überlieferung des tibetischen Schrifttums" spricht, so ist daran zu erinnern,
daß zwischen der Überlieferung als solcher und der technischen Verfassung,
in der sie in den Büchern geboten wird, zu unterscheiden ist. In deutschen
Landen haben wir zahlreiche schlecht (und wahrscheinlich mehr schlecht als
gut) gedruckte Ausgaben von Schiller und Goethe, aber wir werden aus
dieser ökonomischen Erscheinung nicht schließen, daß die Überlieferung
unserer klassischen Dichter schlecht ist. Über die Güte der tibetischen
Überlieferung ist sich alle Welt einig. Die Schreibfehler — denn um solche
handelt es sich, nicht um Druckfehler, da im Holzdruckverfahren nur der
dem Blockschnitzer das Manuskript liefernde Kopist, nicht aber der Drucker
sich irren kann — brauchen uns nicht zu irritieren; der Zustand der tibetischen Orthographie, obwohl vielen Schwankungen ausgesetzt, ist lange nicht
so verworren als der der meisten europäischen Sprachen. Dann darf man
nicht vergessen, daß die uns jetzt zur Verfügung stehenden Kanjur- und
Tanjurdrucke verhältnismäßig modernen Ursprungs sind und auf ältere Ausgaben zurückgehen, die uns vielleicht eine glücklichere Zukunft noch bescheren wird; wenn Pelliot's tibetische Funde bearbeitet und zugänglich gemacht sein werden, wird sich vieles in neuem Lichte zeigen. Wir müssen
auch menschlich und billig urteilen; das Kopieren eines ganzen Kanjur und
Tanjur für den Druck war eine herkulische Arbeit, und wem würden da
nicht Schreibfehler, Irrtümer und Unterlassungssünden unterlaufen? Ich
zweifle sehr, ob man bei uns die Arbeit viel besser gemacht hätte als in
Tibet. Natürlich mußte das ewige Kopieren eine allmähliche Verschlechte-
rung der ursprünglichen Verfassung der Texte herbeiführen, deren Gefahr
noch dadurch vergrößert wurde, daß seit dem Mittelalter der lebendige
Kontakt Tibets mit Indien verloren ging, als der Buddhismus dort allmählich
erlosch. Vom siebenten bis zum zwölften Jahrhundert war die große Epoche
der Übersetzungen, als die aus Indien und den Nachbarreichen nach Tibet
strömenden Gelehrten indische Sprache dort verbreiteten und lebendig er-
hielten, während böse Zeiten für die tibetischen Editoren begannen, als sie
vom Verkehr mit Trägern indischer Rede abgeschnitten und allein auf sich
selbst angewiesen waren. Aber auch so haben sie ihre Sache ganz gut
gemacht und verdienen unsere Anerkennung und Bewunderung. Wie überall,
gibt es in Tibet gute und schlechte Drucke, kostbare und billige Handschriften, und an Güte und philologischer Qualität der Editionen steht in
Asien das tibetische Schrifttum nur hinter dem chinesischen zurück. — Und

original getreulich wiederspiegeln. Es sind nur einige ver-
einzelte Fälle da, wo ich in die jetzt vorhandenen Lesarten
gelinde Zweifel setze, und auch diese dürften wohl mit den
Fortschritten unserer Kenntnis der tibetischen Lexikographie
verschwinden. Alles in allem darf man mit der Form, in der
das Werk im Tanjur erhalten ist, sehr wohl zufrieden sein.

IX.

Über die Stellung des Werkes im Tanjur[1] ist noch ein
Wort zu sagen. Man wird aus HUTH's Verzeichnis ersehen,
daß das Citralakshaṇa im Index der Narthang-Ausgabe nicht
genannt wird, ebensowenig die oben S. 2 unter Nr. 1 zitierte
Abhandlung, während Nr. 2 und 4 im Index angeführt sind.
Huth hat diese merkwürdige Tatsache nicht beleuchtet, aber
sie hat ihre Ursache in der komplizierten Literaturgeschichte
des Tanjur. Die Auslassung dieser beiden Abhandlungen wie

---

noch eins. Seit etwa drei Generationen liegt in unseren Bibliotheken das
Material in allen Sprachen des nördlichen Buddhismus aufgespeichert. Was
haben wir bisher daraus übersetzt? Liegt hier nicht ein beschämendes
Armutszeugnis für den Stand unserer Wissenschaft vor? Welche Berech-
tigung eines überlegenen Stolzes haben wir gegenüber Tibetern, Chinesen,
Uiguren, Mongolen und andern Völkern Zentralasiens, die den ganzen
buddhistischen Kanon mit unendlicher Mühe und Sorgfalt in ihre Sprachen
übersetzt haben und mit weit größeren philologischen Schwierigkeiten zu
kämpfen hatten als wir? Sind diese Leistungen in ihrer Gesamtheit nicht
kolossal, und gibt es in unserer ganzen europäischen Kultur irgend etwas,
das ihnen als meßbare Größe zur Seite gestellt werden könnte? Bei uns
ist die Götterdämmerung ganz gewiß noch nicht angebrochen.

[1] Das Wort Tanjur བསྟན་འགྱུར་ ist eine Abkürzung (Binom) für
བསྟན་བཅོས་འགྱུར་རོ་ཆག d. h. die [o-cog diese, nämlich die bekannten] Übersetzungen der Çāstra, definiert als བདེ་བར་གཤེགས་པའི་བཀའི་དགོངས་འགྲེལ་
(Skr. *Sugataçāsanavṛtti* oder -*ṭīkā*) Kommentar der Aussprüche des Sugata
(Buddha).

4

mancher anderen im Index von Narthang ist kein Zufall, denn genau dasselbe ist in dem in meinem Besitz befindlichen Index des Tanjur des Klosters Co-ne ཙོ་ནེ་[1] der Fall. Die weitere Tatsache, die HUTH (*l. c.*, S. 279) nur konstatierte, ohne sie zu erklären, daß im Index Werke zitiert werden, die in der Sammlung selbst fehlen, hätte schon den Gedanken anregen müssen, daß dieses Mißverhältnis zwischen Index und den tatsächlich vorhandenen Werken auf geschichtliche Ereignisse in der Entwicklung der Tanjursammlung zurückzuführen ist, — mit anderen Worten, daß der Index, wie er jetzt vorliegt, zu der Sammlung, wie sie uns jetzt vorliegt, nicht paßt, sondern zu einer älteren Sammlung gehört haben muß. Das ist denn auch tatsächlich der Fall.

Der früheste Katalog der tibetisch-buddhistischen Literatur, von dem wir Kunde haben, wurde im achten Jahrhundert zur Zeit des Königs K'ri-sroṅ lde-btsan von dPal-brtsegs[2] und Nam-mk'ai sñiṅ-po[3] verfaßt, der zufolge dem Index des Tanjur von Co-ne (fol. 302) in Band 125 der Sūtra enthalten sein soll. Es werden dort zunächst zwei von dPal-brtsegs verfaßte Schriften erwähnt: ཆོས་ཀྱི་རྣམ་གྲངས་ཀྱི་བརྗེད་བྱང་དང་ཆོས་ཀྱི་རྣམ་གྲངས་

---

[1] Dieser Index führt den Titel འཇིག་རྟེན་གསུམ་གྱི་བདེ་སྐྱིད་པད་ཚལ་ འབྱེད་པའི་ཉིན་བྱེད་ „Die Seligkeit der drei Welten, die die Lotusblüten öffnende Sonne (*divākara* „Tagmacher"). Das Kloster Co-ne, wo ich einige Tage im Januar 1910 zubrachte, liegt einige Meilen östlich von der Stadt T'ao chou im südwestlichen Teile der jetzigen chinesischen Provinz Kan-su, welchen die Tibeter von ihrem Standpunkt noch zu Amdo rechnen.

[2] HUTH (Verzeichnis, S. 274, 279) setzt ihn irrtümlich ins neunte Jahrhundert. In allen mir zugänglichen historischen Werken (auch in *dPag bsam ljon bzaṅ*, p. 173) und in der Padmasambhavaliteratur (z. B. „Roman einer tibetischen Königin", S. 4, 131, 216) wird er zu einem Zeitgenossen des Königs K'ri-sroṅ gemacht.

[3] „Roman", S. 14, 130.

ཀྱི་བརྟེན་བྱུང་གི་རྩ་བ་ལྲུ་ཚོ་རྣ་དཔལ་བརྩེགས་ཀྱིས་མཛད་པ་ Eine klassifizierte Liste religiöser Schriften, und die Grundlage dieser Liste. Dann heißt es: འབྲུག་གི་ལོ་ལ་པོ་ཐང་སྟོང་ཐང་ལྷན་དཀར་གྱི་བཀའ་དང་བསྟན་བཅོས་ འགྱུར་རོ་ཚོག་གི་དཀར་ཆག་ལོ་ཚོ་བ་དཔལ་བརྩེགས་དང་ ། ནམ་མཁའི་སྙིང་པོ་ལ་ སོགས་པས་མཛད་པ་རྣམས་བཞུགས་སོ ། d. h. Der in einem Drachenjahre im Palaste lDan-dkar in sToṅ-t'aṅ von den Lotsāva dPal-brtsegs, Nam-mk'ai sñiṅ-po und anderen verfaßte Index (oder Katalog) des Kanjur und Tanjur sind [in diesem Bande] enthalten. Diese Angabe findet ihre Bestätigung im *dPag bsam ljon bzaṅ* (p. 173), wo gleichfalls das Drachenjahr und der Palast lDan-dkar erwähnt werden, in welchem der Übersetzer Ban-dhe dPal-brtsegs, der Ban-dhe kLui dbaṅ-po und andere heilige Schriften übersetzten; auch fertigten sie eine Liste der Titel der Sammlung an und verfaßten einen Katalog, in dem die Zahl der Bände und Verse angegeben war, wodurch die Lehre damals eine weite Verbreitung fand. Von da müssen wir einen weiten Sprung bis in den Anfang des dreizehnten Jahrhunderts machen.

Der Index des Tanjur (wie der des Kanjur) wurde im Mittelalter aufs neue von Bu-ston (1288—1363) festgestellt. Die Liste der von ihm festgesetzten Schriften findet man im *dPag bsam ljon bzaṅ* (ed. CHANDRA DAS, p. 410) aufgezählt[1].

---

[1] Man darf nur sagen, daß der Kanon zu Bu-ston's Zeit in großen Zügen abgeschlossen war, aber bis ins achtzehnte Jahrhundert hinein befand er sich in beständigem Flusse. Es wurden neue Werke aufgefunden, alte wurden revidiert und neu übersetzt, Kollationen mit dem chinesischen Tripiṭaka veranstaltet, und Übersetzungen ins Tibetische aus dem Chinesischen, dem Uigurischen, der Sprache von Khotan, und sogar aus dem Tocharischen hinzugefügt. Ich muß mich an dieser Stelle mit einigen wenigen konkreten Beispielen begnügen. Kanjur Nr. 446 (Index ed. SCHMIDT, p. 67) war von Bu-ston nach einem unvollständigen Sanskritoriginal übersetzt worden, man

4*

Aber schon zu Bu-ston's Zeit gab es Werke, die früher übersetzt worden waren, die sich aber damals nicht auffinden
ließen und deshalb nicht in den offiziellen Index aufgenommen
wurden. Zu diesen Werken gehörte eben unser Citralakshaṇa
(*ibid.*, p. 416 Z. 6). Daraus geht hervor, daß die Übersetzung

requirierte daher später von Peking (*Mahātsinai rgyal-k'ab*) das im dort
befindlichen Kanjur (d. h. dem chinesischen Tripiṭaka) vorhandene, von dem
Mahāpaṇḍita Danarakshita ins Chinesische übersetzte Werk, das ein aus dem
Geschlechte des Chinggis Khan stammender tibetischer Gelehrte, namens
mGon-po skyabs, ins Tibetische übertrug. Derselbe lebte zur Zeit der Ts'ing-
Dynastie und ist gleichfalls Übersetzer von Kanjur Nr. 502 (Index ed.
SCHMIDT, p. 76) auf Grund einer chinesischen Vorlage. Manche Werke des
Kanjur wie Nr. 199 (Index ed. SCHMIDT, p. 33), Nr. 351 (*ibid.*, p. 53),
Nr. 555 (*ibid.*, p. 81) und Nr. 688 (*ibid.*, p. 96) sind ausschließlich aus
dem Chinesischen übersetzt und haben außer dem tibetischen meist nur
einen in tibetischer Schrift transkribierten chinesischen Titel; andere sind
auf Grund indischer und chinesischer Versionen übertragen wie Nr. 254
(*ibid.*, p. 42), Nr. 352 (*ibid.*, p. 53) und Nr. 1083 (*ibid.*, p. 134); in einigen
Fällen sind sogar chinesische Übersetzer mitbeteiligt wie bei Nr. 119 (*ibid.*,
p. 20) und Nr. 213 (*ibid.*, p. 35). Aus Tocharistan wurde Kanjur Nr. 513
(*ibid.*, p. 78) beschafft, und zwar von dem Bhikshu Ner-ban-rakshita. Will
man die ganze Literaturgeschichte einzelner Werke feststellen, so findet man
die ausführlichsten Angaben in den neueren Peking-Ausgaben. Ein interessantes
Beispiel wiederholter Überarbeitung ist mir im Kolophon einer in Peking
gedruckten tibetischen Ausgabe der Ashṭasāhasrikāprajñāpāramitā begegnet.
Hier finden wir zunächst das im Index des Kanjur (Nr. 12) enthaltene
Kolophon, dann aber Nachrichten über acht weitere Bearbeitungen des Werkes
zu verschiedenen Zeiten, bei denen nicht nur Handschriften in Sanskrit,
sondern auch in Māgadhī benutzt wurden, bis schließlich der Übersetzer
von Ža-lu, dPal Rin-c'en c'os-skyoṅ bzaṅ-po (1439—1525), der Verfasser
des grammatischen Werkes Za-ma-tog, „durch eine Konkordanz der zahlreichen chinesischen und tibetischen Ausgaben den Text vollkommen ins
Reine brachte", worauf eine nochmalige Revision, gestützt auf zwei chinesische und die bereits gedruckten tibetischen Editionen, durch dPal-ạbyor
rgyal-mts'an und Kun-dga c'os-bzaṅ stattfand. Wir besitzen also den Kanjur
nicht mehr in seiner ursprünglichen Gestalt, und Vergleiche der tibetischen
mit den chinesischen Versionen, um an deren Hand die Güte der indischen
Tradition zu prüfen, sind völlig illusorisch, wenn man sich nicht über die
Literaturgeschichte der benutzten Vorlagen Klarheit verschafft hat.

dieses Werkes vor dem Zeitalter des Bu-ston stattgefunden haben muß, also der älteren Periode der tibetischen Übersetzungen angehört, womit der Stil des Werkes übereinzustimmen scheint. Der von Bu-ston redigierte Tanjur wurde unter der Regierung des Mongolenkaisers Buyantu Khan (1311—1319) samt dem Kanjur in Tʻugs-kʻu[1] gedruckt[2] und im Kloster der Goldhalle von Ža-lu ཞུ་ལུ་གསེར་ཁང་གི་གཙུག་ལག་ཁང་ aufbewahrt[3]; denn Ža-lu[4], einige Meilen südwestlich von bKrašis lhun-po gelegen, war der Sitz von Bu-ston's literarischer Tätigkeit. Darin ist der Hauptgrund zu suchen, warum Narthang in der Geschichte der Kanjur und Tanjur Editionen eine führende Rolle spielt, und warum die meisten Klöster für ihre Ausgaben auf die von Narthang zurückgreifen.

Neben der Tradition des Bu-ston scheint aber, dieser parallel laufend, noch eine andere Rezension des Tanjur bestanden zu haben, die sich auf den Ort aPʻyoṅ-rgyas[5] bezieht. Nähere Angaben über dieselbe sind im Index von Co-ne leider nicht vorhanden; wir vernehmen nur, daß der Regent Saṅsrgyas rgya-mtsʻo (1652—1703) die Aufmerksamkeit des Fünften Dalai Lama Ṅag-dbaṅ bLo-bzaṅ rgya-mtsʻo[6] (1617—1680)

---

[1] Wahrscheinlich Name eines Tempels in Narthang, wo ja im Tempel des Mañjughosha ein Exemplar untergebracht wurde.

[2] Huth, Geschichte des Buddhismus in der Mongolei, Band II, S. 165.

[3] Nach dem Index des Tanjur vom Kloster Co-ne.

[4] Wird poetisch als „die Quelle kostbarer Aussprüche" (*legs bšad rincʻen ąbyuṅ gnas* = Skr. *subhāshitaratnasambhava*) bezeichnet.

[5] Vgl. „Roman einer tibetischen Königin", S. 243. Das geographische Moment verdient in der tibetischen Literaturgeschichte die sorgfältigste Beachtung. Nicht ohne Grund hat die zentrale Provinz gTsaṅ von jeher eine Vormachtstellung besessen und stets auf ihre politische Selbständigkeit gesehen. Sie war der Mittelpunkt des geistigen Lebens und literarischen Schaffens.

[6] Sein vollständiger Name lautet: རྒྱལ་དབང་ཐམས་ཅད་མཁྱེན་ཅིང་གཟིགས་

auf diese Ausgabe lenkte, und daß der Dalai Lama auf Grund derselben einen neuen Index des Tanjur festlegte. Diese revidierte Redaktion bildet die Grundlage der von der orthodoxen Kirche (dGe-lugs-pa) adoptierten Tanjur Rezensionen. Sie wurde für die Palastausgabe des Kaisers K'ien-lung zugrunde gelegt, wie der Index von Co-ne ausdrücklich erzählt, ebenso von den großen Klöstern bei Lhasa dPal-ldan ạBras-spuṅs, Se-ra t'eg-c'en gliṅ (Mahāyānadvīpa von Se-ra, gegründet 1418) und Ras-ạp'rul-snaṅ adoptiert, die mit Nachdrucken des Tanjur hervortraten, auch neue Übersetzungen der in der Edition von ạP'yoṅ-rgyas fehlenden Werke hinzugefügt haben sollen.

Einen weiteren Schritt tat man im großen Kloster Derge im östlichen Tibet, indem man die beiden Rezensionen von Ža-lu und ạP'yoṅ-rgyas, d. h. also die des Bu-ston und des Fünften Dalai Lama, mit einander kollationierte und verschmolz. Diesem Vorbild folgte man bei Drucklegung des Tanjur im Kloster Co-ne im Jahre 1772[1].

Diese Vorgänge verleihen uns einen Einblick in die Ursachen der Abweichungen der einzelnen Tanjur Kataloge und des sonderbaren Mißverhältnisses, das in den Editionen des achtzehnten Jahrhunderts zwischen Katalog und Sammlung klafft. Im allgemeinen liegt allen der von Bu-ston aufgestellte

པ་ཆེན་པོ་དགེ་གི་དབང་ཕྱུག་རྡོ་བརྦང་རྒྱ་མཚོ་འཇིགས་མེད་གོ་ཆ་ཕྱབ་བསྲན་ལང་ཚོའི་སྟེའི་ཞབས་པད་

[1] Die Geschichte der Drucklegung wird ausführlich erzählt. Dreihundert Graveure für die Druckplatten und ein Heer anderer Arbeiter waren erforderlich. Die Summe von 13 937 Taels (Unzen Silber), wovon allein 13 000 Taels auf einen Fürsten in der Mongolei fielen, wurde von frommen Gläubigen für das Werk gestiftet, die mit Beiträgen von 48$^{1}/_{2}$, 50, 60, 100, 103, 110 bis zu 150 Taels beteiligt waren. Auch im Index von Narthang finden sich interessante Mitteilungen über Herstellung und Kosten der Arbeit.

Katalog zugrunde, aber seit Bu-ston's Zeit sind einzelne Werke verloren gegangen, andere damals verloren geglaubte wieder aufgefunden und hinzugefügt worden. Die Narthang-Ausgabe, — und darin liegt ihr besonderer Wert — weil sie an dem Orte der Wirksamkeit des Meisters haften blieb, scheint seine Tradition am treuesten bewahrt zu haben. Die Palastausgabe des Kaisers K'ien-lung entfernt sich weiter davon und schließt sich der Autorität des Dalai Lama an, während Derge und Co-ne den Ausweg des Kompromisses suchen.

Machen wir die Anwendung auf unser Citralakshaṇa, so sehen wir, daß sein Titel in den Indexbänden von Narthang und Co-ne (also auch Derge) fehlt, weil es in Bu-ston's Index nicht verzeichnet war, und man hält an der einmal geheiligten Über-lieferung fest. Das Werk war von Bu-ston's Index ausge-schlossen, weil es damals zufällig abhanden gekommen oder in Ža-lu unerreichbar war. Die Tatsache, daß das Werk im Tanjur von Narthang des achtzehnten Jahrhunderts abgedruckt ist, beweist, daß es später wieder entdeckt worden war, und der Abdruck in der Palastausgabe K'ien-lung's beweist, daß es bereits in der Rezension des Fünften Dalai Lama vor-handen gewesen sein und somit auf die Edition von ạP'yoṅ-rgyas zurückgehen muß. Damit ist die Tatsache festgestellt, daß an diesem für die Literaturgeschichte so bedeutsamen Orte Übersetzungen aus der Sanskritliteratur aufbewahrt worden sind, die Bu-ston und seiner Schule unzugänglich waren.

Während sich auf diese Weise manches Rätsel des Tanjur löst, knüpft sich auch manches neue an. Was ist aus den Werken geworden, die im Index, aber nicht in der Samm-lung enthalten sind?

Es wird darüber im Index von Co-ne zu Band 123 am

Schlusse des ersten Teils, der die Werke über Technik (*bzo
rig-pa*) aufzählt, gesagt:

རྒྱལ་དབང་ལྔ་པའི་དཀར་ཆག་ཏུ་བྱས་ཏེ་ཆེན་པོ་དྲ་ར་ཧ་མིས་མཛད་པའི་དུས་བསྟན་
པའི་མེ་ལོང་རྒྱ་གར་ཤར་ཕྱོགས་ཀྱི་པཎྜི་ཏ་ཛྙཱ་ན་ཤྲཱི་དང་ ། ལོ་ཙཱ་བ་ཉི་མ་རྒྱལ་མཚན་
གྱི་འགྱུར་ཞེས་བྱ་བ་དང་ ། དྲང་སྲོང་ཕུར་བུས་མཛད་པའི་ལོ་དྲུག་ཅུའི་འབྲས་རྩིས་
ཕུར་བཏགས་བལ་ཡུལ་གྱི་མཁས་པ་བི་ན་སིང་གིའི་འགྱུར་ ། ཞེས་པ་བྱུང་ཡང་དཔེ་མ་
འཚོར་པས་སླར་བཙལ་དགོས་སོ ༎

„Die im Index des Fünften Dalai Lama aufgeführten
(beiden) Werke, nämlich der von dem großen Brahmanen
Dharahami verfaßte Kālaçāsanādarça nach der Übersetzung
des Paṇḍita Jñānaçrī aus dem Osten Indiens und des Lotsāva
Ñi-ma rgyal-mts῾an, sowie die von dem Ṛishi Bṛihaspati verfaßte Berechnung des Jupiter-Zyklus nach der Übersetzung
des nepalesischen Gelehrten Vinasiṁha sind zwar erschienen,
aber die Bücher sind nicht mehr vorhanden und müssen
wieder aufgesucht werden".

Die beiden Werke werden von HUTH, Verzeichnis, S. 279,
Noten, Nr. 19 und 20, als im Index von Narthang zitiert, aber
als in der Sammlung fehlend bezeichnet. Nun wissen wir den
Grund, warum sie darin fehlen; aber es bleibt doch merk-
würdig, daß diese Werke noch um die Mitte des siebzehnten
Jahrhunderts vorhanden gewesen und innerhalb kaum eines
Jahrhunderts verschwunden sein sollen. Oder waren sie im
Index des Dalai Lama nur ex officio katalogisiert und in seiner
Ausgabe des Tanjur nicht vorhanden? Wir wissen es nicht
oder noch nicht.

Es wird ferner im Index von Co-ne (fol. 304) gesagt, daß
größere und kleinere Fragmente von Çāstras im Index des
Fünften Dalai Lama erscheinen, die gegenwärtig den ge-

druckten Ausgaben nicht mehr einverleibt sind, aber handschrift-
lich zirkulieren; die interessantesten von diesen sind ein Kommentar zu den Sūtra von dem erwähnten dPal-brtsegs, betitelt
མི་ཤེས་མདུད་འགྲོལ་ „Der Knotenlöser der Unwissenheit" und
ein Werk ཉེར་བསྒྱུར་གྱི་མཚན་ཉིད་ (Skr. *upaplavalakshaṇa*) „Vorzeichen der Unfälle" mit erklärendem Kommentar, verfaßt
von Ācārya Indradatta (tib. *dBaṅ-pos byin*) aus Magadha, dem
Verfasser eines historischen Berichts über den Buddhismus in
Indien, der Tāranātha als Quelle gedient hat (*dPag bsam ljon
bzaṅ*, p. 123); die Übersetzung dieses Werkes stammt von dem
Lotsāva Nam-mk'a dpal-bzaṅ. Zu Band 129 (fol. 306) heißt
es, daß der Index des Bu-ston von dem des Dalai Lama in
der Anordnung des Stoffes hier geringe Abweichungen zeige,
daß man aber in Co-ne den vom Kloster Derge gedruckten
Index befolge und sich nach dessen System richte.

Noch auffallender ist, daß im Index von Co-ne in Band 123,
dessen Materie vom Narthang Index verschieden angeordnet ist,
ein Werk registriert wird, das in letzterem fehlt und den Bu-ston
zum Übersetzer hat: མཆོད་རྟེན་གྱི་མཚན་ཉིད་སྟོན་པ་བུ་སྟོན་ལོ་ཙའི་འགྱུར་
d. i. „Lehrbuch der Merkmale der Caitya (also wohl *caitya-
lakshaṇaçikshā*), Übersetzung des Bu-ston, des Lotsāva". Dieses
Caityalakshaṇa muß ein Gegenstück zum Citralakshaṇa gewesen sein, und die Existenz dieses indischen Werkes macht
es uns klar, warum, wie wir oben (S. 39) gesehen haben, in
den ikonographischen Werken der Lamen auch die Caitya behandelt werden, ein Thema, das ihnen gewiß aus diesem oder
einem verwandten indischen Werke zugeflossen ist. Und wenn
Bu-ston der Übersetzer war, so wirft diese Tatsache auch
Licht auf die Schriftstellerei des bCom-ldan Rig-pai ral-gri im
Fache der Ikonographie (S. 41); denn dieser war ja sein Zeit-

genosse und Mitarbeiter an dem großen Werke. Diese für
die buddhistische Kunstgeschichte bedeutsame Abhandlung
muß also im Tanjur des Bu-ston vorhanden gewesen sein;
man darf nicht mit dem Argument operieren, daß Bu-ston sie
nach dem Abschluß und Druck des Kanons übersetzt haben
könnte, denn dem widerspricht ihre Erwähnung im Index von
Co-ne, die nur dadurch gerechtfertigt sein kann, daß sie bereits im Index des Bu-ston erwähnt war. Leider wird die
Hoffnung, daß das Werk in der Ausgabe von Co-ne vorhanden sein könnte, grausam vernichtet durch die kurze Glosse
der Herausgeber མ་བྱུང་ „nicht erschienen“, ein Leichenstein,
der hier auf das Grab jedes in der Sammlung fehlenden
Werkes gesetzt ist. Was ist also aus dem Tanjur des Bu-ston
geworden? Sollten alle Exemplare[1] desselben, sollten die

---

[1] Der Begriff der Auflage fehlt dem tibetischen Buchwesen gänzlich.
Es werden von den geschnittenen Holztafeln nur wenige Exemplare für die
unmittelbaren Bedürfnisse des Klosters und befreundeter Nachbarklöster ab-
gezogen, ausgenommen wenn es sich um volkstümliche Erbauungsschriften,
Legenden, Gebete usw. handelt, die einen größeren Absatz versprechen.
Kanjur- und Tanjurexemplare werden aber nie auf Vorrat, sondern nur auf
Subskription oder besser Bestellung gedruckt. Die Platten werden in einem
besonderen, mit dem Siegel des Abtes verschlossenen Raume aufbewahrt,
und gedruckt werden darf nur mit seiner ausdrücklichen Genehmigung. Das
Drucken geschieht auch nicht zu einer beliebigen Zeit, sondern nur in der
dafür vorgeschriebenen Jahreszeit, einem der Sommermonate (gewöhnlich
August), falls Bestellungen vorliegen. Dies erklärt die große Seltenheit von
Büchern und die Schwierigkeit sie zu erlangen. Eine andere Schwierigkeit
bibliographischer Natur ergibt sich daraus für uns in der Datierung eines
tibetischen Buches. Die Holztafeln mögen z. B. nach Angabe des Kolophons
im Jahre 1740 geschnitten sein, aber der Abzug von denselben mag im
Jahre 1905 gemacht worden sein; man müßte also streng genommen beide
Daten haben. Nur der Charakter des Papiers kann das Datum oder relative Alter des eigentlichen Druckes entscheiden, und solche Bestimmungen
nach dem Papier sind möglich, wenn man ein möglichst großes Vergleichs-
material gesammelt und sich praktische Erfahrungen erworben hat. In China

Drucktafeln selbst der Vernichtung anheimgefallen sein? Man wird es wohl annehmen müssen, denn sonst wären nicht in Narthang zweimal nach Bu-ston, im siebzehnten und achtzehnten Jahrhundert, neue Editionen veranstaltet worden.

Da wir hier einmal von Bu-ston reden, so mag von den anderen Überraschungen, die der Index von Co-ne bietet, die folgende als eine die Sanskritisten gewiß interessierende Episode an dieser Stelle eingeschaltet werden. In Band 123 (nach der Zählung von Co-ne Nr. 36) lesen wir dort folgenden Titel, der im Narthang Index und im Verzeichnis von HUTH fehlt: གཞོན་ནུ་འབྱུང་བའི་བསྟན་བཅོས་ཇི་ལྟར་བྱུང་བའི་གཏམ་རྒྱུད་པཎྚི་ཏ་སུ་མ་ན་ཤྲཱིའི་ངག་བཞིན་བུ་སྟོན་ལོ་ཙྭ་བས་བསྒྱུར་བ ། d. h. „Geschichte davon, wie das Çāstra von Kumārasambhava (tib. *gŽon-nu ạbyuṅ-ba*) entstanden ist", nach den Worten des Paṇḍita Sumanaçrī von Bu-ston, dem Lotsāva, übersetzt. Da nun Sumanaçrī, wie wir wissen, an der tibetischen Übersetzung von Kālidāsas Meghadūta beteiligt war, so ist es aus diesem Grunde wahrscheinlich, daß es sich hier um Kālidāsas Kumārasambhava handeln wird[1]. Da ferner im Index von Co-ne

---

haben wir ganz ähnliche Dinge: Bücher von Holztafeln aus der Zeit der Sung- und Yüandynastie zur Zeit der Mingdynastie gedruckt und zahlreiche im Zeitalter der Manjudynastie von Druckplatten der Ming abgezogen. Gewiegte Buchkenner unter den Chinesen entscheiden auf Grund langer Routine die Sachlage sofort nach den Qualitäten des Papiers und können oft genug eine genaue Zeitperiode fixieren. In China wie in Tibet ist Feuer die häufigste Ursache für die Zerstörung aufbewahrter Holzdrucktafeln.

[1] Ich möchte diese Gelegenheit nicht vorübergehen lassen, ohne meiner Freude über Dr. BECKH's prächtige Meghadūtatrilogie (wie bereits früher *The Monist*, 1907, p. 635) noch einmal Ausdruck zu geben. Diese Arbeit ist von grundlegender Bedeutung für die tibetische Philologie und wird in Zukunft dazu beitragen, das Verständnis der einheimischen Poesie zu fördern. Ohne Kenntnis des indischen Kavi Stils kann man z. B. die gesammelten Schriften (*gsuṅ ạbum*) der lamaischen Kirchenväter gar nicht

diesem Titel das erwähnte verhängnisvolle Totenmal nicht beigesetzt ist, so wäre es möglich, daß diese Schrift im Tanjur von Co-ne wirklich abgedruckt ist. In St. Petersburg soll ein Exemplar dieser Ausgabe vorhanden sein, und es wäre erfreulich, wenn diese Zeilen einen der dortigen Gelehrten veranlassen sollten, einmal nachzusehen[1], wie es damit steht, und was für eine Bewandtnis es mit dieser Geschichte hat. Sie folgt im Co-ne Index auf die kurzen bei HUTH unter Nr. 35—40 aufgezählten Geschichten. Da sie in der Narthang- und Palastausgabe fehlt, so stehen wir hier wiederum der auffälligen Tatsache gegenüber, daß ein von Bu-ston übersetztes Werk [wenigstens zeitweise] verloren gegangen sein muß[2].

verstehen, und es ist ganz wunderbar, mit welcher Eleganz und Beherrschung sie ihn selbst in moderner Zeit pflegen. Man vergleiche die interessanten Ausführungen GRÜNWEDEL's (Der Lamaismus in „Die orientalischen Religionen", Kultur der Gegenwart, S. 155) über das Gedicht eines Literaten aus Zentraltibet auf St. Petersburg, dessen blumenreiche, rituelle Sprache, wenn die Anspielungen nicht verloren gehen sollen, nur durch eine Übersetzung ins Sanskrit verständlich würde. Oder man lese das hübsche Gedicht auf den Ackerbau in der kleinen Anthologie *Legs bsdus sna lua* (ed. CHANDRA DAS, Calcutta, 1890) und vergleiche dann die verfehlte Übersetzung von G. SANDBERG (Tibet and the Tibetans, p. 142, London, 1906), der eben aus Unkenntnis des Sanskrit keine der indischen Anspielungen verstand. Mir scheint, daß Sprache und Stil der Meghadūta-Übersetzung einen starken Einfluß auf das einheimische Schrifttum ausgeübt haben, und daß auch innere Beziehungen zwischen dieser und den Liedern des Marpa und Milaraspa bestehen, wie ja inbezug auf Milaraspa auch BECKH erkannt hat. Aus allen diesen Gründen erscheint es mir von großer Bedeutung, das Datum der Meghadūta-Übersetzung genauer festzulegen. Sie ist vielleicht doch früher anzusetzen als Dr. BECKH geneigt ist, und könnte eventuell ins elfte Jahrhundert zurückgehen, — doch davon ein andermal.

[1] Und zwar in Band ༄ཅོ་ནེ = 124 der Co-ne-Ausgabe, der dem Band 123 der Narthang-Ausgabe entspricht.

[2] Es seien mir einige weitere Bemerkungen zu HUTH's Verzeichnis auf Grund einer Nachprüfung im Index von Co-ne gestattet.

Während der Abfassung seiner Abhandlung sprach Dr. HUTH wieder-

Diese Beispiele führen uns vor Augen, daß wir uns nicht damit begnügen dürfen, den Tanjur als eine gegebene Tat-

---

holt mit mir über den seltsamen Titel „Das kostbare Zwietracht-Rad" (124, 9), und mancher wird sich seither gewundert haben, was darunter zu verstehen sei. Jetzt löst sich das Rätsel ebenso einfach als glücklich dank dem Index von Co-ne: es ist nicht རྩོད་, sondern རྩེད་ zu lesen. Der Titel lautet hier ausführlicher: སློབ་དཔོན་བེ་རོ་ཙ་ (so!) ནས་རྒྱལ་པོ་ཁྲི་སྲོང་ལྡེ་བཙན་ལ་གདམས་ པ་རིན་པོ་ཆེ་རྩེད་པའི་འཁོར་ལོ། „Die von dem Meister Vairocana an den König K'ri-sron lde-btsan gerichteten Ratschläge (Skr. *avavāda*), das kost-bare Spielrad (*krīḍācakra*)". Also wohl ein Buch in usum delphini, das sich spielend erlernen läßt.

Der von HUTH zu Band 123, 15 erschlossene tibetische Übersetzer *dGos-ạdod t'ams-cad ạbyuṅ-ba* existiert leider nicht und beruht auf einem Mißverständnis der Stelle, die im Kolophon der Palastausgabe (fol. 139b) lautet: ལོ་ཙ་བ་དགེ་སློང་རྣམས་པའི་གཉེན་པོ་དགོས་འདོད་ཐམས་ཅད་འབྱུང་བས ། d. h. als die erforderlichen Ausgaben für die Lotsāva und Bhikshu, die geist-lichen Ratgeber der Unwissenden [eine häufige Titulatur], bereitgestellt waren usw., wurde das Werk übersetzt und redigiert. In diesem Sinne er-scheint die Phrase *dgos ạdod t'ams-cad ạbyuṅ-ba*, die überhaupt keinen Eigen-namen repräsentieren kann, recht häufig in der Literatur, z. B. im Co-ne-Index des Tanjur, wenn es sich darum handelt, die für den Druck nötigen Mittel aufzubringen.

Band 123, 2. Im Co-ne-Index lauten Titel und Kolophon: གསེར་ འགྱུར་གྱི་བསྟན་བཅོས་བསྡུས་པ་མཐུག་མ་ཆད་པ་འདི་གྲུབ་ཆེན་ཨོ་རྒྱན་པའི་ འགྱུར ། གོང་གི་བྲ་ལི་པས་མཛད་པ་དང་གཉིས་འགྱུར་ཆད་མ་གདོགས་དོན་གཅིག་ པ་ཡིན་པ་འདོན ། „Dieses *mjug-ma tsʻad-pa,** ,Auszug aus dem Çāstra über die Goldbereitung' wurde [zuerst] von dem Siddha, dem von Udyāna [also wohl Padmasambhava], übersetzt. Der oben [unter Nr. 1] erwähnte Bhalipa ist der Verfasser; die zweite Übersetzung, in der die ausgeschiedenen Be-standteile interpoliert waren, dürfte einen einheitlichen. Sinn enthalten". In der Palastausgabe des Kaisers K'ien-lung (fol. 5a) ist am Schlusse der Ab-handlung weder Verfasser- noch Übersetzername genannt. Sie endet mit der Wiederholung des Titels und dem Zusatz „soviel davon vorhanden ist" (*ji sñed-pao*).

* Dürfte wohl ein alchemistischer Terminus sein. In CHANDRA RAY's History of Hindu Chemistry kann ich nichts finden, was zur Erklärung beitragen könnte.

sache hinzunehmen und zu analysieren, sondern, daß es auch unsere Aufgabe sein muß, zu erforschen, wie diese große literarische Sammlung entstanden und geworden ist.

Band 123, 26. མིའི་མཚན་ཉིད་བསྟན་པ་རྒྱ་མཚོ་ཞེས་བྱ་བ་སྡེ་སྣོད་འཛིན་པ་ཆེན་པོ་བྷོ་ཊ་པཎྜི་ཏའི་གསུང་ལ་བརྟེན་ནས། པཎྜི་ཏ་དྷརྨ་དྷ་རའི་ཞལ་སྔ་ནས། ཡར་ལུངས་པ་གྲགས་པ་རྒྱལ་མཚན་གྱིས་དཔལ་ས་སྐྱའི་གཙུག་ལག་ཁང་དུ་བསྒྱུར་བ་ „Nach dem Diktat des Mahāpiṭakadhara Bhoṭapaṇḍita von seiner Ehrwürden, dem Paṇḍita Dharmadhara, und Grags-pa rgyal-mts'an aus Yar-luṅs im Kloster dPal Sa-skya übersetzt". In der Palastausgabe (fol. 193b) wird der tibetische Übersetzer als བོད་ཀྱི་ལོ་ཙྪ་བ་སྤངས་པ་ „der tibetische Lotsāva, der von Spaṅs" bezeichnet. HUTH's Übersetzung scheint auf einer Lesart *gsuṅ-gi bskul-nas* zu beruhen, wie sie sich in der Palastausgabe findet; die hier gegebene Lesart *gsuṅ-la brten-nas* bedeutet auf Grund eines mündlichen Vortrags. Es war kein Manuskript da, und das Werk wurde daher aus dem Gedächtnis rezitiert. Bhoṭapaṇḍita ist der Name und daher nicht als „der große tibetische Gelehrte" zu übersetzen; es war offenbar ein Inder, und der Name mag etwas ganz anderes bedeuten. Tib. *sde-snod ạdzin* ist nicht Piṭakabekenner, sondern ein offizieller Titel, ein literarischer Grad, der den Kennern des Tripiṭaka verliehen wird.

Band 123, 27. Im Index von Co-ne lautet der Titel wie im Index von Narthang: མི་དཔྱད་ཀྱི་བསྟན་བཅོས་རྒྱ་མཚོས་བསྟན་པའི་སྐྱེས་པའི་མཚན་བཅོ༔ ལྔ་པ་དང་བུད་མེད་ཀྱི་མཚན་བཅུ་པ་པཎྜི་ཏ་ཤྲཱི་ཀ་ར་དང་། བོད་ཀྱི་ལོ་ཙཱ་བ་དགེ་སློང་ཨོ་ཌྱཱ་ན་པའི་རིན་ཆེན་དཔལ་གྱི་འགྱུར། Als Übersetzer erscheinen die von HUTH unter 27 genannten, wo statt „der bhikshu aus U-rgyan" der Bhikshu Rin-c'en dpal von Udyāna einzusetzen ist. Der Ortsname ist von HUTH nicht recht erfaßt; verbessere: die Übersetzung wurde in Ri-bo spud-tra in dPal bde-c'en gliṅ verfaßt. Letzteres ist Name eines Klosters (*Çrīsukhadvīpa*), ersteres Name eines darin gelegenen Tempels. Am Schluß des Bandes begegnet der Titel: དྲང་སྲོང་ཆེན་པོ་ཤ་ལི་ཧོ་ཏྲས་ཡང་དག་པར་བསྡུས་པའི་རྟའི་ཚེའི་རིག་བྱེད་མཐའ་དག་རྗེས་སུ་བསྟན་པ་ཞེས་བྱ་བ་རྒྱས་པ་རྒྱ་གར་གྱི་མཁན་པོ་ཨཱ་ཙཱརྱ་པདྨ་ཤྲཱི་བྷ་དྲ་དང་། བནྡེ་ཤྲཱི་ཤཱནྟི་དང་། ཞུ་ཆེན་གྱི་ལོ་ཙཱ་བ་ཆེན་པོ་

X.

Auch in der chinesischen Literatur sind uns Nachrichten über indische Malerei aufbewahrt. Die wichtige Stelle über Malerei im Kloster Nālanda, die im *Hua-ki* des Têng Ch un[1] enthalten ist, ist bereits zweimal übersetzt worden, von HIRTH[2] und von GILES[3], und beweist, daß im Mittelalter eine buddhistische Malschule in Nālanda geblüht hat, die sich sogar einer in China unbekannten Farbentechnik bediente. Weiteres Material läßt sich in der großen buddhistischen Enzyklopädie des siebenten Jahrhunderts, *Fa yüan chu lin*[4], und wohl in man-

---

དགེ་སྤྱོང་རིན་ཆེན་བཟང་པོས་བསྒྱུར་བ ། „Der von dem großen Ṛishi Çālihotra verfaßte, vollständig gesammelte (*saṁsañcaya*) Açvāyurveda (Veterinärkunde), betitelt der alles lehrende (? *sakalānuçāsana*), von dem indischen Gelehrten Ācārya Dharmaçrībhadra, Buddhaçrīçānti und dem großen Redaktor-Lotsāva Bhikshu Rin-c'en bzaṅ-po übersetzt". Vgl. HUTH, S. 279, Noten, und seine Bemerkung über die Zeit der Übersetzung in der ersten Hälfte des elften Jahrhunderts in ZDMG, Bd. XL, 1895, S. 281; JOLLY, Medicin, S. 14 und P. CORDIER, BEFEO, Vol. III, 1903, p. 620.

123, 28 ist hier nicht erwähnt; 123, 29 eröffnet den Band 125 mit dem Titel བོ་པ་ཆེ་མང་པོས་མཛད་པའི་སྐྱ་བྱེ་བྲག་ཏུ་རྟོགས་བྱེད་ཆེན་མོ „Die von vielen Übersetzern und Paṇḍita verfaßte Mahāvyutpatti."

[1] Publiziert 1167, gewöhnlich als Anhang zum *Süan ho hua p'u*, dem Katalog der Gemäldesammlung der Sungkaiser, gedruckt. Vgl. HIRTH, Über die einheimischen Quellen zur Geschichte der chinesischen Malerei, S. 20, und Scraps from a Collector's Note-Book, p. 111.

[2] Fremde Einflüsse in der chinesischen Kunst, S. 51; dazu GRÜNWEDEL, Buddhistische Kunst in Indien, S. 151.

[3] An Introduction to the History of Chinese Pictorial Art, p. 135. Statt „linen of the West", wie hier übersetzt ist, muß es heißen „indische Baumwollstoffe" (*Si-t'ien pu*), wie bei HIRTH. Nach der Übersetzung HIRTHS „der Körper einfach aufrecht sitzend" könnte es scheinen, als hätte man in Nālanda nur sitzende Buddhafiguren dargestellt; im Texte steht aber 坐 立 而 已 „sitzend oder stehend", wie auch GILES korrekt übersetzt.

[4] D. i. Perlenhain im Garten des Dharma. Vgl. A. WYLIE, Notes on Chinese Literature, p. 207 und PAUL PELLIOT, *T'oung Pao*, 1912, p. 354.

chen anderen Werken auftreiben. Doch dieses Thema muß im Zusammenhang mit der Frage nach dem Einfluß der indischen auf die chinesische Malerei behandelt werden, betreffs deren HIRTH in der zitierten Schrift „Fremde Einflüsse" bereits bemerkenswerte Aufschlüsse gegeben hat. Die ganze Frage ist aber sehr kompliziert geworden und durch die Funde altbuddhistischer Malereien in Turkistan in ein neues Stadium getreten, so daß erst die sachgemäße Publikation derselben abgewartet werden muß, bevor das Problem aufs neue aufgenommen wird. Mir scheint, die Zukunft wird uns vor allem zwei Dinge lehren, einmal daß der Einfluß der indischen auf die chinesische Malerei tiefer und nachhaltiger gewesen ist als wir bisher annehmen durften und sich nicht nur auf die buddhistischen Sujets, sondern auch auf Komposition und Technik, besonders auf das Kolorit, erstreckte; sodann daß die echten Traditionen der indischen Malerei treuer in Tibet und in der Mongolei als in China und Japan erhalten sind. Die Ursache für diese Erscheinung ist nicht schwer zu erkennen. In Tibet war die Malerei vor der Berührung mit der indischen Kultur ein weißes unbeschriebenes Blatt, und die Malerei als wirkliche Kunst ist erst von Indien eingeführt worden. Vor dieser Zeit hatte es in Tibet nur eine primitive Ornamentik gegeben, die sich an die der Türkvölker Zentralasiens anschloß und sich noch jetzt teilweise aus der volkstümlichen Kunstbetätigung, besonders der tibetischen Nomadenstämme, rekonstruieren läßt. Die religiöse Kunst wurde daher erst vom Buddhismus großgezogen, und die Geistlichen Tibets waren leicht empfängliche und dankbare Schüler, die, eben weil sie noch keine fertigen Ideen von Malerei im Kopfe trugen, um so williger die indischen Traditionen genau in der Form und Technik in sich aufnahmen, wie sie ihnen von den

indischen Kollegen beigebracht wurden. Von der unantast-
baren religiösen Überlieferung geheiligt, wurde denn der indische Stil bis auf den heutigen Tag gewissenhaft bewahrt.
Dasselbe dürfen wir wohl von Nepal sagen, und in Nepal wie
in Tibet haben wir die unmittelbare Fortsetzung der buddhistischen Malerei Indiens zu erblicken. Auch dürfte die Hoffnung
gegründet sein, daß in den großen Klöstern Tibets noch alt-
indische Malereien behütet werden. Anders in China. Hier
war, wie wir jetzt wissen, zu der Zeit als der Buddhismus
eintraf, eine künstlerisch bedeutende Malerei bereits in voller
Entwicklung begriffen, die im wesentlichen unter der Inspiration des Taoismus stand und schon eine ganze Reihe
hervorragender Künstler - Persönlichkeiten aufzuweisen hatte.
Diese Künstler nahmen, jeder in seiner Eigenart, zu der
neuen indischen Malerei Stellung und nahmen ihre Lehren
in sich auf, jeder in seiner Weise. Von vornherein
muß man in der chinesischen (wie auch japanischen)
Kunst zwei Dinge scharf voneinander trennen, buddhistische
Bilder, die nach indischen oder turkistanisch-indischen Vor-
bildern getreulich kopiert wurden und den Kultuszwecken der
Tempel dienten, und künstlerisch-individuelle Schöpfungen der
großen Meister. Wie aber die Chinesen mit ihrem phäno-
menalen Assimilationsvermögen alle fremden Elemente in ihrer
Kultur aufgelöst und zu Blut von ihrem Blute gemacht haben,
·so haben sie auch allmählich den Buddhismus aufgesogen und
ihrem universalen System des Lebens entsprechend chinisiert.
Der chinesische Buddhismus ist nicht mehr indischer, sondern
chinesischer Buddhismus, und die chinesisch - buddhistische
Kunst, ob rein religiös, volkstümlich oder künstlerisch, ist nicht
mehr die indisch-buddhistische Kunst, sondern eine auf Grund
indischer Elemente aufgebaute, freie Übertragung ins Chine-

5

sische, und damit eine Erscheinung *sui generis*. Die köstlichen Arhat eines Li Lung-mien sind ebenso chinesisch als Dürers Apostel germanisch sind. Wie bei uns die geistige Macht des Christentums herrliche und freie Kunstwerke ausgelöst hat, so hat in China die von Indien ausstrahlende Offenbarung eine große religiöse Kunst von kräftiger Eigenart geschaffen. In Tibet ist im großen und ganzen die Malerei ihren hierarchischen Zwecken treu geblieben. Der Gedanke des Kunstwerks um seiner selbst willen ist dort wohl nie aufgetaucht, aber trotz ihrer ausgesprochen kirchlichen Bestimmung sind besonders in den Porträts der Kirchenfürsten Arbeiten entstanden, die einen Ehrenplatz in der Geschichte der Kunst beanspruchen dürfen.

Tibetischer Text des Citralakshaṇa

༄༅། རྒྱ་གར་སྐད་དུ། ཙི་ཏྲ་ལཀྵཎ། བོད་སྐད་དུ། རི་མོ་འི་མཚན་ཉིད །

1. Kapitel.

ལྷ་ཆེན་ཚངས་པ་སྲིད་མེད་བུ །

དབངས་ཅན་མཆོག་སྟིན་མཛད་པ་ལ །

བདག་གིས་རབ་ཏུ་ཕྱག་འཚལ་ཏེ །

རྒྱལ་བ་དང་ནི་བཀྲ་ཤིས་བགྱི །

5 ལྷ་ཆེན་བྱུང་བའི་རི་རྒྱལ་གྱི །

སྲས་མོ་ཞལ་གྱི་པད་མ་ཅན །

དེ་ནས་དེ་གསུང་དེ་འོག་ཏུ །

མཁས་པ་ཐམས་ཅད་རྒྱལ་བར་འོག །

བཀྲ་ཤིས་པར་གྱུར་ཅིག །

དེའི་འོག་ཏུ་རི་མོ་འི་མཚན་ཉིད་བཤད་པར་བྱའོ ༎ ལྷ་ཆེན་པོ་རིག་པ་ཐམས་ཅད་སྟོན པར་མཛད་པ་ལ་ཕྱག་འཚལ་ལོ ༎

10 ཐོག་མར་སྐྲ་བ་ལ་བཏུད་ནས །

སྐྲ་བ་ཚོས་པ་གཏུག་རྒྱན་ནས །

ལྷ་ཆེན་དེ་ནས་ཁྱབ་འཇུག་དང་ །

དབང་ནི་ཅུ་བླ་མེ་ཀྲུང་བླ། །

དེ་ནས་སྐྱེ་དགུ་མཛའ་མཛོད་པ། །

15 ལས་ཀུན་པ་ལ་ཕྱུག་འཚལ་ཏེ། །

དེ་ནས་གཉེར་ཐུལ་ཀང་པ་དང་། །

སྐྱོབ་དཔོན་དགའ་ནི་ཐམས་ཅད་ལ། །

ཐལ་མོ་སྦྱར་ཏེ་ཕྱུག་འཚལ་ནས། །

རི་མོ་དེ་ལས་ཀྱི་ཚོན་དགའ་ལ། །

20 མཉམ་པར་གཞག་ནས་ཅི་དྲན་དང་། །

ནུས་པ་ཅུང་ཟད་བཙོད་པར་བྱ། །

ལས་ཀུན་པ་དང་རབ་སིམ་དང་། །

གཉེར་བྱུ་ཐུལ་བྱས་མཆན་ཉིད་དང་། །

ཀུན་བདུས་ སྙིང་པོ་ཞེས་ བྱས་ནས། །

25 བརྗོན་པའི་རྫོ་ཅན་རྣམས་ཀྱི་སྣོ། །

སྐྱེད་ཕྱིར་རི་མོ་དེ་མཆན་ཉིད་དགག །

བདག་གིས་བསྐུལ་ནས་བྱས་པ་སྟེ། །

རི་མོ་དེ་མཆན་ཉིད་མཁས་པ་ཨི། །

ཤེས་རབ་བཟང་རྣམས་ང་ལ་ཉིན། །

30 རི་མོ་དེ་མཆན་ཉིད་འདི་ཉིད་སྟོན། །

མི་ཨི་ཀྲུལ་པོ་ས་འཛིན་པ། །

_____

བློ་དང་ལྷན་ཞིང་གྲགས་པ་ཆེ །

ཚོས་ཤེས་བདེན་པའི་རྩལ་ལྡན་པ །

རྣམ་གྲགས་འཇིགས་ཐུལ་ཤེས་བུ་བྱུང་ །

35 རྒྱལ་པོ་དམ་པ་དེ་རིང་ལ །

ཚོས་དང་མཐུན་པས་ས་བསྐྱངས་[1]པས །

མི་རྣམས་ཀྱི་ནི་ཚོ་ཨེ་ལོ །

གུངས་ནི་སྟོང་ཕྲག་བརྒྱ་ཕྲབ་གྲག །

དེ་ཚོ་ནད་རྣམས་མེད་གྱུར་ཅིང་ །

40 དུས་མིན་འཆི་བ་གང་ཡང་མེད །

ཡིད་ནད་མེད་ཅིང་གསོ་[2]བྱེད་ན །

ཁྲི་དང་ཆགས་སོགས་ག་ལ་ཡོད །

རྒྱུ་ནི་ལེགས་པར་ལྷང་གྱུར་ཅིང་ །

དབང་པོས་ཆར་པ་ལེགས་པར་འབེབས །

45 ས་བོན་རྩ་བ་འཐུས་བུ་རྣམས །

མཌངས་དང་རོ་དང་ལྡན་པར་འགྱུར །

རིགས་བཞི་རྣམས་ཀྱང་མ་འཚོལ་ཅིང་ །

དེ་བཞིན་རང་གི་ཚོས་ལ་བརྟོན །

འགྲོ་བ་རྒྱས་ཤིང་དཔལ་དང་ལྡན །

50 རྟ་འཕུལ་ཡོན་དན་ཀུན་ལྡན་འགྱུར །

---

[1] B: སྐྱངས་   [2] B: གསོད་

ཡུལ་འཁོར་དེ་སྐྱུར་གྱུར་པའི་ཚེ། །

རྒྱལ་པོ་བདག་ཉིད་ཆེན་པོ་ཡི། །

ཨེད་ནེ་དགའ་ཐུབ་ཆེར་ཞུགས་པས། །

དེ་ཨིས་དགའ་ཐུབ་མཚོག་ཀྱང་བྱས། །

55 དགའ་ཐུབ་དེ་ཨིས་རྒྱལ་པོ་དེ། །

ཉིད་ཀྱི་འཇིག་རྟེན་མེས་པོ་ལས། །

ཨོན་དུན་བཀྱུར་ཕུག་སྟེད་དགའ་བ། །

མཚོག་རྣམས་ཀྱང་ནི་ཐོབ་པར་འགྱུར། །

ལྷ་ཨི་མཚོན་དང་དགྲ་རྣམས་ཀྱིས། །

60 མི་ཡུལ་ཐེབས་ པར་མི་ནུས་དང་། །

བསྟུན་བཙོས་ མ་ལུས་རྣམས་ལ་ཡང་། །

ཐོགས་པ་མེད་པའི་བློ་མཚོག་ཐོབ། །

ཨོན་དུན་ཀུན་གྱིས་རབ་བཀྱུན་པ། །

མི་ཨི་བདག་པོ་དེ་ལྷ་བུ། །

65 ནོར་བུའི་སྲས་པོ་མཐའ་ཐང་ཆེ། །

ལྷ་མཐུས་པ་རོལ་གནོན་མཛད་པ། །

རིག་པ་རྣམས་ཀྱི་རྟེན་བཟང་པོ། །

ཚོས་ཀྱི་ལུས་འདུར་ གྱུར་པ་ན། །

ཐུམ་ཟེ་ཞིག་ནི་དུ་བཞིན་དུ། །

70 རྒྱལ་པོ་འི་གན་དུ་འོངས་པ་དང་། །

₁ B: ཐེབས་   ₂ A: ཚོས་   ₃ B: འགྱུར་

མི་དགར་གྱུར་པ་དེ་ལ་ནི ། །

ཅི་ཉེས་ཞེས་ནི་རྒྱལ་པོས་དྲིས ། །

དེ་ཡང་ཁྲོམས་པས་རྒྱལ་པོ་ལ ། །

གདུག་པའི་ཚིག་ནི་སྨྲས་པ་ནི ། །

75 གང་ཕྱིར་ཁྱོད་ཀྱི་ཡུལ་དུ་ནི ། །

དུས་མ་ཡིན་པར་འཚེ་བྱུང་ཡང་ ། །

ཞེས་པར་མི་རྗེ་ཁྱོད་ཀྱིས་ནི ། །

ཚོས་མིན་སྐྱེ་དགུ་བསྐྱངས་པ་འདྲ ། །

སྨྲོན་ཆད་མ་བྱུང་པོ་མཚར་གང་ ། །

80 དེ་ཡང་དེ་རིང་འབྱུང་འགྱུར་ཏེ ། །

དུས་མིན་འགུམ་པ་བདག་གིས་ཐུ ། །

མཚན་དང་ག་གཟུགས་མ་དོན་པར ། །

བྱིས་པ་རིགས་ཀྱི་རྒྱུད་འཚོབ་པ ། །

སྐྱིང་དུ་སྤྱུག་པ་བཀུམ་མོ་ལྟ ། །

85 རྗེ་ཁྱོད་གལ་ཏེ་བྲམ་ཟེ་གཅེས ། །

ཁྱིད་ནི་ཐབས་ཅད་མཐིན་ལགས་ན ། །

སྤྲོག་པས་ཆེས་འཕངས་བདག་བུ་ནི ། །

བདག་གི་ཐད་དུ་ཁྱིད་དེ་ག་ཞིགས ། །

རྒྱལ་པོ་མཐུ་ཆེན་གལ་ཏེ་ཡང་ ། །

༩༠ བདག་ལ་ཕྱུགས་བཅུར་མ་མཛད་ན། །

དེང་ཁྱོད་སྨྱུན་སྤྱོར་བབ་པ་[1]ཡང་། །

ཕྱུག་ནི་རྩ་བཞིན་བདག་གིས་དོར། །

བྲམ་ཟེ་དེ་ཡིས་དེ་སྐད་སྨྲས། །

དེ་ཚེ་དེ་ཡི་བུ་དགུག་ཕྱིར། །

༩༥ རྒྱལ་པོ་དགའ་བ་སློ་ཚོན་ནེས། །

ཕྱུགས་ལ་དགོངས་པ་མཛད་པར་གྱུར། །

ང་ཡི་ཞེས་འབྱིང་འདོད་[2]གསུངས་པས། །

བྲམ་ཟེ་དེ་ཡིས་དབུགས་ཕྱུང་སྟེ། །

ག་ཉིན་རྗེའི་རྒྱལ་པོ་ཅི་མ་དང་། །

༡༠༠ འདུ་བར་ཤར་བ་དེ་མཐོང་ནས། །

འགྲོ་བས་[3]ཕྱུག་བུའི་ངོས་དེ་ལ། །

རྒྱལ་པོ་འཛིགས་ཕྱུལ་ཕྱུག་འཚལ་ཏེ། །

བྲམ་ཟེ་ལ་ནི་ཕན་བཙུན་པས། །

རབ་ཏུ་གུས་པས་ཚིག་[4]འདི་སྨྲས། །

༡༠༥ ལེགས་སྨྲན་བྲམ་ཟེ་འདི་ཡི་བུ། །

ཕྱོག་པས་ཚིམ་འཐངས་སྐྱིང་ཕུག་པ། །

འཆི་བ་ལ་ནི་མ་ཕྱུག་པར། །

ཁྱོད་ཀྱི་མཆག་གཞུག་བགྱིད་པས་ཕྱོགས། །

སྐྱིང་རྗེ་བྲམ་ཟེ་ལ་དགའ་ནས། །

༑༡༠ འཛིག་རྟེན་གསུམ་གྱི་མཛད་མཛོད་ཅིང་། །

འཛིག་རྟེན་ལས་ཀྱི་དབང་པོ་ཡིས། །

བྲམ་ཟེ་འདི་ལ་བུ་སྦྱལ་གསོལ། །

དེ་སྐད་སྨྲས་ནས་རྒྱལ་པོ་ལ། །

ཚོས་ཀྱི་རྒྱལ་པོས་འཛུམ་དམུལ་ནས། །

༑༡༥ མཆོད་གནས་དེ་ལ་ཚིག་འཛམ་དང་། །

བཛིད་བག་དུ་ནི་སྨྲ་སྨྲས་པ། །

བདག་གིས་རང་དགར་འགག་ཡང་ནི། །

འགུགས་པའམ་གདུང་བར་མི་ནུས་ཏེ། །

རང་གི་ལས་ཀྱི་འཕྲས་བུའི་མཐུས། །

༑༡༢༠ སྐྱེ་པོ་དགའ་གི་དབང་དུ་འགྱུར། །

དུས་དང་ལས་ཀྱི་དབང་འགྲོ་བ། །

བདེ་དང་སྡུག་བསྔལ་ཕྲད་པར་འགྱུར། །

དེ་ལྟར་རྒྱལ་པོ་བདག་གིས་ནི། །

མ་ལན་པར་ནི་གོ་ཕྱིར་རིགས། །

༑༡༢༥ བདག་གི་ཐད་དུ་ཕྱིན་ནས་ནི། །

སྨྲ་ནི་འགའ་ཡང་ཐར་མི་འགྱུར། །

དུས་ནི་ཐོབས་དང་སྨྲན་གྱུར་པས། །

བྲམ་ཟེའི་ཕྱིར་ནས་གྱིས་དགུག །

འབྱུང་པོ་ཀུན་གྱིས་མི་ཐུབ་པའི། །

130 དགེ་དང་མི་དགེ་འདིར་སྐྱོང་ངོ་། །

མྱང་ནི་ལས་ཀྱིས་གཞི་སྟེ། །

འདི་ནི་འཕྲས་བུའི་ས་གཞིར་བཤད། །

གཉིན་རྗེས་དེ་སྣང་སྨྲས་པ་དང་། །

མི་ཡི་བདག་པོས་ཕྱིར་སྨྲས་པ། །

135 ཡང་དང་ཡང་དུ་གཉིན་རྗེ་ལ། །

བྱིན་ཅིག་ཅེས་པའི་ཚིག་སྨྲས་སོ། །

དེ་ལ་གཉིན་རྗེས་ཡང་དང་ཡང་། །

མི་རུང་མི་རུང་ཞེས་སྨྲས་སོ། །

རྒྱལ་པོས་བྱིན་ཅིག་བྱིན་ཅིག་སྨྲས། །

140 གཉིན་རྗེས་ཡང་ཡང་མི་རུང་སྨྲས། །

དེ་དག་དེ་ལྟར་ཞེན་པ་ཡིས། །

ཉིན་དུ་འཁྲུགས་པ་དག་ཏུ་འགྱུར། །

དེ་ལྟར་དེ་དག་འཁྲུགས་པས་ན། །

འཐབ་མོ་ཆེན་པོ་བྱུང་གྱུར་ཏེ། །

145 གཉིན་རྗེ་ལ་ནི་རྒྱལ་པོ་དེས། །

སྟོད་སྟོད་ཅེས་ནི་ལན་མང་སྨྲས། །

གཡུལ་དངོས་དག་ཏུ་གཉིན་རྗེས་ཀྱང་། །

དེ་ལས་ང་སྟོད་འདི་ཡིན་ཞེས་སྨྲས་སོ། །

དེ་ནས་རྒྱལ་པོ་དེ་ཡིས་ནི། །

150 མཛའ་ཡང་དྲག་ཅིང་རྩོབ་དང་ [1] །

གྲངས་མེད་པ་ཡི་ཆར་དག་ནི། །

གཉིན་རྗེ་ཡིན་སྟེང་དག་ཏུ། །

ཞིན་ཏུ་དབབ་པ་དག་བྱས་སོ། །

སྨྲིན་ལས་ཆར་དག་རྒྱུན་འདྲ་བའི། །

155 ཕོགས་མེད་གོ་བར་བྱེད་པ་ཡི། །

སྤུགས་དང་སྲན་པའི་ལྷ་ཡི་མཚོན། །

ཀུན་ནས་གཉིན་རྗེས་མགུར་ [2] ཕབ་པས། །

དེ་ཡིས་རྒྱལ་པོ་དེ་ཁྲིས་ཏེ། །

ལྷ་ཡི་མཚོན་ཆ་མཐུ་ཆེན་གྱིས། །

160 གཉིན་རྗེ་མདག་གཞུག་བརྡགས་པ་དང་། །

ཡི་དགས་ཀྱིས་ཀྱང་དེ་འཕངས་ན། །

གཏོད་པ་རྩོམ་ [3] པར་བྱེད་པ་ཡི། །

རལ་གྲི་སོགས་ལ་སྐྲམ་པ་དང་། །

ཕོབ་འཇིགས་པ་འབྱིན་བྱེད་པ། །

165 མ་ལུས་པ་དག་དག་ཏུ་བཙམ། །

དེ་ནས་ཡི་དགས་མི་བཟད་པའི། །

གཟུགས་ཅན་ཕན་ཚོན་ཀུན་ཏུ་རྒྱུག །

_____

¹ B: དག་   ² B: མྱུར་   ³ A und B: ཚོམ་

གཉིན་རྗེའི་དམག་དཔོན་བཙམ་པ་དང་། །
ཕྱུགས་ཕྱུགས་སུ་ནི་རིངས་པར་འགྲོ །

170 རྒྱལ་པོ་ཡིས་ནི་གཡུལ་བྱི་ཙོར། །
ཐམ་པར་བྱུས་པ་མཐོང་ནས་ནི། །
གཉིན་རྗེའི་རྒྱལ་པོ་དེ་ཁྲོས་ནས། །
རྗེ་བཞིན་བྱེད་པའི་དབྱུག་པ་གཟས། །
རྒྱལ་པོ་བསྐལ་པའི་མེ་འདྲ་བའི། །

175 དབྱུག་པ་གཟས་པ་དེ་མཐོང་ནས། །
ཁྲོས་དེ་ཚངས་པའི་མགོ་མཚན་ཅན། །
མཚན་ཆ་ལག་ཏུ་བླངས་པ་དང་། །
དེ་ནས་འབྱུང་པོ་ཐམས་ཅད་དང་། །
གནས་རྣམས་ཀུང་ནི་བྱེད་པར་གྱུར། །

180 ཀུན་ནས་འབྱུང་པོ་ཆེན་པོ་ཡང་། །
སྲས་ངན་ཆེན་པོ་གཏེར་བར་འགྱུར། །
དེ་ནས་འགྲོ་བ་ཀུན་གདངས་པར། །
ཚངས་པ་ཡིས་ནི་མཐོང་ནས་སུ། །
ཚངས་པས་ལྷ་རྣམས་སྐྱེན་ཅིག་ཏུ། །

185 ས་ཕྱོགས་དེར་ནི་ཕྱིན་པར་གྱུར། །
དེ་ནས་ཚངས་པ་མཐོང་ནས་ནི། །
མཚན་བཅུལ་ཐལ་མོ་སྦྱར་ནས་སུ། །

རེས་སྐྱེས་འདི་ནི་ཕོག་དུ་སྐྱེས། །

དེ་དག་གིས་ནི་འཁྲུག་པའི་རྒྱུ། །

190 དེ་ཀུན་མ་ལུས་ཕོས་ནས་སུ། །

དེ་དག་ལ་ནི་ཚངས་པ་ཡི། །

གཡུལ་ལས་རྣོག་ནས་ཚོག་སྐྱེས་པ། །

གཤིན་རྗེའི་བདག་ཉིད་ཆེན་པོ་འདི༌། །

ཉིས་པ་མེད་བདེན་མིའི་རྒྱལ་པོ་ཡིས༌། །

195 འཚེ་བདག་མ་ཉེས་དུས་མ་ཉེས། །

འོན་ཀྱང་རང་གི་ལས་ཀྱིས་ཉེས། །

ཕྱིན་ཆད་དགེ་དང་མི་དགེ་བའི། །

ལས་དེ་བྱེས་པ་དེ་ཡིས་བྱུས། །

མི་ཡི་སྐྱེ་བ་ཡང་དག་ཐོབ། །

200 འཚེ་བའང་སྱུར་དུ་བྱུང་བར་གྱུར། །

འོན་ཀྱང་བྲམ་ཟེ་མཆོད་པའི་ཕྱིར། །

ཁྱོད་ཀྱི་དལ་བ་འབྲས་བཅས་བཤིག །

ང་ཡི་དྲིན་ལ་དེ་ལ་ཡང་། །

ཐབས་ཀྱི་རིག་པ་འདི་སྣང་སྟེ། །

205 གཟུགས་དང་ཁ་དོག་སྒོ་ནས་ནི། །

བྲམ་ཟེའི་བུ་ནི་ཅི་འདྲ་བ། །

ཁྱོད་ཀྱིས་ལེགས་པར་རྟིས་ཤིག་དང་། །

དེ་ནས་ཁྱོད་ཕན་དེ་ཛེས་ཱབུ།

ཚངས་པས་སྣང་སྟེ་སྣུས་པ་དང་།

210 རྒྱལ་པོ་རྡོ་སྣན་དེ་ཨིས་ཀྱང་།

བྲམ་ཟེའི་ཁྱིད་དེ་མཐོང་བར།

གུར་དུ་ཟེར་ཀྱང་ཉེས་སྣུས་སོ།

དེ་ནས་བྱེས་ནས་ཚངས་པས་ཀྱང་།

• དེ་ལྟ་བུ་ནི་བསྐང་ནས་སུ།

215 བྲམ་ཟེ་ལ་ནི་གསོན་པོར་ཕྱིན།

བྲམ་ཟེ་མགུ་བས་ཨིག་རྒྱས་ཏེ།

ཡུ་ཐུལ་གསལ་བའི་ཨིག་སྣན་ནི།

གཞིན་ལ་ཨིན་དུ་གཞིན་མརོག་ཚན།

ཚངས་པ་ལ་ནི་ཕྱག་བྱས་ནས།

220 རང་གི་བུ་ནི་སྣང་བ་བྱས།

དེ་ནས་ཚངས་པས་རྒྱལ་པོ་ལ།

ཁྱིད་ཀྱིས་གསུལ་དུ་བྲམ་ཟེའི་ཕྱིར།

མངག་གཞུག་དེ་དག་མཐུ་ཨིས་ནི།

བཏུལ་བ་དེ་ནི་ལེགས་སོ་ལེགས།

225 ཚངས་པས་དེ་སྣད་སྨྲས་པ་དང་།

རྒྱལ་པོ་དེ་ནི་མགུ་བར་གྱུར།

_____

¹ A: ངས་

ཐམས་ཅད་སྟོམ་པར་བྱེད་པའི་སྐྱ། །

གཉེན་རྗེའི་ཡེད་ནི་མི་དགར་གྱུར། །

གཉེན་རྗེའི་རྒྱལ་པོ་མི་དགའ་བ། །

230 དེ་ཡང་ཚངས་པས་མཐོང་ནས་ནི། །

བཤེས་ཚིག་[1]མང་པོས་དབུགས་ཕྱུང་ནས། །

མི་ཨེ་རྒྱལ་པོ་ལ་སྨྲས་པ། །

རྒྱལ་པོས་ཚོས་ལུགས་རྣམ་[2]ཤེས་ནས། །

ཁྱེད་ཀྱིས་སྐྱེས་པར་མི་བྱའི། །

235 མི་ལ་དགའ་བ་བྱུང་བ་བཞིན། །

མཆོན་པའི་ང་རྒྱལ་བྱེད་པ་བཅག །

དམ་པ་རྣམས་ཀྱིས་བསྟེན་མི་འགྱུར། །

ཤེན་དུ་སྐྱེད་པ་དག་དུ་འགྱུར། །

དག་དུ་ག་ཤུང་བར་འགྱུར་བ་དང་། །

240 བདེ་བ་ཆུང་དུའང་ཐོབ་མི་འགྱུར། །

གཏབས་པ་རྣམས་ཀུང་སྡུང་བར་འགྱུར། །

མཐོ་རིས་ལས་ཀུང་སྒྱུར་དུ་ཆགས། །

རྒྱལ་པོ་ང་རྒྱལ་བྱེད་པ་ནི། །

ཀུན་གྱིས་སྤང་བར་སྒྱུར་དུ་བརྩི། །

245 དེ་བས་དག་པ་ཁོ་ནར་ནི། །

རྒྱལ་པོས་མཛེས་པར་བྱ་བ་དང་། །

རིག་པ་ནོར་དང་རིག་སྟོབས་དང་། །

ལྷུན་ཞིང་ང་རྒྱལ་མེད་པར་བྱ། །

ཏེ་ཕུག་ཏུ་ནི་རྒྱལ་པོ་ཡིས། །

250 ལྷ་རྣམས་དང་ནི་ཕྲམ་ཟེ་ལའང་། །

ལེགས་པར་ཡང་ནི་གུས་པར་བྱ། །

བརྐུས་པའི་ཐབས་ནི་ཡང་མི་བྱ། །

བཀུར་བ་དང་ནི་མགལ་མི་བྱ། །

མི་བཟོད་མི་བྱ་ཁྱད་མི་བསད། །

255 ཀུན་ཉེས་ང་རྒྱལ་མི་བྱ་ཞིང་། །

ཡོན་ཏན་ཅན་ལ་སྤང་མི་བྱ། །

མི་གཞན་མི་སྨྲ་དཀྲུ་མི་བྱ། །

མིག་སྤུར་བབལ་པ་ནོན་ཡོད་བྱ། །

ལྷ་རྣམས་དང་ནི་མི་མཐུན་དང་། །

260 ཕྲམ་ཟེ་དང་ཡང་མི་མཐུན་ལ། །

གང་ནའང་བའི་མེད་དེ་ཡི་ཕྱིར། །

རྒྱལ་པོ་གཉིས་རྗེའི་མི་དགའ་སྟོངས། །

ཚངས་པས་དེ་ལྟར་གདམས་པ་དང་། །

ཕྲམ་ཟེ་ལ་དགའ་རྒྱལ་པོ་དེ། །

265 གཉིས་རྗེ་ལ་ནི་ཕྱག་འཚལ་ནས། །

གཉིས་རྗེ་ལ་ནི་དགའ་བར་བྱས། །

གཉེན་རྗེ་ཡང་ནི་དགའ་གྱུར་ཞིང་། །

ཚངས་པ་ཡང་ནི་དགའ་གྱུར་ཏེ། །

དེ་ཙོ་འགྲོ་བ་ཐམས་ཅད་ཀྱང་། །

༢༧༠ སྐྱ་ངན་མེད་ཅིང་དགའ་བར་གྱུར། །

དེ་ནས་ཚངས་པས་རྒྱལ་ལ་སྨྲས། །

ཁྱོད་ཀྱིས་ཡི་དགས་གཉེར་ཐུལ་འོག །

གཉེན་རྗེའི་མདག་གལུག་དེ་དག་གིས། །

དག་ཏུ་འགྲོ་བར་མི་ཡིབས་འོག །

༢༧༥ ཐོབས་དང་གཉི་བཞིད་དགའ་ཐུབ་ཀྱིས། །

འདི་འདྲ་རྒྱལ་པོ་གཞན་དག་ལའང་། །

ཐུག་པ་མེད་པར་ཐལ་བ་ནི། །

གྱུར་ན་མི་རུང་དེ་ཡི་ཕྱིར། །

དགེ་མེད་བུ་མིན་མི་བྱེད་བཞིན། །

༢༨༠ བདག་གིས་རྒྱལ་པོ་ཁྱོད་བརྐྲག་སྟེ། །

འཇིག་རྟེན་དུ་ནི་གྲགས་པ་ཆེ། །

བྱང་བ་དང་ནི་སྨན་པར་གྱུར། །

གཉེར་བུ་ཞེས་བུའི་རྣམ་གྲངས་འདི། །

ཡི་དགས་རྣམས་ལ་བྱས་པས་ན། །

༢༨༥ སྐྱེས་བུ་སྐྱེས་མཆོག་གང་ཕྱིར་ཁྱོད། །

གཉེར་བུ་ཐམས་ཅད་ཡེབས་པས་ན། །

དེ་ཕྱིར་གཅེར་བུལ་ཞེས་བྱ་བ། །

མ་ཆོད་སྦྱིན་གྱིས་ནི་གྲགས་མ་ཆོག་འགྱུར། །

གྲགས་མ་ཆོག་བྲམ་ཟེར་འཛིན་ཕྱིར་དང་། །

290 ང་ཡང་དགྱེས་པར་རེ་ཕྱིན་ཆད། །

སྐྱེ་དགུའི་བདག་པོ་གཞན་བཞིན་དུ། །

ས་ཨེ་སྟེང་དུ་གྲགས་ལྡན་འགྱུར། །

བདུལ་ལུགས་དགའ་ཕྱུབ་རིག་བྱེད་རིག །

སྐྱིག་མེད་སྐྱེ་དགུ་བདེ་བར་སྐྱོངས། །

295 ང་ཡིས་གནང་བས་ཁྱོད་ཀྱིས་ཀྱང་། །

བྲམ་ཟེའི་བུ་འདི་ཐྱིས་པས་ན། །

རེས་ཁྱོད་འཚོ་བའི་འཇིག་རྟེན་འདིར། །

རི་མོ་དང་པོ་འབྱིན་བྱེད་ཡིན། །

འཇིག་རྟེན་རྣམས་ལ་ཕན་འདོགས་ཕྱིར། །

300 བརྗེད་པའི་ངོས་སུ་འགྱུར་བ་ཡིན། །

དེ་རིང་ཕྱིན་ཆད་རི་མོ་དེ། །

མཛེན་པར་ཡང་ནི་མཆོད་པར་ཤོག །

མི་རྣམས་ཀྱི་ཡང་ཨིན་འཐྱོག་ཅིང་། །

གདན་དུ་དགན་དང་བདེ་སྦྱིན་ཤོག །

305 བཀྲ་ཤིས་དཔལ་ཡོན་སྐྱིག་སྟོང་ཤོག །

སྦྱིན་པོ་གསོད་ཅིང་དགྲ་འཛོམས་ཤོག །

འདི་ལྟར་ང་ཨི་ཆོག་གིས་བྱིན། །

དང་པོར་བྲིས་པ་དེ་ཡི་ཕྱིར། །

དང་པོར་བྲི་བ་ཡིན་པས་ན། །

༣༡༠ སིང་ཡང་རེ་མོ་ཞེས་བྱར་འགྱུར། །

དེ་ལྟར་ཚངས་པ་དེ་ཡིས་སུ། །

གཡེན་རྟེ་གཅེར་བུལ་དེ་གཉིས་ཀྱིས། །

ཐུམས་མཛད་ནས་ནི་དེ་གཉིས་དང་། །

བྲམ་ཟེས་ཀྱང་ལ་ཕྱུག་འཚལ་ཏེ། །

༣༡༥ དེ་དག་ཀུན་ལ་ཡིས་བྱས་ནས། །

དེ་ནས་རང་གི་གནས་ཉིད་དུ། །

བསྡུན་པ་འཇིག་རྟེན་གསུམ་གྱི་བླ་ །[1]

ཧ་ཡི་ཚོགས་དང་ལྷན་ཅིག་དང་། །

རྒྱལ་པོས་ཀྱང་ནི་ཚོས་རྒྱལ་ལ། །

༣༢༠ མཆོད་གནས་དགའ་སྟེ་གནས་སུ་དོང་། །

བྲམ་ཟེ་རང་གི་གྲོང་ཁྱེར་དུ། །

གང་དུ་འོངས་པར་སོང་བར་གྱུར། །

བྲམ་ཟེ་བུ་དང་བཅས་སོང་ནས། །

རྒྱལ་པོ་དེ་ཡང་ཡིད་རངས་ཏེ། །

༣༢༥ སྐྱིང་དུ་ཕྱག་དང་གཉིས་དང་བཅས། །

རེ་མོ་ལ་ནི་དྲག་དུ་བརྩོན། །

_____

དེ་ནས་མི་ཡི་དབང་ཕྱུག་དེ། །

ཚངས་པ་ཡི་ནི་གནས་སུ་དོང་། །

གཟུགས་ཀུན་མཐུན་པའི་ཆད་དེ་ཡང་། །

330 ཚངས་པ་ལ་ནི་ཚུལ་བཞིན་རིས། །

ཚངས་པ་ཚད་ནི་གང་གིས་སུ། །

རི་མོ་བདག་གིས་བྲི་བ་ཡི། །

མཚན་ཉིད་དག་གིས་སྐུ་ཚོགས་པས། །

དེང་སྐྱེད་དེང་འདིར་བདག་གསུང་འོས། །

335 རི་མོ་ཡང་ནི་རྗེ་ལྟར་བྱུང་། །

དེ་ཡི་ཚད་ཀྱི་ཚོ་གགང་། །

དེ་ལྟར་ཐལ་མོ་སྦྱར་ནས་ཞུས། །

ཚངས་པས་མི་དབང་ལ་སྨྲས་པ། །

དེ་ལ་མཆོག་ཏུ་གསང་བ་འདི། །

340 ཁྱོད་ལ་ང་ཡིས་བཤད་པར་བྱའོ། །

རྒྱལ་པོ་དང་པོར་འཛིག་རྟེན་གྱི། །

རིག་བྱེད་རྣམས་དང་མཆོད་སྦྱིན་བྱུང་། །

མཆོད་རྟེན་བྱེད་པར་རི་མོ་འདི། །

དེ་ཕྱིར་རི་མོ་རིག་བྱེད་བཅེ། །

345 ང་ཡིས་དང་པོ་སྐྱེ་དགུ་བྲི། །

འདི་ལྟར་སྐྱེ་དགུ་རྣམས་ལ་བསྟན། །

---

¹ A und B: བདག ² A und B: རིགས ³ B: ཡི

དེ་བས་དང་པོར་བྱིས་པའི་ཕྱིར། །

རི་མོ་ཞེས་བྱ་བར་བརྗོད་དེ། །

འགྲོ་བ་དང་ནི་གེལ་བའི་བར། །

350 སྐྱུ་དང་མི་སྐྱུར་བཅས་པ་དག །

དེ་ཨེ་ཚལ་བཞིར་བྱིས་བྱས་པས། །

དེ་ཕྱིར་འདི་ནི་རེ་ཤོར་བ་ཡད། །

རི་པོ་འི་མཚག་ནི་རེ་རབ་ཡིན། །

སློང་ལས་སྐྱེས་རྣམས་ཀྱི་ནི། །

355 གཙོ་པོ་ནམ་ཁ་ལྟིང་བཞིན་ནོ། །

མི་ཨེ་གཙོ་པོ་ས་བདག་བཞིན། །

དེ་བཞིན་རེ་ཤོ་སྐུ་ཚལ་གཙོ། །

ཇི་ལྟར་ཆུ་པོ་ཐམས་ཅད་ཀྱང་། །

རྒྱ་མཚོ་ཆེན་པོར་འབབ་པ་བཞིན། །

360 རྒྱ་མཚོ་ལ་ནི་རིན་ཆེན་བརྗེན། །

ཉི་མ་ལ་ཡང་གཟའ་རྣམས་བརྗེན། །

དང་སྲོང་དམ་པ་ཚངས་པ་ལ། །

ཇི་ལྟར་ལྷ་རྣམས་བསྟེན་པ་བཞིན། །

རྒྱལ་པོ་དེ་བཞིན་སྐུ་ཚལ་ཀུན། །

365 རི་མོ་དེ་ལས་ལ་ཡང་དག་བརྗེན། །

དེ་རྒྱལ་རི་རྣམས་ཆུ་༼ རྣམས་ནི། །

---

¹ So A und B statt ནམ་མཁབ་ ² A und B: ཀྲུ་; vielleicht ist ཀྲུང་ zu lesen.

གང་གྲུ་ཊི་མ་ གཟན་གཏོ་གཉིས། །

བྱུ་ཨི་རྒྱལ་པོ་ནས་ཁ་སྟེང་། །

སྐྲ་ནམས་གཏོ་བོ་དབང་ཆེན་ཡིན། །

370 དེ་བཞིན་རེ་མོ་སྐྱུ་རྒྱལ་གཏོ། །

དེ་བས་ལས་ཀུན་སྐྲ་ གན་པོང་། །

གཅེར་ཕྱུལ་དེས་ནི་རེ་མོ་ཨི། །

མཚན་ཉིད་ཚོ་ག་ཆད་ནམས་ཀུང་། །

ཐམས་ཅད་ཁྱོད་ལ་སྟོན་པར་འགྱུར། །

375 དེ་ནས་ཆངས་པས་སྐྱོ་བ་བཞིན། །

མི་ཨི་རྒྱལ་པོ་དགའ་བཞིན་དུ། །

ལས་ཀུན་པ་ཨི་ཐད་དུ་སོང་། །

སྟང་བ་ཆེན་པོ་དེས་ མཐོང་རོ ༎

མཐོང་ནས་རྒྱལ་པོས་ཕྱག་འཚལ་ཏོ ༎

380 དེས་ ཀུང་དེ་དང་སྐྲན་ཆིག་ཏུ །

བསྟན་མཆོག་བཀོད་ནས་བཀུར་སྟི་བྱས། །

ཞེ་སྡའི་ཚིག་ཀུང་སྐྲ་བར་འགྱུར། །

ཁྲབ་བདག་འཕྲིན་ལས་ཐམས་ཅད་པས། །

ཁྱོད་ཀྱིས་ཐད་དུ་ཚངས་པའི་བཀའ། །

_____

¹ A: irrtümlich མ་སྟེ་. B: bietet གི་ག་མ་སྟེ; die Verbesserung ergibt sich aus གཟན་   ² B: ལུ་   ³ A: ཆད་   ⁴ B: ང་   ⁵ A: དེས་   ⁶ B: ཟེས་

385 བདག་ཕྱིན་རི་མོ་འི་འཕྲིན་ལས་ཀྱི།

མཚན་ཉིད་བདག་ལ་བ་ཤད་དུ་གསོལ།

ཅི་རིགས་པར་ཡང་ཚད་དང་ནི།

རྣམ་པ་ཚོ་གར་བཅས་པ་གསུངས།

རྒྱལ་པོས་དེ་སྐད་སྨྲས་པ་དང་།

390 ལས་ཀུན་པས་ནི་བཙོ་བ་འདང་པ།

མི་བདག་དམ་པ་རྗེ་གཅིག་ཏུ།

ང་ཡིས་བ་ཤད་པ་ཐམས་ཅད་ཉིན།

རྗེ་ལྟར་རི་མོ་བྱུང་བ་དང་།

ཚད་དང་གནས་ཐབས་མདོག་ལྡན་པ།

395 བློ་དང་སྤྲན་པ་ཀུན་གྱི་གཙོ།

རི་མོ་འི་མཚན་ཉིད་སྤྲས་བཀུར་བ།

པ་ཧྲ་ལས་བྱུང་ལེགས་སྤྲན་གྱིས།

རྗེ་ལྟར་བསྟན་པ་བཞིན་དུ་ཉིན།

དངོས་པོ་ཀུན་གྱི་བྱད་གཟུགས་ཀུན།

400 འཇིག་རྟེན་རྣམས་ནི་དང་གྱུར་པ།

དགེ་བའི་མཚན་ཉིད་ཚངས་པ་ཨོ།

ཕྱིས་ནས་བདག་ལ་དང་པོར་བྱིན།

ཚད་ནི་གང་དང་གང་གིས་ནི།

གནས་དང་ཐབས་ནི་མཛེས་གང་ཡིན།

405 ཚངས་པ་དེ་ལས་དེ་དག་ནི། །

ཐོབ་ནས་རི་མོ་མཚོག་རྣམས་ཀུན། །

མཐའ་བདག་དམ་པའི་བཀའ་སྟིན་ལས། །

བརྩོ་རྣམས་ཐམས་ཅད་བདག་གིས་བགྱིའི། །

རྒྱལ་པོ་སྐྱེ་དགུ་འདི་དག་ནི། །

410 རྣམ་པ་དེ་ལྟར་བདག་གིས་སྒྲུལ། །

མཚོན་ཉིད་དག་དང་སྲན་པ་ཨི། །

རི་མོ་སྣ་ཚོགས་སྟེད་ནས་ནི། །

དེ་ནས་ཕྱིན་ཆད་ལྷ་རྣམས་ཀྱིས། །

རི་མོ་དག་འཕེལ་བྱུང་བ་ཡིན། །

415 ཚད་དང་མཚོན་ཉིད་ཕྱད་བཞིན་དང༌། །

འདུ་དང་སྐུ་དང་མཛེས་པ་དག །

ཅི་འདུ་དེ་འདུ་དེ་ནི་དང༌། །

བདག་ལ་ཤེས་ནས་མ་ལུས་པར། །

སྐུ་ཚལ་ཀུན་སྣན་རི་མོ་འདི། །

420 མ་ལུས་པར་ནི་ཙྲོ་སྣན་མཚོག །

ཁྱོད་ཀྱིས་མི་རྣམས་དག་དང་ནི། །

ལེགས་པར་གཟུང་ལ་མ་ཁབས་པ་དང༌། །

གཟུང་བར་སྐྱོ་བའི་སེམས་སྤྲན་དང༌། །

རི་དག་བྱ་བར་དགའ་བ་དང༌། །

425 དག་ཏུ་ཡིད་གཞུངས་མ་དུན་དག་ཏུ། །

རབ་ཏུ་བསྟན་པར་བྱིས་ཤིག་དང་། །

ཐུབ་པ་རྣམས་པ་སྐྲ་ཚོགས་དང་། །

གྲུ་དང་གཟོར་སྦྱིན་སྦྱིན་པོ་དང་། །

ཡི་དགས་ལྷ་མིན་ཁ་ཟར་བཅས། །

430 མཚན་ཉིད་ལ་སོགས་ཚོ་ག་བཞིན། །

བྱིས་ནས་ཁྱོད་ལ་འངས་བསྟན་ཏོ། །

གཅེར་བུ་ཐུལ་གྱི་རེ་མོ་ནི་མཚན་ཉིད་བསྟན་པ་ལས། གཅེར་བུ་བཏུལ་བ་ཞེས་བུ་བའི་

ལེའུ་སྟེ་དང་པོ་ན ༎

_____

2. Kapitel.

དེ་ནས་བཅོད་ཀླུན་དེ་གུས་པས། །

རྒྱལ་པོ་བདག་ཉིད་ཆེ་ནེ་ལ། །

ཚངས་པས་རེ་མོ་ཅི་འདྲ་བར། །

435 བྱི་བ་མཛོན་བསྟན་སྐྱས་པ་ནི། །

སྤ་མ་འཇིག་པའི་མཐར་གྱུར་ཅིང་། །

གཏན་དང་འགྲོ་བ་བཞིག⁎པ་ན། །

གསེར་གྱི་སྐོ་ངས་སྐུན་བཅོམ་སྟེ། །

རྒྱ་ཡི་ནང་ནས་ཀུན་ཏུ་བྱུང་། །

440 སྐོང་དེ་ལས་འཇིག་དེན་གྱི། །

མེས་པོ་རང་ཉིད་ཀུན་ཏུ་བྱུང་། །

_____

¹ B: གྱིས་   ⁎ A und B: ཞིག་

དེ་ནས་ཚངས་དེ་ཡོ་ལ་སོགས། །

རིག་བྱེད་དང་ནི་རིག་པ་དང་། །

ཏོག་པར་བྱ་དང་སྐྱེ་དགུ་ནི། །

445 རྣམ་པ་བཞི་དང་དེ་དག་གིས། །

གཟུགས་དང་མིང་ལ་སོགས་པ་ནི། །

ཚེ་དང་བཅས་དང་ཚངས་པ་ཡིས། །

རིགས་དང་གནས་ཀྱི་སྦྱོར་པ་དང་། །

ཚོས་དང་ལུགས་རྣམས་ཤ་སྲག་བྱུང་། །

450 ཚངས་པས་དེ་ལྟར་བྱས་ནས་ནི། །

ཇེས་པར་ཚངས་པས་འགྲོ་ཀུན་ལ། །

ཕན་པར་བྱ་ཕྱིར་རང་ཉིད་ཀྱི། །

ཡིད་ལ་ཀུན་ཏུ་སེམས་པར་གྱུར། །

དེ་ནས་སེམས་པར་གྱུར་པ་ནས། །

455 དེ་ཡི་བློ་ནི་འདི་འདྲ་བྱུང་། །

ཇི་ལྟ་བུར་ན་སྐྱེ་བོ་རྣམས། །

རྒྱལ་པོ་དང་ནི་ལྷ་རྣམས་ཀྱིས། །

མི་དག་ཤེས་ཤིང་གཡེལ་མེད་པར། །

དག་ཏུ་གུས་པ་བྱེད་པར་འགྱུར། །

460 ཚངས་པས་དེ་ལྟར་བསམས་ནས་དེས། །

ལྷ་ཆེན་ཁྱབ་བདག་བརྒྱ་བྱིན་དང་། །

---

¹ B: དེས་པ་

ཤླ་ནྲམས་དག་ནི་ཐབས་ཅད་དང་། །

སྐྱལ་པ་ཆེན་པོར་རང་གི་ཡང་། །

ཆད་དང་ཡོན་དན་མཛེས་པ་དང་། །

465 མཚན་སྟོན་ཤིན་ཏུ་འདྲ་བ་དང་། །

གཟུགས་བཟང་ཤིན་ཏུ་ལེགས་ཕེབས་དང་། །

ཡན་ལག་ཉིང་ལག་རྫོགས་པ་དང་། །

ཀུན་ནས་ཤིན་ཏུ་འབྱེས་པ་དང་། །

དབྱིབས་ནི་རྣམ་པ་སྣ་ཚོགས་དང་། །

470 རྒྱུན་དང་གོས་ནི་ཤིན་ཏུ་མཛེས། །

མཚན་ཆ་སྣ་ཚོགས་འཆང་བ་ཡི། །

བུད་གཟུགས་རེ་མོ་ནི་མཚན་འཕེལ་བས། །

རང་ཉིད་ཀྱིས་ནི་རྣམ་པར་བྱེས། །

དེ་དག་ལྷ་རྣམས་ཀྱིས་མཐོང་ནས། །

475 མིག་རྩེ་རངས་ཤིང་སྐྱིང་དགའ་སྟེ། །

ཚངས་པ་ལེགས་སོ་ལེགས་སྐུས་ནས། །

གུས་པ་ཡིས་ནི་བསྟོད་བྱས་ཤིང་། །

དེ་རྣམས་ཀྱིས་ནི་རང་རང་གི །

བུད་གཟུགས་ཕྲིན་བརྐྱབས་མཐུ་ཆེན་བྱས། །

480 དེ་ནས་ཚངས་པ་དགར་གྱུར་ནས། །

ལྷ་བདུན་ལ་ནི་འདི་སྐད་སྨྲས། །

འཇིག་རྟེན་གསུམ་དུ་དེང་ཕྱིན་ཆད། །

ཁྱེད་ཅག་གིས་ནི་ཕྱད་གཟུགས་ལ། །

བསོད་ནམས་རྟེན་པ་ཕྱེ་ཚོམ་མེད། །

485 དྲག་ཏུ་མཚོད་པ་བྱེད་པར་འགྱུར། །

གང་ཞིག་ཡང་དག་ཡིད་བཟང་བྱས། །

རིག་[1]་པ་ཡིས་ནི་བསྒྲུབས་པའི་ནོར། །

གཙང་དང་དེ་ལ་གཞིལ་བ་ཡིས། །

མི་ཡི་ལྱུལ་དུ་མཚོད་བྱེད་ན། །

490 དེ་རྣམས་དེ་དག་ལ་དག་ཏུ། །

བསྐུབ་དོན་ནད་མེད་འདོད་པ་སྟེར། །

འཇིག་རྟེན་རྣམས་ནི་སྒྲིག་སེལ་ཞིང་། །

སྐྱེ་ལམ་འན་དང་སྲུང་རྣམས་ཀྱིས། །

བྱིན་པ་དག་ནི་སེལ་བྱེད་ཅིང་། །

495 བསྱུང་བ་ཉིད་ནི་བྱེད་པར་འགྱུར། །

ཚོས་བཞིན་བྱེད་པ་མཐའ་དག་གིས། །

སྟིན་པོ་འརྫོམས་ཤིང་གྲགས་པ་འཕེལ། །

འཇིག་རྟེན་རྣམས་ཀྱི་མཚོད་བྱའི་ཕྱིར། །

ལྷ་ཡི་བྱེད་གཟུགས་རྣམས་ལ་ནི། །

500 མིང་ནི་བརྫོད་བྱ་བསྒོད་བྱ་དང་། །

མཚོད་བྱ་ཕྱུག་བྱར་གྲགས་[2]་པར་འགྱུར། །

_____

¹ A und B: རིགས་   ² B: རགས་

སྐྱེས་བུ་བསོད་ནམས་བརྩོན་གང་གང༌། །

ཁྱེད་ཀྱིས་བྱད་གཟུགས་གང་གང་ལ། །

ཉིན་ཅིག་བཞིན་དུ་མཆོད་བྱེད་ན། །

505 དེ་ལ་དེ་དག༌ྋ ཞི་བ་སྟེར། །

དེ་སྐད་སྨྲས་ནས་དེར་གྱུར་ཅེས། །

ལྷ་རྣམས་ཡིད་ནི་རབ་དགའ་ནས། །

རང་གི་གནས་བཟྗེད་ཆས་བྱིན་རྫབས། །

རང་རང་གི་ནི་ྋ གནས་སུ་དོང༌། །

510 དེ་ལྟར་མཆོད་པ་བྱ་བྱུང་ཞིང༌། །

འགྲོ་བས་མཆོད་པ་བྱ་གྱུར་པ། །

དེ་དག་གིས་ནི་ཚད་ལ་སོགས། །

བདག་ལས་མཆན་ཉིད་ཐོས་ནས་ནི། །

མི་དབང་ཁྱེད་ཀྱི་དེ་རིང་དུ། །

515 མི་ཡི་འཛིག་རྟེན་མ་ལུས་སྒྲོགས། །

མཆན་གྱི་རིག་པ་དམ་པ་གང༌། །

བདག་གིས་ཚངས་ལས་སྙེད་དེ་དང༌། །

ཚད་ཅི་འདུབ་དེ་ལྟ་བུར། །

མ་ལུས་པར་ནི་ཤེས་པར་བགྱི། །

520 བགྱར་འོས་འཛིག་རྟེན་གསུམ་དག་ན། །

ལུས་ཅན་གྱིས་ནི་ལུས་ཀུན་གྱིས། །

ཡན་ལག་ཆད་ནི་གང་ཡིན་པ །

བདག་གིས་བཤད་ཀྱིས་མཉན་པར་གྱིས །

རྒྱལ་པོ་ཆེན་པོ་རེ་མོ་འི་ཆད་ནི་སྐྲ་ཀུན་གྱི །

525 མཚོད་ཅིང་བྲགས་འཕེལ་སྟེག་དང་འཇིགས་པ་མེལ་ཕྱེད་དང་ །

སེག་སྤུར་སྤུག་ཅིང་ཕོག་པར་བྱུང་ལ་སྐུ་ཚོགས་གཞི༔ །

བཙོད་པ་སྐྱོན་དང་ཕྲལ་བ་འདི་ནི་གོ་བར་གྱིས །

རེ་མོ་ནི་མཚན་ཉིད་ལས་མཚོད་པར་བྱ་བ་འབྱུང་བ་ཞེས་བྱ་བ་སྟེ་ལེའུ་གཉིས་པ་རྫོགས་

སོ ॥

3. Kapitel.

རྒྱལ་པོ་འགྲོ་བའི་ལུས་ཚད་ཚད་ནེ་ནི །

ཚངས་པ་རང་གི་བདག་ལ་བཤད་པ་སྟེ །

530 བདག་ཀྱང་ཁྱོད་ལའི་ཁྱོད་ཀྱང་སྐྱེ་དགུ་ལ །

ཚད་ཀྱི་ཚུལ་ནི་རབ་ཏུ་བཤད་པར་གྱིས །

སྐྲ་དང་སྐྲ་མིན་སྐྲ༔ དབང་དང་ །

སྤྱིན་པོ་ཏེ་ཟ་མི་འམ་ཙེ །

གུབ་དང་རོལ་མ་ཁབ་མཚོན་སྒྲོགས༔ དང་ །

535 ཕ་ཟ་ཨི་དགས་གྲུལ་བུམ་དང་ །

རེག་སྤྲགས་འཆང་གི་ཚོགས་དང་ནི །

མི་དང་མི་མ་ཡིན་རྣམས་དང་ །

. ᵃ B: བཞི་   ᵇ B: སྐྲ   ³ B: སྒྲོགས་

རྒྱལ་པོ་དགའ་ནི་ཐབས་ཅད་ཀྱི། །

ཆད་དུ་བྱ་བ་འདི་ཡིན་ནོ། །

540 དྲལ་ཕུ་རབ་དང་སྟབ་འི་ཙེ་་མོ། །

སྲོ་མ་ཤིག་དང་ནས་དང་སོར། །

ཐབས་ཅད་རིམ་བཞིན་བཀྲད་འགྱུར་དུ། །

སྐྱེད་པ་ཡིན་པར་ལོངས་སུ་བསྟན། །

དྲལ་ཕུ་རབ་ནི་བཀྲད་པོ་ལ། །

545 སྐྲ་ཨི་ཙེ་མོ་ཞེས་བསྟན་ཏོ། །

ཆད་ཤེས་པས་ནི་སྐྲ་ཙེ་ཡང་། །

བཀྲད་ལ་སྲོ་མ་ཞེས་བསྟན་ཏོ། །

སྲོ་མ་བཀྲད་ལ་ཤིག་ཡིན་ཏེ། །

ཤིག་བཀྲད་ལ་ནི་ནས་སུ་བཤད། །

550 ནས་བཀྲད་ལ་ནི་སོར་ཡིན་ཏེ། །

སོར་ཞེས་བྱ་བ་སོར་མོར་བཤད། །

སོར་གཉིས་ལ་ནི་སོར་ཆད་གཉིས། །

སོར་བཞི་ལ་ནི་སོར་ཆད་བཞི། །

ཆད་ཤེས་པས་ནི་ནས་བཞི་ལ། །

555 སོར་གྱི་ཕྱེད་དུ་བསྟན་པ་ཡིན། །

ཕྱེད་ཅེས་བྱ་བ་མཚུ་ཡིན་ཏེ། །

ཞིང་ནི་ཐད་ཀ་ཞེས་བྱ་བ། །

རྒྱལ་པོ་མི་ནི་འདི་དག་གི །

ཆད་ནི་ཡོངས་སུ་བསྟན་པར་བྱ། །

560 འཁོར་ལོ་བསྒྱུར་བའི་རྒྱ་ཞིང་ནི། །

རྫི་སྟེར་ཡིན་པ་བཤད་ཀྱིས་ཉིན། །

མཚུ་དང་ཞིང་ནི་ཁབ་པ་སྟེ། །

ནུ་གྲོ་རྣ་དང་འདྲ་བ་ཡིན། །

འཁོར་ལོ་བསྒྱུར་བའི་སྟེད་དུ་ནི། །

565 རང་གི་སོར་གྱི་གཚལ་བ་ཡིས། །

བཀྱུ་དང་ཚ་ནི་བཀྱུད་ཡིན་ཏེ། །

གང་དུའང་གཞན་གྱི་ སོར་གྱིས་མིན། །

རྒྱལ་པོ་གང་དུ་སྤྱོས་པ་ཡི། །

འཁོར་ལོས་བསྒྱུར་བའི་རྒྱལ་པོ་ཡི། །

570 སྤྱིད་ཀྱི་ཚད་དུ་བྱ་བ་དག །

འདིར་ནི་དངོས་སུ་བཤད་པར་བྱ། །

གཏོང་ལ་སོགས་པའི་ཚད་གང་ཡིན། །

དེ་ནི་བདག་གིས་བཤད་ཀྱིས་ཉིན། །

གཏོང་ལ་དཔུལ་བ་སྟ་དང་ནི། །

575 གོས་ཀི་སོར་བཞི་ཚད་གསུམ་དབྱེ། །

གདོང་གི་ཞིང་དུ་སོར་མོ་དགུ །

བཅུ་བཞི་ཨེན་པར་རབ་དུ་བསྟན །

དེ་ཉིད་ཀྱི་ནི་སྟོད་དམད་ཞིང་ །

སོར་མོ་དགུ་ནི་བཅུ་གཉིས་ཨེན །

580 ཚད་ཤེས་པ་ཨེས་མཚུར་ཡང་ནི །

སོར་ནི་བཅུ་གཉིས་ཨེན་པར་འདོད །

གཙུག་ཏོར་མཚུར་ནི་སོར་བཞི་སྟེ །

ཞིང་དུ་སོར་ནི་དྲུག་ཨེན་ནོ ༎

མགོ་ནི་གདུགས་ཀྱི་ཚད་དུ་བྱ །

585 མགོ་བོ་འི་ཞིང་དུ་སོར་བཅུ་གཉིས །

དེ་བཞིན་མགོ་བོ་འི་སོར་ཡུག་ཏུ །

སོར་མོ་སུམ་ཅུ་རྩ་གཉིས་ཨེན །

རྣ་བའི་ཞིང་ལ་སོར་གཉིས་ཏེ །

སྲིད་དུ་སོར་ནི་བཞི་ཨེན་ནོ ༎

590 རྣ་བའི་བུ་ག་སོར་ཕྱེད་དེ །

རྣ་བའི་བུ་ག་སོར་ཅིག་བའད །

རྣ་མཆོག་དང་ནི་སྨིན་མཛག་མཚམས །

མིག་སྤུབས་རྣ་བའི་བུ་གར་མཚམས །

རྣ་ག་ཁལ་ལ་ནི་ཙིས་པ་མེད །

595 སྨིན་མའི་སྲིད་ལ་སོར་བཞི་སྟེ །

ཞིང་ལ་ནས་ནི་གཉིས་ཨེན་ནོ ༎

ཞི་བ་ཀུན་གྱི་སྐྱེན་མ་ནི། །

སྨྲ་བ་ཚོས་པ་འདྲ་བ་ཡིན། །

ཚད་ཤེས་པས་ནི་གར་བྱེད་དང་། །

༦༠༠ ཁྲོས་དང་དུ་ལ་གཞུ་འདྲར་དགུག །

སྐྱག་དང་ཀྲུ་འན་བྱེད་པ་ལ། །

མཐུག་སྐྱད་མགོ་ནི་བསྒྱོར་བྱ་ཞིང་། །

སྐྲ་ཡི་སྨྱུབས་ལས་སྐྱེན་མ་དག །

བསྐྱེད་དེ་སྒྱལ་བའི་ཕྱིར་དུ་བཤག །

༦༠༥ འབྱོར་ལོ་སྒྱུར་བའི་མཛོད་སྐྱུ་ཡིས། །

སྐྱེན་མའི་བར་ནི་སོར་ཅིག་བུ། །

སྐྱེན་མའི་སྐྱེད་ནས་སྐྱ་འཚམས་ཀྱི། །

ཞིང་དུ་སོར་ནི་ཕྱེད་དང་གསུམ། །

སྐྱེན་མའི་མགོ་ནས་དཔལ་དཔྱིངས་ལ། །

༦༡༠ སོར་བཞི་པ་ཡི་ཚད་དུ་བྱ། །

མིག་གི་དབྱབས་ལ་སོར་གཉིས་ཏེ། །

མིག་གི་བར་ཡང་དེ་བཞིན་ནོ། །།

མིག་གི་དཀྱུས་སུ་སོར་གཉིས་ཏེ། །

ཞིང་དུ་སོར་ནི་ཅིག་ཏུ་བཤད། །

༦༡༥ མིག་གི་འབྲས་བུ་མིག་གསུམ་སྟེ། །

མིག་ལ་འབྲས་བུ་གསུམ་འགྱུར་བཤད། །

གདོང་པ་ཡི་ནི་དཀྱུས་ལ་ནི། །

སྨིག་རྒྱུ་བུ་བར་ཡོངས་སུ་བཤད། །

སྨིག་ནི་འོན་པའི་གཞུ་འདྲའི་ཞིང་། །
620 ནས་གསུམ་གྱི་ནི་ཆད་དུ་བྱ། །
ཁྱུ་མྲུལ་འདབ་འདྲའི་སྨིག་ལ་ནི། །
ནས་དྲུག་གིས་ནི་ཆད་དུ་བཤད། །
ཅི་སྟེ་འདྲ་བའི་སྨིག་གང་ཡིན། །
ནས་བཅུད་ཀྱི་ནི་ཆད་དུ་བཤད། །
625 པཊྚའི་འདབ་མ་འདྲ་བའི་སྨིག །
ནས་དགུ་ཡི་ནི་ཆད་ཡིན་ནོ ༎
མགྲོན་བུ་འདྲ་བའི་སྨིག་གང་ཡིན། །
དེ་ནི་ནས་དག་བཅུ་ཡིན་ནོ ༎
རྒྱལ་པོ་དེའི་སྨིག་དག་གི །
630 དཀུས་དང་ཞིང་དུ་བསྟན་པ་ཡིན། །
རུལ་འགྱུར་པ་ལ་རྟོག་མེད་པའི། །
སྨིག་ནི་འོན་པའི་གཞུ་འདྲ་བྱ། །
བྱད་མེད་རྣམས་དང་གཡེམ་པོ་ཡི། །
ཅི་ཡི་སྟོ་འདྲར་ཤེས་པར་བྱ། །
635 ཐ་མལ་འདུག་པའི་སྨིག་དག་ནི། །
ཁྱུ་མྲུལ་འདུ་བར་འཐོར་པ་ཡིན། །
སྨག་དང་དུ་བ་ཤིང་ཀྱི་ནི། །

པ་རྣའི་འདབ་མ་འདྲ་བར་བསྟན །

ཁྲིས་དང་སྨུག་བསྨལ་གཟེར་བའི་སྨིག །

640 མགྲོན་བུ་འདྲ་བར་བྱི་བར་བུ །

ཁྲུ་རྩལ་འདབ་མ་འདྲ་བའི་སྨིག །

མཐན་དམར་འབྲས་བུ་གནས་ལ་དང་ །

གསལ་ཞིང་རྗེ་མ་རྩེ་རིང་དང་ །

ཡིད་འོང་མདོག་དབྱུང་འཇམ་འཇམ་ལ །

645 སྐྲ་རྣམས་ལ་ནི་བྱིས་བྱས་ནས །

རྒྱལ་པོ་སྐྱེ་དགུ་ཐན་བྱེད་འགྱུར །

བ་ཡི་ནུ་མའི་མདོག་གསལ་འདྲ །

སྐྱུམ་ཞིང་རྗེ་མ་རང་རོང་མེད །

གསལ་བས་པ་རྣའི་འདབ་འདྲ་ལ །

650 མཐིང་ཤུན་གྱིས་ནི་བསྒྱུར་བྱས་ཤིང །

འབྲས་བུ་གནས་ལ་རྒྱུ་ཆེ་བའི །

སྨིག་བྱིས་དཔལ་དང་བདེ་བ་སྟེར །

འདི་ནི་སྨིག་གི་ཞིང་གི་ཚད །

ཡིན་པར་འོངས་སུ་བསྟན་པ་ཡིན །

655 ལྷ་སྲུངས་སྐྱམ་ཅུ་རྩ་དྲུག་གང་ །

དེ་དག་མཚན་ཉིད་ཕྱིས་བཤད་དོ ༎

སྐྲ་ལ་སོར་ནི་བཞི་ཡིན་ཏེ །

ཙེ་མོ་ནི་འཕང་དུ་སོར་གཉིས་སོ ། །

སྣ་སྲུབས་ལོ་གར་བཙས་པ་ཨེ ། །

660 ཞིང་དུ་སོར་ནི་གཉིས་ཨེན་ནོ ། །

སྣ་ཨེ་བུ་གའི་ཞིང་ནས་དྲུག །

འཕང་དུ་ནས་ནི་གཉིས་སུ་བཞད །

སྣ་ཨེ་བུ་གའི་བར་ཞིང་ནི །

ནས་གཉིས་དཀྲུས་སུ་ནས་དྲུག་གོ ། །

665 ཨ་མཚུའི་ཊམས་སུ་སོར་ཙིག་སྟེ ། །

མ་མཚུ་ལ་ནི་སོར་ཕྱེད་བཞད །

ཡ་མཚུ་དང་ནི་མ་མཚུའི་དཀྲུས །

སོར་བཞི་དག་ཏུ་ཤེས་པར་བྱ །

ཡ་མཚུ་ཨེ་ནི་སྟེང་མ་ཨེ །

670 ཕྱེད་དུ་སོར་ནི་ཕྱེད་ཨེན་ནོ ། །

མཚུ་མཐའ་ཕིམ་བ་ལྱར་དམར་ཞིང[1] །

དེ་བཞིན་གཞུ་ཨེ་དུ་ཤིང་འདྲ །

ཁ་ཨེ་གྲུ་ནི་ཆུང་ཟད་བསྡོར །

དེ་བཞིན་འཇོམ[2] བག་ཅན་དུ་བྱ །

675 གོས་ཀོ་འཕང་ནི་སོར་གཉིས་ཏེ ། །

[1] B: ཞིག་   [2] B: མཇོམ་

ཞིང་དུ་སོར་ནི་གསུམ་དུ་བཞད།

མགལ་པ་དག་གི་སྙིང་ཚད་ནི།

སོར་བཞི་དག་དུ་བསྟན་པ་ཡིན།

གནམ་ལྔང་པོ་ནི་སྙིད་སོར་དུ།

680 སྤུམ་པོ་ནི་ཞིང་ལ་སོར་བཅུ་སྟེ།

ཕུ་མོ་ཉིས་ལ་སོར་བཅུད་དོ།།

སྤྱོམས་དག་དུ་ནི་སུམ་འགྱུར་རོ།།

གཏིར་མ་གསུམ་གྱིས་མཛེས་བྱས་པ།

དུང་དང་འདྲ་བ་དག་དུ་བྱ།

685 རྒྱབ་ལོགས་སུ་ནི་མི་དམན་ཞིང་།

འཁོར་བར་རྩམ་པོ་ཉིད་དུ་བྱ།

མཁར་ཚོས་ལ་ནི་སོར་ལྔར་བྱ།

འགྱམ་པ་ལ་ནི་སོར་བཞིར་བྱ།

གྱུར་འགྱམ་ཕོགས་ཀྱི་ཚད་དག་ཉི།

690 སོར་བཞི་དག་དུ་ཤེས་པར་བྱ།

འདི་ནི་བདག་གིས་རྡོ་མཆུ་དང་།

ཞིང་དག་བསྟན་པར་བྱས་པ་ཡིན།།

གྲུ་བོ་ཉི་སྐམ་དང་ལེགས་རྟོགས་དང་།

གསལ་དང་མཛེས་པའི་མཚན་ཉིད་དུ། །

695 གྲུ་གསུམ་མི་བྱ་ལོ་མི་བྱ། །

ཁྲིས་བག་མི་བྱ་རྣུམ་མི་བྱ། །

གདོང་འདིའི་འདྲབ་ཐིས་བྱས་ནས། །

དག་དུ་འབྲོར་བ་བྱེད་པར་འགྱུར། །

སྐྱེ་དགུ་རྣམས་ལ་ཞི་འདོད་པ། །

700 རིང་པོ་རྣུམ་པོ་ཡོན་པོ་དང་། །

ཟུར་གསུམ་ལ་སོགས་གང་ཡིན་ནས། །

དེ་དག་རྣམས་ནི་སྣ་ལ་ཡང་། །

རྒྱལ་པོ་འདི་ནི་གདོང་དག་གི །

ཆད་དུ་བྱ་བ་བསྟན་པ་ཡིན ༎

705 ལུས་ཀྱི་རབ་དུ་བརྗོད་པར་བྱ། །

འདི་ནི་མ་ཡིངས་བཟུང་བར་གྱིས། །

སྐྱེད་བྱོ་བར་ཞིང་སོར་གཉིས་ཏེ། །

སྲུང་ཡའི་ཞིང་དུ་སོར་དྲུག་གོ ༎

ཕྱག་པ་ལ་ནི་སོར་བཅུད་དོ ༎

710 མཐོང་ག་གདོང་ནི་གཉིས་འགྱུར་བའད། །

པོ་མཚན་སྙེད་ནི་སོར་དྲུག་བྱ། །

སྣ་རྣར་དཀྱུས་སུ་སོར་བཙོ་བཀྱད། །

སྨྲག་དུས་བར་ནས་སྙིང་གར་གདོང་། །

སྐྱིང་ནས་སྟེ་བ་ཁུང་དུ་འདུང་མིན །

715 སྟེ་ནས་པོ་མཆན་བར་ལའང་གཏོང །

སྐྱིད་པ་ལ་ནི་སོར་བཅུ་བཞི །

སྟེ་བ་དག་ནི་སོར་ཕྱེད་དེ །

 རྩབ་ཅིང་གཡས་སུ་འཁྱིལ་བར་བཤད །

ནུ་མཛོར་ལ་ནི་ནས་གང་སྟེ །

720 ཕོར་ཕོར་ཡུག་སོར་གཉིས་སོ ༎

གང་དུ་ཉམ་ཐབས་སྐོས་ལ་ནས །

སྐ་རགས་དག་ནི་བཅིངས་ནས་སུ །

སྟེ་བའི་འོག་གི་སྐྲོ་ཡི་བར །

སོར་བཞི་དག་དུ་བྱ་བ་ཡིན །

725 པོ་མཆན་ཞིང་དུ་སོར་གཉིས་ཏེ །

ཅུ་སོ་ནི་སྟེད་དུ་སོར་དྲུག་གོ ༎

སྟིག་པ་ཏ་ཅང་མི་འཕྱང་དང །

དོ་མཉམ་རྣམ་ཞིང་གཤིས་ཡིབས་དང །

ཕོམས་སུ་སོར་བདུན་ཡིན་ཏེ[2] །

730 སྟིད་དུ་སོར་ནི་བཞི་ཡིན་ནོ ༎

པོ་མཆན་དག་གི་ཕོམས་སུ་ནི །

སོར་དྲུག་དག་དུ་ཤེས་པར་བྱ །

---

[1] B: བྲླུ་   [2] So in A und B; eine Silbe fehlt; vermutlich hat
ནི་ nach སོར་ gestanden wie im folgenden Verse.

ཕོ་མཚན་ནས་ནི་ལྟོ་མཚམས་ཀྱི། །

བར་ལ་སོར་ནི་དྲུག་ཡིན་ནོ། །

735 དེ་ནས་གཞུང་སྟེ་བཀྲ་དག་གི །

སྲིད་ཀྱི་ཚད་དུ་བྱ་བནི། །

སོར་མོ་ཉི་ཤུ་རྩ་ལྔར་བུ། །

ཕྱས་མོ་ལ་ནི་སོར་བཞིར། །

གོང་བུའི་ཕྱོགས་ལ་སོར་བཞིར། །

740 རྗེ་ངར་ལ་ནི་གདོང་གཉིས་བ་འད། །

གོང་བུའི་ཕྱོགས་ཀྱི་རྗེ་ངར་ཁྱི། །

ཞིང་དུ་སོར་ནི་བཞི་ཡིན་ནོ། །

རྗེ་ངར་དག་གི་དབུས་དག་གི །

ཞིང་དུ་སོར་ནི་དྲུག་དུ་བ་འད། །

745 ཕྱས་གདོང་ཞིང་ནི་སོར་གསུམ་བུ། །

སྲིད་ཉིད་དུ་ནི་མ་ཡིན་ནོ། །

ཕྱས་མོ་གོང་གི་བཀྲ་གཉིས་ཀྱི། །

ཞིང་དུ་སོར་ནི་བཀྱུད་ཡིན་ནོ། །

བཀྲ་དྲུང་སྤྱོམ་པོའི་ཞིང་དུ་ནི། །

750 སོར་ནི་བཅུ་གཉིས་དག་དུ་བྱ། །

བཀྲ་ཡི་མདུན་ནི་མཐོན་པོ་སྟེ། །

གཉིས་གའང་ཀུན་ནས་འཁ་རྒྱས་བུ། །

འཇམ་ཞིང་༌འདང་༌གོང་མེད་པ་དང་། །

སྒྲང་པོ་ཆེ་ལ་སྐྲ་ལྤར་བྱ། །

755 བོང་བུ་ཕྱུས་མོ་མི་མཛིན་དང་། །

རྟ་རྣམས་ཀྱང་ནི་མི་མཛིན་དང་། །

ཕྱིན་གྱིས་རླམ་པོ་དག་ཏུ་བྱ། །

སྤང་ནི་ཅུང་ཟད་མཐོན་པོར་བྱ། །

རྫིང་པའི་འཕང་ལ་སོར་སྤུར་བྱ། །

760 ཞིང་དུ་སོར་ནི་གསུམ་དུ་བ་འད། །

ཀང་པ་གཉིས་ལ་སོར་བཅུ་བཞི། །

མཐེ་བོང་ལ་ནི་སོར་བཞིན། ॥

པཙ་དམར་པོ་ནི་རེ་འདྲ་དང་། །

རྒྱུ་སྐྱིགས་ཁ་ལྤར་ཟང་བག་ཅན། །

765 ཀང་རྒྱུ་མ་ཀྲོངས་ས་ལ་འཛར[3]། །

འཕྲོར་ལོ་མཚན་གྱིས་སྤུས་པ་འདྲ། །

རྫིང་པ་དང་ནི་མཐེ་བོང་གཉིས། །

ས་ལ་རེག་ན་འཇར་བར་བྱ། །

རྒྱལ་པོ་འཕྲོར་ལོ་སྤྱུར་བ་ཡི། །

770 ཀང་པའི་དུ་བ་ངང་པའི་འདྲ། །

ཀང་པོ་ལ་སྤྱིད་སོར་བརྒྱུད་ཞེས་བྱ[4]། །

---

1 B: ཤིང་   2 B: བ་འད་

3 A: འཛར་, B: མཛར་   4 So in A und B; eine Silbe zu viel

དུས་སྨལ་རྒྱབ་ལྟར་མཐོ་བ་དང་། །

མཇེས་པའི་མཚན་ལྟར་ཞིང་དུ་ཡང་། །

སོར་ནི་ལྟ་དང་ལྡུག་པར་བྱ། །

༧༧༥ མཐེའུ་ཆུང་སོར་ཀྱི་དྲུག་པ་ནས། །

ཞེང་ནི་སོར་མོ་དྲུག་ཡིན་ནོ། །

མཐེ་བོང་ཞེང་དུ་སོར་གཉིས་ཏེ། །

སྐྱོམས་སུ་སོར་ནི་དྲུག་ཡིན་ནོ། །

སྦྱིད་དུ་སོར་ནི་བཞི་ཡིན་ཏེ། །

༧༨༠ དུང་མ་རྩེ་མོ་¹ མཐོར་པོར་²བྱ། །

འཇོབ་མོ་དཀ་ནི་ཆུང་རབ་ཅིག །

མཐེ་བོང་³ ལ་ནི་སྣག་པར་བྱ། །

དེ་ཉིད་ཀྱི་⁴ ནི་སྙིད་དང་སྐྱོམས། །

སོར་མོ་དཀ་ནི་གསུམ་དུ་བྱ། །

༧༨༥ སོར་མོ་སྣག་མ་རྣམས་དང་ནི། །

ཆད་ནི་སོར་མོ་རེ་རེས་དབྱེ། །

མཐེའུ་ཆུང་གི་ནི་སྙིད་དང་སྐྱོམས། །

སོར་མོ་དཀ་ནི་གཉིས་གཉིས་སོ། །

སོར་མོ་བར་ཕགས་བཟང་བ་དང་། །

༧༩༠ ཚིགས་སྤུག་འབར་འབུར་མེད་པ་དང་། །

<hr>

¹ B: རུམ་ ² B: པོ་ ³ A: མཐེ་འོང་ ⁴ A: ཀྱིས་

ཤིན་དུ་རྣམ་དང་ཙ་མི་མཛེན། །

རྒྱས་པ་རུས་པ་མི་མཛེན་དང་། །

སེན་མོ་ཧྲ་གས་འདྲ་བ་དང་། །

མདོག་དམར་ཡང་ཞིང་ཐང་བག་ཅན། །

795 མིག་གི་འབྲས་བུ་ལྗང་མདོག་སྲུམ། །

གསལ་ཞིང་ཐང་བག་ཅན་དང་ནི། །

ལེབ་ཀྲན་རྩི་ཡིས་བསྐུ་བར་བྱ། །

ཀེ་ཀེ་རུ་ལྟར་ཐང་བག་བྱ། །

ཧྲ་གས་དབྱིབས་འདྲ་མདོག་དམར་ཞིང་། །

800 མདོག་སྲུམ་སྟེ་མེད་འཚམ་པར་བྱ། །

ནས་ཀྱི་འབྲིང་བས་གང་བ་དང་། །

ཚིགས་རྣམས་འབྱོར་ཞིང་ཤིན་དུ་གྲིམས། །

སོར་མོ་དགའ་ནི་ཚིགས་གསུམ་སྟེ། །

མཐེ་བོང་ནི་ལྟར་སྐྱིམས་ནས་བྱ། །

805 མཐེ་བོང་དང་ནི་མཛུབ་མོ་ནི་བར། །

ཞིང་ནི་སོར་གྱི་ཕྱེད་དུ་བྱ། །

སོང་བུའི་ཕོག་ནི་ཀང་པར་བ་འདྲ། །

འཕང་དུ་སོར་ནི་བཞི་ཡིན་ནོ། །

འདིའི་ཀང་པའི་མཚུ་ཞིང་གི །

---

¹ A und B: མེ་ཡི་   ² A und B: སྐུ་   ³ A und B: སྐྲ་

⁴ A: ཚིགས་   ⁵ A und B: ཐེ་   ⁶ B: མཛུབ་   ⁷ B: པས་

810 མཚན་ཉིད་བདག་གིས་བཤད་པ་ཡིན་ ༎

མི་ཡི་བདག་པོ་ལག་པ་ཡི །
མཚན་ཉིད་དག་ནི་ཉན་པར་གྱིས །
ལག་པའི་མཐིལ་ལ་སོར་བདུན་ཏེ །
ཞིང་དུ་སོར་ནི་ལྟར་བྱུའི ༎

815 གུང་མོ་ཡི་ནི་སོར་མོ་ལ །
སོར་ལྟ་དག་གི་ཚད་དུ་བྱ །
མཐེབ་མོ་མེན་མོ་འི་ཕྱེད་དུ་ཕྱིན །
སྲིན་ལག་ཀུང་ནི་དེའི་ཚད་དོ ༎

མཐེའུ་ཆུང་ཡང་ནི་ཐུང་དུར་བྱ །
820 མཐེ་བོང་ལ་ནི་སོར་བཞིར་བསྟན །
མཐེ་བོང་ལ་ནི་ཚིགས་གཉིས་ཏེ །
ནས་འཐེང་དགང་ཞིང་འབར་འབྱུར་མེད །
མཐེ་བོང་དྲུང་གི་ཉེད་ ཅ་ལ །
སོར་མོ་དག་ནི་གསུམ་དུ་བཤད །

825 མཐེ་བོ་འི་རྩེ་ཡི་ཚད་དག་ནི །
ནས་དགུ་དགུ་དུ་བྱ་བ་ཡིན །
རྩེ་ཡི་ཞིང་དུ་ནས་བཅུད་དོ ༎
སྲིད་དུ་ནས་ནི་དགུ་ཡིན་ནོ ༎

_____

¹ **A und B:** མཐེ་   ² **A:** སྲིད་, **B:** སྲི་

མཁེ་པོ་ཡི་ནི་བཞི་ཆ་ནས །

830 དུ་བའི་རིམ་པས་འཁེལ་བར་བྱ །

སེན་མོ་མདོག་དམར་མང་ཞིང་སྤུག །

ཅུ་ཕྱིས་འདུ་བར་ཆོ་མཆར་བྱ །

མཁེ་པོ་འི་དུང་གི་ནད་་ཁ་ཡེ །

ཞིང་དུ་སོར་ནི་གསུམ་ཞེས་་བྱ །

835 མཁྲིག་མའི་དུང་ནས་སྤྱིད་དུ་ནི །

སོར་མོ་བདུན་དུ་བྱ་བ་ཡིན །

འདི་ནི་མཁེ་པོ་ནི་སྲིད་དང་ཞིང་ །

ཆད་དུ་བྱ་བ་བསྟན་པ་ཡིན ༎

སོར་མོ་རྣམས་ཀྱི་སེན་མོ་ནི །

840 ཚིགས་ཀྱི་ཕྱེད་ཀྱི་ཆད་ནས་བྱ །

ཅི་ཕྱུ་་ཀེད་པ་ནས་འདུ་དང་ །

སྐྱིན་མེད་ཚིགས་ནི་འཁྱོར་པ་དང་ །

ཚིགས་སྤུག་རྐུལ་ཞིང་རིང་བ་དང་ །

རྒྱལ་པོ་སོར་མོ་རྣམས་སུ་བཤད །

845 ལག་པ་ཡི་ནི་མཁྲིལ་་དག་ཀྱང་ །

བཙ་དམར་པོ་འདུ་བར་བྱ །

---

1 A und B: ནི་   2 B: ཤེས་   3 A und B: འཕུ་   4 B: འཁྲིལ་

རེ་མོ་ཊབ་ཙིང་ཡ་ལོ་མེད །

མཐའ་གཉིས་ཊབ་ཆུང་ཕྱུ་བ་དང༌ །

རེ་བོང་ཁྲག་གིས་མདོག་འདུ་བར །

850 རེ་མོ་གསུམ་གྱིས་རྣམ་མཛེས་བྱུ །

དཔལ་བེའུ་གཡུང་དྲུང་འཁྱིལ་པ་དང༌ །

འཁོར་ལོ་ནི་མཚན་ཉིད་བཀྱུས་པ་དང༌ །

ཤིང་བལ་འདའ་བ་རེག་པ་དང༌ །

དར་སྐུད་ཆུན་པོ་ལྡར་འཛམ་བྱུ །

855 ཀུན་ནས་ཁ་ནི་རྒྱས་པ་དང༌ །

རྩ་སྡུང་བ་དང་མི་མཛིན་དང༌ །

བོལ་མཐོ་མཐིལ་ནི་ག་ཤིང་བ་དང༌ །

སོར་མོ་ནི་དུ་བ་སྲུབ་ཆིང་མཛེས །

སེན་མོ་ནི་ཕྱོ་བ་ཨྱུ་རྒྱལ་དམར །

860 ཀྱུ་ཨི་རྒྱལ་པོ་གནིངས་ཀ་འདྲ །

སེན་མོ་དོས་ནི་འཛམ་པ་དང༌ །

སྲུབ་ཆིང་མདོག་སྐྱམ་དཀ་པ་དང༌ །

མདོག་དམར་ཡངས་ཤིང་འཚེར་བ་དང༌ །

མཐོ་ཞིང་གསལ་བས་ལག་མཛེས་བྱུ །

865 ལག་པའི་མཚུ་དང་ཞིང་དག་གི །

མཚན་ཉིད་རབ་ཏུ་བསྟན་པ་ཨིན ༎

དཔུང་པ་ཨེ་ནི་སྟིད་དང་ཞིང་། །

རྒྱལ་པོ་བདག་ལ་མཆན་པར་གྱིས། །

ལག་པའི་སྟིད་དུ་སོར་མོ་དག །

870 སྐུམ་ཚུ་རྩ་དྲུག་དགུ་བ་འདད། །

དཔུང་པ་ལ་ནི་སོར་བཙོ་བཅུད། །

ལག་ངར་ལ་ཡང་དེ་བཞིན་ནོ། །

དཔུང་མགོ་ལ་ནི་སོར་དྲུག་སྟེ། །

དཔུང་པའི་ཞིང་ལ་སོར་ལྔར་བུ། །

875 མཁྲིག་མ་ལ་ནི་སོར་གསུམ་མོ། །

ལག་ངར་གཉིས་ལ་སོར་བཞིན། །

ཕྱག་པའི་ཕྱོགས་ནི་ཡངས་པ་དང་། །

བྱིན་གྱིས་རྐྱམ་པོ་དག་ཏུ་བུ། །

ཞིང་དག་ཏུ་ནི་སོར་བཙོ་བཅུད། །

880 རྩ་དང་ཚིགས་ནི་མི་མངོན་བུ། །

ལག་པ་གཏོགས་པར་སྟིད་དུ་ནི། །

བཞི་བཅུ་རྩ་བཅུད་དག་བུ་སྟེ། །

ལག་པ་དག་གིས་ཕུས་མོ་ལ། །

སྐྱིབས་པ་དག་ཏུ་རིང་བ་དང་། །

885 ཀུན་ནས་ཕ་ནི་རྒྱས་པ་དང་། །

བྱིན་གྱིས་ཕོ་ཞིང་ཕྱུག་ལེགས་བུ། །

མི་ཨི་དབང་ཕྱུག་དཔུང་པ་ནི། །

བ་སྐྱང་མཇུག་ལྱེར་བྱིན་ཀྱིས་ཕུ །

གང་ཚོ་དྲང་པོར་བཞིང་བ་ན །

890 ལག་པ་དག་གིས་ཕུས་མོ་རིག །

དེ་ཕྱིར་རྒྱལ་པོ་ལག་པ་ཡིས །

ཕུས་མོ་སྣེབས་པ་ཞེས་བཤད་དོ །།

འདི་ནི་དཔུང་པ་ལག་པ་ཨེ' །

ཕྱེད་ཞིང་མཚན་ཉིད་བསྟན་པ་ཨེན །།

895 དངུང་དུ་ནི་མ་བསྟན་པ །

གང་ཨེན་ཁྱོད་ལ་བའད་པར་བྱ །

སྨ་ཡི་མཚམས་ནས་སྐྱག་ཙུས་པར །

ཚད་ནི་'ཚོར་མོ་དྲུག་ཨེན་ནོ །།

རྒྱབ་ཀྱི་ཕོགས་པ་གཉིས་ལ་ནི །

900 ཕོར་མོ་བཅུ་དྲུག་དག་དུ་འདོད །

ཕོགས་པའི་བར་ཀྱི་ཕྱེད་དུ་ནི །

ཚད་ཤེས་རྣམས་ཀྱི་ཕོར་བཅུ་སྟེ །

ཞིང་དུ་ཕོར་ནི་དག་ཨེན་པར །

རྣམ་པར་ཕྱེ་ནས་ཡོངས་སུ་བསྟན །

905 ཕོགས་པའི་བར་ནི་རྒྱུ་ཞིབ་དབྱིབས །

འདུན་ཞིང་ཤེན་དུ་སྨུག་པར་བྱ །

རྒྱབ་ཀྱི་ཤུལ་ ¹ ལ་སོར་རྡུག་གོ ༎

རོ་སྟོད་གཞུང་ལ་སོར་ཏེ་ཤུ །

གཞུང་གི་ཞིང་དུ་སོར་གཉིས་ཏེ །

910 ལྱང་བུ་ཚན་དང་སྱུག་གུར་བུ །

སྐྱེས་པ་ ² དེ་ལྱར་ཤེས་པར་བུ །

བུད་མེད་ལ་ནི་ལྱང་ ³ བུ་ཆུང་ །

དཔྱི་ཡི་བར་ན་འདུག་པ་ཡི །

སྐུ་དོ་གུ་བཞི་སྐྱམ་པ་སྟེ །

915 དམར་ཞིང་ཚད་དུ་བྱ་བ་ཡང་ །

སོར་མོ་དག་ནི་རྡུག་དུ་བཕད །

སྐུ་དོ་ནས་ནི་སྲུ་རྱར་འཚམས །

ཞིང་དུ་སོར་མོ་བཞི་ཡིན་ནོ ༎

ཀྲུབ་ཀྱི་ལྱང་བུ་སོར་གཉིས་སོ ༎

920 ཀྲུབ་ཚོས་ ⁴ སོར་བརྒྱད་མཛེས་པ་དང་ །

དུ་ཅང་མི་ཉུམ་མི་འབུར་དང་ །

ལེགས་པར་སྐྱམ་ཞིང་ཤེན་དུ་མཛེས །

ཀྲུབ་ཚོས་ཞིང་དུ་སོར་བདུན་གྱི །

ཚད་དུ་བྱ་བ་བསྟན་པ་ཡིན །

925 འདི་ནི་མཆུ་དང་ཞིང་དག་གི །

མཚན་ཉིད་དག་དུ་བསྟན་པ་ཡིན ༎

_____

¹ A und B: གཤུལ་   ² A: པའི་   ³ B: ལོང་   ⁴ B: ཙོས་

དུད་དུ་ཡང་དེ་བ་ཤད་ཀྱིས། །

དེ་ནི་མ་ཡིངས་གཞུང་༡བར་ཀྱིས། །

སོ་དང་སྐྱ་དང་སྨུ་རྣམས་ཀྱི། །

930 སྐྱེ་བའི་ཐབས་དང་ཁ་དོག་གི །

མཚན་ཉིད་སྣ་ཡི་གདོང་དགག་གི །

མེད་པ་འང་བ་ཤད་པར་བྱ་ཡིས་ཉིན ༎

མཉམ་ཞིང་ཐགས་ནི་དམ་པ་དང་། །

མདོག་སྩམ་གཙང་༢ཞིང་རྣོ་དགར་བ། །

935 སྐྱི་ཏིག་བ་ནོ་པ་བཞིའི་ཙ། །

ཁ་བའི་ཕུང་པོ་སྐྱར་དགར་བ། །

བཞི་བཅུ་ཡོད་ཅིང་སོ་བཟང་ལ། །

མཆེ་བ་རྣམས་ཀྱིས་ལེགས་པར་མཛེས། །

སྙིད་དུ་ནས་གསུམ་ཤེས་པར་བྱ། །

940 དེ་བཞིན་ཞིང་དུ་ནས་གཉིས་སོ ༎

སོ་ཡི་སྐྱིལ་དང་ཀན་དང་ནི། །

ཐྱེ་ཡི་མཚམས་ནི་དམར་པོར་བྱ། །

མཆེ་བ་ནས་ཀྱི་ཐྱེད་ཀྱི་བསྐྱེད། །

སྣ་མའི་མེ་དོག་ཁ་འབུས་འདྲ། །

945 པདྨའི་ཙ་ལྟར་འཛམ་པ་དང་། །

རྗེ་མོ་རྡོ་ལ་ཤེན་དུ་རྩུམ། །

མཆེ་བ་ཤེས་ཀྱི་སང་ཚེར་བུ། །

ལྗེ་ནི་པདྨའི་འདབ་འདྲ་ཞིང་། །

སྒྲོག་འགྱུ་བ་ལྟར་གསལ་བ་དང་། །

950 མདོག་སྔམ་ཁྲག་ནི་འདྲ་བར་དུ། །

ཨ་རྩའི་འདབ་མ་གཞིན་ནུ་འདྲ། །

གདོང་ལ་ཀུན་དུ་ཁེབས་པ་ཡིན། །

སྐྲ་ཀྱི་གདངས་ནི་སྣར་པ་དང་། །

སྣང་ཆེན་རྒྱགས་པའི་འདྲ་བ་དང་། །

955 ཏེ་ཨི་རྒྱལ་པོ་ནི་སྐྲད་འབྱིན་འདྲ། །

འབྱུག་སྨྲ་སྒྲོགས་ལྷར་སྐྲད་བརྗོད་དོ། །

མཆུ་རིས་འདྲ་བར་ཕུ་བ་དང་། །

རང་བཞིན་ཀྱིས་ནི་སྣམ་བག་རྒྱན། །

ལྷུག་ཅིང་ཡ་ཛ་རྐྱལ་དང་འདྲ། །

960 ལྗང་མདོག་མིག་སྐྱན་འདྲ་བ་དང་། །

མེ་ཕུའི་མགྱིན་པ་འདྲ་བ་དང་། །

ཁ་ལྷུག་གི་ནི་མཐོང་ཀ་འདྲ། །

ཅི་ལ་གནས་པའི་མདོག་ལྟར་འཆོར། །

---

1 A: ཟང་ B: བཟང་ཚེར་   2 B: ན་   3 B: ཕུག་པ་   4 A und B: ཕྱིང་
5 A und B: ལེ་

ཤོགས་པའི་སྟེང་དུ་སྐྱེབས་པ་དང་། །

965 འཁྱིལ་བག་ཅན་ལ་གཡས་སུ་འཁྱིལ། །

ཆུ་རིས་མེང་གིའི་རལ་པ་འདྲ། །

ཕྱིར་ཚུགས་རལ་པ་གཡོན་དུ་འཁྱིལ། །

འདི་དག་སྨ་ཡི་རྣམ་པ་ཡིན ༎

བདག་ཉིད་ཆེན་པོ་འི་ཕྱིན་པ་དང་། །

970 བཙ་དང་འོ་རྡོམས་དང་མཆན་ཁུང་དང་། །

རྣ་བ་སྨ་ཡི་བུ་ག་དང་། །

འགུལ་པ་ཁ་དང་འགྲམ་པ་ལ། །

སྤུ་མེད་གྱེན་ཕྱོགས་འཇམ་པ་དང་། །

ཕ་སྤུམ་མིག་སྨན་སྤྱན་པོ་འདུ། །

975 ཆུ་རིས་འདུ་བར་སྤུག་པ་དང་། །

རྣམ་པོ་དག་དུ་འཁྱིལ་བ་ཡི། །

བ་སྤུ་རྣམས་ཀྱིས་རྒྱལ་པོ་ཡི། །

མཐོང་ཀ་མཛེས་བྱས་ཡིད་འོང་དུ། །

མི་རྣམས་ཡིན་གྱི་སྨ་འབྲེལོ་དང་། །

980 ཁ་ལ་ཁ་སྤུ་བ་སྤུ་མེད། །

ལུས་ལ་སྤུ་ནི་མེད་པ་དང་། །

བཅུ་དྲུག་ལོ་ལོན་འདྲར་ཤེས་བྱ། །

<hr />

[1] A und B: ཕྱོང་

མི་ཡི་བདག་པོ་ལྷ་རྣམས་ནི། །

སྐུ་ཚོགས་ཕུ་ཞིང་འཕྲིལ་[1] བག་ཆན། །

985 མཐོན་མཐིང་གིས་ནི་རྣམ་པར་མཛེས། །

སེམས་ཅན་ཀུན་གྱི་ཡིད་འོང་སྟེ ༎

འཛམ་[2] བུ་ཆུ་བོ་འི་གསེར་ལུན་འདྲ། །

པད་རྒྱས་པའི་སྤུབས་འོབས་ཅན། །

ཙམ་པ་ཀ་མདོག་སྟེལ་[3] འོད་ཅན། །

990 སྐྱེས་མཆོག་འཁོར་ལོ་སྐྱུར་བ་དུ། །

སྨྲང་ཆེན་རྒྱལ་པོ་ཁྱུ་མཆོག་དབང་། །

ངང་པ་སེང་གེ་རྣམས་ཀྱི་དང་། །

འཁོར་ལོ་བསྐྱུར་བའི་འགྲོས་འདྲ་བར། །

མི་ཡི་རྒྱལ་པོ་ཤེས་པར་གྱིས། །

995 སྨྲང་ཆེན་རྒྱལ་པོ་དི་གོ་སྟུབས་ཆེ། །

ཁྱུ་མཆོག་དབང་པོའི་རྟེན་བག་ཆན། །

སེང་གེ་རྒྱལ་པོ་བཞིན་བག་ཆན། །

ངང་པའི་རྒྱལ་པོ་ཛོམ་[4] བཛིད་ཅན། །

མི་ཡི་དབང་ཕྱུག་ཆས་འདུ་ཞིང་། །

1000 འགྲོས་རྣམས་ཐམས་ཅད་སྟེལ་གཅན་ནས། །

བློས་[5] གར་དག་ནི་སྟོན་པ་འདུ། །

སེམས་ཅན་རྣམས་ལ་སྤྱོབ་པ་འདུ། །

¹ B: འཕྲིལ ² B: མཛོམ ³ A: སྟེལ ⁴ A: ཛོམ ⁵ A und B: བློ

འཁོར་ལོ་སྒྱུར་བའི་ཤ་མདོག་ནི། །

དུལ་དང་རྡི་མས་མི་གོས་ཤིང་། །

1005 སྒྲུབ་ལ་མདོག་རྣམ་དྲི་ཞིམ་དང་། །

མེ་ལོང་ལྟར་རེག་འཇམ་པ་ཡིན། །

ནོར་བུ་འབར་བའི་འོད་ཅན་དག །

གོས་སྒྲུབ་དཀར་པོར་བཞག་པ་ལྟར། །

འཁོར་ལོ་སྒྱུར་བའི་རྒྱལ་པོ་ནི། །

1010 ཀུན་ནས་འོད་ཀྱི་འཁོར་བར་བྱེ ། །

བླ་བ་སྤྲིན་བྲལ་ལྟག་པར་མཚེས་གྱུར་ཅིང་། །

གཙུགས་ནི་འོད་ཀྱི་སྐོར་བ་ཅེ་འདུ་བར། །

ས་བདག་འཁོར་ལོ་སྒྱུར་བའང་མཚེས་པའི་འོད། །

དཀྱིལ་འཁོར་ལུས་ལ་གནས་པའང་ཌེས་པར་བྱ། །

1015 གདོང་ཡང་བླ་བ་འདུ་བའི་འོད་ལྟར་དཀར། །

མི་ཡི་འཇིག་རྟེན་བླ་དགོང་བླ་གཉིས་འདུ། །

སྤྲིན་ལེགས་མགལ་ལེགས་དབུལ་བ་གདོང་པར་ལེགས། །

སྐུ་མདོག་ལེགས་རྣམ་འཇམ་ཞིང་ཇེ་མོ་འཕྱིལ། །

སྐུ་མཐོ་དང་ཞིང་མཚུ་ནི་ཁྲག་ལྟར་སྨུ ། །

1020 པོ་ནི་དག་ཅིང་མུ་ཏིག་གསར་པ་འདུ། །

རྫི་མའི་མདོག་རྣམ་མཆོག་གི་འདུ་ཞིང་རིང་། །

¹ A und B: པར་

མཐོན་ཕྱིང་¹གསལ་བར་མིག་ཀྱང་དཀྱུས་རིང་མཛེས། །

མཛོད་སྤུ་འབར་བ་སྟེན་པའི་བར་གནས་པ། །

དཀར་པོ་ལུས་ནི་རྣམ་པར་མཛེས་པར་བྱ། །

1025 ནུ་བ་བཀྲ་ཀྱིས་པའི་ལྲུབས་མདོག་འདྲ། །

དེ་ཡི་ནུ་ག་ཤལ་ཕྱིན་ཀྱིས་ཕྲོ་²བར་བྱ། །

མགྲིན་པ་དུང་འདྲ་ཕྱལ་གོང་ཀྲུས་³པ་ཡིན། །

ཕྲག་པ་ལེགས་འབྲེལ་ཡན་ལག་ཉེས་པར་སྣུམ། །

མཐོང་ཀ་ཤ་ཀྲུས་མཆུ་དང་ཞིང་དུ་གབ། །

1030 སེང་གེའི་ཀྲལ་པོ་ལྟོ་ཕྲ་ཕྱིན་ཀྱིས་རྫོང་། །

ཀེང་པ་ལེགས་སྣུམ་ལུས་ནི་ཤ་ཀྲུས་ཤིང་། །

ཕྱེ་བ་གཡས་སུ་འཁྱིལ་བ་ཟབ་པ་ཡིན། །

སྐྱང་ཆེན་ཀྲལ་འདྲར་འདྲོམས་ཀྱི་སྲ་བ་སྤུབས་སུ་ཉུབ། །

ཀུན་ནས་ལེགས་སྣུམ་ཚིགས་ཀྱང་མི་མཛེས་ཡིན། །

1035 བཀྲ་ནི་སྐྱང་ཆེན་སྲ་འདྲ་ཕྱིན་ཀྱིས་སྣུམ། །

བཀྲ་གདོང་མཐོ་ཞིང་ལོང་བུ་མི་མཛེན་ལ། །

སོར་མོ་ལེགས་འབྲེལ་མཛེས་ཤིང་ཀྲང་པ་དག །

ཕོལ་ནི་རྣས་སྣལ་འདྲ་བར་ཤ་ཀྲུས་དེ། །

ཏེ་མོ་སྣ་གས་འདྲ་བའི་སེན་མོས་སྤྲ། །

1040 ཀྲང་འཐམ་འཁོར་ལོ་འི་མཚན་ལྡན་སྤྲ་ལ་ལྟུང་། །

ལག་འདར་སྣུམ་ཞིང་ཀྱང་⁴ན་རིང་བ་དང་། །

ཁྱུ་མཆོག་མཐུག་ ༷ མ་འདྲ་བར་ཕྲིན་གྱིས་ཕུ །

སོར་མོ་རིང་ཞིང་ལེགས་རྒྱས་ལག་མཐིལ་གཡོང་ །

སེན་མོ་འི་འོད་ཀྱིས་གསལ་བྱས་འཁོར་ལོས་རྒྱན །

1045 དབྱིབས་ལེགས་ཚམ་པའི་ཕྱང་པོ་འདྲ་བར་དགར །

ངང་སྤྱུར་འཁོར་ལོ་སྐྱུར་བ་བཟིན་འགྲོར་བའད །

དེ་ལྟར་འདི་ནི་འཁོར་ལོ་སྐྱུར་བ་ཨེ །

མཆུ་ཞིང་མཐའན་དག་གི་ནི་ཚད་དུ་བའད །

ཇི་ངར་བརྒྱུས་ལུས་ནི་ཞིན་དུ་རྒྱུས །

1050 ཕྱུག་དང་ཞབས་ལེགས་ཚོགས་རྣམས་མི་མཚོན་ནོ ॥

རྒྱབ་དང་རྫིབ་ལོགས་ལེགས་ཞིང་ལག་པ་ཨེ །

ཕྱག་ལེགས་གཡོང་ལེགས་ལུས་ཀྱང་གྱ་བཞི་ལེགས །

ཞིན་དུ་ཕ་རྒྱུས་ཡན་ལག་ཐམས་ཅད་མཛོས །

ལུས་ཀུན་མདོག་སྤྱུག་མཐིལ་ ༷ བག་ཆན་ཡང་ཨེན །

1055 ཨོན་དན་སྤྱུག་པ་རྒྱ་ཆེན་དག་སྤུན་ཞིང་ །

ངང་པ་ལྟར་འགྲོ་འཁོར་ལོ་སྐྱུར་བར་བའད ॥

མཁས་པར་གྲུགས་པ་དག་གིས་ནི །

མཛོར་ནེ་མི་ལ་ཡན་ལག་དུ །

མགོ་བོ་མགལ་པ་ལག་པ་གཉིས །

1060 བརྒྱ་དང་རྗེ་ངར་ཨེན་པར་བའད །

འདི་སྐད་བསྟན་པ་གང་ཡིན་ནས། །

ཀུན་ལ་རང་གིས་སྲོར་མོས་ཡིན། །

ཞིང་ལག་ཆད་ནི་གང་ཡིན་ནས། །

ཆད་དུ་བྱ་བ་དང་བཤད་པ། །

1065 ས་གཞི་མཚམས་དང་རྒྱ་དང་ནི། །

ཞིང་དང་སྦྱོམས་ལ་སོགས་པ་ཡི། །

ཆད་དང་ཚུལ་གྱིས་ཞིང་ལག་དང་། །

ལེགས་པ་ཕུན་སུམ་ཚོགས་པར་བྱི། །

དེ་ཕྱིར་ཐོག་མར་ཁ་བས་རྣམས་ཀྱིས། །

1070 ཆད་ནི་འབད་དེ་ཤེས་པར་བྱ། །

ཆད་དང་གནས་ཐབས་མི་སྐྱེན་གཟུགས། །

ལྷ་རྣམས་ཀྱིས་ནི་རྣམ་སྤྱང་ཤིང་། །

ཁ་ཟ་འབྱུང་པོ་སྤྱིན་པོ་རྣམས། །

དེ་དག་ལ་ནི་གནས་བྱེད་ཅིང་། །

1075 སྒྱུར་དུ་མི་ཤེས་འཛིགས་པ་འབྱུང་། །

ལེགས་པ་རྣམས་ཀྱི་བགེགས་ཀྱང་བྱེད། །

གཙོ་བོ་མཆུ་དང་ཞིང་དག་གི །

མཚན་ཉིད་རབ་ཏུ་བསྟན་པ་ཡིན། །

རྒྱལ་པོ་བཞི་པོ་དག་གིས་ནི། །

༡༠༨༠ རང་གི་བློ་ཨིས་རྣམ་ཕྱེ་ཤིག །

སྟོབས་ཅན་དང་ནི་སྣང་བྱེད་དང་། །

རམ་ཤིང་དུ་བཅུ་པའི་བུ། །

མེ་རོ་དེ་བུ་ནི་ཚད་བཞིན་དུ། །

ཚད་ཤེས་དགའ་གིས་བྱ་བ་སྟེ། །

༡༠༨༥ བཟང་པོ་ལ་ནི་སོར་གཉིས་དབྲེ། །

མ་ལ་བར་སྐྱེས་སོར་བཞིན༔

གསལ་བྱེད་སོར་ནི་བརྒྱད་ཀྱིས་དབྲེ། །

རེ་པོ་འཚོ་ནི་སོར་བཅུས་སོ༔

ཇི་ལྟར་མི་སྣག་མི་འགྱུར་བར། །

༡༠༩༠ རང་གི་བློ་ཨིས་ཚད་བྱི་བུ། །

ལུས་ལ་མཇེས་པ་མ་ཚང་ན། །

ཚད་དགའ་གིས་ཉི་ཙེ་ཞིག་བྱ། །

རྒྱལ་པོ་དག་ནི་བཞི་རྣམས་ཀྱི། །

ཚད་ཀྱི་མཚན་ཉིད་སོ་སོ་སྟེ། །

༡༠༩༥ ཚངས་པས་བདག་ལ་བསྟན་པ་བཞིན། །

གལ་ཏེ་ཁྱོད་ལ་བའད་པ་ན། །

རེས་ནི་གཞུང་ནི་མཛེས་པ་དང་། །

མཚན་ཉིད་ཕུན་སུམ་ཚོགས་ཐོབ་ན། །

སྐྱེ་པོ་བློ་ནི་ཆུང་བ་དག །

¹ B, མི་

༑༑༠༠ མི་ཕོས་པས་ཀྱང་སྨུག་པར་འགྱུར །

བྱུང་མེད་སྐྱེས་པའི་མཚན་ཉིད་ཀྱི །

བསྟན་བཅོས་སྟོང་ཕྲག་བཅུ་གཉིས་ཏེ །

རྣམ་པ་ལྔ་ཡང་ཚངས་པ་ཡིས །

རང་ཉིད་ཀྱིས་ནི་བསྟན་པ་ཡིན །

༑༑༠༥ མི་ཨི་མཚན་ཉིད་སྟན་པ་ཨི །

རྣམ་པ་ལྔ་པོ་བསྟན་དེ་དག །

དེ་ཁྱོད་གལ་ཏེ་བསྟན་བྱས་ན །

དེས་ནི་གཞུང་ནི་མངས་གྱུར་བསྟན །

མི་ཨི་རྒྱལ་པོ་ཁྱོད་ལ་ནི །

༑༑༡༠ བསྟན་པ་དེས་ནི་ཅི་ཞིག་བྱ །

འཁོར་ལོ་སྒྱུར་བའི་ཆད་བསྟན་པ །

མི་ཨི་སྐྱེས་མཆོག་ལྟ་དང་ནི །

ཐམས་ཅད་དག་གི་ཆད་དག་གི །

མཚན་ཉིད་རབ་ཏུ་བསྟན་ཡིན་པས །

༑༑༡༥ རང་གི་བློ་ཨི་ནུས་པ་ཡིས །

ཅེས་བྱས་བཟང་ལ་སོགས་བཞི་ཐི །

འདི་ལ་བཟང་པོ་ནི་སྒྲིད་དུ་ནི །

བཀྲ་དང་ཙ་ནི་རྣུག་ཏུ་བྱ །

གསལ་བྱེད་ལ་ནི་བཀྲུར་ཞེས་བྱ །

༑༑༢༠ མ་ལ་བར་སྐྱེས་བཞི་ཨི་སྐྱེད །

རི་བོ་འཚང་གི་སྟེང་དུ་ནི། །

སོར་མོ་དགུ་བཅུ་རྩ་བཅུར་བུ། །

དེ་ལས་གཞན་པའི་ཚད་ཀྱི་ཚུལ། །

མ་བྱུང་འབྱུང་བར་མི་འགྱུར་རོ། །

1125 མི་ཡི་དཔའ་ཕྱུག་གཙོ་བོ་དང་། །

འབྲི་དང་ཐ་མ་དག་གི་གཟུགས། །

སྙིད་དང་ཞིང་དུ་བུ་བ་དང་། །

ཐམས་ཅད་བསླས་ཏེ་བསླས་པ་ཡིན། །

བུད་མེད་བཞིན་དུ་སྐྱེས་པ་ལའང་། །

1130 དེ་ལྟར་རང་གི་བློས་སྤྱད་ནས། །

རྒྱལ་པོ་འི་སྙིད་དུའང་སོར་རོ་དང་། །

སོར་དེ་དག་གིས་དབྱེ་བར་བྱ། །

མི་གྱིར་པོ་མགོ་བོ་བླེ་བ་དང་། །

བྱང་དང་ཀེད་པའི་བར་དངེ། །

1135 བཅུ་དང་རྗེ་ངར་ཕྱུས་མོ་དང་། །

ཀང་པ་གཉིས་ཀྱི་སྙིད་དག་དང་། །

མི་ཕྱུག་པ་ནི་མི་འགྱུར་བར། །

ཚད་ཀྱི་ཚུལ་གྱིས་དབྱེ་བར་བྱ། །

གཙོ་བོ་བར་མ་ཐ་མ་ཡི། །

---

¹ Wiederhergestellt nach V. 1088. A und B: རི་བོང་མཚོག; vielleicht ist རི་བོ་མཚོག in beiden Fällen korrekt.   ² A: བྱུང   ³ B: བས

1140 ཆད་དུ་བྱ་བ་གོ་རིམ་བཞིན།

མཚུ་དང་ཞིང་ནི་རབ་ཏུ་བསྟན།།

བྱད་མེད་ཆད་ཚུལ་མཚན་པར་གྱིས།

རྒྱལ་པོས་རྗེ་སྐྱེད་བསྟན་འདི་རང་གི་བློས་སྤྱད་ནས།

ཆད་དང་འཚམ་པར་བྱ་ཞིང་སྒྱོད་པར་མི་འགྱུར་དང་།

1145 གནས་ཐབས་མང་པོ་དང་ནི་ས་གཞི་མཚམས་གཉིས་དང་།

ཤེན་དུ་གཞན་པ་ལ་ཡོ་མེད་པར་རེ་མོ་བྱིས།།

རེ་མོ་འི་མཚན་ཉིད་ལས། ཆད་ཀྱི་ལེའུ་སྟེ་གསུམ་པའོ།། ཆངས་པའི་ཆད་ཡིག་རྗེ་
སྟེད་པ་རྫོགས་སོ།། ༈   ༈

ÜBERSETZUNG DES CITRALAKSHAṆA

DIE MERKMALE DES GEMÄLDES[1].

Auf Sanskrit: *Citralakshaṇam.* Auf Tibetisch: *Ri-moi mts'an-ñid.*

1. Kapitel.

1—9. Dem\ Mahādeva[2], Brahma und Nārāyaṇa[3], der die trefflichsten Gaben spendenden Göttin Sarasvatī bezeige ich innige Verehrung. Möge Sieg und Segen daraus erwachsen! Der Gemahlin des Mahādeva, der Parvatarājaputrī[4], und der Padmāvatī[5] (Verehrung). Darauf mögen, nachdem ich dieses

---

[1] Die von GEORG HUTH vorgeschlagene Übersetzung „Theorie der Malerei" ist nur bedingt richtig. Skr. *citra* und tib. *ri-mo* bedeuten wohl nicht Malerei im abstrakten Sinne als die Malkunst, sondern das konkrete Gemälde. Der Titel lautet zweifellos in genauer Übertragung: die wesentlichen Kennzeichen oder die charakteristischen Merkmale des Gemäldes. *Lakshaṇa* ist hier in demselben Sinne gebraucht wie z. B. in den zweiunddreißig körperlichen Abzeichen (*lakshaṇa*) des Mahāpurusha. Es wird ja auch in dem ganzen Werke keine Theorie vorgetragen, sondern es werden die Abzeichen gelehrt, die der Cakravartin und die anderen großen Figuren in den Werken der Malerei haben sollen (vgl. EINLEITUNG, S. 28).

[2] Tib. *Lha c'en*, ein Beiname Çivas.

[3] Tib. *Sred med bu*, ein Beiname Vishṇus. Die tibetische Übersetzung „der Sohn des Wunschlosen" beruht auf falscher Analyse des Namens.

[4] Tib. *Buṅ-bai* (= *spuṅ-bai*) *ri rgyal-gyi sras-mo*, Patron. der Durgā (*pw*).

[5] Tib. *Žal-gyi padma-can*, „die Lotusgesichtige", Beiname der Lakshmī, (Vishṇus Gemahlin) = Skr. *Vaktrāmbujā.*

gesprochen, alle Weisen siegreich sein! Möge es zum Segen gereichen!

Weiter unten sollen „die Merkmale des Gemäldes" (das Citralakshaṇa) erklärt werden. Dem alle Wissenschaft lehrenden Mahādeva Verehrung!

10—29. Zuerst verneige ich mich vor Candra[1], dann vor Mahādeva, der die Mondsichel als Scheitelschmuck trägt, vor Vishṇu und Indra, vor Sūrya und Varuṇa, vor Agni und Pavana[2]. Dann verneige ich mich vor dem Prajāpati-Viçvakarman[3], dann verneige ich mich dem Nagnajit[4] zu Füßen, und vor allen (anderen) Meistern (der Malerei) mit gefalteten Händen. Was die Farben in den Werken der Malerei anbelangt, soll ihre gleichmäßige Anordnung nach meinem Erinnerungsvermögen und meiner Befähigung ein wenig besprochen werden. Wenn ich die Theorien[5] des Viçvakarman, des Pra-

---

[1] Tib. *Zla-ba*, der Mondgott.

[2] Tib. *Ñi c'u lha, me rluṅ lha*, die Götter der Sonne, des Wassers, des Feuers und des Windes.

[3] Tib. *Las kun-pa*, wörtliche Übersetzung des Sanskritnamens Viçvakarman, der alles schaffende, ein weltbildender Genius, ähnlich dem Prajāpati, später Name des Baumeisters und Künstlers der Götter (*pw*). Der Name ist den Tibetern auch in der Transkription des Sanskritwortes bekannt (s. HUTH, Geschichte des Buddhismus in der Mongolei, Band II, S. 89, 359). Tib. *sKye-dgu mña mdsad-pa* ist offenbar Übersetzung von Skr. *prajāpati*, und es ist klar, daß in dieser Stelle Prajāpati, der Schöpfer, mit Viçvakarman identifiziert wird (A. A. MACDONELL, Vedic Mythology, p. 118). Wären sie als zwei verschiedene Götter aufgefaßt, so hätte dies im tibetischen Texte durch *mdsad daṅ* (statt *mdsad-pa*) leicht zum Ausdruck gelangen können.

[4] Tib. *gcer t'ul* „der Bezwinger der Nackten". Die Erklärung des Begriffes wird V. 271 (S. 138) gegeben. Unter Nagnajit haben wir einen alten Künstler zu verstehen, der nackte Götterfiguren malte und wohl als Inkarnation des Viçvakarman gedacht wurde. Näheres siehe S. 4—13; vgl. auch Anmerkung 1, S. 129.

[5] Tib. *mts'an-ñid*. Hier behalte ich der Kürze wegen die Übersetzung „Theorie" bei, da man sonst eine zu schwerfällige Umschreibung wählen

hlāda[1] und des Nagnajit[2] in einer alles zusammenfassenden Übersicht im Auge habe, so habe ich damit, um den Einsichtigen, die sich bemühen wollen, Einsicht zu verschaffen, „die Merkmale des Gemäldes" (*citralakshaṇa*) übersichtlich dargestellt. Daher vernehmet nun von mir das Citralakshaṇa, ihr in der Weisheit der Gelehrten bewanderte!

30—50. Folgendes ist die Vorgeschichte[3] „der Merkmale des Gemäldes". Einst lebte der Männerbeherrscher, der

---

müßte wie „die Merkmale der Malereien, wie sie von ... aufgestellt werden". Man muß natürlich den Gedanken im Auge behalten, daß es sich auch hier nicht um Theorien handelt, sondern um bestimmte Beobachtungen und Regeln.

[1] Tib. *Rab-sim* = Skr. *prahlāda, prahrāda*. Tib. *sim-pa* entspricht auch Skr. *sāta* (S. Lévi, Mahāyāna-Sūtrālamkāra, p. 188). Prahlāda war ein Asura oder Daitya, der Sohn des Hiraṇyakaçipu, der von Vishṇu in der Form des Narasiṁha erschlagen wurde, während Prahlāda ein gläubiger Verehrer des Vishṇu war, der zum Range Indras erhoben und schließlich mit Vishṇu vereinigt wurde (vgl. J. Garrett, Classical Dictionary of India, p. 463, und J. Dowson, A Classical Dictionary of Hindu Mythology, p. 238). Prahlāda gilt im obigen Texte als der Urheber einer Abhandlung über Kunst, die, wie es scheint, als von Vishṇu emanierend gedacht wird. Viçvakarman hat, wie wir am Schlusse dieses Kapitels erfahren, sein Wissen von der Kunst von Brahma empfangen, und so ist die Vermutung gerechtfertigt, daß der dritte in der obigen Reihe, Nagnajit, zu Çiva in irgendwelchen Beziehungen steht. Im Anfang des Kapitels ruft der Verfasser die drei großen Götter als die geistigen Schöpfer der Kunst an; hier handelt es sich um drei Praktiker, die jeder an ihrem Geiste teilhaben und von ihnen inspiriert sind, wahrscheinlich ausübende Künstler, die gleichzeitig ihre Gedanken über die von ihnen betriebene Malerei niedergelegt haben. Die Stelle ist sehr interessant, denn sie läßt ahnen, daß es bereits in der alten indischen Malerei mindestens drei verschiedene Richtungen oder Schulen gegeben hat, die durch die drei obigen Namen charakterisiert werden und ihre Anregungen aus verschiedenen religiösen Kulten erhalten zu haben scheinen.

[2] Tib. *gcer-bu t'ul byas*, „der die Nackten bezwungen hat".

[3] Tib. *sñon* ist vermutlich in *sñon rabs* = Skr. *purāṇa* zu ergänzen und als Sage der Vorzeit, alte Überlieferung aufzufassen, oder aber vom Standpunkt der Jaina mit Skr. *pūrva* zu identifizieren. — Zur Sache vgl. E. Kuhn, Zu den arischen Anschauungen vom Königtum (*Festschrift Thomsen*, S. 214 u. f.).

9

Erdenträger[1], der verständige[2] und hochberühmte, im wahren Wissen vom Dharma bewanderte, mit Namen rNam-grags aJigs-t'ul[3]. Zu Lebzeiten dieses heiligen Königs wurde die Erde im Einklang mit der Religion beschützt, und die Lebensjahre der Menschen erreichten bekanntlich an Zahl die Hunderttausend. Zu jener Zeit gab es keine Krankheiten, und frühzeitiges Sterben gab es überhaupt nicht. Gemütskrankheiten gab es nicht, und zu ihrer Heilung war keine Veranlassung, da für Zorn und Leidenschaft kein Raum war[4]. Günstig erhob sich der Wind, und Indra ließ Regen trefflich herabfallen. Same, Wurzeln und Früchte besaßen ein frisches Aussehen und feinen Geschmack. Die vier Kasten richteten noch keine Verwirrung an, sondern jeglicher war auf seine eigene Sitte[5] bedacht. Die Geschöpfe entfalteten sich und wurden des Glücks teilhaftig, sie erlangten Wohlfahrt[6] und Vorzüge aller Art.

51—72. Während sich das Land eines solchen Zustands erfreute, begann der König, von seinem eigenen hohen Sinne angetrieben[7], eine große Askese (*tapas*) und vollzog die höchsten Bußübungen. Infolge dieser Bußübungen wurde jener

---

[1] Skr. *nripati bhūmidhara*.

[2] Tib. *blo dan ldan* = Skr. *dhīmat* oder *buddhimat* (HUTH, Geschichte des Buddhismus in der Mongolei, Band II, S. 282).

[3] D. i. der berühmte, die Furcht bezwingende; Skr. etwa *Prakīrti* (oder *Vikhyāta-)bhayajit*. Tib. *grags-pa* entspricht auch Skr. *yaças* und Tib. *rnam-par grags-pa* Skr. *vighushṭa* (z. B. in der Liste der Tausend Buddhas, Nr. 926).

[4] Wörtlich: wenn man sie hätte heilen wollen oder können, wo waren Zorn und Leidenschaft?

[5] Tib. *c'os* = Skr. *dharma* im brahmanischen Sinne.

[6] Tib. *rdsu ap'rul* = Skr. *riddhi*, Gedeihen, Wohlstand, Glück.

[7] Tib. *bdag-ñid c'en-po-yi yid* = Skr. *ātmamahāmanas*; *mahāmanas*, hohen Sinnes, großgesinnt (*pw*).

König reicher als seine irdischen Ahnherren an den acht Vorzügen und errang von den schwer zu erlangenden¹ Dingen
die trefflichsten. Weder die Waffen der Deva noch die
(menschlichen) Feinde vermochten in das Land der Menschen
zu gelangen, und da alle Çastra vollständig vorhanden waren,
erlangte er den höchsten alles durchdringenden Verstand. Mit
allen Vorzügen geschmückt, ein solcher Beherrscher der Menschen, ein machtvoller Maṇiputra², übertraf er die Götter an
Kraft. Als er, ein trefflicher Behälter der Wissenschaften,
gleichsam eine Verkörperung des Dharma³ geworden war,
kam einst ein Brahmane weinend vor den König. Ihn, den
übelgelaunten, fragte der König, was ihm widerfahren sei.

73—92. Da geriet er in Zorn und sprach zu dem Könige
mit wütenden Worten: „Weshalb ist in deinem Lande frühzeitiges Sterben⁴ entstanden? Sicherlich gleichest du, Men-
schenbeherrscher, einem, der religionslose Geschöpfe beschützt
hat. Daß solches nicht schon früher geschehen ist, wundert
mich; heute ist es nun eingetroffen: vorzeitig verschieden ist
mein Sohn!⁵ Ohne Vorzeichen und körperlichen Schmerz⁶ zu
zeigen, ist der Knabe, der Stammhalter des Geschlechts, der
meinem Herzen wohlgefällige, dahingeschieden, Majestät! Herr,

---

¹ Tib. *sñed = rñed*; *sñed dka-ba* = Skr. *durlabha*.

² Tib. *nor-bui sras-po*, „Sohn des Juwels".

³ Tib. *cos-kyi lus*, was hier nicht etwa im buddhistischen Sinne als
Dharmakāya (= tib. *cos sku*) zu fassen ist, sondern nur so zu verstehen ist,
daß der König durch das Studium der Wissenschaften, die er ganz in sich
aufgenommen hatte (*rten*), den Dharma vollständig beherrschte und also
leibhaftig (*lus*) vertrat.

⁴ Tib. *dus ma yin-par aci* = Skr. *akālamaraṇa*.

⁵ Man beachte die untibetische Konstruktion des Satzes, dessen Wort-
stellung sich eng an die des Sanskrittextes anzuschließen scheint.

⁶ Tib. *ša-gzugs* = *zug*, wie auch CHANDRA DAS (Dictionary, p. 1228a)
richtig anführt.

9*

wenn dir die Brahmanen lieb sind, und wenn du ein All-
wissender[1] bist, so führe meinen Sohn, der mir werter als das
Leben ist[2], zu mir zurück. Wenn der König trotz seiner
großen Zauberkraft mir keine Gnade erweisen will, so werde
ich, obwohl ich heute vor deine Augen getreten bin, mein
Leben wie einen Grashalm[3] wegwerfen."

93—112. Also sprach der Brahmane. Um seinen Sohn
zurückzurufen, versank da jener heilige, hochverständige König
in Nachdenken. Er sprach das Wort: „Ich verlange nach
meinem Diener!" Und kaum war der Brahmane wieder zu
Atem gelangt, da sah er den König Yama[4] gleich der Sonne
aufgehen. Vor diesem, den die Wesen der Verehrung würdig
halten, verneigte sich der König aJigs-tʻul, und auf das Heil
für den Brahmanen bedacht, sprach er andächtig folgende
Worte: „Den Sohn dieses gütigen[5] Brahmanen, der ihm lieber
als das Leben und seinem Herzen wohlgefällig ist, hast du,
ohne daß er den Tod erreicht hat, entrissen und zu deinem
Boten gemacht. Da du dich an den mildtätigen Brahmanen
erfreust, so geruhe, Machthaber über die drei Welten, Beherrscher der Werke der Welt, diesem Brahmanen seinen Sohn
zu gewähren."

---

[1] Skr. *viçvavid* oder *sarvajña*.

[2] Tib. *srog pas cʻes apʻaṅs* (= *dpaṅs?*) scheint Übersetzung einer
Sanskritphrase wie *prāṇasaṁmita, prāṇādhika, prāṇādhikapriya* zu sein.

[3] Tib. *rtsa* = *rtsva*.

[4] Tib. *gŚin-rjei rgyal-po* (Yamarāja), auffallend statt *rgyal-po gŚin-rje*,
wie z. B. V. 262. Der Genitiv ist gebraucht, um die Identität auszudrücken:
der König, welcher Yama ist (vgl. *Yamárājaḥ*: A. A. MACDONELL, Vedic
Mythology, p. 171).

[5] Tib. *legs-ldan* (entspricht vielleicht Skr. *kalyāṇa*). Es kann nicht als
Anrede an Yama aufgefaßt und etwa mit Mahākāla (GRÜNWEDEL, Mythologie
des Buddhismus, S. 177) identifiziert werden. In V. 397 erscheint das Wort
als Attribut des Brahma, und auch in andern buddhistischen Texten findet

113—140. Als er so gesprochen, lächelte den König der König des Gesetzes[1] an, und jenem Opferpriester erwiderte er mit sanfter Rede und würdevoll:[2] „Meine Selbständigkeit ist nur gering, und ihn zurückzurufen oder frei zu lassen vermag ich nicht. Kraft der Vergeltung ihrer eigenen Werke (Karma) werden die Wesen meiner Macht unterworfen. Der Zeit[3] und dem Karma untertan, werden sie Heil und Unheil erfahren. Daher muß man verstehen, daß nicht ich es bin, der König, der die Vergeltung übt. Selbst wenn du zu mir gelangtest, könntest du ebensowenig befreit werden. Da du mit der Zeit mächtig geworden bist, willst du den Sohn des Brahmanen dringend zurückgerufen haben. Von allen Wesen unübertroffen, mußt du doch von guten und schlechten Taten leiden. Der Name[4] ist die Grundlage des Karma, und dieses wird als die Grundlage der Vergeltung erklärt." Als Yama so gesprochen hatte, nahm der Herrscher der Menschen wieder das Wort, und wiederholt rief er dem Yama zu: „Gib ihn doch zurück!" Und Yama sagte zu ihm wieder und wieder: „Es geht nicht, es geht nicht!" Der König sprach: „Gib ihn, gib ihn doch zurück!" Yama sagte beständig: „Es geht nicht!"

141—169. Da sie sich so leidenschaftlich erregten, entspann sich daraus ein Streit, und aus ihrem Streiten entwickelte sich ein großer Kampf. Da sagte der König viele Male zu

---

es sich in Verbindung mit brahmanischen Gestalten. So wird Çālihotra, der angebliche Verfasser des Açvāyurveda, *legs-ldan Çālihotra* genannt (Tanjur, Sūtra, Narthang-Ausgabe, Vol. 130, fol. 137).

[1] Dharmarāja, d. i. Yama in seiner Eigenschaft als Richter der Toten.
[2] Tib. *brjid bag-tu*; wohl = Skr. *tejasvitā* energisches, würdevolles, majestätisches Wesen (*pw*).
[3] Tib. *dus* = Skr. *kāla*; wohl doppelsinnig: Zeit und Tod.
[4] Skr. *nāma*, das Individuum.

Yama: „Halt ein, halt ein!" Während des Kampfes sagte da
Yama: „So will ich denn einhalten!"[1] Darauf ließ jener König
kraft seiner furchtbaren Macht und seines erfinderischen Geistes
zahllose Regengüsse auf Yama kräftig herabregnen. Die einem
furchtbaren Regenstrom aus den Wolken vergleichbaren, die
mit — das Verständnis des Alldurchdringenden (asaṅga) er-
schließenden — Zaubersprüchen geweihten Götterwaffen (deva-
vadha) ließ Yama von allen Seiten freudig[2] herabfallen. Darüber geriet der König in Zorn und erkannte mit seiner großen
Zauberkraft, daß die Götterwaffen die Diener des Yama seien,
und als die Preta (Pfeile) auf ihn schleuderten, da besiegte er
sie alle gründlich mit dem Verderben bereitenden Schwerte,
mit Zangen und mit der Schrecken entsendenden Streitaxt.
Da zeigten die Preta, daß sie nicht länger standzuhalten
vermochten[3], und eilten der eine hierhin, der andere dorthin
fort. So waren die Heerführer des Yama geschlagen und
gingen schleunigst nach allen Richtungen.

170—220. Als Yama, der König[4], sah, daß er von dem
Könige im Kampfe besiegt worden war, geriet er in Zorn
und schwang die nach Belieben wirkende Keule[5]. Als der
König ihn die dem Weltbrand[6] gleichende Keule schwingen
sah, geriet er in Zorn und nahm die mit dem Abzeichen des

---

[1] Dieser Vers (148) hat neun Silben statt der regelmäßigen Zahl von
sieben Silben.

[2] So nach der Lesart von **A** (mgur); nach **B**: schleunigst (myur).

[3] Tib. mi bzad-pa = Skr. asaha, asahant, asahishṇu. Tib. gzugs can
= Skr. rūpavant, die Gestalt von — habend, erscheinend als.

[4] Tib. gŚin-rjei rgyal-po, wie oben V. 99.

[5] Die Keule ist ein Attribut des Yama, der infolgedessen der Keulen-
träger, Daṇḍin, Daṇḍapāṇi, genannt wird. Tib. ji-bźin byed-pai dbyug-pa =
Skr. yathākaradaṇḍa?.

[6] Tib. bskal-pai me = Skr. kalpāgni, das alles vernichtende Feuer am
Ende eines Kalpa (pw).

Brahma-Kopfes[1] versehene Waffe in die Hand. Da gerieten
alle Wesen und die Höllenbewohner in Schrecken. Allenthalben wurden die großen Wesen (*mahābhūta*) von großen
bösen Vorzeichen gequält. Als darauf Brahma wahrnahm,
daß alle Geschöpfe bekümmert waren, begab sich Brahma
samt den Göttern an jenen irdischen Ort. Alsdann, da sie
des Brahma ansichtig wurden, streckten[2] sie die Waffen,
falteten die Handflächen und tauschten Rede und Gegenrede
miteinander aus. Als sie nun die Ursache, die zu ihrem Streit
geführt hatte, und alles das vollständig vernommen hatten, da
hielt sie Brahma von weiterem Kampfe zurück und sprach zu
ihnen die Worte: „Den hochherzigen[3] Yama trifft wahrlich
keine Schuld. König der Menschen, den Herrn des Todes
(*mrtyupati*) trifft keine Schuld, die Zeit (*Kāla*) trifft keine
Schuld; die Schuld liegt vielmehr **an** dem eigenen Karma.
Durch die früheren guten und bösen Werke, die jener Knabe
vollbracht hat, hat er die Geburtsform eines Menschen erlangt, und sein Tod ist frühzeitig **eingetreten.** Jedoch, um
den Brahmanen zu ehren, hast du deinen inneren Frieden mit
seinen guten Früchten zerstört. Nun soll ihm meine Huld zuteil werden, und folgendes ist die Kenntnis von diesem Mittel.
Entsprechend seiner Gestalt und mittels der Farben sollst du
den Sohn des Brahmanen in einer ihm ähnlichen Weise trefflich malen[4]; alsdann soll dein Heil gewiß sein." Nachdem
Brahma so[5] gesprochen, sagte der verständige König: „Möge
es mir beschieden sein, jenes Knaben des Brahmanen ansichtig

---

[1] Tib. *T'saṅs-pai mgo* = Skr. *Brahmaçiras* oder *-çīrshan*, eine bestimmte mythische Waffe (*pw*).

[2] Wörtlich: bändigten sie.

[3] Tib. *bdag ñid c'en-po* = Skr. *mahātman*, edel, hochherzig (*pw*).

[4] Tib. *dris* (in **A** und **B**) = *bris.*

[5] Tib. *snaṅ-ste.*

zu werden!" Darauf malte er ihn; Brahma ließ ihn als eben
solchen (wie er gemalt worden war) wieder auferstehen und
schenkte ihn dem Brahmanen als einen Lebenden[1]. Dem
Brahmanen weiteten sich die Augen vor Freude; mit Augen
leuchtend wie Utpala Lotus, verjüngt und mit frischen Farben,
verneigte er sich vor Brahma und nahm den eigenen Sohn
in Empfang.

221—270. Darauf sagte Brahma zu dem Könige: „Daß
du im Kampfe um des Brahmanen willen durch deine Zauberkraft die Boten (des Yama) bezwungen[2] hast, ist ganz vortrefflich." Nachdem Brahma diese Worte gesprochen, wurde
der König erfreut. Aber der alle bändigende[3] Gott, Yama,
war gar nicht im Herzen froh. Da nun Brahma wahrnahm,
daß sich der König Yama nicht freute, sprach er zu dem
Könige der Menschen mit vielen freundlichen Worten, indem
er ruhig Atem holte: „Da der König mit den Formen des
Dharma wohl vertraut ist, so wird er ihm keine Verachtung[4]
zeigen. Sobald die Freude an den Menschen zum Durchbruch
gekommen ist, wird sichtbarer Stolz von selbst gebrochen.
Die Heiligen werden einem Stolzen nicht vertrauen, sondern

---

[1] Das Bild wurde beseelt und als etwas Lebendiges empfunden; es
war der Stellvertreter des Toten und schloß seine Seele ein.

[2] Tib. *brtul-ba*, offenbar = *btul-ba*, *t'ul-ba*, wie denn dieser Text und
auch andere *brtul* statt *btul* schreiben; *brtul-ba* kann an der obigen Stelle
nicht das nominale „Benehmen" sein, sondern läßt sich nur verbal fassen
wegen des vorausgehenden Objekts und des Instrumentalis in V. 223. Und
dann wird ja im folgenden ausgeführt, daß der König, eben weil er die
nackten Preta bezwungen hat, den Titel Nagnajit erhält; es kann daher an
der obigen Identifikation kein Zweifel obwalten.

[3] Tib. *sdom-pa* = Skr. √*yam*; Yama, der Bändiger; das Epitheton gibt
also die Etymologie der nachvedischen Zeit wieder (s. A. A. MACDONELL,
Vedic Mythology, p. 172).

[4] Tib. *sñas* = *brñas*; vgl. V. 252. Wörtlich: *du* wirst ihn nicht
verachten.

ihn sehr schelten und beständig tadeln; auch das geringste Glück wird er nicht erlangen. Die Weisen werden ihm zürnen, denn gegen den Himmel[1] hat er sich vorschnell verfehlt. Von einem Könige wird erwartet, daß er den Stolz gänzlich ablegt, und ein König soll sich daher gerade in beständigem Gleichmut weise zeigen. Wer im Besitze der Schätze des Wissens und der Kraft des Wissens ist, wird sich nicht stolz zeigen. Insbesondere sollen die Könige den Göttern und Brahmanen Ehrfurcht erweisen, Verachtung[2] aber nicht bezeigen, sondern es an den schuldigen Ehrbezeigungen nicht fehlen lassen[3]. Was andere nicht ertragen können, soll man nicht tun noch andere verachten. Der Gebildete wird sich nicht stolz zeigen, und dem Tugendhaften wird der Zorn nicht erregt. Unpassendes soll man nicht sprechen noch jemandem Kränkungen zufügen; die vor den Augen liegenden nutzbringenden Werke soll man tun[4]. Da er sich weder mit den Göttern noch mit dem Brahmanen im Einklang befindet, so möge der König Yama, weil ja darin kein Heil liegt, seine Verstimmung aufgeben." Nachdem Brahma so geraten hatte, verneigte sich der König, der die Partei des Brahmanen genommen hatte, vor Yama und versetzte damit Yama in freudige Stimmung. Yama wurde auch wirklich froh gelaunt, und auch Brahma wurde froh gelaunt. Da empfanden alle Geschöpfe eine ungetrübte Freude.

271—326. Darauf sagte Brahma zum Könige: „Mögest

---

[1] Tib. *mt'o-ris* = Skr. *svarga*.

[2] Tib. *brñas-pai t'abs*; *t'abs* dient hier zur Bildung eines Abstraktums.

[3] Tib. *mgal* = *agal*, entgegenhandeln.

[4] Die Kenner der Nītiçāstra werden sicher in der Lage sein, diese Sprüche auf ihre Quellen zurückzuführen. Boehtlingks „Indische Sprüche" ist leider nicht in meiner Bibliothek.

du die nackten Preta bezwingen![1] Jene Boten[2] des Yama
sollen nicht länger an das Licht der Sonne kommen![3] Da es
nicht passend gewesen wäre, daß deine so mächtigen und
glänzenden Bußübungen zu Ende gegangen wären, ohne sich
anderen Königen mitzuteilen, so habe ich dich, o König, damit du keine Sünde und Untat begehen sollst, (vom Kampfe)
zurückgehalten[4], und du bist in der Welt hochberühmt geworden. Mit dem Terminus[5] der Nackten (*nagna*) verhält es
sich so; er bezieht sich auf die Preta. Wenn alle Nackten,
weil du der vortrefflichste aller Männer[6] bist, zu dir gekommen
sind, wirst du aus diesem Grunde Bezwinger der Nackten
(*Nagnajit*)[7] genannt und durch Opferspenden den höchsten

---

[1] Oder (der Satz ist mit Absicht doppelsinnig): mögest du der Nacktbezwinger (*nagnajit*, d. i. der Maler) der Preta werden, d. h. möchtest du
auch in Zukunft ein Maler bleiben!

[2] Instrumentalis, auffallend als Subjekt mit intransitivem Verbum.

[3] Tib. *rtag-tu ḥgro-ba* kann hier nur als ein Nomen gefaßt werden
und entspricht wohl einem Skr. *sadāgati*, „was in steter Bewegung ist; Wind,
Windgott; die Sonne" (*pw*). Vielleicht kommt hier auch die von MONNIER
WILLIAMS beigebrachte Bedeutung in Frage: *everlasting happiness, final eman-
cipation.* Der Sinn ist wohl der: die Preta, die Geister der Abgeschiedenen,
sollen nicht mehr, wie es hier der Fall gewesen ist, an die Oberwelt kommen,
um die Menschen zu beunruhigen, sondern in der Hölle verbleiben; von
den Menschen sollen sie in Zukunft nur gemalt werden; ihre Bilder sind
ein ausreichender Ersatz für ihre nicht wünschenswerte Anwesenheit auf
der Erde.

[4] Tib. *bzlog*, nämlich *ɤyul-las*, wie in V. 192.

[5] Tib. *rnam-graṅs* = Skr. *paryāya* (CHANDRA DAS).

[6] Tib. *skyes-bu skyes mc'og* = Skr. *purushottama, purushapuṁgava.*

[7] Tib. *gcer t'ul.* Hier liegt also die Idee zugrunde, die aus dem
Buddhismus übernommen die geistige Grundlage der religiösen tibetischen
Malerei bildet und auch in der Geschichte der chinesischen Malerei auftritt,
daß der Künstler Gottheiten und Dämonen bannt und in seine Gewalt
bringt, indem er sie malt. Der Maler muß zu diesem Zweck hervorragende
sittliche Eigenschaften besitzen wie im vorliegenden Falle der König. Unter
den Opferspenden sind seine Malereien zu verstehen, die seinen Namen

Ruhm erwerben. Weil du den höchsten Ruhm für den Brahmanen beanspruchtest, wirst du zu meiner Freude von jetzt ab, wie die anderen Beherrscher der Menschen, auf der Erde berühmt sein. Du kennst die religiösen Pflichten (*vrata*), die Askese (*tapas*) und den Veda, und herrschest glücklich über sündlose Menschen. Da ich es gewährte, hast du den Sohn des Brahmanen gemalt und in dieser Welt der Lebenden[1] das erste Bild hervorgebracht. Wegens des Nutzens, der dadurch den Welten erwächst, hast du dir Ansprüche auf Verehrung erworben. Heute und später soll dieses Bild hervorragend geehrt werden. Möge es das Herz der Menschen gefangen nehmen und beständig Freude und Heil gewähren! Möge Segen und Glück von ihm ausgehen, mögen die Sünden ver-

---

berühmt machen werden; denn wie im zweiten Kapitel gezeigt wird, waren die Malereien für den Kultus bestimmt.

[1] Tib. *ạtso-bai ạjig-rten* = Skr. *jīvaloka*, „die Welt der Lebenden", oder vom Standpunkt der Jaina „die Welt der Seelen". Nach der Lehre der Jaina besteht die Welt aus Seelen (*jīva*) und Nichtseelen (*ajīva*), d. h. aus belebter und unbelebter Schöpfung. Im Gegensatz zu den Buddhisten fassen die Jaina die Seele als eine Realität auf. "A wise man believes in the existence of the soul, he avoids the heresy of the non-existence of the soul" (H. JACOBI, Jaina Sūtras, Part II, p. 84). Vom jainistischen Standpunkt wird die obige Stelle vollkommen verständlich. Die Seele des toten Knaben ist in das Bild zurückgekehrt, während der Tote selbst in der Unterwelt gehalten wird, und weilt somit wiederum in der Welt der Seelen. Das Bild ist die neue körperliche Form, die seine Seele aufgenommen hat; die Malerei wird damit zu einer Welt der Beseelung. Es ist daher auch klar, daß die Buddhisten den Gedanken, daß das Bild die Seele der Gottheit darstelle, von den Jaina empfangen haben, und daß die Entlehnung nicht etwa auf dem umgekehrten Wege stattgefunden haben kann. Die buddhistische Anschauung von diesen Dingen beginnt nicht etwa mit den Bannungen der Götter und Dämonen im Tantra- und Yoga-system, welche den letzten Ausläufer dieser Entwicklung darstellen, sondern setzt bereits bei den ersten Buddhastatuen ein, die nach dem Text Nr. 1 des Tanjur die irdischen Stellvertreter des im Tushitahimmel abwesenden Buddha vorstellen.

mieden werden! Möge es die Rākshasa vernichten und die
Feinde besiegen! Weil du also auf mein Geheiß hin zuerst
gemalt hast, wird nun auf Grund dieses ersten gemalten
Werkes das Wort *citṛa*[1] (in dem Sinne eines Gemäldes) gebildet." Als auf diese Weise Brahma zwischen den beiden,
dem Yama und dem Nagnajit, eine Aussöhnung zustande gebracht hatte, verneigten sich beide und der Brahmane zu den
Füßen (der Anwesenden) und segneten sie alle. Darauf begab
sich zu seinem Wohnort der Gott der Erde[2] und der Dreiwelt
(Brahma) samt den Scharen der Deva. Der König[3] voll
Freude darüber, dem Dharmarāja (Yama) ein Opferpriester zu
sein, begab sich an seinen Ort. Der Brahmane kehrte in
seine Stadt zurück, aus der er gekommen war. Der Brahmane
nahm seinen Sohn mit sich, und in herzlicher Freude über
den König, war er beständig um das seinem Herzen wohlgefällige und ihm hilfreiche Bild bemüht.

327—374. Darauf begab sich jener Herrscher der Menschen zu dem Wohnort des Brahma und erkundigte sich bei
ihm nach der rechten Methode, was die zu den Körperformen
passenden Maße anbelangt: „Brahma, welche[4] Maße kommen
in Betracht? Da es bei dem von mir gemalten Bilde der
Abzeichen[5] verschiedene gibt, so ziemt es sich, daß du sie

---

[1] Tib. *ri-mo.*

[2] Tib. *bstan-pa* = *brtan-pa*, Beiname der Erde (CHANDRA DAS, Dictionary,
p. 558 b); vgl. *bstan-ma* und *brtan-ma*, die Göttin der Erde, = Skr. *prithivī*.

[3] Wiederum Instrumentalis als Subjekt mit intransitivem Verbum.

[4] Das Relativpronomen *gaṅ* hängt von dem folgenden *gsuṅ* ab: durch
welche Maße, sollst du mir sagen.

[5] Statt *mtsʿan-ñid bdag* der beiden Drucke ist wohl zweifelsohne *dag*
(Pluralpartikel) zu lesen, da *bdag-gis* keinen Sinn hat und nur durch das
im vorhergehenden Verse vorkommende *bdag-gis* in die Feder des Kopisten
geflossen ist.

mir[1] hier heute ein für alle Male erklärst. Wie sind die Malereien entstanden? Welches sind die Regeln für ihre Maße?" So fragte er mit gefalteten Handflächen. Brahma sagte zu dem Herrscher der Menschen: „Das darin liegende vortreffliche Geheimnis will ich dir also erklären. König, zuerst sind in der Welt die Veda[2] und die Opferspenden entstanden. Um ein Caitya[3] zu machen, muß man Bilder malen. Daher wird die Malerei als Wissenschaft (*Veda*) gerechnet. Ich bin der erste, der Menschen gemalt, und so habe ich den Menschen [das Malen] gelehrt. Daher besteht mit Rücksicht auf die von mir zuerst gemalten Werke der Sprachgebrauch des Wortes *citra* in dem Sinne von „Gemälde", und wenn die Wesen in Übereinstimmung damit aus einem bestimmten Anlaß oder ohne Anlaß im Stil der von mir geschaffenen Bilder malen, so wird mit Rücksicht darauf auch ein solches Bild als „Gemälde" (*citra*) definiert[4]. Der trefflichste der Berge ist der

---

[1] Tib. *bdag* ist hier als Dativ zu fassen.

[2] Tib. *rig-byed-rnams*.

[3] Tib. *mc'od-rten*, ist hier wohl einfach als Grabmal zu fassen, vielleicht auch in dem Sinne von Höhlentempel, dessen Wände dekorativ bemalt wurden.

[4] Die Konjekturen, die wir hier leider zu machen gezwungen sind, werden hoffentlich Billigung finden. Der Vers *agro-ba dan ni gel-bai bar,* wie er in beiden Drucken steht, ist sinnlos. Die Partikel *dan* weist darauf hin, daß *agal-ba* zu lesen ist; die Schreibung *agel-ba* statt *agal-ba* ist mir übrigens verschiedentlich begegnet. Ist diese Vermutung richtig, so ist *ni* in *mi* zu verbessern. Folgerichtig wäre dann zu übersetzen: während man sich mit den Wesen nicht in Widerspruch setzt, mit ihnen harmoniert; wenn man, um den Wesen zu gefallen, malt usw. Das hat indessen keinen Sinn, da der Gedanke Brahmas dadurch nicht zum Ausdruck kommt. Brahma gibt zweimal eine Definition des Wortes *citra,* das von jetzt ab in der neuen Bedeutung „Gemälde" gebraucht werden soll. Zunächst wird es auf seine eigenen, die ersten Bilder, angewandt, sodann auch auf die von Menschen gemalten. Folglich kann *agro-ba* nur das Subjekt zu *bris byas-pas* sein, aber nicht von *agal-ba* abhängen, während sich *mi agal-ba* nur

Sumeru, und wie von den aus dem Ei Geborenen[1] der am Himmel schwebende (Garuḍa) der erste ist, wie der erste unter den Menschen der König[2] ist, so ist die Malerei die erste unter den Fertigkeiten. Wie sich die Ströme alle in den großen Ozean ergießen, wie vom Ozean die Edelsteine abhängen, wie von der Sonne die Planeten abhängen, wie die heiligen Ṛishi von Brahma, wie die Götter (von Brahma) abhängen, ebenso, o König, hängen alle Fertigkeiten in Wahrheit von den Werken der Malerei ab. Wie der Sumeru der erste unter den Bergen, wie unter den Strömen die Gaṅgā die erste und die Sonne der erste unter den Himmelskörpern[3] ist, wie der am Himmel schwebende (Garuḍa) der König der Vögel, wie unter den Göttern Indra[4] der erste ist, so ist die Malerei die erste der Fertigkeiten. Daher geh zu Viçvakarman, dem Gotte, und dieser Nagnajit[5] wird dich über die Merkmale, Regeln und Maße der Malereien[6], kurz, über alles belehren."

---

auf die vorher genannten Bilder des Brahma beziehen kann; ich zweifle daher nicht, daß der Vers ursprünglich gelautet hat: *agro-ba bdag daṅ mi agal-bar.* Im folgenden Verse ist *bcas-pa* als Perfektum des Verbums *aç'a-ba* „machen" zu fassen, und *bcas-pa dag de* bezieht sich auf Brahmas Malereien. Ob ich *rgyu daṅ mi rgyur* „aus einem bestimmten Anlaß oder ohne Anlaß" richtig auffasse, weiß ich nicht; vielleicht liegt dem Ausdruck ein mir unbekanntes, besonderes Sanskritäquivalent zugrunde. Ich denke dabei an religiöse und weltliche Malereien, die einen, die ihre Veranlassung im Kultus, in der Götterverehrung haben, wie im zweiten Kapitel auseinandergesetzt wird, die andern, die keinen bestimmten Zweck verfolgen.

[1] Skr. *aṇḍaja.*
[2] Tib. *sa bdag* = Skr. *bhūpati, kshitipati.*
[3] Tib. *gza* = Skr. *graha*, Planet. Die Sonne (*āditya*) wurde unter die Planeten gerechnet, deren man sieben annahm: Sonne, Mond, Mars, Merkur, Venus, Jupiter und Saturn.
[4] Tib. *dbaṅ c'en* = Skr. Mahēndra.
[5] Tib. *gcer t'ul.*
[6] Man beachte, daß hier der Titel des ganzen Werkes auftritt.

375—431. Darauf begab sich, entsprechend der von Brahma empfangenen Weisung, der König der Menschen freudig zu Viçvakarman und schaute die große Erscheinung[1]. Als er ihn erschaute, verneigte sich der König. Beide im Verein stellten die trefflichsten Lehren auf und erwiesen einander Verehrung. Mit ehrerbietigen Worten sprach er: „Da du der alles Durchdringende und der alles Wirkende bist[2], bin ich zu dir auf das Geheiß des Brahma gekommen. Erkläre mir bitte die Merkmale in den Werken der Malerei und sage mir, in welcher Weise die Maße und Formen samt den Methoden (zur Anwendung kommen)." Nachdem der König diese Worte gesprochen hatte, erklärte ihm Viçvakarman die Kunst: „Heiliger Herrscher der Menschen, (richte deine Aufmerksamkeit) auf einen Punkt und höre auf alle von mir gegebenen Erklärungen. Wie die Malereien entstanden sind, wie das von den Göttern geehrte Citralakshaṇa, das die Maße, Kompositionen[3] und Farben beschreibt und bei allen Verständigen als das Hauptwerk gilt, von dem lotusgeborenen gütigen Gotte[4] gelehrt worden ist, vernimm! Alle Formen sämtlicher Körper hat Brahma gemalt zum Wahrzeichen[5] der Wohlfahrt der

---

[1] Tib. *snan-ba* = Skr. *āloka*.

[2] Tib. *k'yab-bdag* = Skr. *vibhu* (CHANDRA DAS); *qprin-las t'ams-cad-pa*, worin offenbar eine Anspielung auf den Namen Viçvakarman steckt.

[3] Tib. *gnas t'abs*, die Art und Weise der Gegenstände, die in der Malerei dargestellt werden, Komposition. Vgl. weiter unten V. 404 *gnas dan t'abs*, und gegen Schluß des dritten Kapitels S. 178, 183.

[4] Brahma. Das Wort *Padma-las byun* fasse ich als Übersetzung von Skr. *Padmayoni*, Beiname des Brahma.

[5] Tib. *mts'an-ñid* bedeutet wörtlich die charakteristischen Merkmale, die Quintessenz. Die Bilder sollten der Erbauung der Gläubigen dienen und stellten somit das Symbol, das Wahrzeichen ihres frommen Lebenswandels und ihres Heils dar. Dieser Gedanke wird im zweiten Kapitel dieser Schrift ausführlich entwickelt.

Gläubigen in den Welten und sie mir zuerst übergeben. Mit was für Maßen dabei zu verfahren ist, welche Gegenstände und Mittel schön sind, das alles habe ich von Brahma erlangt: die trefflichsten Malereien alle verdanke ich der Huld des heiligen Gebieters, und dank derselben habe ich alle Kunstwerke geschaffen. Könige und Menschen, wie du sie hier siehst[1], sind in ihrer äußeren Erscheinung so von mir verwandelt worden. Da es der Malereien mit ihren Merkmalen so viele Arten gibt, werden sich in Zukunft die Gemälde der Götter vermehren. Wenn du dir in der Beschaffenheit der Maße und der Merkmale, der Proportionen und Formen der Ornamente und Schönheiten bei mir Kenntnisse erwirbst, wirst du vollständig in allen Fertigkeiten bewandert und in der Malerei allseitig ein hervorragender Sachverständiger sein. Du sollst darüber belehrt werden, wie man die Menschen gut erfaßt, und wie man die Weisen erfaßt, wie frohsinnige Menschen und solche ein schwieriges Thema darstellen und eine konzentrierte Aufmerksamkeit verlangen. In den verschiedenen Arten der Muni, der Nāga, Yaksha, Rākshasa, Preta, Asura samt den Piçāca, will ich dich, wie man sie nach ihren Abzeichen und nach den Regeln malen soll, unterweisen."

Aus der Lehre von „den Merkmalen des Gemäldes" (*Citralakshaṇa*) des Nagnajit[2] das erste Kapitel, betitelt Nagnavrata[3].

---

[1] *aḍi-dâg-ni;* das Pronomen *aḍi* kann sich nur auf etwas Gegenwärtiges, Vorliegendes beziehen; es scheint daher, daß die Auffassung besteht, als ob der Gott auf seine Kunstwerke hinweise, was durch den Zusatz von *de-ltar* bestätigt wird.

[2] Tib. *gcer-bu t'ul.*

[3] Tib. *gcer-bu brtul-ba.* An dieser Stelle darf man *brtul-ba* wohl nicht gleich *t'ul* auffassen wie oben S. 136. Beide Drucke bieten die gleiche

2. Kapitel.

432—527. Darauf sprach jener Sachkundige[1] vertrauens-voll[2] zu dem hochherzigen[3] Könige, indem er ihn über die Beschaffenheit der von Brahma gemalten Bilder belehrte: „Vor alters, als der Vernichtung ein Ziel gesetzt, und die Festigkeit und Beweglichkeit[4] (des Welteneis) beseitigt war, wurde durch das goldene Ei[5] die Finsternis überwunden, und aus dem Wasser entstand alles. Aus jenem Ei ging der Stammvater der Welt selbst hervor[6]. Darauf schuf Brahma das Oṁ usw., den Veda, die Wissenschaft, die Spekulation[7] und die Kreatur[8]. Aus diesen vier entstanden Formen und Namen[9] usw., und

---

Lesart, in der ein von dem kurz vorhergehenden *t'ul* verschiedenes Wort gesucht werden muß. Tib. *brtul-ba* entspricht Skr. *vrata*, und *nagnavrata* (vgl. auch *nagnacaryā*, *pw* Nachträge, S. 351 c) bedeutet „das Gelübde des Nacktgehens", hier aber wohl mit Bezug auf den Inhalt des Kapitels: die Art und Weise, das Gebahren der Nackten; mit großer Wahrscheinlichkeit wohl: die Praxis des Nackten in der Malerei. Tib. *gcer-bu* scheint hier gleichzeitig subjektiver und objektiver Genitiv zu sein: Tätigkeit der Nackten und in bezug auf die Nackten oder das Nackte.

[1] Tib. *bzod-ldan* = Skr. *saha*, nämlich Viçvakarman.

[2] Tib. *gus-pas*. Vgl. H. BECKH, Beiträge zur tibetischen Grammatik, S. 42 (*Abhandlungen der Berliner Akademie*, 1908).

[3] Tib. *bdag ñid c'e* = Skr. *mahātman*, wie oben S. 135, Anm. 3.

[4] Tib. *gtan dañ agro-ba*; *gtan* ist hier Nomen „Stärke, Solidität, Dauer-haftigkeit" (mong. *nuta*) und steht im Gegensatz zu *agro-ba* „das Gehende, sich Bewegende". Die harte Schale des Welteis wurde zersprengt, und es vertauschte die Bewegung mit dem Zustand der Ruhe.

[5] Skr. *hiraṇyagarbha*, das goldene Embryo, das kosmische goldene Ei, aus dem nach der vedischen Vorstellung der Geist des Schöpfers und das ganze Weltall hervorgingen. Vgl. z. B. A. A. MACDONELL, Vedic Mytho-logy, p. 14; J. GARRETT, Classical Dictionary of India, p. 203; E. C. SACHAU, Alberuni's India, Vol. I, p. 221.

[6] Tib. *rañ byuñ* = Skr. *svayambhū*, was zum Beinamen Brahmas ge-worden ist.

[7] Tib. *rtog-par bya* = Skr. *tarka*.

[8] Tib. *skye-dgu* = Skr. *praja*.

[9] Tib. *gzugs dañ miñ* = Skr. *rūpanāma*.

10

mit der Schaffung des Lebens traten durch Brahmas Wirksamkeit Kasten und Rangstufen in Tätigkeit, und Recht und Sitte
formten sich. Nachdem Brahma so gewirkt hatte, richtete er
in seinem Geiste alle Gedanken auf das Heil der Geschöpfe.
Als er in diesen Gedanken versunken war, brachte sein geistiger
Zustand die Wirkung hervor, daß die Wesen, Könige und
Götter, infolge der Kenntnis der Namen, unaufhörlich und beständig gläubige Verehrung bezeigten. Durch solche Gedanken Brahmas erlangten Mahādeva, Vishṇu, Indra und sämtliche Götter großen Segen[1]. An Maßen und Vorzügen wurden
sie schön aus eigener Kraft und glichen sehr den mit guten
Merkmalen Ausgestatteten[2]. Schöne Formen entwickelten sie,
die Haupt- und Nebenglieder wurden vollendet und blühten
allseits zur vollen Entfaltung auf[3]. Ihre Gestalten nahmen

---

[1] Tib. *skal-pa c'en-po* = Skr. *mahābhāga* ausgezeichnet, hervorragend.

[2] Tib. *mts'an-ldan* = Skr. *lakshaṇya*.

[3] Tib. *ạbyes*, scheint perfektische Bildung zu dem intransitiven Verbum
ạbye-ba, „sich öffnen" zu sein (= *bye*). In der Liste der Tausend Buddhas
wird der Name *Vibhaktatejas* (Nr. 486) durch tib. *gZi-brjid rnam-par ạbyes-
pa* wiedergegeben. In der Liste der achtzig kleineren Körpermerkmale
(*anuvyañjana*) des Mahāpurusha, wie sie in der Mahāvyutpatti enthalten ist,
findet sich *yan-lag daṅ ñiṅ-lag šin-tu ạbyes-pa* (Schiefner, Triglotte, fol. 4b),
während in der Ausgabe von Csoma und Ross (Part I, p. 95, Nr. 31) an
dieser Stelle *ạbyed-pa* geschrieben ist. Die Phrase entspricht im Sanskrit
suvibhaktāṅgapratyāṅga, von Csoma „mit wohlproportionierten (oder an-
geordneten) Gliedern" übersetzt (ebenso Dharmasaṁgraha ed. Max Müller
und Wenzel, p. 57, Nr. 33). Ferner findet sich ebenda (Triglotte, fol. 7b)
spyan dkar nag ạbyes šiṅ padmai ạdab-ma raṅs-pa (Csoma: *mdaṅs*) *lta-bu*,
wo bei Csoma (p. 97, Nr. 63) *dbyes* geschrieben ist (Jäschke hat nur die
Form *dbye*); im Sanskrit *sitāsitakamaladalanayana* läßt sich nichts entdecken,
was diesem *ạbyes* oder *dbyes* entsprechen könnte. Es bezieht sich zunächst,
ebenso die Beiworte „weiß" und „schwarz", wie auch aus der mongolischen
Übersetzung hervorgeht, auf die Entfaltung der Blütenblätter des Lotus.
Des weiteren stimmen Triglotte (fol. 8a) und Csoma (*Ibid.*, Nr. 71 und 72)
in der Phrase *dpral-ba dbyes c'es-pa* in der Schreibung *dbyes* überein, das
hier Skr. *supariṇata* oder *prithu*, also „entwickelt, voll, groß", entspricht.

mannigfache Formen an, verschönert durch Schmuck und Ge-
wandung. Verschiedene Waffen, die sie in der Hand hielten,
trugen zur Vermehrung ihrer Attribute auf den Bildern bei,
und so wurden ihre Figuren von ihnen selbst gemalt. Als
diese die Götter erschauten, gingen ihnen die Augen über[1],
und sie waren voll Herzensfreude. Brahma sprach: vortrefflich! und erhob andächtig Lobpreisungen. Da segneten sie
ihre eigenen Gestalten und erfüllten sie mit großer Macht[2].
Darauf sprach Brahma freudig zu den sieben Göttern folgendermaßen: „In den drei Welten wird man von jetzt ab im Vertrauen auf eure Heiligkeit keinen Zweifel an euern[3] Göttergestalten hegen und euch beständig Opfer darbringen. Wer
wahrhaft Herzensgüte[4] erworben, ist im Besitze des durch das
Wissen erreichten Schatzes. Wer der Reinheit[5] beflissen im
Lande der Menschen Opfer darbringt, der hat dadurch immer-
dar seinen Lohn dahin[6], ist frei von Krankheit und empfängt

---

Jedenfalls beweisen diese Stellen, vor allem die erste, deren Wortlaut *yan-lag ñiṅ-lag .. šin-tu ạbyes* sich ebenso in unserem Texte wiederholt, daß
die Zugehörigkeit der Form *ạbyes* zu dem Stamme *ạbye(d)* gesichert ist. —
Tib. *ñiṅ-lag* ist aus *ñid-lag* entstanden; *ñid* ist Übersetzung von Skr. *prati*
in *pratyaṅga* (*prati* + *aṅga*).

[1] Wörtlich: sie freuten sich in den Adern der Augen. *Mig rtsa* wird
von JÄSCHKE als die Blutgefäße in der harten Haut des Auges (Sklerotica)
erklärt. Die Freude wird also hier als physische Erregung, dann als innere
Befriedigung ausgedrückt.

[2] Tib. *mt'u* = Skr. *prabhāva*.

[3] Tib. *k'yed-cag-gis* ist als Genitiv zu fassen.

[4] Tib. *yid bzaṅ* = Skr. *hṛidayabhadra*.

[5] Tib. *gtsaṅ daṅ*, Synonymkompositum (*daṅ* = *daṅ-ba*, „rein"), scheint
Skr. *çuddhi, çucicarita* zu entsprechen. Tib. *gžol-ba* ist Äquivalent von Skr.
abhirakta (s. Çikshāsamuccaya, ed. BENDALL, p. 33) oder *parāyaṇa* (Mahāvyutpatti ed. CSOMA und Ross, Part I, p. 89, Nr. 3).

[6] Tib. *bsgrub dọn*, in Nachahmung eines Bahuvrīhi: einer, der sein
Ziel erreicht hat, ein Gesegneter.

10*

die Gewährung seiner Wünsche. Die Sünden aus der Welt verbannend, böse Träume und das Besessensein von den Leidenschaften aus dem Wege räumend, wird er sein eigener Schützer sein. Alle, die dem Dharma gemäß handeln, werden die Rākshasa bezwingen[1] und ihren Ruhm mehren. Um den Opferkult[2] der Welt einzurichten, sollen den Göttergestalten Namen verliehen werden, Lobpreisungen, Opfer und Anbetung, und so soll ihnen Ruhm zuteil werden. In welcher Richtung auch immer sich ein Mensch in der Ausübung tugendhafter Handlungen anstrengen mag, wenn er euch, unter welcher euerer Gestalten auch immer, an einem einzigen Tage Opfer darbringen wird, dem werden jene Bilder[3] inneren Frieden bescheren." „So soll es sein!" Mit diesen Worten segneten die Götter hoch erfreut ihre eigene majestätische Erscheinung[4] und kehrten jeder an seinen Ort zurück. So sind denn die Opferhandlungen entstanden. Wenn du die Maße usw. der Bilder[5], welche von den Menschen zum Gegenstand des Opferkults gemacht werden, und wenn du die Definitionen davon von mir vernommen hast, dann wird noch heute dein Name, o Herrscher der Menschen, in der ganzen Welt der Menschen, verkündet werden. Das heilige Wissen von den Merk-

---

[1] Vielleicht Reminiszenz an das vedische *rakshohan* „Vernichter der Unholde" (A. A. MACDONELL, Vedic Mythology, pp. 95, 110, 164). Vgl. auch oben S. 140. Möglicherweise ist hier ein Wortspiel zwischen *rakshitar* und *rakshas* beabsichtigt.

[2] Tib. *mc῾od-bya* scheint an dieser Stelle einem Skr. *yajusya* zu entsprechen.

[3] Tib. *de-dag*. Vgl. V. 512 und unten Anmerkung 5.

[4] Tib. *c῾as*, Kostüm mit allen Paraphernalia, dann Form, Gestalt, Erscheinung; bezieht sich hier wohl auf die Götterbilder.

[5] Tib. *de-dag-gis* ist genitivisch in Abhängigkeit von *ts῾ad* zu fassen und kann sich nur auf die Götterbilder beziehen, die bei den Opfern gebraucht werden.

malen¹, soweit ich es von Brahma gelernt habe², und die Beschaffenheit der Maße sollen vollständig verstanden werden. In den drei verehrungswürdigen Welten sind die Körperlichen alle mit einem Körper begabt: so höre du denn auf meine Erklärungen darüber, welches die Maße der Glieder sind. Großer König³, die Meßkunst der Malerei beruht auf der Opferverehrung aller Götter, der dadurch bewirkten Vermehrung des Ruhmes und der Verbannung der Sünden und der Furcht; indem der Nachdruck auf das, was unserem Auge schön erscheint, zu legen ist⁴, sollst du die fehlerlose Darlegung der verschiedenen Grundgesetze⁵ folgendermaßen erfassen.

Aus den „Merkmalen des Gemäldes" (*Citralakshaṇa*) ist das zweite Kapitel, betitelt die Entstehung der Opferhandlungen, hiermit beendet.

---

3. Kapitel.

528—539. Was die einzelnen Körpermaße der Könige und anderen Wesen betrifft, so hat sie mir Brahma selbst erklärt; ich werde sie dir, und du wirst den Menschen die Methode der Maße genau erklären. Für die Götter, Asura, Nāga-Gebieter (*nāgendra*), Rākshasa, Gandharva, Kinnara, Siddha⁶, Vādana und Jaritar⁷, Piçāca, Preta, Kumbāndha und

---

¹ Tib. *mts'an* = Skr. *lakshaṇa*, nämlich von den Bildern.
² Tib. *sñed de*. Die Partikel *te* zeigt an, daß *sñed* verbal ist, und daß es ein Verbum *sñed-pa* gibt. Vgl. am Schluß des dritten Kapitels *ji sñed-pa*.
³ Die vier Verse 524—527 bestehen aus je elf Silben.
⁴ Tib. *t'og-mar byuñ*, an den Anfang gestellt, in den Vordergrund gerückt, vorangesetzt.
⁵ Tib. *gźi*. B liest dafür *bźi*, „vier" (vier Arten von Regeln?).
⁶ Tib. *grub*.
⁷ Tib. *rol mk'an mts'an sgrogs*, „Musiker und Namenrufer". Die

die Scharen der Vidyādhara, für die Menschen, Amanusha und Könige, für alle diese kommen die folgenden Maße in Betracht.

540—543. Atom[1], Haarspitze[2], Niß[3], Laus[4], Gerstenkorn[5], Finger[6], wachsen progressive durch Multiplikation[7] mit der Zahl acht, wie genau gezeigt werden soll.

544—555. Acht Atome bilden eine Haarspitze, so wird gelehrt. Weiß man dieses Maß, so gelangt man zu dem Satze, daß eine Haarspitze gleich acht Nissen ist. Acht Nisse bilden eine Laus, und acht Läuse werden als ein Gerstenkorn erklärt. Acht Gerstenkörner bilden einen Finger: die Bezeichnung Finger ist als ein Zoll zu erklären, so bedeuten z. B. zwei Finger soviel als zwei Zollmaße und vier Finger soviel als vier Zollmaße. Aus der Kenntnis dieses Maßes folgt es, daß vier Gerstenkörner einen halben Zoll machen.

556—573. Was den Ausdruck „Ausdehnung" betrifft, so bedeutet er die Länge; die Breite bedeutet die Richtung geradeaus. Nun sollen die Maße der Könige und Menschen vollständig gelehrt werden. Was die Flächenausdehnung des

---

Zurückübersetzung ins Sanskrit ist hypothetisch. In Çikshāsamuccaya (ed. BENDALL, p. 330) gibt tib. *rol-mo mk'an* Skr. *jhullakamalla* wieder. In tib. *sgrogs* steckt vielleicht Skr. *nādin* (*nāmanādin?*).

[1] Tib. *rdul p'ra rab* = Skr. *paramāṇu* (Atom) oder *trasareṇu* (Sonnenstäubchen) (vgl. WASSILJEW, Der Buddhismus, S. 306). Über diese und die folgenden Maße vgl. H. T. COLEBROOKE, On Indian Weights and Measures (Miscellaneous Essays, Vol. II, p. 538) und E. C. SACHAU, Alberuni's India, Vol. I, p. 162.

[2] Tib. *skrai tse-mo* (= *rtse-mo*) = Skr. *aṇu*.

[3] Tib. *sro-ma* = Skr. *likshā* oder *liksha*, Ei einer Laus.

[4] Tib. *śig* = Skr. *yūkā* oder *yūka*.

[5] Tib. *nas* = Skr. *yava*.

[6] Tib. *sor* = Skr. *aṅgula*.

[7] Tib. *agyur*.

Cakravartin betrifft, höre, wie sie erklärt wird. Er ist wohlproportioniert wie der Nyagrodha Baum[1]. Was die Länge

---

[1] Skr. *nyagrodhaparimaṇḍala*, „kreisförmig wie der Nyagrodha", erscheint unter den zweiunddreißig Schönheitsmerkmalen (*lakshaṇa*) des Mahāpurusha (Mahāvyutpatti ed. Csoma und Ross, Part I, p. 93, Nr. 20; Schiefner,
Triglotte, fol. 4b; vgl. auch Schiefner, Berichtigungen und Ergänzungen zu
Schmidt's Ausgabe des Dsanglun, S. 31). Wir ersehen hier, daß der Vergleich mit dem Nyagrodhabaum ursprünglich vom Cakravartin gebraucht und
später auf Buddha übertragen wurde. In der ersten die Buddhastatuen be-
handelnden Schrift des Tanjur (S. 2) bildet dieses Gleichnis den Ausgangspunkt
der Betrachtung, wenn Buddha selbst die Anweisung gibt, daß sein Bild in
einer der Größe des Nyagrodhabaumes entsprechenden Proportion dargestellt
werden soll. Dem Kommentar zufolge bezieht sich das Gleichnis darauf,
daß die Äste des Baumes wohlproportioniert sind; nach der Meinung anderer
wird es jedoch auf den Nimbus gedeutet, während wieder andere an das
Längenmaß denken und die vergleichsweise Höhe des Baumes (er erreicht
20 bis 30 Meter) und der Buddhastatue im Auge haben.

Der Nyagrodha ist der Banyanbaum (*Ficus indica* oder *bengalensis*),
einer der kolossalsten und schönsten Bäume Indiens, in der Subhimalaya-
region und in Zentralindien einheimisch. Die Gesetze des Manu schreiben
vor, unter andern Bäumen den Nyagrodha und Açvattha (pippala, *ficus reli-
giosa*) auf die Grenzen der Felder zu pflanzen. Açoka ließ, wie er in einem
seiner Edikte meldet, die Straßen mit Nyagrodha bepflanzen, um Menschen
und Tieren Schatten zu spenden, und diesem Zwecke dient er noch jetzt
in den Dörfern (*grāmadruma* des Rāmāyana). Die große Zahl seiner Luft-
oder Adventivwurzeln verschaffte ihm den Beinamen *bahupada*. Er errregte
die Bewunderung der Hellenen, und seine Schilderung ist nach Bretzl das
Glanzstück der botanischen Mitteilungen über ostindische Pflanzenwelt, die
unter Alexander erschienen. Vgl. W. Roxburgh, Flora Indica, pp. 639—642
(Calcutta, 1874); Hugo Bretzl, Botanische Forschungen des Alexander-
zuges, S. 158—190 (Leipzig, 1903); Charles Joret, Les plantes dans l'antiquité et au moyen âge, II L'Iran et l'Inde, pp. 291—293 (Paris, 1904).
Wie der Açvattha den Lebensbaum versinnbildlicht, wie Vishnu mit dem
Nyagrodha-udumbara-açvattha identifiziert wird (Hopkins, JAOS, Vol. XXX,
1910, p. 349), so dürfte auch die Beziehung des Cakravartin-Mahāpurusha-
Buddha zu dem Nyagrodha seine Ursache in der Baumverehrung haben, die
durch die vielen merkwürdigen botanischen Eigenschaften dieses Baumriesen
verständlich wird. Daß der Nyagrodha von einer Gottheit bewohnt gedacht
wurde, ist aus der Erzählung von Mahākāçyapa und Bhadrā (Schiefner,
Mélanges asiatiques, Bd. VIII, S. 297) ersichtlich, wo ein kinderloser Brah-

des Cakravartin angeht, so beträgt sie entsprechend der Messung seiner eigenen Finger hundertundacht Zoll; in keinem Falle sind die Zollmaße anderer auf ihn anwendbar[1]. Die Längenmaße der oben genannten Könige und der Cakravartin Könige sollen nun hier in der Tat erläutert werden. Welches die Maße des Gesichts und der übrigen Körperteile sind, darüber höre nun meine Erläuterung.

574—618. Das Gesicht wird in drei Teile geteilt, Stirn, Nase und Kinn, deren Maß je vier Zoll beträgt. Was die Anzahl der Zoll in der Breite des Gesichts anbelangt, so wird sie auf vierzehn angegeben; die obere und untere[2] Partie des Gesichts betragen zwölf Zoll an Breite; auf Grund dieser Maße ergibt sich für die Länge des Gesichts die Annahme von zwölf Zoll. Der Scheitelknorren (ushnīsha) beträgt vier Zoll an Länge und sechs Zoll an Breite. Der Kopf muß sich den Maßen des Baldachins (chatra) anpassen und ist zwölf Zoll an Breite; so beläuft sich der Umfang des Kopfes auf zweiunddreißig Zoll. Die Breite des Ohres beträgt zwei Zoll, und seine Länge vier Zoll. Die Ohröffnung wird auf einen halben Zoll (an Breite) und auf einen Zoll (an Länge) angegeben. Die Ohrspitzen und die Enden der Augenbrauen

---

mane die Baumgottheit unter Zeremonien um Verleihung eines Sohnes bittet; ebenda die Angabe nach dem Kanjur, daß, wenn das Fällen eines Baumes für religiöse Bauzwecke notwendig war, die Baumgottheit unter Blumenopfern und Segenssprüchen ersucht werden mußte, sich einen andern Wohnsitz zu suchen, wozu ihr sieben bis acht Tage Zeit zu lassen waren; war dann keine Veränderung an dem Baume sichtbar, durfte er gefällt werden; war eine Veränderung vorhanden, durfte er nicht niedergehauen werden.

[1] Sie haben also gleichsam ihre eigenen Königszolle. Entsprechend der größeren Länge ihrer Finger muß auch ihre Körperlänge die gewöhnlicher Wesen überragen.

[2] Tib. dmad = smad.

liegen in gleicher Höhe; die Augenhöhlen befinden sich mit den Ohröffnungen in gleicher Höhe. Für die Ohrlappen gibt es keine bestimmten Angaben[1]. Die Länge der Brauen beträgt vier Zoll, ihre Breite zwei Gerstenkörner ($^1/_4$ Zoll). Bei allen von mildem Wesen[2] sind die Brauen gleich der Sichel des Mondes. Auf Grund der Kenntnis der Maße sind sie bei den Tanzenden[3], den Zornigen und Weinenden wie ein Bogen gekrümmt. Bei den Erschreckten und Klagenden, deren Kopf in seinem hinteren und unteren Teile gehoben[4] ist, erscheinen die Brauen[5] wie aus den Nasenlöchern herausgewachsen und bedecken die Hälfte der Stirn[6]. Die Stirnlocke (*ūrṇā*) des Cakravartin, die ihre Stelle zwischen den Brauen hat, soll einen Zoll gemacht werden. Der Raum von der Mitte der Brauen bis zur Grenzlinie[7] der Haare beträgt an Weite $2^1/_2$ Zoll. Von

---

[1] Dies ist interessant im Hinblick auf die lang gezogenen Ohrlappen, die sich an Bildern und Statuen der Buddha, Bodhisatva, buddhistischen Heiligen usw. befinden. Schon FOUCAUX (*Annales du Musée Guimet*, Vol. XIX, p. 29) macht darauf aufmerksam, daß diese Eigenschaft unter den Abzeichen des Buddha im Lalitavistara nicht erwähnt ist, wahrscheinlich weil diese Verlängerung des Ohres auf künstliche Weise hervorgerufen werde. Jedenfalls ist aus unserem Text ersichtlich, daß dies auch kein offiziell anerkanntes Merkmal des Cakravartin gewesen ist, und daß dem Künstler in diesem Punkte freie Hand gelassen war.

[2] Bei *ži-ba* und dem folgenden *k'ros* handelt es sich natürlich nicht um die bekannte Unterscheidung von milden (*çānta*) und zornigen (*krodha*) Formen von Gottheiten, da hier nicht von diesen, sondern von Königen die Rede ist. Die betreffenden Ausdrücke beziehen sich lediglich auf die Darstellung von Gemütsaffekten.

[3] Tib. *gar byed* = Skr. *naṭ* (vgl. GRÜNWEDEL, Mythologie d. Buddhismus, S. 97).

[4] Tib. *stod-pa* (von *stod*) hat hier die wörtliche Bedeutung „erhöhen".

[5] Die Pluralendung *dag* dient hier zum Ausdruck des Duals des Sanskrit, wie weiter unten *mig-dag* (vgl. SCHIEFNER, Über Pluralbezeichnungen im Tibetischen, § 16).

[6] Tib. *spral-ba* = *dpral-ba*. Tatsächlich begegnet man dieser Art der Darstellung in den grotesken chinesischen und japanischen Figuren der Arhat.

[7] Tib. *ats'ams* = *mts'ams*.

den Anfängen der Brauen über die Ausdehnung der Stirne hin[1] soll ein Maß von im ganzen vier Zoll gemacht werden. Die Augenhöhlen sind zwei Zoll (lang), und ebenso der Raum zwischen den Augen. Die Augen betragen zwei Zoll an Länge und einen Zoll an Breite, so wird erklärt. Die Augäpfel betragen ein Drittel des Auges, wird erklärt[2]. Im Verhältnis zur Länge des Gesichtes sollen die Augen gemacht werden, so ist die genaue Erklärung.

619—656. Was die Weite der Augen betrifft, die einem Bogen von Bambus gleichen, so ist das Maß auf drei Gerstenkörner ($^3/_8$ Zoll) festgelegt. Die dem Blatte des Utpala[3] gleichenden Augen werden als das Maß von sechs Gerstenkörnern ($^3/_4$ Zoll) erklärt. Die Augen, welche einem Fischmagen ähnlich sind, werden als das Maß von acht Gerstenkörnern (ein Zoll) erklärt. Die dem Blatte des Padma[4] Lotus gleichenden Augen haben das Maß von neun Gerstenkörnern ($1^1/_8$ Zoll). Die Augen, die Cowrie-Muscheln[5] ähnlich sind, betragen zehn Gerstenkörner ($1^1/_4$ Zoll). Was die Augen der Könige anbelangt, so ist ihre Länge und Weite bereits gelehrt worden. Für die Yogin sollen die Augen der Herzenseinfachheit gleich einem Bogen von Bambus gemacht werden. Bei Frauen und Buhlern sollen die einem Fischmagen gleichenden zur Anwendung gelangen. Was die Augen der gewöhnlichen Menschen betrifft, so ist das dem Utpala gleichende Auge zugrunde zu legen.

---

[1] D. i. die Breite der Stirn.

[2] Sie sind also $^2/_3$ Zoll lang und $^1/_3$ Zoll breit.

[3] Der blaue Lotus, *Nymphaea stellata*.

[4] Der weiße Lotus, *Nelumbium speciosum*.

[5] Tib. *mgron-bu = agron-bu*. CHANDRA DAS führt *mgron-bu t'al-ba* an als Name einer Medizin, die Blutungen hemmen soll, was offenbar mit dem von JÄSCHKE notierten *agron-bui t'al* (*Med.*) „die pulverisierte Asche der verbrannten Muschel" identisch ist.

Zum Ausdruck des Schreckens und des Weinens bedient man sich des dem Blatte des Padma Lotus gleichenden Auges, wird erklärt. Die Augen der von Zorn und Schmerz Geplagten sollen gleich der Cowrie-Muschel[1] gemalt werden. Die dem Utpala Blatte gleichen Augen sind an den Rändern rot, die Augäpfel sind schwarz und glänzend, die Wimpern sind mit langen Spitzen versehen und anmutig, der Glanz ihrer Farbe in weichen Tönen. Wenn man den Göttern die Augen gemalt hat, wird den Königen und Geschöpfen Heil daraus erwachsen. (Die Augen der Götter) sind glänzend wie die Farbe der Kuhmilch und fettig[2], mit Wimpern ohne Rauheit, in ihrem Leuchten wie das Blatt des Padma Lotus, und durch die blaue Bindehaut im Farbenspiele wechselnd[3], mit Augäpfeln schwarz und groß: das Malen solcher Augen bringt Reichtum und Glück ein. So ist also vollständig gezeigt worden, wie die Maße der Weite der Augen sind. Welches die sechsunddreißig Arten von Mienen sind, die Theorie davon wird später gelehrt werden[4].

657—664. Die Nase ist vier Zoll (lang), die Spitze beträgt an Höhe zwei Zoll. An der Stelle, wo die Nasenlöcher schräg gemacht werden, beträgt die Weite zwei Zoll. Die Weite der Nasenlöcher ist sechs Gerstenkörner ($^3/_4$ Zoll), ihre Höhe wird als zwei Gerstenkörner ($^1/_4$ Zoll) erklärt. Der Raum zwischen

---

[1] Siehe Anm. 5, S. 154.

[2] Tib. *snum* = Skr. *snigdha*.

[3] Tib. *bsgyur byas*, besonders von der Veränderung der Farbe gebraucht. In der Mahāvyutpatti kommt einmal die Gleichung Skr. *lilāyita* „spielend, sich belustigend" = tib. *lus bsgyur-ba* vor, letzteres also „einer, dessen Körper sich spielend verändert oder dreht, wendet." In ähnlichem Sinne ist hier von dem veränderlichen Farbenspiele der Augen die Rede: bald erscheinen sie weiß wie der Lotus, bald himmelblau.

[4] Dieser Abschnitt ist nicht in diesem Werke enthalten, muß also wohl verloren gegangen sein.

den Nasenlöchern beträgt an Breite zwei Gerstenkörner ($^1/_4$ Zoll), an Länge sechs Gerstenkörner ($^3/_4$ Zoll).

665—674. Die Oberlippe beträgt an Höhe einen Zoll, die Unterlippe einen halben Zoll, wird erklärt. Die Länge der Oberlippe und Unterlippe beträgt je vier Zoll, soll man wissen. Der über der Oberlippe liegende Teil beträgt an Ausdehnung einen halben Zoll. Der Rand der Lippen ist rot wie die Bimbafrucht[1], ebenso sind sie dem Rücken eines Bogens gleich[2]. Die Mundwinkel sind ein wenig emporgezogen und verleihen ihm so den Ausdruck eines lieblichen Lächelns[3].

675—686. Das Kinn beträgt an Höhe zwei Zoll, an Breite drei Zoll, wird erklärt. Was das Längenmaß des Halses betrifft, so wird gelehrt, daß es vier Zoll beträgt. Bei Gestalten, die gen Himmel schweben, ist der Hals gleichmäßig lang und dick, nämlich zehn Zoll. Bei den mit schlanker Gestalt Gesegneten ist er acht Zoll lang, bei den Dicken nur ein Drittel davon. Bei diesen tritt eine Verschönerung von drei Hautfalten ein, die gleich einer Muschel geformt sind; auf der Rückseite ist der Hals nicht so niedrig und soll rund im Umfange gemacht werden.

687—692. Die Wangen sollen fünf Zoll gemacht werden, die Kinnbacken vier Zoll[4]. Die Maße der Konturen der Backen[5] betragen vier Zoll, soll man wissen. So habe ich

---

[1] Die rote Frucht einer Cucurbitacee (*Momordica monadelpha*), mit der die Lippen häufig verglichen werden (*pw*).

[2] D. h. so gekrümmt oder geschwungen. Es handelt sich hier um den zusammengesetzten indischen Bogen, der aus Schichten von Horn und Holz (*ru śiṅ*) gebildet ist.

[3] Der poetische Ausdruck *ḍdsum bag-can* findet sich auch bei Milaraspa.

[4] Nach der Mahāvyutpatti entspricht tib. *mkʻur tsʻos* Skr. *gaṇḍa* und tib. *ḍgram·pa* Skr. *hanu.*

[5] Tib. *mur* (oder *mu*) *ḍgram* = Skr. *çaṅkha* (Mahāvyutpatti). Nach

denn die Länge und Breite des Gesichts auseinandergesetzt.

693—704. Viereckig, scharf umrissen[1], schön vollendet,
mit glänzenden und schönen Abzeichen versehen, soll das Gesicht gemacht werden. Dreieckig soll es nicht gemacht werden, auch nicht schief; zornig soll es nicht gemacht werden,
auch nicht rund: wer ein solches Gesicht gemalt hat, wird beständig Güter erwerben. Für die gewöhnlichen Menschen
kommt ein nach Ruhe verlangendes, langes, rundes, schiefes,
dreieckiges usw. Gesicht in Betracht. Die vorher erwähnten
Eigenschaften dagegen passen auf die Gesichter der Götter
und Könige, deren Maße oben dargelegt worden sind.

705—734. Nun sollen die Körperteile dargelegt werden.
Hier wende deine ganze Aufmerksamkeit[2] an! Der Raum
zwischen Lenden und Bauch ist zwei Zoll breit. Die Breite
der Schultern[3] beträgt sechs Zoll. Die Schultern[4] sind acht
Zoll (lang). Die Brust wird in zwei Flächen eingeteilt, lautet
die Erklärung. Der Penis soll an Länge sechs Zoll gemacht
werden. Die Hüfte beträgt achtzehn Zoll an Länge. Von
der Stelle zwischen den Schlüsselbeinen[5] bis auf die Brust ist

---

dem viersprachigen Wörterbuch des Kaisers K'ien-lung = manju šakšaha
= chin. 腮 sai.

[1] Tib. *lhám* = *lham-me*, *lham-pa*; vgl. V. 914.

[2] Tib. *yeñs* = *ryeñs*.

[3] Tib. *spuñ ya* ist offenbar = *dpuñ mgo* = Skr. *skandha* (Mahāvyutpatti ed. Csoma und Ross, p. 93, Nr. 14).

[4] Tib. *p'rag-pa* = Skr. *aṁsa*.

[5] Tib. *sgrog rus* wird in unseren Wörterbüchern und in der Mahāvyutpatti nicht aufgeführt. Das tibetisch-mongolische Wörterbuch *Zla-bai od-
snañ* (Pekingdruck, 1838) erklärt es durch mong. *tobčilaghur*, *tobčilūr*, was
nach Kowalewski 1. Kropf der Vögel, 2. Schlüsselbein, Seitenknochen des
Halses, 3. Agraffe bedeutet. Als tibetische Äquivalente zitiert Kowalewski
og-sgrog und *sgrag* (offenbar für *sgrog* verdruckt)-*rus*. Als Sanskritäquivalent
läßt sich also *jatru* ansetzen, ein Wort, von dem freilich Hoernle (Studies

eine ebene Fläche, aber nicht von der Brust bis zur Nabelhöhle; dagegen ist wieder eine Fläche vom Nabel bis auf den Penis. Die Lenden betragen vierzehn Zoll (an Länge). Der Nabel ist einen halben Zoll (im Umfang), ist tief und nach rechts gewunden[1]. Die Brustwarzen betragen ein volles Gerstenkorn

in the Medicine of Ancient India, p. 159) nachgewiesen hat, daß es nicht, wie gewöhnlich angenommen wird, das Schlüsselbein, sondern die Luftröhre, wenigstens in der medizinischen Literatur, bezeichnet. In der Tat könnte *sgrog rus*, wenn die Ableitung von *sgrog-pa* „Töne ausstoßen" (*rus* „Knochen") berechtigt wäre, in letzterem Sinne gedeutet werden. Doch das in unserem Texte auf *sgrog rus* folgende Wort *bar* „Zwischenraum" gibt zu erwägen, daß es sich an dieser Stelle um ein Knochenpaar handelt, und daß daher die Annahme der Bedeutung „Schlüsselbein" hier vorzuziehen ist. Auch paßt dies besser zu dem Sinn der Stelle, in der von der einheitlichen Oberfläche der Brust die Rede ist, während „von der Luftröhre bis auf die Brust" sinnlos wäre. Für die Bezeichnung der Luftröhre haben wir im Tibetischen die beiden Wörter *lkog-ma* und *mid-pa*. In der Mahāvyutpatti wird unter den Namen der Körperteile Skr. *jatru* durch tib. *nam ąts'oň* (so im Druck der Palastausgabe; ed. Csoma, p. 39, Nr. 40: *nam ts'oň*) interpretiert, und das Wörterbuch der katholischen Missionare, offenbar einem einheimischen Wörterbuch folgend, setzt *nam ts'oň* (verlesen als *ts'od*) gleich *sgrog rus* (erklärt als ein Knochen des Körpers, mit Fragezeichen). Daraus scheint zu folgen, daß Skr. *jatru* den Tibetern wirklich als „Schlüsselbein" von den Indern erklärt oder von ihnen wenigstens so verstanden worden ist. Eine andere Gleichung erlaubt uns, die Bedeutung von *sgrog rus* in dem Sinne von „Schlüsselbein" zu bestätigen und auch die Etymologie richtig zu erkennen. In dem großen viersprachigen Wörterbuch des Kaisers K'ien-lung (s. meine Skizze der mongolischen Literatur, S. 177, *Keleti Szemle*, 1907) finden wir nämlich unter den Namen der Körperteile (Kapitel 10, p. 81) die Reihe tib. *sgrogs rus* = manju *alajan* = mong. *omoro* = chinesisch *sotse ku* 鎖子骨. Letzteres bezeichnet das Schlüsselbein und bedeutet wörtlich „Kettenknochen"; eine Erklärung lautet: „Knochen, die wie Ketten rasseln" (Giles, Chinese-English Dictionary, 2. A., p. 1264 c). Danach ist also wohl der tibetische Ausdruck als wörtliche Übersetzung aus dem Chinesischen zu verstehen (*sgrog* oder *sgrogs* „Fesseln", *rus* „Knochen"), und darin wird der Grund zu suchen sein, warum diese chinesische Bildung nicht in der Vyutpatti und anderen tibetisch-indischen Wörterbüchern Aufnahme gefunden hat.

[1] Diese beiden Eigenschaften entsprechen Nr. 37 und 38 in der Liste

('/₈ Zoll) und im Umfang ringsherum zwei Zoll. Bei denen, die mit einem Untergewand bekleidet sind und einen Gürtel umgebunden haben, soll der unterhalb des Nabels liegende Teil des Bauches vier Zoll gemacht werden. Der Penis ist zwei Zoll breit, der Hodensack sechs Zoll lang; die Hoden sollen nicht zu sehr herabhängen und beide gleichmäßig rund sein. Im Zustand der Erektion[1] betragen sie an Dicke sieben Zoll, an Länge vier Zoll. Was die Dicke des Penis betrifft, so ist sie sechs Zoll, soll man wissen. Vom Penis bis zur Grenzlinie des Bauches beträgt die Entfernung sechs Zoll.

735—784. Von da kommen wir zu den Oberschenkeln, und was deren Längenmaße anbelangt, sollen sie fünfundzwanzig Zoll betragen. Die Knie sind vier Zoll (lang), die Seiten der Fußknöchel vier Zoll. Die Unterschenkel haben zwei Flächen, wird erklärt. Da, wo sie die Seiten der Knöchel berühren, sind sie vier Zoll breit; in der Mitte sind die Unterschenkel sechs Zoll breit. Die Flächen der Knie sollen drei Zoll breit gemacht werden; es gibt für sie kein Längemaß.

---

der achtzig Anuvyañjana, *lte-ba zab-pa* (*gambhīranābhi*) und *lte-ba ɣyas-su aḱyil-ba* (*pradakshiṇāvartanābhi*), s. Mahāvyutpatti ed. Csoma und Ross, pp. 95, 96; Dharmasaṁgraha ed. Max Müller und Wenzel, p. 57, Nr. 39, 40. Es ist bekannt, daß *ɣyas-su aḱyil-ba* von den nach rechts gewundenen Trompetenmuscheln (*duṅ*) gesagt wird, die als kostbarer gelten als die gewöhnlichen nach links gewundenen (vgl. Pander-Grünwedel, Das Pantheon des Tschangtscha Hutuktu, S. 105, Note 4); Skr. *nandāvartaya* oder *nandyāvarta* (Eitel, Handbook of Chinese Buddhism, p. 105). Diese Anschauungen erklären sich aus der Physiognomik. Die Schrift Sāmudravyañjanāni (Tanjur, Sūtra, Vol. 123, Nr. 27) sagt im achten Abschnitt darüber: „Wer einen weiten, tiefen und innen gerundeten Nabel hat, wird im Glück dahinleben; wessen Nabel im Inneren nicht gerundet ist, der wird ein Dasein des Unheils führen." Ebenda heißt es: „Wenn bei einer Frau die Linien der Finger nach rechts gewunden sind, so wird sie einen Sohn erlangen; wenn sie nach links gewunden sind, wird eine Tochter geboren werden."

[1] Tib. *gšis pʻebs*.

Die Breite der beiden Oberschenkel zusammen, über der oberen Seite der Knie gemessen, beträgt acht Zoll. In der Ausdehnung der Dicke sollen die Oberschenkel zwölf Zoll gemacht werden. Die Vorderseite der Oberschenkel ist hoch, und beide sollen vom Muskelfleisch geschwellt sein, zart und ohne Unebenheiten wie der Rüssel eines Elefanten. Knöchel und Knie sollen nicht sichtbar sein, ebensowenig die Adern[1]. Die Waden sollen gerundet werden. Die Fußbiege[2] soll ein wenig hoch gemacht werden. Die Ferse soll fünf Zoll an Höhe betragen und drei Zoll an Breite, wird erklärt. Die beiden Füße betragen vierzehn Zoll, die großen Zehen vier Zoll, den Spitzen des roten Lotus gleich, und wie der Saft des Lacks[3] glänzend. Die Fußsohlen sind ausgestreckt und haften fest an der Erde; sie erscheinen mit dem Abzeichen des Rades geschmückt[4]. Ferse und die großen Zehen sollen,

---

[1] Dieselben Eigenschaften in bezug auf Knöchel und Adern werden in der Liste der Lakshana und Anuvyañjana hervorgehoben (Mahāvyutpatti ed. Csoma und Ross, Part I, p. 94, Nr. 26, 7 und 9), ebenso bei den Merkmalen der Frauen in Tanjur, Sūtra, Vol. 123, Nr. 26, mit denselben Worten: *rtsa ni mi mṅon daṅ, loṅ-bu-dag ni mi mṅon žiṅ*; eine solche Frau wird die Herrin der Männer werden.

[2] Tib. *sgaṅ* bezeichnet einen vorspringenden Teil am Körper (vgl. das Wörterbuch der französischen Missionare, p. 239b); hier ist *rkaṅ-sgaṅ*, der hohe Teil des Fußes, gemeint. Die wörtliche Bedeutung von *sgaṅ* ist „kleine Erhöhung, hoch gelegener Ort"; vgl. das viersprachige Wörterbuch (Kap. 2, p. 46): tib. *sgaṅ* = manju *nuhu* = mong. *ghurbi* = chin. 高地 *kao ti*.

[3] Tib. *rgya skyegs*. Vgl. Heinrich Laufer, Beiträge zur Kenntnis der tibetischen Medicin, S. 64. Die Frauen im alten Indien pflegten sich die Zehen und Nägel der Füße mit Lack zu färben. Im Drama Mālavikā und Agnimitra schmückt eine Dienerin den Fuß der Mālavikā mit künstlerisch gezeichneten Mustern in Lack (Charles Joret, Les plantes dans l'antiquité II L'Iran et l'Inde, p. 378).

[4] Ebenso in der Liste der Lakshana (Mahāvyutpatti ed. Csoma und Ross, p. 94, Nr. 29).

wenn sie die Erde berühren, aneinander haften[1]. Die Netzhaut der Füße des Cakravartin gleicht der der Wildgans[2]; diese Füße sind acht Zoll lang, hoch wie der Rücken einer Schildkröte[3] und mit Schönheitsabzeichen versehen. An Breite betragen sie fünf Zoll und sollen hübsch gemacht werden. Die kleinen Zehen betragen den sechsten Teil eines Zolles und ein Gerstenkorn (an Länge) und sind $^1/_6$ Zoll breit. Die große Zehe ist zwei Zoll an Breite und sechs Zoll dick; an Länge ist sie vier Zoll, die vordere Spitze[4] in die Höhe gerichtet. Die anderen Zehen sollen den Daumen ein wenig überragen. Was Länge und Dicke derselben anbetrifft, so sollen sie drei Zoll gemacht werden.

785—810. Die überragenden Zehen sollen in ihren Maßen um je einen Zoll verringert werden. [Länge und Dicke der kleinen Zehe betragen je zwei Zoll.][5] Das Haut-

---

[1] Hier haben wir wieder Entlehnung aus der Physiognomik anzunehmen. In der Abhandlung Sāmudravyañjanāni (*l. c.*) heißt es: „Eine Frau, deren Fußsohlen die Erde berühren (*sa-la reg*, ebenso wie im Texte oben) wird gleichförmig im Glück leben." Die kleine Zehe braucht dagegen den Boden nicht zu berühren; eine Frau, bei der dies der Fall ist, wird ein Alter von hundert Jahren erreichen. Die Füße sollen flach sein, um fest auf dem Erdboden zu stehen. Dasselbe wird in der Reihe der Lakshaṇa durch die Phrase *supratishṭitapāda* (tib. *žabs šin-tu gnas-pa*) ausgedrückt.

[2] In den Lakshaṇa (Nr. 28, nach anderer Zählung Nr. 30, nach Dharmasaṁgraha Nr. 3) sind sowohl die Finger als die Zehen durch eine Netzhaut verbunden (*dra-bas ạbrel-ba*). Von den Händen wird dieselbe Angabe weiter unten S. 164 gemacht. Die Ausgabe der Mahāvyutpatti von Csoma und Ross, zu der Schiefners Triglotte und andere Editionen leider nicht zu Rate gezogen sind, gibt an dieser Stelle ein falsches Sanskritäquivalent; es muß heißen: *jalabandhahastapāda*.

[3] Aus der Physiognomik entlehnt, wo dieser Vergleich häufig auftritt.

[4] Tib. *druṅ-ma rtse-mo*; *druṅ-ma*, das vorn Befindliche.

[5] Dieser Satz scheint eine aus einem anderen Werke entlehnte, hier interpolierte Angabe zu sein, da sie mit der vorherigen Aufstellung in Widerspruch steht.

II

gewebe[1] zwischen den Zehen soll schön, die Gelenke hübsch und ohne Erhöhungen[2] sein, sehr gerundet, ohne daß die Adern hervortreten, ohne daß Muskeln oder Knochen sichtbar werden. Die Nägel sollen dem Halbmond gleichen, von roter Farbe und glänzend sein, wie die Augäpfel leuchtend[3]. Dieses leuchtende Glänzen ist durch den Saft des Safflors[4] hervorgebracht, mit dem sie eingerieben werden. Wie Chrysoberyl[5] sollen sie glänzen. Der Form des Halbmonds gleich, von roter Farbe, von fettiger Farbe, sollen sie ohne Flecke und glatt sein. Erfüllt von einem Kranz von Gerstenkörnern[6], sind

---

[1] Tib. *bar-t'ags*.

[2] Tib. *ạbar ạbur* wird (SCHIEFNER's Triglotte, fol. 36b) durch *šañ šoñ*, *mt'o dman* und Skr. urkūlanikūla (?) erklärt.

[3] Die Nägel werden ähnlich in den drei ersten Merkmalen der achtzig Anuvyañjana charakterisiert, nämlich als kupferfarbig (dem „rotfarbig" unseres Textes entsprechend), als glänzend (*mdog snum-pa* im obigen Texte ist also einfach gleich Skr. *snigdha*) und hoch, d. h. lang. Ebenso werden die Nägel in den beiden Schriften im Tanjur über Physiognomik behandelt. „Kupferfarbene und lange Nägel sind schön" (*sen-mo zañs mdog riñ-ba mdses*) und gelten als Merkmale einer erfolgreichen Frau. Runde Finger mit kupferfarbenen Nägeln und ein Daumen von blutroter Farbe deuten als günstige Vorzeichen auf die Erlangung der Königsherrschaft. Weiße Nägel dagegen bringen Unheil, gespaltene Nägel führen nicht zur Erfüllung der Wünsche, und einer mit krummen, abgebrochenen und dicken Nägeln wird ein armer Schlucker bleiben, der beständig von Almosen leben muß. Man ersieht aus diesem Beispiel die Arbeitsmethode der Physiognomik.

[4] Tib. *leb-rgan rtsi* wird in der Mahāvyutpatti als Skr. *kusumbha* (mong. *küsünük*) erklärt (*Carthamus tinctorius*), noch jetzt in vielen Teilen Indiens als Färberpflanze und des Öles wegen gebaut (W. ROXBURGH, *l. c.*, p. 595).

[5] Tib. *ke-ke-ru* = Skr. *karketaṇa* (vgl. Prākṛit *kakkeraa* in der Mṛicchakaṭikā: L. FINOT, Les lapidaires indiens, p. III) Chrysoberyl (*ibid.*, pp. XVI, 49—51).

[6] Damit sind offenbar die runden Linien gemeint, welche um und über die Fingergelenke laufen. Vgl. unten V. 822, wo sich die Variante *nas ạp'reñ dgañ žiñ* findet. Tib. *nas-kyi ạp'reñ-ba* würde einem Skr. *yava-māla* oder *yavāvali* entsprechen. *Yava* ist in der Chiromantie eine dem Gerstenkorn ähnliche Figur an der Hand (*pw*). MONNIER WILLIAMS: (*in*

die Gelenke der Finger wohl gefügt[1] und schön verbunden[2]. Die Zehen bestehen aus drei Gliedern, und die große Zehe soll so weit seitwärts gestellt werden, daß der Raum zwischen dem großen und dem vierten Zeh einen halben Zoll an Breite beträgt. Was unterhalb des Knöchels liegt, wird als Fuß definiert; an Höhe beträgt dieser Teil vier Zoll. Die Merkmale von der Länge und Breite des Fußes habe ich bereits erklärt.

811—838. Höre nun die Merkmale von der Hand des Menschenbeherrschers[3]. Die Handfläche soll sieben Zoll (lang) und fünf Zoll breit gemacht werden. Der Mittelfinger soll auf das Maß von fünf Zoll gebracht werden. Der Zeigefinger erreicht die Hälfte des Nagels (des Mittelfingers), und der vierte Finger hat eben dieses Maß. Der kleine Finger soll am kürzesten gemacht werden. Dem Daumen werden vier Zoll zugeschrieben. Der Daumen besteht aus zwei Gliedern, die mit einem Kranz von Gerstenkörnern erfüllt sind, aber keine

---

palmistry) *a figure or mark on the hand resembling a barley-corn, a natural line across the thumb at the second joint compared to a grain of barley and supposed to indicate good fortune.*

[1] Tib. *ḁbyor-ba* ist hier Intransitivum zu *sbyor-ba* in dem Sinne von *adhaerere*, angefügt sein. Vgl. V. 842.

[2] Tib. *šin-tu grims*. Diese Phrase erscheint unter Nr. 30 in der Liste der achtzig kleineren Schönheitsmerkmale eines Buddha: *sku šin-tu grims-pa* = Skr. *susaṁhatagātra* (Mahāvyutpatti, ed. Csoma und Ross, p. 95), wo indessen die Übersetzung *a very patient or subdued body* verbesserungsbedürftig ist. Gemeint ist offenbar „ein Körper, in dem alle Teile schön untereinander verbunden sind". In der tibetischen Übersetzung des Lalitavistara (ed. Foucaux, p. 99) wird dafür *sku rim-gyis legs-par grub-pa*, „ein wohl proportionierter Leib" gesagt. Verschieden von *šin-tu grims* scheint *šin-tu ḁgrims* zu sein; die Phrase *sku šin-tu ḁgrims-pa* wird im tibetisch-mongolischen Wörterbuch *Zla-bai od-snaṅ* als „mit einem sehr geraden, aufrechten Körper" (mong. *maši čikärän bäyä-tü*) erklärt. Schiefner's Triglotte (fol. 6 b) hat gleichwohl die Schreibung *ḁgrims* in der obigen Stelle.

[3] Tib. *mi-yi bdag-po* = Skr. *nṛipati*.

11*

Erhöhungen zeigen. Was den Daumenballen[1] betrifft, so werden ihm drei Zoll zugewiesen. Die Maße des unteren Teiles an der inneren Seite des Daumens[2] sollen neun Gerstenkörner (1⅛ Zoll) betragen, und zwar acht Gerstenkörner (1 Zoll) an Breite, und an Länge neun Gerstenkörner. In vier Abschnitten soll, vom Daumen angefangen, die Hand von einer Netzhaut durchzogen sein[3]. Die Nägel sind von roter Farbe, glänzend und hübsch; der Perlmutter gleich sollen sie wunderbar sein. [Der Daumenballen soll an Breite drei Zoll gemacht werden.][4] Das Handgelenk soll an der Vorderseite sieben Zoll an Länge betragen. Damit sind die Längen- und Breitenmaße des Daumens erklärt.

839—844. Was die Nägel der Finger anbelangt, so soll ihr Maß die Hälfte eines Fingergliedes betragen; die Spitzen sind fein und gleichen (in ihrer Schlankheit) der Taille des Leibes. Fehlerlose Fingerglieder sind wohl gefügt; die Glieder sind hübsch, rund und lang. Damit sind die Finger der Könige erklärt.

845—866. Die beiden Handflächen sollen dem roten Lotus gleichen; die Linien in denselben sind tief[5], doch nicht ge-

---

[1] Wörtlich: das dicke Fleisch vorne am Daumen. Meine Konjektur *rbad śa* ist nur ein Notbehelf; vielleicht ist die **A** und **B** gemeinsame Lesart *rbe śa* in V. 833 doch richtig. Der Umstand, daß in V. 823 **A** *sbed* und **B** *sbe* schreibt, beweist, daß es sich um ein ungewöhnliches, wohl veraltetes Wort handelt. An dem Sinn der Phrase kann indessen kein Zweifel sein.

[2] Wörtlich: der Bauch des Daumens.

[3] Tib. *dra-bai rim-pa* = Skr. *jālakrama*. Vgl. oben S. 161.

[4] Wiederholung des oben Gesagten, offenbar Interpolation.

[5] Skr. *gambhīrapāṇilekha*, ist Nr. 44 in der Reihe der achtzig Schönheitsmerkmale eines Buddha (Mahāvyutpatti ed. CSOMA und Ross, Part I, p. 96).

krümmt[1]. An beiden Enden sind sie nur ein wenig tief und fein; an Farbe dem Blute des Hasen gleichend[2], werden sie durch dreifache Linien verschönert. Die Handflächen sind mit den Abzeichen des Çrīvatsa, des Svastika, Nandyāvarta, und des Rades geschmückt[3], fühlen sich wie Baumwolle an[4] und sollen wie ein Bündel[5] Seidenfäden weich sein. Allenthalben ist das Muskelfleisch üppig; die Adern sind vorhanden, aber werden nicht sichtbar. Der Handrücken[6] ist hoch, aber die Handfläche hat eine leichte Vertiefung[7]. Die Netzhäute zwischen den Fingern sind zart und schön. Die innere Seite[8] der Nägel ist rot wie der Utpala, der Haube des Nāga-Königs

---

[1] Tib. *ya-yo med-pa*. Die Angabe JÄSCHKES, der DESGODINS und CHANDRA DAS folgen, daß *ya-yo* nur der Umgangssprache angehöre, besteht also nicht zu Recht. Im Lalitavistara (FOUCAUX's Text, p. 99, 8) wird derselbe Ausdruck gebraucht: *p'yag-gi ri-mo ya-yor ma gyur-pa* (*il a les lignes de la main non tortueuses*); weiter unten: *sku ya-yo ma mc'is-pa* und *lte-ba ma yo-ba*. Vgl. auch Avadānakalpalatā (Tib. Prosaausgabe), p. 38, 8.

[2] Skr. *çaçaçrij*. Dabei ist wohl an eine rubinrote Farbe zu denken, denn in der Ratnaparīkshā des Buddhabhaṭṭa wird der Rubin mit dem Hasenblut verglichen (L. FINOT, Les lapidaires indiens, p. 26).

[3] Nr. 80 der Schönheitsmerkmale. Tib. *ak'yil-pa* = Skr. *nandyāvarta*. Vgl. FOUCAUX's Übersetzung des Lalitavistara (Paris, 1848), p. 110. Eine Abbildung dieses Symbols bei J. ANDERSON, Catalogue of the Archæological Collections in the Indian Museum, Part II, p. 197. Es findet sich häufig auf tibetischen Petschaften.

[4] Nr. 42 der Schönheitsmerkmale.

[5] Oder vielleicht wie ein Seidenband, da in der Mahāvyutpatti einige Komposita angeführt werden, in denen tib. *c'un-po* = Skr. *dāma* ist, z. B. *mu-lig-gi c'un-po* = *muktādāma* „Perlenschnur".

[6] Tib. *bol*, hat nicht ausschließlich die Bedeutung Fußrücken, wie JÄSCHKE angibt. Das viersprachige Wörterbuch des Kaisers K'ien-lung hat die folgenden beiden Gleichungen: tib. *lag-pai bol* = mong. *ghar-un aru* = manju *gala-i huru* = chin. 手背 „Handrücken"; tib. *rkan bol* = mong. *ülmäi* = manju *umuhun* = chin. 脚面 „Fußrücken".

[7] Tib. *gšon-ba*, Höhlung, Vertiefung; Gegensatz: das oben (S. 160) erwähnte *sgan*. Derselbe Ausdruck kehrt in V. 1043 wieder: *lag mc'il gšon*.

[8] Tib. *lto-ba* „Bauch".

vergleichbar. Die Seiten der Nägel sind weich, zart und leuchtend, rotfarbig, groß und glänzend. Wenn sie hoch und leuchtend sind, verleihen sie der Hand eine besondere Schönheit. So sind die Merkmale von den Größenverhältnissen der Hand dargelegt.

867—894. Was die Länge und Breite des Oberarms betrifft, höre, o König, mir zu! Das Längenmaß des ganzen Arms wird auf sechsunddreißig Zoll angegeben, wovon auf den Oberarm achtzehn Zoll kommen, und somit ebensoviel auf den Unterarm. Der Schulter[1] kommen sechs Zoll zu. Der Oberarm erreicht an Breite fünf Zoll; dem Armgelenk[2] stehen drei Zoll zu, den beiden Unterarmen vier Zoll. Die Seiten der Schultern sollen breit und völlig abgerundet gemacht werden, an Weite achtzehn Zoll[3]; Adern und Gelenke sollen nicht sichtbar sein[4]. Die Arme mit Einschluß der Hände betragen in ihrer ganzen Länge achtundvierzig Zoll[5], indem sie bis auf die Knie herabreichen[6]; sie sind sehr lang, allenthalben

---

[1] Tib. *dpuṅ-mgo* = Skr. *bāhuçikhara* oder *skandha*. Nach der oben (S. 157) gemachten Angabe sind die Schultern sechs Zoll breit und acht Zoll lang.

[2] Tib. *mk'rig-ma*, kann, wie aus dem Zusammenhang hervorgeht, an dieser Stelle nur das Armgelenk bezeichnen. JÄSCHKE und CHANDRA DAS geben nur die Bedeutung „Handgelenk" (Vyutpatti: Skr. *maṇibandha*); das Wörterbuch der französischen Missionare (p. 126) erklärt mit Recht: *junctura manus et brachii*. Das viersprachige Wörterbuch hat die Gleichung: tib. *ạk'rig-ma* = mong. *šu* (KOWALEWSKI: Knochen, der die Schulter mit dem Arm verbindet) = manju *sabta* (v. D. GABELENTZ: Knochen des Oberarms, Armspindel) = chin. 連細骨 *lien si ku*.

[3] Vgl. Nr. 14 der Lakshaṇa (*dpuṅ-mgo šin-tu zlum-pa*, Mahāvyutpatti ed. CSOMA und ROSS, Part I, p. 93). Das Breitenmaß ist vom Ende der einen bis zum Ende der anderen Schulter zu verstehen.

[4] Vgl. Nr. 7 der Anuvyañjana (*rtsa mi mṅon-pa*), wie bereits oben S. 160.

[5] Die Zahl ergibt sich durch die Addition der Längen des Armes, der Handfläche und des Mittelfingers $36 + 7 + 5 = 48$.

[6] Tib. *pus-mo la slebs-pa*. Dieselbe Phrase kehrt in Nr. 18 der zwei

vom Muskelfleische geschwellt; symmetrisch[1] und hübsch sollen sie gemacht werden. Der Oberarm des Herrschers der Menschen ist symmetrisch wie der Schweif eines Rindes; wenn er aufrecht dasteht, berühren beide Hände die Knie. Daher werden die Hände eines Königs als „die Knie erreichende" (*jānudaghna*) definiert. So sind denn die Merkmale der Längen- und Breitenmaße des Oberarms und des Arms dargelegt.

895—926. Nunmehr soll dir, was noch nicht dargelegt worden ist, erklärt werden. Von der Grenzlinie des Haupthaars bis zum Halsgelenk[2] sind es sechs Zoll an Maß. Für die beiden Schulterblätter des Rückens werden sechzehn Zoll angenommen. Die Länge des Raumes zwischen den Schulter-

---

unddreißig Schönheitsmerkmale des Mahāpurusha wieder, dessen Arme so lang sind, daß, wenn er aufrecht steht, die Hände bis zu den Knien reichen (Mahāvyutpatti, *l. c.*, p. 93; Lalitavistara, ed. Foucaux, p. 98, 7). Über die Bedeutung der Langarmigkeit vgl. vor allem Grünwedel's interessante Ausführungen (Buddhistische Kunst in Indien, S. 139). In dem bereits zitierten Werke über Physiognomik wird gesagt, daß derjenige, dessen Arme die Knie erreichen, wenn er gerade aufgerichtet ist, den mit allen Vorzügen geschmückten Arm besitzt, während ein Mann mit sehr kleinem Arm ein Untertan wird.

[1] Tib. *byin-gyis p'ra*, ist in der Mahāvyutpatti Äquivalent für Skr. *anupūrva* „regelmäßig, symmetrisch".

[2] Tib. *ltag rus-pa*. Beide Drucke haben den gemeinsamen Fehler རྔ, was leicht für རྔ versehen werden konnte. Einen Knochen *lhag* gibt es aber nicht. Die Lesart kann, wie auch der Zusammenhang lehrt, nur *ltag* sein. Das Wort *ltag-pa* wird in der Mahāvyutpatti durch Skr. *krikāṭikā* „Halsgelenk" (von *krikāṭaka* „Nacken") erklärt. Und das ist offenbar, was der Zusammenhang verlangt. Denn hier beginnt, wie die folgenden Sätze zeigen, die Beschreibung des Rückens. Unter der Grenzlinie des Haupthaars ist also das Nackenhaar zu verstehen. Jäschke erklärt *ltag-pa* als „Rückseite des Halses, Nacken"; das tibetisch-mongolische Wörterbuch *Zla-ba od-snań* schlechtweg durch „Rücken" (mong. *aru*), ebenso das Wörterbuch der katholischen Missionare („hintere Seite, Rücken, Rückseite eines Gegenstandes").

blättern beträgt zehn Zoll, die Breite neun Zoll, wird von den Kennern der Maße gelehrt und erklärt. Der Raum zwischen den Schulterblättern gleicht der Figur eines feinen Netzes und soll sehr hübsch gemacht werden. Das Rückgrat[1] hat sechs Zoll. Die Mittellinie des Brustkastens[2] ist zwanzig Zoll lang und zwei Zoll breit. Sie ist durch einen Einschnitt[3] bezeichnet und soll hübsch gemacht werden. Diese Eigenschaft muß man sich für die Männer merken, bei den Frauen dagegen ist der Einschnitt klein. Der zwischen den Hüften liegende Teil des Bauches[4] ist viereckig, scharf umrissen, eingezogen und wohl proportioniert; er wird auf sechs Zoll angegeben. Der Raum von diesem Teil des Bauches bis zum Hüftbein ist vier

---

[1] Tib. *rgyab-kyi šul.* Die tibetischen Bezeichnungen für die Körperteile bereiten uns oft Schwierigkeiten, da sie keinen anatomisch korrekten Begriffen entsprechen und bald weiter, bald enger als unsere Definitionen zu fassen sind. Das Wort *šul* wird weder in der Mahāvyutpatti noch im viersprachigen Wörterbuch erklärt. Nach JÄSCHKE, der auf CSOMA zurückgreift, bedeutet es „Rückgrat, Rücken, Hintern"; nach dem Wörterbuch der katholischen Missionare ist es gleichbedeutend mit *rgyab* „Rücken". Aus unserem Texte geht hervor, daß es hier nur einen Teil des Rückens bezeichnet. Wahrscheinlich ist unter dem Maße von sechs Zoll einfach die Breite der Mittelpartie des Rückens zu verstehen. Die Tibeter haben natürlich keine rechte Vorstellung von der Wirbelsäule; das in der Mahāvyutpatti dafür aufgenommene Wort *sgal ts'igs* ist eine künstliche Bildung und Übersetzung von Skr. *prishṭavaṃça.*

[2] Tib. *ro-stod* = Skr. *pūrvārdha* (Mahāvyutpatti ed. CSOMA und ROSS, Part I, p. 93, Nr. 19). Nach dem viersprachigen Wörterbuch des Kaisers K'ien-lung (Kap. 10, p. 76) bezeichnet es den Brustkasten: tib. *ro-stod* = mong. *tsäghäji* = manju *cejen* = chinesisch 胸膛.

[3] Tib. *luṅ-bu* = *luṅ-pa* Furche, Vertiefung, Kerbe.

[4] Tib. *sgu-do,* wird in unseren Wörterbüchern nicht erklärt, fehlt auch in der Mahāvyutpatti. Die im Texte selbst gegebene Definition läßt keinen Zweifel an der Bedeutung und der Gleichsetzung von *sgu* mit *dku;* offenbar ist *sgu-do* = *dku-zlum,* Skr. *kukshi.* Das tibetisch-mongolische Wörterbuch *Zla-bai od-snaṅ* (fol. 27) definiert *sgu-do* als einen Teil der Brust (*äbčigün sabar saba*), was auf die obige Stelle nicht paßt. Zur Wortbildung vgl. *mjug-do* = Skr. *trika* (Vyutpatti).

Zoll breit. Die Arschkerbe beträgt zwei Zoll. Hinterbacken von acht Zoll Länge sind schön und sollten nicht zu flach[1] noch zu weit vorspringend sein. Wenn wohl gerundet, sind sie sehr schön. Ihre Breite beträgt sieben Zoll, und damit sind diese Maße gelehrt. Hiermit sind die Definitionen der Proportionen dargelegt und erledigt.

---

[1] Tib. *žum* wird im tibetisch-mongolischen Wörterbuch *Zla-bai od-snan* („Das Mondlicht", gedruckt in Peking, 1838) durch mong. *khamsaku* oder *khamsiku* „abgeflacht sein" erklärt. Vgl. bei KOWALEWSKI mong. *khamsighar* = tib. *sna žom* = Skr. *avabhraṭa, avaṭīṭa* „flachnasig". Die Schreibung *žom* habe ich im Mahāçītavanasūtra (*bsil-bai ts'al c'en-poi mdo*, fol. 8b) gefunden, wo die Amanusha als *mig ser sna žom-pa* „gelbäugig und flachnasig" beschrieben werden (ebenso im viersprachigen Wörterbuch des Kaisers K'ien-lung, Kap. 11, p. 11). Tib. *žum* oder *žom* scheint demnach hauptsächlich von Körperteilen gesagt zu werden. Nach dem viersprachigen Wörterbuch des Kaisers K'ien-lung scheint tib. *žum-pa* auch die Bedeutung von „eng, spitz zulaufend" zu haben. In Kap. 11, p. 22 finden wir tib. *smad žum* = mong. *šubtung* = manju *sibsihôn* = chin. 臉下窄 „einer, dessen Gesicht oben breit und unten spitz ist; spitzkinnig". KOWALEWSKI zitiert auch mong. *šubtughur* = tib. *žum-pa*. Ferner bedeuten tib. *lus žum* und *žum ạdug* „mager, abgemagert" (*ibid.*, p. 23) = mong. *ghardsaiksan* = manju *gebsehun, gebserehebi*. Eine weitere in unseren Wörterbüchern nicht verzeichnete Bedeutung von tib. *žum-pa* ergibt sich aus der Gleichsetzung mit mong. *mitarakhu* „sich biegen" (von Eisen und Holz); das große viersprachige Wörterbuch des Kaisers K'ien-lung (Kap. 16, p. 2) verzeichnet das Kausativum tib. *žum ajug* = mong. *mitarighulumui* (KOWALEWSKI: *mitara-*) = manju *musembumbi* „beugen, krümmen". Daraus wird nun die abgeleitete Bedeutung von *žum-pa*, die einzige in unseren Wörterbüchern notierte, klar, „niedergebeugt, niedergeschlagen, bekümmert"; auch im Mongolischen bedeutet das erwähnte Kausativum „einschüchtern", ferner mong. *mital* „Betrübnis, Schüchternheit", *mitakhu* „betrübt sein, sich fürchten". Im Tibetischen zeigt die negative Form am besten die ursprüngliche Bedeutung. Vgl. Nr. 28 der Anuvyañjana (bei CSOMA und ROSS, p. 95): *sku žum-pa med-pa* = Skr. *adīnagâtra*, wo offenbar noch die Idee eines ungebeugten Körpers vorschwebt, wenn auch der Sinn „unverzagt, unerschrocken" sein mag. Tib. *žum-žum-por byed-pa* bedeutet „verscheuchen" (von Vögeln gesagt). Etymologisch gehört auch das Verbum *ajum-pa* (fut. *gžum*) 'zum Stamme *žum*.

927—932. Nunmehr, da ich mit diesen Erklärungen zu Ende bin, und du sie aufmerksam in dich aufgenommen hast, höre meine Auseinandersetzung über das Verfahren bei der Komposition[1] der Zähne, Haupt- und Körperhaare und die Merkmale der Farben, welche für die Gesichter der Götter nicht in Betracht kommen[2].

933—947. Die Zähne sind gleichförmig, dicht geschlossen[3], glänzend, rein, scharf und weiß, weiß wie Perlen, wie Kuhmilch, wie der Stengel des Lotus[4], wie ein Haufen Schnee. Vierzig[5] sind es, treffliche Zähne. Die Eckzähne sind sehr schön, an Länge drei Gerstenkörner ($\frac{3}{8}$ Zoll), soll man wissen, ebenso an Breite zwei Gerstenkörner ($\frac{1}{4}$ Zoll). Das Zahnfleisch, der Gaumen und die Umrisse der Zunge sollen rot gemacht werden. Die Eckzähne gleichen, wenn man sie ein halbes Gerstenkorn ($\frac{1}{16}$ Zoll) größer macht (also $\frac{7}{16}$ Zoll), der sich entfaltenden Blüte des Jasmins[6], sind weich wie ein Lotus-

---

[1] Tib. *skye-bai t'abs*, die Methode der Entstehung oder Bildung, nämlich in der Malerei; scheint fast unserem Begriffe „Stil" nahezukommen.

[2] Sondern für den Cakravartin ausschließlich.

[3] Diese Eigenschaften der Zähne werden auch in der Liste der zweiunddreißig (Nr. 6—9) und der achtzig Körpermerkmale (Nr. 53—57) aufgeführt. Der Ausdruck im Texte *t'ags ni dam-pa* (= *dam-po* „fest, dicht, eng") ist dort mit *t'ags bzaṅ-ba* (Skr. *aviraladanta*) wiedergegeben.

[4] Tib. *padmai rtsa* = Skr. *bisa*.

[5] Auf diese Anomalie macht bereits FOUCAUX (*Annales du Musée Guimet*, Vol. XIX, p. 29) aufmerksam. Natürlich haben die indischen Ärzte und wohl auch das Volk gewußt, daß die Zahl der menschlichen Zähne zweiunddreißig beträgt (HOERNLE, Studies in the Medicine of Ancient India, Part I, p. 182). Die typische Zahl der Zähne bei den Säugetieren ist vierundvierzig, bei den Zibetkatzen vierzig; der Löwe, mit dem ja der Cakravartin auch in bezug auf Körperform verglichen wird, hat aber nur dreißig. Die vierzig Zähne sind daher wohl nur durch die mystische Bedeutung der Zahl vierzig zu erklären.

[6] Tib. *sna-mai me-tog* = Skr. *sumanapushpa* (vgl. L. FEER, Le Karma-Çataka, p. 28, *Extrait du Journal asiatique*, 1901) *Jasminum grandiflorum* (HOERNLE, The Bower Manuscript, General Sanskrit Index, p. 339). Vyut-

stengel, von scharfer Spitze und sehr gerundet. Die Eckzähne sollen überaus[1] fleckenlos sein.

948—956. Die Zunge gleicht dem Lotusblatt, schimmert wie ein züngelnder Blitz und soll in ihrer leuchtenden Farbe dem Blute gleichen. Sie gleicht dem jungen Blatte der Adha-Pflanze[2] und wird vom Gesicht völlig verdeckt. Wohllautend ist der Klang der Stimme, dem mächtigen Elefanten gleich; wie der König der Pferde[3], stößt sie Töne aus und äußert Töne wie die Stimme des Donners.

957—968. Das Haupthaar ist fein, einer Spiralfigur[4] gleichend, von Natur in fettigem Glanze schimmernd, in seiner Schönheit dem Saphir[5] gleich, von der Farbe der Biene[6], dem

---

patti gibt *sumana* und *jātikusuma*; beide Namen repräsentieren dieselbe Spezies.   [1] Tib. *šas-kyi* = *šas c'er*.

[2] W. Roxburgh (Flora Indica, p. 47, Calcutta, 1874) erwähnt eine Pflanze mit dem Bengālīnamen *Adha-birni*, identifiziert mit *Gratiola Monnieria* (Blätter: *sessile, long, obovate, entire*). Vielleicht ist auch das von Hoernle (The Bower Manuscript, General Sanskrit Index, p. 249) erschlossene *aḍhaki, Cajanus indicus*, zur Erklärung heranzuziehen.

[3] Tib. *rta-yi* (= *rtai*) *rgyal-po* = Skr. *açvarāja*. So wird auch Çākyamuni an drei Stellen im Lalitavistara bezeichnet.

[4] Tib. *mc'u-ris*, weiter unten zweimal *c'u-ris*. Das Wort ist weder in unseren noch in einheimischen Wörterbüchern erklärt; *ris* ist Zeichnung, Figur, Ornament; vgl. *sdoṅ ris* (Lalitavistara, p. 100, Z. 19) = Skr. *vardhamāna*. An dieser Stelle — es handelt sich um das letzte der achtzig kleineren Schönheitsabzeichen — werden die Formen des Haares mit den Figuren des Çrīvatsa, Svastika, Nandyāvarta und Vardhamāna verglichen, mit anderen Worten, es handelt sich um das spiralisch gewellte Haar, wie es z. B. auf den Köpfen der Buddhastatuen dargestellt wird, und ich glaube daher *c'u-ris* in dem Sinne von Spiralornament fassen zu müssen.

[5] Tib. „*anda-rñil* (auch „*andra-rñil* oder -*sñil*), volkstümliche Schreibung für Skr. *indranīla* (*Li-šii gur k'aṅ*, fol. 20b: *indranīla žes-pa dbaṅ siion-te, zur-c'ag-pas (Apabhraṃça) anda-rñil žes-pa daṅ*) Saphir (R. Garbe, Die indischen Mineralien, S. 83; L. Finot, Les lapidaires indiens, pp. 39—42, 118—122, 162—164).

[6] D. i. schwarz. Vgl. Nr. 74 der achtzig Schönheitsabzeichen: *dbu skra buṅ-ba ltar gnag-pa*.

Antimon[1] gleich, dem Nacken des Pfaus gleich, der Brust des Kuckucks gleich, an Glanz einer unveränderlichen Farbe gleich, bis auf die Schulterblätter herabwallend, in Locken gedreht[2] und nach rechts gewunden. Die Spiralen gleichen der Mähne des Löwen[3]. Der Haaraufsatz auf dem Kopfe (çikhābandha) ist nach links gewunden. So ist die Erscheinung des Haupthaars.

969—986. Bei dem großen Manne[4] sind die Waden, Oberschenkel, Pubes, Armhöhlen, Ohren und Nasenlöcher, Hals, Gesicht und Wangen unbehaart. Mit Haarflaum dagegen, der nach oben gerichtet[5], glatt, fein und fettig ist und an Farbe dem Antimon[6] gleicht, der, wenn spiralförmig, schön ist, der in runden Windungen gedreht ist, soll die Brust der Könige verschönert und anmutig gemacht werden. Bei den Göttern, die als Menschen dargestellt werden, ist im Gesichte weder Barthaar noch Flaum; ihr Körper ist unbehaart und gleicht dem eines Sechzehnjährigen, soll man wissen. Bei dem Herrscher der Menschen und den Göttern ist das Haupthaar fein

---

[1] Tib. *mig sman* (wörtlich: Augenmedizin) = Skr. *añjana* (mit derselben Bedeutung), nach R. GARBE (Die indischen Mineralien, S. 54): Schwefelantimon (s. auch P. CHANDRA RAY, A History of Hindu Chemistry, 2nd ed., vol. I, p. 119). Der Vergleich bezieht sich auf die schwarze Farbe des Stoffes.

[2] Die wie Weidenblätter herabfallen, wie es Lalitavistara p. 100, Z. 17 heißt: *dbu skra kaṅ lor aḳʿril-ba daṅ ldan-pa*, woraus sich auch die Identität von *aḳʿril bag can* und *aḳʿril-ba daṅ ldan-pa* ergibt.

[3] Die Löwenmähne wird bekanntlich in Spiralen dargestellt, eine Stilart, die sich aus der indischen in die chinesische Kunst verpflanzt hat.

[4] Tib. *bdag ñid cʿen-po* = Skr. *mahātman, mahāpurusha*.

[5] Tib. *gyen pʿyogs*; *pʿyogs* ist verbal zu fassen, wie sich aus Nr. 22 der zweiunddreißig Schönheitsabzeichen ergibt.

[6] Tib. *mig sman sṅon-po* = Skr. *nīlāñjana*.

und gelockt[1] und wird durch eine himmelblaue[2] Farbe verschönert. So sollen die anmutigen Züge aller Wesen gemalt werden.

987—1010. Gleich dem geschmolzenen Golde des Jāmbū-Flusses[3], gleich dem ausgehöhlten Stengel des voll entfalteten[4] Lotus, gleich der farbensatten, leuchtenden Magnolia[5], soll der das Rad drehende große Mann[6] dargestellt werden[7]. Wie der

---

[1] So nach **A** (*aḱril bag can*); nach **B** (*aḱyil bag can*): gedreht, gewunden.

[2] Tib. *mḱon mḱin* = Skr. *atinīla*.

[3] Tib. *aDsam-bui ću-boi gser* = Skr. *Jāmbūnadasuvarṇa*. Vgl. Nr. 17 der zweiunddreißig Schönheitsmerkmale. Das Gold vom Jāmbūfluß wird auch in der einheimischen Literatur, besonders in Vergleichen, häufig erwähnt. So sagt ein Spruch:

མཁས་པ་གང་ཞིག་རབ་ཏུ་སེམས་དང་བས །

སངས་རྒྱས་མཆོད་རྟེན་ལ་ནི་གོམ་འདོར་ན །

ཇམྦུའི་ཆུ་བོའི་གསེར་གྱི་སྲང་ཚད་ནི །

སྟོང་ཕྲག་བརྒྱ་ཡང་དེ་དང་མཚམས་པ་མིན །

„Wenn ein Weiser gläubigen Sinnes seine Schritte zu einem Caitya Buddhas lenkt, so kommen hundertmal tausend Unzen Goldes vom Jāmbūfluß dem nicht gleich."

[4] Tib. *rgyas-pa*, vgl. H. BECKH, Beiträge zur tibetischen Grammatik, S. 43 (*Abhandlungen der Berliner Akademie*, 1908).

[5] Tib. *tsam-pa-ka* = Skr. *campaka*, *Michelia Champaka* (W. ROXBURGH, Flora Indica, p. 453). Vgl. HEINRICH LAUFER, Beiträge zur Kenntnis der tibetischen Medicin, II. Teil, S. 59—60; CH. JORET, Les plantes dans l'antiquité, II L'Iran et l'Inde, p. 305 (Paris, 1904). Nach J. S. GAMBLE (List of the Trees, Shrubs, and Large Climbers found in the Darjeeling District, p. 3, Calcutta, 1896) kommt der Baum bis zu einer Höhe von dreitausend Fuß vor, und sein Stamm erreicht in einem Alter von hundert bis hundertzwanzig Jahren einen Umfang von acht Fuß. Die Blüten sind groß, gelb und angenehm duftend.

[6] Skr. *purusha cakravartin*.

[7] Nämlich in bezug auf seine Körperfarbe, die einem Goldgelb entspricht.

Gang des Elefantenkönigs, des Führers der Herde, wie der Gang der Wildgans[1], des Löwen und des Cakravartin, so, soll man wissen, ist der König der Menschen darzustellen[2]. Er hat den Rang und die große Kraft des Elefantenkönigs, den Scharfsinn des Stierführers[3], des Herrschers, die Stärke des Löwenkönigs, die Majestät des Königs der Wildgänse: so ist die äußere Erscheinung des Gebieters der Menschen. Alle Gangarten übertrifft er, einem Lehrer der Schauspielkunst[4] gleich, einem Lehrer der Wesen gleich. Was die Fleischfarbe des Cakravartin betrifft, so ist sie unbefleckt von Staub und Schmutz, fein, leuchtend, wohlriechend und wie ein Spiegel[5]

---

[1] Tib. *ñañ-pa* = Skr. *haṁsa*.

[2] Vgl. die achtzig Schönheitsmerkmale, Nr. 11—13.

[3] Tib. *k'yu mc'og* = Skr. *vrishabha*.

[4] Tib. *zlos-gar-dag ston-pa* = Skr. *naṭyācārya*.

[5] Über Spiegel (Skr. *ādarça*) im alten Indien scheint leider wenig bekannt zu sein. FOUCHER (L'art gréco-bouddhique, Vol. I, p. 465) beschreibt eine Reliefskulptur von Hidda, auf der eine Frau bei ihrer Toilette dargestellt ist; sie betrachtet sich in einem kreisförmigen Standspiegel, der zusammen mit einer Parfümbüchse auf einem Tische steht. In einer fragmentarischen Replika dieser Szene im Museum von Kalkutta hält sie den Spiegel in der linken Hand. R. FICK (Die sociale Gliederung im nordöstlichen Indien, S. 189) schließt auf das Vorhandensein von Spiegeln aus der Beschreibung einiger Gauklerkunststücke. Eine greifbarere Stelle findet sich in den Vedānta-Sūtra (G. THIBAUT, *Sacred Books of the East*, Vol. XLVIII, p. 67): "The mirror is only the cause of a certain irregularity, *viz.* the reversion of the ocular rays of light, and to this irregularity there is due the appearance of the face within the mirror; but the manifesting agent is the light only." Aller Wahrscheinlichkeit nach waren die altindischen Spiegel aus Metall verfertigt (RHYS DAVIDS, Dialogues of the Buddha, p. 24, London, 1899). Im *mDsaṅs-blun* (SCHMIDT's Ausgabe, S. 151, Z. 1; Übersetzung, S. 188; SCHIEFNER's Ergänzungen, S. 37) werden Spiegel aus Bronze oder Eisen erwähnt (*mk'ar-ba am lcags-kyi me-loṅ*); auch im alten China hat es eiserne Spiegel gegeben. Bhāvamiçra, Verfasser des medizinischen Werkes Bhāvaprakāça (16. Jahrhundert) empfiehlt den Gebrauch des Spiegels bei der Pflege des Haares, Bartes und der Nägel (JOLLY,

sanft anzufühlen. Wie Licht ausstrahlende Edelsteine, ist sein Gewand weißfarbig und lose angeordnet. Der das Rad drehende König soll rings von Licht umflossen gemalt werden.

1011—1056.[1] Der wolkenlose Mond ist überaus schön. Will man die von Licht umflossene Gestalt mit etwas vergleichen, so ist der das Rad drehende Beherrscher der Erde das schöne Licht. Darum soll gewiß der Nimbus an seinem Körper haften. Sein Antlitz ist weiß wie das Licht der Mondscheibe. Die Welt der Menschen ist daher an einem Mondscheinabend gleichsam eine, die zwei Monde hat[2]. Seine Brauen sind schön, sein Hals ist schön, seine Stirn im Antlitz ist schön. Die Farbe des Haupthaars ist schön, glänzend, weich, die Spitzen gewunden. Die Nase ist hoch und gerade, die Lippen sind wie mit Blut bedeckt[3]. Die Zähne sind rein, gleich frischen Perlen. Die Wimpern gleichen dem besten Ölglanz und sind lang. Himmelblau leuchtend sind die Augen, lang gezogen und schön. Die Stirnlocke (ūrṇā), Licht ausstrahlend, hat ihre Stelle zwischen den Brauen. Der weiße Körper soll sehr schön dargestellt werden. Die Ohren gleichen

---

Medicin, S. 38). In den eigentümlichen Metallspiegeln von Java, von denen sich einige gute Exemplare im ethnographischen Museum zu Leiden befinden, dürfen wir wohl Ausläufer der altindischen Spiegel erblicken.

[1] Diese 45 Verse haben sicher als Interpolation oder als eine von unserem Autor aus einer anderen Quelle geschöpfte Entlehnung zu gelten. Äußerlich charakterisieren sie sich durch das veränderte Versmaß, indem hier statt des bisherigen siebensilbigen ein neunsilbiges trochäisches Maß auftritt (in V. 1033 elf Silben). Die Merkmale des Cakravartin werden zusammengefaßt, der Mehrzahl nach Wiederholungen des schon vorher Gesagten, aber auch Widersprüche zu den vorhergehenden Angaben enthaltend (vgl. EINLEITUNG, S. 24).

[2] Tib. zla gñis scheint mir Übersetzung von Skr. dvicandra „zwei Monde habend" zu sein. Der Tibeter kann es doppelsinnig verstehen, denn zla gñis bedeutet auch „zwei Gefährten".

[3] Tib. sba = Skr. gupta. Vgl. V. 1039 und 1040.

an Farbe dem ausgehöhlten Stengel des voll entfalteten Lotus; die Ohrlappen sind symmetrisch darzustellen. Der Nacken gleicht der Muschel (çankha)[1], und der Zwischenraum zwischen den Schultern ist ausgefüllt[2]. Die Schultern sind schön zusammengefügt, die Glieder völlig rund. Die Brust ist muskulös und wohl proportioniert. Wie der Bauch des Löwenkönigs ist der Körper lang ausgestreckt[3]. Die Lenden sind wohl gerundet, und der Körper strotzt von Muskeln. Der Nabel ist nach rechts gewunden und tief[4]. Gleich dem Elefantenkönig, hält er seine Geschlechtsteile in einer Höhlung eingezogen[5]. Allseits ist er wohlgerundet, so daß die Gelenke nicht sichtbar sind[6]. Die Oberschenkel sind wohlgerundet wie der Rüssel des Elefanten. Die obere Fläche des Oberschenkels ist hoch, der Knöchel ist nicht sichtbar[7]. Die Zehen sind durch eine Netzhaut wohl verbunden und schön; der Rücken beider Füße gleicht der Schildkröte[8] und ist fleischig.

---

[1] Nämlich in der weißen Farbe.

[2] Tib. *f'al gon rgyas-pa*, ebenso Lalitavistara p. 98, Z. 6 (FOUCAUX: l'épaule bien arrondie) und Mahāvyutpatti (ed. CSOMA und ROSS, p. 93, Nr. 16; CSOMA's Übersetzung ist schwerlich korrekt) = Skr. *citāntarāṁsa*. Vgl. Dharmasaṁgraha (ed. MAX MÜLLER und WENZEL, p. 54, Nr. 20): having the place between the shoulders well filled out (ebenso GRÜNWEDEL, Buddhistische Kunst in Indien, S. 138). Die tibetische Übersetzung ist unerklärbar. In der Liste der Körperteile in der Mahāvyutpatti wird tib. *f'al gon* durch Skr. *skandha* „Schulter" gedeutet.

[3] Wie das Wörterbuch der katholischen Missionare (p. 391 b) richtig erklärt, ist *rñon* eine alte Form für *rkyon*. Vgl. V. 765.

[4] Nr. 37 und 38 der achtzig Anuvyañjana. Vgl. oben S. 158, Anm. 1.

[5] Tib. *ądoms-kyi sba-ba sbubs-su nub* (ebenso Lalitavistara, p. 98, Z. 10 und Mahāvyutpatti, p. 93, Nr. 23) = Skr. *koṣagatavastiguhya*. Dharmasaṁgraha, p. 53, Nr. 13.

[6] Vgl. oben S. 166.

[7] Vgl. Mahāvyutpatti, p. 94, Nr. 26 und Nr. 9, und oben S. 160.

[8] Vgl. oben S. 161, Anm. 3.

Sie werden von Nägeln geschützt, deren Spitzen die Form des Halbmonds haben. Die Fußsohlen sind weich, mit dem Abzeichen des Rades versehen[1], geschützt und nicht gewölbt[2]. Der Vorderarm ist rund, und wenn er ausgestreckt wird, lang, und wie der Schweif des Stieres symmetrisch[3]. Die Finger sind lang und wohlgerundet; die Handfläche ist hohl und mit dem Rade geschmückt, das von dem Lichte der Nägel erstrahlt. Die Gestalt ist schön und weiß wie ein Büschel Campa-Blumen[4]. Der Wildgans wird die Majestät des Cakravartin in seinem Gang verglichen. So habe ich denn sämtliche Proportionsmaße des Cakravartin erklärt[5]. Unter- und Oberschenkel sind üppig, der ganze Körper ist sehr üppig. Hände und Füße sind schön, doch die Gelenke sind nicht sichtbar. Rücken und Rippen sind schön; vom Arme ist der Oberarm schön, das Gesicht ist schön, und der Körper, viereckig, ist schön und sehr muskelkräftig; kurz, alle Glieder sind schön. Der ganze Leib ist von gefälliger Farbe und

---

[1] *Ibid.*, p. 94, Nr. 29, und oben S. 160, Anm. 4.

[2] Tib. *ljaṅ* fasse ich im Sinne von *ljaṅ-duṅ* „solide, nicht hohl". Die Lesart *sba-la ljaṅ*, obwohl übereinstimmend in **A** und **B**, erscheint mir etwas zweifelhaft.

[3] Vgl. oben S. 167.

[4] Das Petersburger Wörterbuch identifiziert Skr. *campa* mit *Bauhinia variegata*. W. Roxburgh (Flora indica, p. 344) hat indessen unter dieser Spezies einen anderen Sanskritnamen, und die Beschreibung der Blüten „von lebhaftem Purpurrot" paßt nicht auf den obigen Fall; dagegen hält er für *campa* den Baum *Liriodendron grandiflora*, dessen Blüten weiß und duftig sind (p. 452).

[5] Es macht den Eindruck, als sei dieser interpolierte Abschnitt hier abgeschlossen, und als sei das Folgende bis Vers 1056 wiederum einer anderen Quelle entlehnt. Die Wiederholung am Schluß betreffend den Gang der Wildgans weist notwendig darauf hin.

12

rhythmischer Bewegung[1]. Die körperlichen Vorzüge sind schön und weit ausgedehnt. Ein Gang wie der der Wildgans wird für den Cakravartin in Anspruch genommen.

1057—1076. Was die anbelangt, die den Ruf des Weisen (*paṇḍita*) haben, so haben sie, kurz gesagt, die allen Menschen zukommenden Hauptgliedmaße, Kopf, Hals, zwei Arme, Ober- und Unterschenkel, so lautet die Erklärung. Die Lehre davon gipfelt in dem Satz, daß jedem seine eigenen Zollmaße zukommen[2]. In betreff der Maße der Nebenglieder besteht die Erklärung, daß sie proportioniert sein sollen. In gleicher Höhe mit dem Boden[3], mittels Länge, Breite, Dicke und der übrigen Maße und Methoden, sollen sie in bezug auf die Nebenglieder schön und vortrefflich gemalt werden. Daher soll das Studium der Maße bei den Weisen den Anfang nehmen, soll man wissen. Wenn man beim Studium der Maße und Komposition[4] die menschlichen Formen und die Götter beiseite setzen und nur Piçāca, Bhūta, Rākshasa und solcherlei Platz einräumen[5] wollte, so würden diese bald das Glück der Menschen

---

[1] So fasse ich die Bedeutung von *mk'ril* (= *ak'ril*)-*bag-can* (vgl. Mahāvyutpatti, p. 95, Nr. 18) = Skr. *vṛttagātra*. Die Übersetzungen *having round members* (Dharmasaṁgraha p. 56, Nr. 18) und *a chosen body* (CSOMA) sind nicht zutreffend. Die tibetische Übersetzung zeigt, daß es sich um eine Bewegung des Körpers handelt.

[2] D. h. es kommt kein bestimmtes Maßschema in Frage, sondern jeglicher ist nach seiner Individualität zu behandeln.

[3] Tib. *sa gži mñam*, ist vielleicht nicht wörtlich zu fassen, sondern von dem Boden, dem Hintergrund der Malerei zu verstehen und zu übersetzen: im Verhältnis zu dem Gesamtbilde, im Maßstab der Komposition. Vgl. unten V. 1145 und S. 183.

[4] Tib. *gnas t'abs*. Vgl. oben S. 143, Anm. 3.

[5] Tib. *gnas byed*.

vernichten[1] und den trefflichen Bestrebungen[2] Hindernisse bereiten.

1077—1141. Die Merkmale von den Proportionen der Gebieter[3] sind bereits gelehrt worden. Die vier Könige mögen nach eigenem Ermessen dargestellt werden. Dies sind Balin[4], Bhāskara[5], Rāma[6], der Sohn des Daçaratha, und der Sohn des Agnidhāra[7], die je nach ihren entsprechenden Maßen von den

---

[1] Bei dieser Auffassung ist statt *ajigs* „sich fürchten" *ajig* „vernichten" zu lesen; es ist bekannt, daß die Orthographie beider Wörter vielfach verwechselt wird. Möglicherweise ließe sich *mi* als Negation und *mi-śis* als „Unheil" auffassen (vgl. *mi śis-pai ltas* bei JÄSCHKE; den nominalen Gebrauch kann ich freilich nicht belegen, doch vgl. *bkra mi śis*); dann ließe sich *ajigs-pa* halten und übersetzen: dann wird bald Unheil und Furcht entstehen. Doch scheint mir die obige Auffassung sinngemäßer zu sein, indem den Dämonen eine unmittelbare Veranlassung der bösen Einflüsse durch Vermittlung ihrer Bilder, in denen sie lebendig werden, zugeschrieben wird.

[2] Oder einfacher: den Guten (*legs-pa-rnams*). Darunter sind die Maler zu verstehen, die ja in ihrem Bestreben, die Dämonen zu malen, nur von den besten Absichten geleitet waren, ohne an das Unheil zu denken, das diese etwa über die Menschen bringen könnten.

[3] Tib. *gtso-bo*, nämlich die Cakravartin. Wie in einem Buddhatempel die größte und vornehmste Statue des Gottes, dem der Tempel geweiht ist, *gtso-bo* genannt wird, so ist der Cakravartin die Hauptperson unter den Figuren der Malerei.

[4] Tib. *stobs-can* (= *stobs-ldan*). Vielleicht ist Balin, der König der Unterwelt Pātāla, gemeint(?).

[5] Tib. *snan byed*. Das Sanskritäquivalent wird von CHANDRA DAS gegeben. Ich weiß nicht, um welchen König es sich handelt. Vermutlich ist ein anderer Sanskritname anzusetzen.

[6] Im Tib. *ra-ma* transkribiert. "From various references in Hindu dramatic literature we may conclude that the history of Rāma and Sītā and of the Pāṇḍava heroes from whom many of the Hindu kings claimed descent were frequently illustrated in the fresco paintings of the royal *citra-çāla*, or picture-halls, which have now entirely disappeared" (E. B. HAVELL, The Ideals of Indian Art, p. 138).

[7] Tib. *me rnoi bu*; *me* „Feuer" entspricht Skr. *agni*, *rno-ba* „scharf" Skr. *tīkshṇa* oder *dhāra* „Schärfe, Schneide". Agnidhārā ist nach *pw* der

12*

Kennern der Maße darzustellen sind. Die Sādhu[1] sollen um zwei Zoll kleiner gemacht, die Mālava[2] um vier Zoll; die Vyañjana[3] sollen um acht Zoll verringert werden, die Giridhara[4] um zehn Zoll. Um häßliche Wirkungen zu vermeiden, sollen diese in ihren Maßen nach freiem Ermessen gemalt werden. Wenn einem Körper die Schönheit nicht vollständig vorhanden ist, wie soll man dann in bezug auf die Maße verfahren? Die Definition von den Maßen[5] der vier Könige nimmt deshalb

---

Name eines Tīrthaka. Möglicherweise liegt ein anderer Sanskritname zugrunde.

[1] Tib. *bzaṅ-po* = Skr. *sādhu*, ist vielleicht in dem Sinne der Heiligen (Arhat) der Jaina zu fassen. Mögen die Indologen entscheiden; mein Standpunkt ist nur der des Jainaphilosophen *syādvādin*.

[2] Tib. *ma-la-bar skyes-pa*, in *Ma-la-ba* geboren, würde einem Skr. *malavajāta* oder *mālava* entsprechen. Vielleicht entspricht aber ein anderes Sanskritoriginal, denn das Tibetische läßt sich auch analysieren: *ma-la*, Dativ zu *ma* „Mutter", *bar* „zwischen" (Skr. *antara*), *skyes-pa* „geboren" (der in der Mutter geborene? *mātariçvan?*). Diese Deutung soll natürlich nur dem Sanskritisten den Weg ebnen, um in der Aufspürung des Richtigen zu helfen·

[3] Tib. *gsal-byed*.

[4] Tib. *ri-bo ạc'oṅ* (= *ạc'aṅ*) = Skr. *giridhara, parvatadhara*. In V. 1121 ist die Lesart *ri-bo mc'og* „trefflichster Berg" (*parvatottara?*). Bei diesen vier Namen handelt es sich jedenfalls nicht um Eigennamen bestimmter Personen, sondern um generelle Termini einer Gruppe. Denn wie aus den weiter unten angegebenen Längenmaßen hervorgeht, folgen sie unmittelbar auf den Typus des Cakravartin; ihre Längenmaße sind 106, 104, 100, 98 zu 108 des letzteren, so daß anzunehmen ist, daß sie gleichfalls gewisse Typen darstellen, vielleicht vier Rangstufen von Königen.

[5] Tib. *ts'ad-kyi mts'an-ñid* = Skr. *mānalakshaṇa*. Nicht alle Körper besitzen die Schönheitsmerkmale in vollständiger Zahl, wie sie nur bei der Idealgestalt des Cakravartin vorkommen. Es läßt sich daher für das Malen der gewöhnlichen Könige (Könige der Menschen, tib. *mii rgyal-po*) kein bestimmter Kanon von Regeln aufstellen. Es läßt sich nicht mit Sicherheit sagen, welche individuellen Schönheitsabzeichen einem gewissen Könige eignen oder fehlen; nur die allgemeine Regel steht fest, daß sie gegenüber den des Cakravartin ein Minus bedeuten. Die Art der Darstellung muß also den besonderen Fall im Auge haben und sich vom Gefühl und von

eine Sonderstellung ein. Wenn ich entsprechend der mir von Brahma gegebenen Lehren dir die Erläuterungen gegeben habe, und du Grantha[1] in großer Zahl und die Lehre von den vortrefflichen Merkmalen erlangt hast, werden die Wesen geringen Verstandes, die noch nicht davon gehört haben, in Schrecken geraten[2]. Çâstra über die Merkmale der vom Weibe Geborenen gibt es zwölftausend; in fünf Klassen (eingeteilt), sind sie von Brahma und mir selbst gelehrt worden. Diese Lehren von den fünf Klassen, die sich mit den körperlichen Merkmalen der Menschen befassen, werden, wenn ich sie dir vorgetragen habe, zahlreiche Grantha zur Folge haben. Wie sind nun die Könige der Menschen, nach den dir gegebenen Lehren, zu behandeln? Die Maße des Cakravartin habe ich ja vorgetragen, und da ich dir die Definition der Maße sämtlicher großer Männer[3] unter den Menschen und der Götter gelehrt habe, so bist du kraft deines eigenen Verstandes am sicheren Ziel[4] und kannst die Sâdhu usw., jene vier, malen.

ästhetischen Gesichtspunkten leiten lassen. Feste Regeln oder schematische Behandlung würden ein Zerrbild ergeben und den ästhetischen Eindruck zerstören. Daher steht die Behandlung der vier Könige für sich allein (tib. *so-so*), ist individuell.

[1] Tib. *gžuṅ*.

[2] Man muß im Auge behalten, daß die körperlichen Merkmale und die Malereien, die sich darauf gründen, sozusagen identisch sind, was wir in der Übersetzung kaum zum Ausdruck bringen können. Unter dem Erschrecken des Volkes ist wohl der überwältigende Eindruck zu verstehen, den die Werke der Malerei auf den Laien hervorbringen werden; die große Mannigfaltigkeit der Abzeichen, die vielen Formen der Schönheit, die sie vorführen, und die große Zahl der Schriften, zu denen sie Veranlassung bieten, werden den Unbefangenen, der mit dieser Materie noch nicht vertraut ist, auf den ersten Blick verwirren.

[3] Tib. *skyes mc'og* = Skr. *purushottama*.

[4] Tib. *ṅes byas*. Du stehst dann auf eigenen Füßen und bist gesichert, kannst auf eigene Hand arbeiten; denn die Kenntnis der Malerei des Cakravartin bildet die Grundlage für alle anderen Figuren.

Was hierbei die Länge der Sādhu betrifft, so ist sie hundert-
undsechs (Zoll) zu machen[1], bei den Vyañjana hundert, soll
man wissen. Die Mālava haben ein Plus von vier[2]. Die
Länge der Giridhara ist achtundneunzig Zoll zu machen. Eine
Methode der Maße für die anderen als diese hat es nicht gegeben, und wird es nicht geben. Die Gebieter der Menschen,
die Herren, werden gemalt (in derselben Weise wie) die Gestalten der niedrigsten Menschen: auf ihre Längen- und Breitenmaße werden sie alle geprüft, und so werden die Maße durch
Schätzung festgestellt. Daher läßt man so bei den vom
Weibe Geborenen sein eigenes Urteil walten, und in ihrer Länge
sollen sie, indem man den Königen je einen Zoll mehr gibt,
entsprechend um einige Zollmaße verringert werden. Nacken,
Kopf, Nabel, Brust, Lenden, Ober- und Unterschenkel, Knie
und beide Füße sollen in ihren Längenmaßen nicht häßlich
wirken. Nach der Methode der Maße sollen Verkürzungen
eintreten: die Maße der adligen, mittleren und niederen Klassen
sind je nach ihrem Rang in einer Folge von Proportionen ab-
gestuft, so wird gelehrt[3].

1142—1146. Nun höre die Methode der Maße für die

---

[1] Die Länge des Cakravartin beträgt, wie im Anfang dieses Kapitels
angegeben, hundertundacht Zoll; die Sādhu nehmen also die zweite Stelle
nach ihnen ein.

[2] Sind also hundertundvier Zoll lang; *skyed* ist Nomen: Zuwachs, Wachs-
tum, Zinsen.

[3] Der Sinn ist wiederum, daß es keine besondere in Regeln gefaßte
Lehre für das Malen der Könige und gewöhnlichen Sterblichen gibt. Sie
sind je nach Rang und Stellung in ihren Längenmaßen verschieden, die
vom Könige ab gerechnet entsprechend niedriger werden. Der Unterschied
in der Länge bedingt eine entsprechende Verkürzung der Breite, und daher
fällt der Nachdruck auf die richtige Proportion, deren Vernachlässigung
eine häßliche Wirkung zeitigen würde. Man beachte das Maß persönlicher
Freiheit, das auf Grund dieser Anschauung dem Künstler gelassen war.

Frauen[1]. Auch hier läßt man, wie es für die Könige ausein-
andergesetzt wurde, sein eigenes Urteil walten; in harmonischem
Maßstab[2] sollen sie dargestellt werden und züchtig erscheinen[3].
In zahlreichen Gruppen[4], im Verhältnis zur Gesamtdarstellung[5],
von jugendlichem Fleisch[6], in aufrechter Haltung[7] sollen ihre
Bilder gemalt werden.

Aus den „Merkmalen des Gemäldes" (*Citralakshana*) das
dritte Kapitel, das von den Maßen handelt.

Brahmas Schrift von den Maßen, so viel davon vorhanden, ist hiermit beendet.

---

[1] Die vier folgenden Verse bestehen aus je elf Silben.

[2] Wörtlich: in Übereinstimmung mit den (rechten) Maßen.

[3] Tib. *smod-par mi qgyur*; *smod-pa* abgeleitet von *smad*, also „hinunter-
ziehen, herabsetzen", daher „tadeln" (Gegensatz *stod-pa*). Vgl. JÄSCHKE's
Zitat: *smod-par qgyur-bas* „weil es despektierlich wäre, seiner Ehre Eintrag
täte", und die im Wörterbuch der katholischen Missionare angeführten
Synonyme.

[4] Tib. *gnas t'abs*. Vgl. oben S. 143, Anmerkung 3.

[5] Tib. *sa gži mñam*. Vgl. S. 178, Anmerkung 3.

[6] Tib. *šin-tu gžon ša*. Derselbe Ausdruck kommt in Nr. 27 der zweiunddreißig Lakshaṇa (Mahāvyutpatti ed. CSOMA und ROSS) als Übersetzung
von Skr. *taruṇa* vor.

[7] Tib. *ya-yo med-par*, nicht gekrümmt, nicht gebückt; zu verstehen im
Sinne von Skr. *avakragāmitā*, Nr. 17 der achtzig Anuvyañjana (Mahāvyutpatti, p. 95; Dharmasaṁgraha, p. 56: *having a straight step*).

NACHRICHTEN ÜBER MALEREI

Aus dem *dPag bsam ljon bzaṅ* des Lama Sum-pa mk'an-po
Ye-šes dpal-ạbyor (1702—1775; ed. SARAT CHANDRA DAS,
Calcutta, 1908, p. 136).

Da die Kunst aus Werk, Wort und Gedanke[1] besteht, so
muß sie in der Tat als die harmonische Vereinigung aller
Wissenschaften erklärt werden. Hier indessen erklären wir
nur die anfängliche Entstehungsgeschichte, wie in Werk, Wort
und Gedanke heilige Bildwerke geschaffen wurden. Was die
Art und Weise der Bildmalerei im allgemeinen betrifft, [so besteht darüber folgende Tradition]. Ehemals als das Lebens-
alter der Menschen noch hunderttausend erreichte, starb einem
Brahmanen, einem Untertanen des Königs Bhayajit (*aǰigs-t'ul*),
sein jugendlicher Sohn. Er ging den König um Rückgabe
seines Sohnes an, worauf sich der König in das Reich des

---

[1] Die drei Sphären menschlicher Betätigung (*lus ṅag yid* oder im
hohen Stil *sku gsuṅ t'ugs*) erhalten in der tibetischen Kunst ihre besondere
Interpretation. Das mit „Werk" übersetzte Wort *lus* oder *sku* bedeutet
eigentlich „Körper" und wird auch zur Bezeichnung einer Statue, besonders
einer Buddhastatue, gebraucht. In dieser Trias wird es auf die Plastik überhaupt bezogen, auf die Kultusfiguren aus Stein, Kupfer, Bronze oder Ton.
Das „Wort" wird auf die heiligen Schriften, in dieser Verbindung auf die
durch Miniaturen illuminierten Bücher gedeutet. Der „Gedanke" (wörtlich:
Herz) bezieht sich auf die Reliquienbehälter, die Caitya (tib. *mc'od-rten*).
Der Sinn der symbolischen Phrase ist also, daß sich die Tätigkeit der Kunst
auf die drei Gebiete der Plastik, Malerei und Architektur erstreckt.

Yama begab und diesen um den Sohn bat. Yama erklärte, daß er ihn nicht zurückerstatten könne, da er infolge seines Karma verschieden sei. Gleichwohl begann der König, ohne auf ihn zu hören, einen Kampf mit ihm; da erschien Brahma und sprach zu dem Könige: „Es ist das Karma; Yama ist unschuldig. Daher male die Gestalt des Sohnes." Er malte ihn, und Brahma machte das Bild lebendig, das er dem Brahmanen überwies. So entstand die Malerei[1]. Später wurde sie durch den Künstler Viçvakarman weit verbreitet; dessen Geschlecht wurde von dem Ṛishi Ātreya[2] fortgesetzt, von dem sich die Kaligraphen ableiten. Obwohl die Tradition so lautet, ist es schwer ihr Glauben beizumessen.

Die früheren Caitya sind das, wie bekannt, zur Zeit des Buddha Kanakamuni errichtete ruhmreiche Dhānyakataka Caitya im Süden Indiens, das Upacaitya Viçuddha[3] im Lande der Malla[4], die in der „Prophezeiung von Khotan"[5] aufgeführten Caitya, Gomasala des Kāçyapa, das große Caitya von Nepal Bya-ruṅ k'a-šor[6] u. a. Auch Handschriften mit Illustrationen

---

[1] Offenbar ist dieser kurze Auszug auf dem ersten Kapitel des Citralakshaṇa basiert und beweist, daß dieses Werk die Lektüre gelehrter Lama gebildet hat.

[2] Tib. *A-trai-bu*, scheint mit *Eṭei-bu* identisch zu sein, der als Maharshi (großer Ṛishi) bezeichnet wird, dem angeblichen Verfasser eines ikonographischen Werkes, das im Tanjur auf das Citralakshaṇa folgt (EINLEITUNG, S. 2). Im Index von Narthang findet sich die Variante *Aṭei-bu*. Eine Übersetzung dieser Abhandlung wird im zweiten Hefte dieser Dokumente folgen.

[3] Tib. *ñe-bai mc'od-rten rnam-dag*.

[4] Tib. *Gyad yul* (mongolisch *Bükä-yin oron*).

[5] Tib. *Li yul luṅ bstan*, eine kurze im Tanjur (Sūtra, Vol. 94) enthaltene Schrift (ROCKHILL, The Life of the Buddha, p. 231).

[6] Die legendenhafte Geschichte der Gründung dieses Klosters wird in einer kleinen Schrift (32 fol.) der padmaistischen Literatur beschrieben, die den Titel führt: མཆོད་རྟེན་ཆེན་པོ་བྱ་རུང་ཁ་ཤོར་གྱི་ལོ་རྒྱུས་ཐོས་པས་གྲོལ་བ; ein Exemplar derselben habe ich in Sikkim erlangt. Die Geschichte wird in

nach dem Lalitavistara, wie der Kronprinz Siddhártha von seinen Lehrern Unterricht empfängt, sind jetzt mehr als früher verbreitet, wie man allgemein weiß, aber in früheren Zeiten waren offenbar schon Malereien und Skulpturen vorhanden. Insbesondere hat man die Gestalt dieses unseres Lehrers auf Caitya und in Büchern dargestellt. Was die Caitya anbelangt, so sind z. B. seine Geburt in Kapila, die Erlangung der Bodhi in Magadha, seine heiligen Umwandlungen in Váráṇasī, seine magischen Verwandlungen in Çrávastī, seine Herabkunft vom Himmel in der Stadt Káçī, seine Versöhnung des Zwiespalts der Geistlichkeit in Rájagriha, seine Segnungen für ein lang dauerndes Leben im Jetavana zu Vaiçálī, sein Nirváṇa in Kuçinagara[1] zu dieser oder jener Zeit in diesen oder jenen Caitya dargestellt worden.

Was die früheren Bilder und Handschriften betrifft, so ließ Utrayana[2], der König von Rávaṇa[3], für den König Bim-

Form eines Dialogs zwischen Padmasambhava und König Kʻri-sroṅ lde-btsan (wie in so vielen Padmaschriften) erzählt. Eine zur Strafe auf die Erde versetzte Göttertochter Indras wird im Distrikt Maguta in Nepal in der Familie eines Geflügelzüchters geboren und übernimmt den Beruf der Eltern. Sie erhält vier Söhne, einen von einem Pferdezüchter, einen zweiten von einem Schweinehirten, einen dritten von einem Hundepfleger und einen vierten von einem Geflügelzüchter. Diese vier erbauen mit ihrer Mutter von ihren zurückgelegten Ersparnissen das erwähnte Kloster, in dessen Name das erste Wort *bya* infolgedessen als „Geflügel" erklärt wird. Man vergleiche den biederen Sanang Setsen (I. J. SCHMIDT, Geschichte der Ostmongolen, S. 39), der diese Tradition gekannt hat, aber drei statt vier Söhne hat und den Tempel bŠarung Gasur nennt (statt „Aufseher über Vögel" in der Übersetzung von SCHMIDT ist „Geflügelzüchter" einzusetzen). Vgl. auch *dPag bsam ljon bzaṅ*, p. 109, Z. 19.

[1] Tib. *rTsa-can* (so auch bei ROCKHILL, *l. c.*, p. 133).

[2] Oder Udayana (s. SCHIEFNER's Übersetzung des Táranátha, S. 2, 71). Es handelt sich hier um das berühmte Sandelholzbild; vgl. GRÜNWEDEL, Buddhistische Kunst in Indien, S. 148—150, und HUTH, Geschichte des Buddhismus in der Mongolei, Band II, S. 408—412.

[3] Tib. *sgra sgrogs*. Die obige Identifikation gibt CHANDRA DAS im

bisāra zum Dank für einen ihm gesandten kostbaren Panzer von großem Werte, ein Bildnis Buddhas malen, indem er die Spiegelung der Gestalt des mit zehn Kräften Ausgerüsteten (*daçabala*) zum Muster nahm. Dieses Bild ist die unter dem Namen *C'u lon-ma* („aus dem Wasser hergeleitet")[1] bekannte Zuflucht zu den drei Kostbarkeiten in der oberen und unteren Welt und der Ursprung der Sitte glücklicher Vorzeichen[2]. Darauf ließ er das *lDog dan bslab gži lna*[3] niederschreiben, sandte diese Schrift [samt dem Bilde] ab, und der König [Bimbisāra] im Vertrauen darauf erschaute die Wahrheit. Da war ferner das Mädchen Muktāhārā[4], das, als es bei dem Jina die heilige Lehre hören wollte, von einem Hausmeister des Çākyamahā-nāma[5] ermahnt wurde, Dienste zu nehmen; reuigen Sinnes erschien sie, und der Jina trug ihr einen Gāthā aus den heiligen

Index. SCHIEFNER (Tibetische Lebensbeschreibung Çākjamunis, S. 101) hatte dafür Roruka mit einem Fragezeichen vorgeschlagen. In anderen Werken wird Udayana König von Vārāṇasī (SCHIEFNER, *l. c.*, S. 43; HUTH, *l. c.*, S. 408), in anderen König von Kauçāmbī genannt (I. J. SCHMIDT, Geschichte der Ostmongolen, S. 313).

[1] Sum-pa mk'an-po folgt also der von Tāranātha (SCHIEFNER's Über-setzung, S. 71) verworfenen Erklärung. Offenbar ist Tāranātha im Recht. Das Wort *c'u-lon* bedeutet „Damm", und die darauf gegründeten Erklärungen von *c'u len* (Tāranātha, Text 56, 17) im aktiven Sinne („das Wasser ab-leitend") in bezug auf die sieben Steinsäulen mit Bildern des Muni, welche die Überschwemmung von Vajrāsana abwehrten, und im passiven Sinne („vom Wasser hergenommen") in bezug auf das Buddhabild des Udayana sind ohne Zweifel spätere volksetymologische Umdeutungen. Im obigen Texte ist *c'u lon-ma* natürlich doppelsinnig; auch dort schwebt die Bedeutung „Damm" vor, die weiter zu der Vorstellung von der Zuflucht zu den drei Kostbarkeiten Anlaß gibt.

[2] Tib. *rten-ąbrel*, in einer Anmerkung im Texte als die Glaubensformel *Yedharma* erklärt.

[3] Skr. *vartamānaçikshāpañcamūla* (?), vermutlich ein im Tanjur ent-haltenes Werk; es ist mir nicht gelungen, den Titel festzustellen.

[4] Tib. *Mu-tig do-šal*, „Perlenhalsband".

[5] Tib. *Šākya miṅ c'en*, „Çākya der hochberühmte".

Schriften vor. Nach ihrem Tode nahm sie die Geburtsform der Tochter des Königs von Siṁgala (Ceylon), Muktālatā[1], an; da entsandte Buddha in eigener Person Lichtstrahlen seines Körpers auf ein Stück Kaliko[2], wodurch eine Malerei seiner Gestalt entstand, und trug den unter dem Namen „Mārīcī"[3] bekannten Gāthā vor. Da erkannte sie im Vertrauen darauf die Wahrheit, und auf jener Insel verbreitete sich die Lehre des Jina.

Was die früheren Skulpturen anbelangt, so wurden, als Buddha nicht mehr zur Mittagszeit[4] den Hauptplatz in den Reihen der Mönchsgemeinde einnahm, von Anāthapiṇḍada viele Statuen des Lehrers an seinem Ehrenplatze, dem obersten Sitze, an Stelle des Jina errichtet. An Stelle des Munīndra, der sich um seiner Mutter willen in die Region der Dreiunddreißig Götter begeben hatte[5], wurde von dem König von

---

[1] Tib. *Mu-tig aḱri śiṅ.*

[2] Kaliko (*ras*) benutzen die Tibeter noch jetzt als Material zum Bemalen; daher *ras bris* „Gemälde". Baumwolle und Baumwollzeuge haben die Tibeter natürlich erst von Indien kennen gelernt. In tib. *ras* steckt unzweifelhaft ein Nachhall von Skr. *karpāsa.* Im Lalitavistara (FOUCAUX's Übersetzung des tibetischen Textes, p. 228) finden wir die interessante Gleichung Skr. *paṭa* (*paḍa*) = tib. *t‘aṅ. Paṭa* oder *paṭikā* waren appretierte Baumwollstoffe, die zum Schreiben für Briefe, amtliche und Privaturkunden verwendet wurden (G. BÜHLER, Indische Palaeographie, S. 88); *paṭa* bedeutet ferner Bild, Malerei. Damit ist eine bestimmtere Definition für tib. *t‘aṅ-ka,* (s. „Roman einer tibetischen Königin", S. 1) gewonnen, das also ursprünglich wohl eine zum Schreiben oder Malen bestimmte Rolle von Baumwollzeug gewesen sein wird.

[3] Die Göttin des Lichts. L. FEER (Fragments extraits du Kandjour, *Annales du Musée Guimet,* vol. V, pp. 430—432) hat eine ihr gewidmete Dhāraṇī übersetzt.

[4] Wann sich die Mönche versammelten, um die Mahlzeit einzunehmen.

[5] Buddha begab sich in den Tushitahimmel, um seiner Mutter den Abhidharma zu predigen (H. KERN, Manual of Indian Buddhism, p. 33; ROCKHILL, The Life of the Buddha, p. 80; I. J. SCHMIDT, Geschichte der

Kaçi[1] in Vāraṇasī, der ein Gabenspender war, der berühmte „Sandelherr" errichtet, der jetzt in die Gefilde der Verdienste Chinas eingetreten ist[2].

Später, als der Jina im Alter von fast achtzig Jahren ins Nirvāṇa eingegangen war, wurden die Mahābodhi genannte Statue von Vajrāsana in Magadha, der Trommelton des Mañjuçrī (*Mañjuçrīdundubhisvara*) und andere wundervolle Bildwerke, im ganzen acht, und andere göttliche Kunstwerke errichtet[3]. Zur Zeit des Königs Açoka wurden die Caitya der acht großen Wallfahrtstätten und die innere Umhegung von Vajrāsana u. a. größtenteils von den Yaksha errichtet. Zur Zeit des Nāgārjuna entstanden vielfach von Nāga-Künstlern verfertigte Werke. Später, zur Zeit des Königs Buddhapaksha[4], lebte in Magadha ein Menschenkünstler, Bimbasara, dessen Malereien und Skulpturen

Ostmongolen, S. 15; SCHIEFNER, Tibetische Lebensbeschreibung, S. 43). Dieses Ereignis wird auch im *Buddhapratimālakshaṇa*, der ersten kunsttechnischen Abhandlung im 123. Bande des Tanjur, als Ausgangspunkt für die Anfertigung von Buddhastatuen genommen, indem Çāriputra an Buddha die Frage stellt, wie sein Bild dargestellt werden soll, wenn er nicht mehr unter seinen Anhängern verweile. So ist auch in der obigen Stelle die Errichtung des Sandelholzbildes durch die Abwesenheit Buddhas im Himmel motiviert; das Bild ist sein irdischer Stellvertreter.

[1] Tib. *gSal-ldan* (SCHIEFNER, *l. c.*, S. 93).

[2] Tib. *bsod-nams-kyi žiṅ* = Skr. *puṇyakshetra*. Eine quellenmäßige Geschichte dieser Statue wird im zweiten Hefte mitgeteilt werden.

[3] Diese und die folgenden Angaben in Übereinstimmung mit Tāranātha, S. 279, dessen Reflexionen aber unterdrückt sind; nur die Traditionen selbst sind hier mitgeteilt. Vgl. die zu Tāranātha's Angaben gemachten Erläuterungen von VINCENT A. SMITH, A History of Fine Art in India and Ceylon, pp. 304—306; auch E. B. HAVELL, The Ideals of Indian Art, p. 81.

[4] So von SCHIEFNER restituiert (vgl. auch sein Vorwort zu Tāranātha, S. IX). Es ist interessant, daß dieser Königsname in der von dem Panc'en rin-po-c'e Blo-bzaṅ dpal-ldan Ye-šes (1737—79) für den Kaiser K'ienlung fabrizierten Genealogie seiner Inkarnationen an dritter Stelle erscheint (HUTH, Geschichte des Buddhismus in der Mongolei, Band II, S. 320), wo statt Buddhadiç ebenfalls Buddhapaksha zu lesen sein dürfte.

denen der göttlichen Künstler gleich waren: dies ist die Schule der Künstler von Madhyadeça[1]. Zur Zeit des Königs Çila lebte in Maru[2] Srigadhari[3], dessen Malereien und Skulpturen den Kunstwerken der Yaksha gleich waren: dies ist die alte Schule des Westens. Zur Zeit des Königs Devapāla und des Çrīmant Dharmapāla[4] lebten in Nalendra[5] Dhimāna und Bitpālo, Vater und Sohn, deren Guß-, Schnitz- und Malarbeiten den Kunstwerken der Nāga gleich waren: in der Gießerei ist dies der Bronzeguß des Ostens (*šar li*)[6]. Im Anschluß an die Malereien des Vaters spricht man von der Malschule des Ostens (*šar ri*), in bezug auf die Malerei des Sohnes, die sich in Magadha verbreitete, von der Schule von Madhyadeça. Beide hatten zahlreiche Nachfolger in Indien und Nepal. Im Süden hat es Nachfolger dreier Schulen, Jaya, Parojaya und Vijaya gegeben: dies ist die Schule des Südens (*lho-ma*)[7]. In Kashmir verbreitete sich später eine Kunstschule in Malerei und Skulptur eines gewissen Hasurāja[8], die als die Schule von Kashmir (*K'a-c'e-ma*) bekannt ist.

Als Ausläufer dieser Schulen[9] erschienen in Tibet Dho-ba

---

[1] Tāranātha gibt eine ausführlichere Begründung.

[2] So nach Tāranātha; Matu im Texte ist wohl Druckfehler.

[3] SCHIEFNER rekonstruiert Çriṅgadhara; die Schreibung des Textes sollte aber beibehalten werden, da der Name offenbar schon der nachsanskritischen Sprachperiode angehört.

[4] Nach der Berechnung von VIDYABHUSANA (History of the Mediaeval School of Indian Logic, p. 148, Calcutta, 1909) wäre das Datum des Devapāla 705—753 und das des Dharmapāla 765—829.

[5] Tāranātha hat Varendra.

[6] Tāranātha hat *šar-gyi lha*, „Götter (d. i. Götterbilder) des Ostens".

[7] Diese Bezeichnung findet sich bei Tāranātha nicht.

[8] Tāranātha macht ihn zum Gründer einer neuen Schule.

[9] Dieser Absatz ist im Werke des Tāranātha nicht enthalten.

bKra-rgyal¹, aJam-dbyaṅs Don-grub (*Mañjughosha Siddhārtha*) von sMan-t'aṅ in Lho-brag, mK'yen-brtse von sGaṅ-stod von der Sekte der Rotgewandigen², Lha-bzo³ byeu von Yar-stod u.a. im Ozean der Sphäre der Religion wirkend. In der Malerei und Skulptur der Bilder des Jina nehmen Indien und Nepal die vorzüglichste Stelle ein, Tibet die mittlere, und China die letzte.

---

¹ D. i. bKra-rgyal von Dho-ba = Do-ba, einem Distrikt in der Provinz Lho-brag im südlichen Tibet, die an Bhūtān grenzt.

² Tib. *gos dmar* = Skr. *tāmraçāṭīya*, eine Schule der Sarvāstivādin (SCHIEFNER, Tāranātha, S. 272).

³ D. i. Götterbildner.

ADDENDA

S. VIII. Das Werk von ANANDA K. COOMARASWAMY, Mediaeval Sinhalese Art, kam mir erst in die Hand, als jene Zeilen geschrieben waren. Er spricht (p. 163) von der schwachen und verweichlichten Skulptur der Gandhāra-Schule und gibt weitere Ausführungen über diesen Gegenstand (pp. 256—261), mit denen ich im allgemeinen übereinstimme. Auf das von ihm behandelte Werk *Sāriputra* (pp. 150—163) werde ich im zweiten Heft zurückkommen. Beachtenswert ist auch der Aufsatz von WILLIAM COHN, Einige Bemerkungen zum Verständnis der indischen Kunst (*Ostasiatische Zeitschrift*, Bd. I, 1912, S. 217—220).

S. 3, Anm. 1. Auch in *Sitzungsberichte der Berliner Akademie*, 1901, S. 209 hat GRÜNWEDEL auf das kunstgeschichtlich wichtige Material in der tibetischen Literatur hingewiesen.

S. 13. Darstellungen nackter Jainamönche bei A. GRÜNWEDEL, Altbuddhistische Kultstätten in Chinesisch-Turkistan, S. 119, 142 (Berlin, 1912). In demselben Werke zahlreiche weitere Beispiele von Nacktdarstellungen im Gebiete der buddhistischen Kunst.

S. 13. Sanskritisten seien auf die richtige Datierung des *Si yü ki*, des Berichts von Hüan Tsang über die Länder des Westens, hingewiesen, die wir PAUL PELLIOT (*T'oung Pao*, 1912, p. 483) verdanken: das Jahr 648, das JULIEN als das Datum der Vollendung des Werkes gegeben und auf das Titelblatt seiner Übersetzung gestellt hat, ist unrichtig; das Werk lag vielmehr im August 646 abgeschlossen vor und wurde damals dem Kaiser mit einer Denkschrift überreicht.

S. 31, Note 3. Gegen die evolutionistische Richtung hat F. GRAEBNER in seinem bemerkenswerten Buche „Methode der Ethnologie" (Heidelberg, 1911) mit vollem Recht Front gemacht. Noch schärfer ist die Sachlage gekennzeichnet von FRANZ BOAS (The Mind of Primitive Man, pp. 174—196, New York, 1911), der die logischen Fehler der Evolutionsmethode in scharfsinniger Weise bloßstellt. Im übrigen akzeptiere ich GRAEBNER's Methoden nur mit den von BOAS (*Science*, Vol. XXXIV, 1911, pp. 804—810) in seiner Kritik des Buches gemachten Einschränkungen; den psychologischen Gesichts-punkt hat GRAEBNER ganz übersehen, und die von ihm vorgeschlagenen historischen Methoden müssen noch beträchtlich modifiziert werden. Der von mir hervorgehobene Fall der verschiedenartigen Entwicklung der Malerei in Indien und China ist ein gutes Beispiel für eine Konvergenzerscheinung, die ROBERT H. LOWIE (On the Principle of Convergence in Ethnology, *Journal of American Folk-Lore,* Vol. XXV, 1912, pp. 24—42) treffend behandelt hat, und dessen Ausführungen ich mich vollkommen anschließe. Mit Recht tadelt LOWIE den Hang zu falschen Klassifikationen und hebt hervor, daß ein Ethnologe, der z. B. ein Zweiklassensystem in Australien mit einem solchen in Amerika oder Totemismus unter den Indianern des Nordwestens mit Totemismus in Melanesien gleichsetzt, auf das Niveau eines Zoologen herabsinkt, der Wale mit Fischen und Fledermäuse mit Vögeln klassifizieren würde.

S. 53. Über aP'yon-rgyas s. WASSILJEV, Tibetische Geographie nach dem Werke des Minčul Chutuktu übersetzt, p. 34 (russisch, St. Pet., 1895).

S. 63. Ein Verzeichnis der indischen Maler, die nach China gekommen sind, findet man im *Li tai hua shi hui chuan,* Kap. 63. Vgl. ferner HIRTH, Scraps, pp. 65, 70—75, 92 (Nr. 37).

S. 161, 164, 176. Die Schwimmhaut zwischen Daumen und Zeigefinger ist an der Hand Buddhas wirklich dargestellt worden (A. GRÜNWEDEL, Altbuddhistische Kultstätten in Chinesisch-Turkistan, S. 89, 154, Berlin, 1912).

13

093

纽伯里图书馆的汉文、藏文、蒙古文及日文书

PUBLICATIONS OF THE NEWBERRY LIBRARY

NUMBER 4

WOODCUT FROM THE ISE MONOGATARI "TALES OF ISE," A
ROMANCE OF THE TENTH CENTURY, FROM THE EDITION
OF 1608, ONE OF THE EARLIEST SPECIMENS OF JAPANESE
WOOD-ENGRAVING.

DESCRIPTIVE ACCOUNT OF
THE COLLECTION OF CHINESE, TIBETAN, MONGOL, AND JAPANESE BOOKS IN THE NEWBERRY LIBRARY

BERTHOLD LAUFER, PH.D.

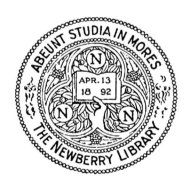

THE NEWBERRY LIBRARY
CHICAGO, ILLINOIS

v

OFFICERS

President
ELIPHALET W. BLATCHFORD

First Vice-President
GEORGE E. ADAMS

Second Vice-President
HORACE H. MARTIN

Secretary and Financial Agent
JESSE L. MOSS

Librarian
WILLIAM N. C. CARLTON, M.A.

vi

LIST OF ILLUSTRATIONS

INTRODUCTORY NOTE

The following brief sketch of the East Asiatic Collection in the Newberry Library was prepared by Dr. Laufer at the request of the Library authorities. His extensive knowledge of Chinese, Japanese, and Tibetan antiquities, art, history, and literature, together with the fact that he himself had gathered the books and manuscripts, marked him as peculiarly well fitted to describe the character and contents of the Collection and to indicate the range and degree of its usefulness to sinologues and all others whose studies require access to such material as this Collection comprises.

W. N. C. C.

ix

EAST ASIATIC COLLECTION

In 1907, in connection with an expedition to be undertaken on behalf of the Field Museum, I was commissioned by the Trustees of the Newberry Library to gather for them a representative collection of East Asiatic works on subjects falling within the field in which that Library specializes, viz., religion, philosophy, history, belles-lettres, philology, and art.[1] The result of this commission was the purchase of 1,216 works in 21,403 volumes. Although a fair and solid foundation, it should not be presumed that any section of this collection can be designated as really complete, in view of the inexhaustible wealth of Oriental literatures, and Chinese in particular; but so much has been attained by including the majority of all important works that the student will be able to carry on serious and profound research work in any of the branches of knowledge enumerated, and it may therefore be considered a truly representative collection of the Chinese, Manchu, Tibetan, and Mongol literatures.

As to language, the Japanese is represented by one hundred and forty-three works, Tibetan by three hundred and ten, Mongol by seventy-two, Manchu by sixty; the rest are in Chinese which is the most extensive and important literature of the East, and the one from which the light of the others radiates. There are eighteen manuscripts, all unpublished and deserving of publication. Of early printed

---

[1] At the same time, a corresponding commission was given me by the Directors of the John Crerar Library to collect for them Oriental works on geography, law, and administration, trade, industries, national economy, sociology, agriculture, mathematics, medicine, and the natural sciences.

I

books there are two fine works printed in the Sung period, dated 1167 and 1172 respectively, one of the Yüan or Mongol dynasty (thirteenth century), forming indisputable proof of the Chinese having antedated Gutenberg by centuries; and fifty-seven from the Ming period (1368–1644), with such early dates as 1395, 1447, 1453, 1467, 1504, 1558, etc.

The Japanese collection was made only incidentally during a short trip from Peking to Tōkyo. Its main object was to search for editions of Chinese works which can no longer be found in China, and to secure a collection of books fairly representative of Japanese art, in which there is at present such a live and intelligent interest in this country. As I made at the same time a collection of Japanese color prints for the Field Museum, it was thought a fit occasion to secure for the Newberry Library, for the benefit of our art students and collectors, a selection of illustrated books bearing on this interesting subject and its history. I purchased in this connection the works of Hokusai, Kuniyoshi, Kyōsai, as far as published in book form, many of them in original editions, color reproductions of the painter Kanō Tanyū (1602–1674), the Masterpieces of Thirty Great Painters of Japan, and the works of Ogata Kōrin (1661–1716) published by the Shimbi Shoin in Tōkyo, and numerous other volumes relating to manners and customs, arts and crafts, costume, textiles, gardening, flower arrangement, architecture, swords, armor, and antiquities, many of them in eighteenth-century editions. There are also several manuscripts on archery, and books on tea, the tea-ceremonies, on the Shintō religion and the objects of its cult. It is hoped that this collection will prove useful to art-designers and art students. Most of the useful books published by the firm Hakubunkwan in

Tōkyo were procured. Among these are eight works on Buddhism in sixteen volumes, and the most extensive catalogue of Japanese literature, the *Kokushōkaidai*, in twenty-six volumes, 1897–1900. From the viewpoint of the development of printing in Japan, the early edition of the novel *Ise Monogatari*, "Tales of Ise" (No. 232), is most interesting. This edition was printed in 1608 during the period Keicho and represents the earliest specimen of a Japanese printed and illustrated book. Unfortunately I was able to obtain only the second of the two volumes of which the work consists. W. G. Aston (*A History of Japanese Literature*, p. 84) characterizes it as "block-printed on variously-tinted paper, and adorned with numerous full-page illustrations which are among the very earliest specimens of the wood-engraver's art in Japan." B. H. Chamberlain (*Things Japanese*, p. 435) defines it as "the earliest illustrated book at present known" (see also E. F. Strange, *Japanese Illustration*, p. 2). An example from the woodcuts of this book is here reproduced as the frontispiece.

The Japanese collection also contains seven very interesting manuscript volumes from the colossal work *Gunshoruiju* (No. 239), by Hanawa Hokiichi (1746–1821), a famous littérateur who grew blind in his seventh year, lost his mother shortly afterward, and was brought up by a Buddhist monk. He first studied music and acupuncture, but later found his proper field in the study of Japanese antiquities and literature. In 1782 he published the collection of rare and ancient works above mentioned which consisted of 2,805, according to others of 1,821 volumes, and is said to have remained in manuscript. The seven volumes in the Newberry collection comprise the Index volume, which will be valuable in studying the contents of the work, and Vols. 66, 214, 494, 503 *a*, 503 *b*, 503 *c*. Another interesting work is an old illustrated edition in forty-one volumes of the famous novel *Taiheiki*, by Kojima who died in 1374, on which Aston (*l.c.*, pp. 169 *et seq.*) gives a great deal of information. A fine manuscript, dated 1804, containing eighteen water-color sketches, illustrates the gradual stages in donning the parts of a coat of mail by a Diamyō.

In Manchu literature, Chicago has one of the richest collections in existence, including as it does many rare early editions, unique Palace editions, and manuscripts for imperial use, of whose existence nothing had before been known. When I published a sketch of Manchu literature in 1908, I was under the impression that I had made as complete a survey of the subject as possible; now I am able to make a series of important additions which will show the character of this literature in a new light. As I expect to publish these notes before long, I need only say here that the majority of Manchu and Manchu-Chinese bilingual prints catalogued in my sketch are now in the Newberry Library. Among the unique works of which no other copies are known may be mentioned a Commentary to the Four Classical Books (*Se shu*) composed in Manchu by the Emperor Kʻang-hi in twenty-six quarto volumes, the Palace edition of 1677 (No. 639); a Manchu commentary to the classical Book of Mutations, *Yi king*, by the same monarch, in eighteen quarto volumes, the Palace edition of 1683 (No. 692); and a commentary to the ancient Book of History, *Shu king*, written by the Emperor Kʻien-lung in thirteen volumes of the same size, Palace edition of 1754 (No. 564). These three works seem never to have been placed on the book-market and to have come out of the Palace in consequence of the panic following the death of the Emperor

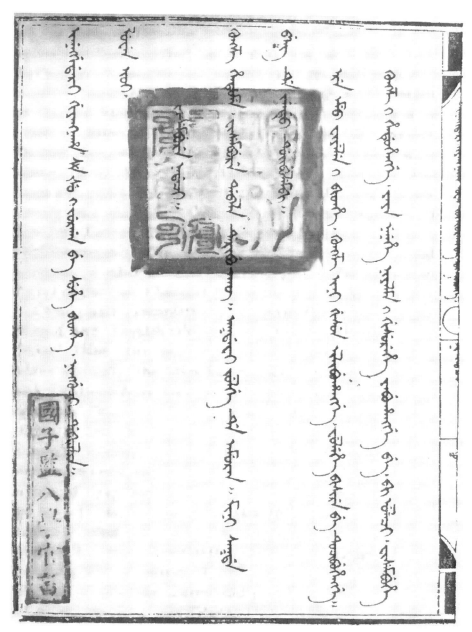

PAGE FROM THE MANCHU COMMENTARY OF THE FOUR CANONI-
CAL BOOKS COMPOSED BY THE EMPEROR KᶜANG-HI IN 1677.

and the Empress-Dowager in the autumn of 1908. It was a curious circumstance that just at that time, for a few weeks, the Peking book-market was flooded with rare Manchu books for sale to foreigners; the Chinese are certainly no customers for Manchu literature. Mention may be, further, made of a Palace edition of the philosopher Mêng-tse (No. 703), in Manchu only, without the Chinese version; the Manchu account of the War against Galdan, 1709 (No. 710); the *Yooni bithe* (Nos. 690 and 693), the oldest Manchu dictionaries of 1682 and 1687; Collection of Words from 120 Old Men (see Laufer, *Sketch of Manchu Literature*, p. 19), a valuable manuscript in eight volumes, written in 1709, and a number of other unedited manuscripts; a Palace edition of 1741 of the Four Classical Books (*Se shu*) in Manchu only (No. 559); and a complete edition of the Genealogies of the Mongol and Turkish Princes in Chinese, Manchu, and Mongol (Nos. 563 and 574, seventy-two vols., quarto, Palace edition); the Manchu translation of the historical work *Tung kien kang mu* (No. 573), which is discussed in the report on Chinese history, a great rarity, in the Palace edition of 1681 in ninety-six volumes; the Ritual of the Manchu Dynasty, written in Manchu, illustrated with wood-engravings describing the objects of the cult, Palace edition of 1747 (see Laufer, *Sketch of Manchu Literature*, pp. 39–40); the *Ku wên yüan kien* (No. 592), an excellent work containing historical extracts and selections in Manchu from the Tso-chuan down to the writers of the Han and Sung dynasties, Palace edition of 1685 in thirty-six volumes; a collection of Buddhist charms and prayer formulas (*dhāranī*) in Chinese, Manchu, and Tibetan (No. 783) in ten volumes, a splendidly printed book with fine large folded wood-engravings executed in the Palace exclusively for imperial use during the Kʿien-lung period (1736–1795). In view of the recent overthrow of the Manchu dynasty, it is the more gratifying to have become heirs to their literary bequest, and to have saved, in the interest of the future historian, so many important monuments inspired by the literary zeal and activity of its illustrious rulers. The time is sure to come when this subject will become one of primary importance for research.

Tibetan books were acquired during three different stages of my expedition, first in Darjeeling and during a journey in Sikkim where books printed in the monasteries of either Sikkim or southern Tibet were secured; secondly in Peking where I gathered all Tibetan and Mongol books, so far as still available, issued from the press near the temple Sung-chu-sze;[2] and thirdly, during a journey in eastern Tibet, in the Tibetan states of Sze-chʿuan, and in Kansu and the Kukunōr region. The richest harvest was obtained in the ancient monastery of Derge in eastern Tibet. Tibetan literature has had but little attention thus far from scholars, and our knowledge of it is still very scanty. The elaboration of a bibliography remains a work for the future. The lists of Tibetan books published by some European libraries usually give no more than the mere titles or are meagre in contents and not entirely reliable. The only serious bibliographical attempt is Csoma's analysis of the two collections of the *Kanjur* and *Tanjur*[3] Copies of both are now in Chicago, in the edition printed in 1742 at the monastery of Narthang near Trashilhunpo in central Tibet.

[2] A description of the activity of this printing establishment will be found in my *Sketch of Mongol Literature*.

[3] Published in *Asiatic Researches*, Vol. XX, Calcutta, 1836. French translation, with indices, by L. Feer, in *Annales du Musée Guimet*, Vol. II. Of P. Cordier's new work, *Index to the Tanjur*, the first part has appeared (Paris, 1909).

FIRST PAGE FROM A VOLUME OF THE TIBETAN KANJUR (EDITION OF NARTHANG), CONTAINING THE VINAYAVASTU, THE FIRST PORTION OF THE SECTION VINAYA, THE DISCIPLINARY RULES FOR THE ORDER OF BUDDHIST MONKS.

The copy in question had been brought by the Dalai Lama from Lhasa to Peking. The Narthang edition has sometimes been described as inferior in make-up. This opinion is largely based on the poor condition of the copy in the Royal Library of Berlin; the reason why this copy is difficult to read is because it was struck off on bad and thin paper. The Newberry copy of the *Kanjur*, however, is printed on a good quality of strong Tibetan paper and perfectly clear and legible.[4] It all depends upon the kind of paper, as can be proved from several practical examples, good and bad, readable and illegible copies of the same work being printed from the same blocks; the different results are due to the different grades of paper used. I am informed by several Tibetan Lamas that the Narthang edition is considered by far the best of all, from the viewpoint of textual criticism; it certainly contains far fewer mistakes than the red-printed imperial editions of Peking, and continuous reading of it is much easier, as the vermilion color of the Peking issues is a great strain on the eye. Nor are the red-printed editions the ideal thing for another reason, viz., the color is liable to fade; in the St. Petersburg copy I have come across many folios where the lettering had faded to a pale white.

The *Kanjur*, which means "Translation of the Word" (i.e., of Buddha), is the adopted canon of the sacred writings of Buddhism translated into Tibetan mostly from the original Sanskrit texts by a trained staff of Buddhist monks from the ninth to the thirteenth century. A few translations go back to the latter part of the seventh century, the time of the first introduction of Buddhism into Tibet; some have

[4] Tibetan books are not kept in stock, and have no ready-made editions. The blocks for the *Kanjur* and *Tanjur* are kept under lock and key in a certain hall of the temple. A copy is printed only when ordered, and requires a permit from the Abbot. There is also but one printing season a year.

been made also from Chinese and from the Turkish language of the *Uighur* in which, as we now know from discoveries made in Turkistan, a translation of the Buddhist scriptures existed. On the other hand, the interesting fact has been brought to light by F. W. K. Müller that the Tibetan version played a prominent rôle in the composition of the Chinese *Tripitaka* which contains a number of terms to be explained only from Tibetan. The Tibetan translations are almost literal and prepared with a great deal of care and accuracy, and as most of the Sanskrit originals are lost, they become a primary authentic source for the study of Buddhism; even in those cases where the Sanskrit texts are preserved, the Tibetan documents always provide considerable assistance in making out the correct Sanskrit reading and facilitating understanding. To one equally versed in Tibetan and Sanskrit and familiar with Buddhistic style and terminology, it is even possible successfully to restore the Sanskrit original from the reading of the Tibetan text. The vast stores of this collection have in part been repeatedly ransacked by scholars interested in the history of Buddhism. A. Schiefner and L. Feer have extracted from its pages a large number of Buddhist legends and stories; the Hon. W. W. Rockhill has skilfully utilized it for a reconstruction of a life of Buddha, and some texts even yielded to him material for a history of Khotan. But the bulk of its contents still remains unstudied; many parts, *e.g.* the *Vinaya* containing the discipline or rules for the orders of the monks, should be translated intact. A task of the first order, the literary history of the collection, remains to be done. This would require a comparative study of all the existing editions. We now know that there are different editions of the *Kanjur* varying in contents and illustrations, and in the arrangement of the matter,[5] and that these differences have sprung up from the midst of different sects. As in China and Japan, so also in Tibet, Buddhism does not form an harmonious unity, but is split up into various sects which came into being at various times and are often bitterly antagonistic to each other, not only on religious but also on political grounds. Only by a thorough investigation of the history of these various sectarian formations can we ever hope to penetrate into the mystery of the history of Lamaism. The history of the collections embodied in the *Kanjur* can be fully understood through the history of the sects only, and the latter subject will shed new light again on the formation of the Canon. Each work in it has had a long and varied life-history, having been translated, corrected, revised, re-edited, and commented upon many times, and this subject is still a *terra incognita*. What is required, therefore, is a critical concordance of the various sectarian editions of the *Kanjur*, the literary history of which is recorded in their lengthy prefaces, and finally also a collation of the works of the Tibetan with those in the Chinese *Tripitaka*, a Tibeto-Chinese concordance.

The bulk of Tibetan literature is of a religious Buddhistic character, but it would be erroneous to believe that it is all secondary matter derived from Indian sources. Native authors have developed a fertile literary talent and produced a quantity of literature relating to theology, logic, metaphysics, rhetoric, grammar, lexicography, medicine, poetry, and history. Tibetan writers have preserved to us the history of India for periods where Indian history presents a

[5] Compare Laufer in *T͑oung Pao*, 1908, p. 8; and Die Kanjur-Ausgabe des Kaisers K͑ang-hsi (*Bulletin de l'Académie des sciences de St. Pétersbourg*, 1909, pp. 567–74).

perfect blank. The poems and legends of Milaraspa are a fascinating production of Tibetan poetic imagination; his works and his biography, in the original Tibetan as well as in the Mongol translations, are to be found in the Newberry Library; also the voluminous literature crystallized around Padmasambhava, the great apostle from Udyāna who played a rôle of great consequence in the establishment of Lamaism in Tibet during the eighth century. "Collected works" of individual authors occupy a prominent place in Lamaist literature. Of all the Dalai Lamas, the Pan-chen rin-po-che, the Metropolitans (Chutuktu) of Peking, and other high church-dignitaries, vast collections of their personal writings embracing all departments of literature have been made, forming a substantial and valuable part of native erudition; a great many of these works, of extraordinary extent and importance, were secured for the Newberry Library. There are also in the collection beautiful Tibetan books printed at the imperial press of Peking under the reigns of the Emperors Kʻang-hi (1662–1722) and Kʻien-lung (1736–1795), as well as some fine specimens of manuscript work in gold and silver writing. Especially noteworthy is an ancient and splendid copy, written in silver on a black polished background, of the famous work *Māni Kambum* (No. 826), in its main portions traceable to the seventh century, written in glorification of the god Avalokiteçvara who incarnates himself in the Dalai Lamas, and containing the laws of the first historical Tibetan king, Srong-btsan-sgam-po of the seventh century. The copying of sacred books is considered a great religious merit; writing in vermilion insures a higher merit than ordinary writing with black ink, while silver and gold writing surpass both.

In connection with the Buddhistic literature of Tibet, the Chinese *Tripitaka* may be mentioned in this place. It is the corresponding Chinese version of the sacred writings of the Buddhist Canon, embracing approximately two thousand works of dogmatic, metaphysical, and legendary character translated from the Sanskrit. The edition in the Newberry Library was formerly preserved in a temple at Wu-ch'ang and is that known under the designation of the Buddhist Canon of the Ts'ing or Manchu dynasty (*Ta T'sing San tsang king*). Until 972 A.D. the Chinese Canon was preserved in manuscript only; in that year, it was printed for the first time by order of the Emperor T'ai-tsu. Thereafter it was printed repeatedly from wooden blocks which were as often destroyed by fire or in the course of wars. During the Sung and Yüan dynasties (960–1367 A.D.) as many as twenty different editions are said to have been issued, but all of them perished in the catastrophe marking the downfall of the Mongols. A few copies of editions coming down from the Ming period have survived in some temples of northern China; one printed in the Yung-lo period (1403–1424), and preserved in a monastery of Shansi Province and alleged to be complete, was once offered to me for ten thousand Mexican dollars. But the K'ien-lung Palace edition in the Newberry collection is certainly just as satisfactory. The plan of this publication was drafted in 1735 by the Emperor Yung-chêng, and on his death in the same year, taken up in 1736 by his son and successor, the Emperor K'ien-lung. The printing of the entire work extended over three years and was completed at the end of 1738. The printing blocks are still preserved in the temple Po-lin-sze, situated east of the Great Lama Temple in Peking. According to an official notice posted there, it required

28,411 blocks to engrave the entire collection, which is composed of 55,632 leaves. The work is arranged in 154 sections and 1,263 chapters. It consists of 7,920 oblong flat volumes bound in 792 wrappers (*t'ao*). The Index, with prefaces and table of contents, makes five volumes. Each volume is illustrated with a fine wood-engraving of delicate tracing. It is bound in brocade, and the wrappers are also mounted on beautiful silk brocades of different designs. This peculiar feature gives the work a great artistic value. These textiles with their variety of ornament and color are rare specimens well authenticated as to origin and date and traceable to the beginning of the K'ien-lung period. Ancient Chinese textiles are rare, and if found, their dating rests on internal evidence only. It should be emphasized that the edition in question is one of the originals actually printed in 1738, and not a later reprint made from the same blocks. In the summer of 1910, when paying a visit to the temple Po-lin-sze, I witnessed there myself the printing of a new edition from the old blocks for the benefit of a temple near Peking.[6] Besides this fundamental work for the study of Buddhism, the Newberry collection has a large number of single editions of Chinese Buddhist works, among them some of the Ming period, and other books bearing on Buddhist subjects. The presence of all the important works of Buddhism in the three principal languages of Northern Buddhism—Chinese, Tibetan, and Mongol—will enable the advanced student to investigate fully and comparatively almost every phase of Buddhist literature.

[6] Regarding the bibliography and contents of the Chinese Tripitaka compare Bunyiu Nanjio, *A Catalogue of the Chinese Translation of the Buddhist Tripitaka, the Sacred Canon of the Buddhists in China and Japan*, Oxford, 1883; and Cl. E. Maitre, Une nouvelle édition du *Tripitaka* chinois (*Bulletin de l'École française d'Extrême-Orient*, Vol. II, 1902, pp. 341–351).

EXAMPLE OF WOOD-ENGRAVING AND INTRODUCTORY PAGE FROM A VOLUME

The engraving represents the attainment of the Buddhaship (*bodhi*) on the part of the four
rounded by Bodhisatvas and Arhats, and the gods of the Brahmanic Heaven are des
doctrine, composed in three langu

TAKA PUBLISHED UNDER THE PATRONAGE OF THE EMPEROR K'IEN-LUNG (1736–95).

in the center, his two disciples Ānanda and Kāçyapa standing at the foot of the altar.   He is sur-
    opening page framed by six five-clawed dragons contains a poem, a eulogy of the Buddhist
etan, by the Emperor K'ien-lung in 1759.

The output of books in China is enormous, and the number of editions, particularly in the department of so-called classical literature, is really bewildering. My primary aim was to secure of all standard works first editions, or whenever this was not possible, the best editions procurable with the idea of permanency in view. Paper and type were carefully examined in each case and stress was laid on obtaining, wherever possible, books printed on Korean paper (*Kao-li chih*) which is the strongest and most expensive, never loses its beautiful and uniform whiteness, and enjoys the same reputation among Chinese scholars that hand-made linen paper does among us. The search for good ancient editions is now beset with more difficulties than ever before, because, as one of the remarkable results of the awakening of China, the practice of establishing public libraries was instituted on a large scale. Until recently only the private library of the scholar and a few more ambitious libraries in the possession of distinguished wealthy clans were known; the latter were guarded with such watchful jealousy that their utilization through a wider circle of students was rendered well nigh impossible. The foundation of universities and colleges has also given an impetus to the establishment of libraries for the benefit of students. At times, the higher Chinese officials became somewhat alarmed at the exportation abroad of valuable libraries through foreigners, and the new national spirit now rapidly asserting itself is inclined to regard old books and manuscripts as national monuments requiring governmental protection. They are placed on the same plane as antiquities, and an export duty *ad valorem* is placed on them, while new books are simply listed as paper and pay a very low amount of duty charged according to weight. The order of the governor of Shantung forbidding the trade and export of all kinds of antiquities within the pale of his province has gone into effect and includes a ban on the exportation of ancient books.   This new movement has naturally resulted in a strong upward tendency of prices which, in some cases, have doubled during the last decade.   When in 1901 I started on my first collection of Chinese books, it was still comparatively easy to secure ancient printed books at reasonable rates.   At Si-ngan fu, I had in 1902 the first edition of Ma Tuan-lin's famous work *Wên hien t'ung k'ao* of 1319 offered to me at 90 Taels (about $63.00) and a Ming edition of the same work of 1524 (period Kia-tsing) at 40 Taels (about $28.00); these editions, no longer procurable, would now cost at least double that rate.   Despite this discouraging situation I was able to secure a good many original and Palace editions of such standard works as form the nucleus of every Chinese library, *e.g.* the famous dictionary of the Emperor K'ang-hi in the original Palace edition of 1716 in forty volumes (No. 34); the great concordance *P'ei wên yün fu* in the Palace edition of 1711 in one hundred and nineteen volumes (No. 42);[7] the *P'ei wên chai shu hua p'u*, a collection of essays on classical and historical books in sixty-four volumes, dated 1705 (No. 41), likewise originating from that great promoter of literature

---

[7] As the wooden blocks used for the printing of these editions have been destroyed by fire, it is impossible to have new impressions struck off from them, as is done with many books out of print the blocks of which are preserved.   Thus, there are books printed with Sung or Yüan blocks under the Ming, and others struck off from Ming blocks under the Manchu dynasty; the paper is then the only means of ascertaining this fact.   The high value placed on the Palace edition of the *P'ei wên yün fu* becomes evident from the fact that the Emperor K'ien-lung presented a copy of it as a reward to persons who sent up a hundred and more rare books to his library, when he had a search made for such throughout the empire for the purpose of compiling a complete bibliography of literature.

and printing, the Emperor Kᶜang-hi, to whom we also owe a fine edition of the collected works of the philosopher Chu Hi (1130–1200), the *Yü chᶜi Chu-tse tsᶜüan shu*, twenty-four volumes, 1713 (No. 31), and an anthology of poetry chronologically arranged, the *Yü chᶜi li tai fu hui* of 1706, in eighty volumes (No. 156).

The marked historical sense of the Chinese is one of their most striking characteristics. Hardly any other nation can boast of such a long and well-authenticated record of a continuous uninterrupted historical tradition extending over a millennium and a half down to 1644, the year of the accession to the throne of the first Manchu ruler. The official history of a dynasty is compiled only after its downfall, and it becomes the duty of the succeeding dynasty to take charge of the archives of its predecessors, and to appoint a commission of scholars to sift and arrange them for the writing of the dynastic history. Some of these histories have been composed by men of high standing in the literary world. Excluding the present one, there are in existence the official records of twenty-four previous dynasties, known as the "Twenty-Four Histories" (*Êrh shi se shi*), comprising altogether 3,264 extensive chapters. With pedantic accuracy, all events are there registered not only year by year, but also month by month, and even frequently day by day. They also contain chapters on chronology, state ceremonial, music, law, political economy, state sacrifices, astronomy, geography, foreign relations, and the condition of literature in that particular period. Of the Twenty-Four Histories, the Newberry collection contains three series of different issues: (1) The complete lithographic edition based on the Palace edition of the Emperor Kᶜien-lung, published in Shanghai, 1884, in seven hundred and eleven volumes, bound in eighty wrappers; of the three Shanghai editions, varying in quality of paper and size of type, this one is the best. (2) Several Palace editions of the Kʻien-lung period of separate Annals, as *e.g.* the *Weishu* or Annals of the Wei dynasty (386–556 A.D.), in thirty volumes, printed in 1739 (No. 730); the *Kiu Wu tai shi* or Old History of the Five Dynasties (907–959 A.D.), in sixteen volumes, published 1775; the *Ming shi*, or the Annals of the Ming dynasty, one hundred and twenty-two volumes, 1739 (No. 646); the *Liao Kin Yüan shi*, *i.e.* the three Histories of the Liao (916–1125 A.D.), Kin (1115–1234 A.D.), and Yüan (1206–1367 A.D.) dynasties, eighty-two volumes, issued in 1740. The Manchu rulers had a special predilection for these three dynasties, with whose representatives they were connected by ties of blood, the Liao representing the Khitan and the Kin the Niüchi, both Tungusic tribes closely allied in speech and culture to the Manchu, while Yüan is the designation under which the Mongols held sway over China. The Newberry collection also includes the important work, first compiled under Kʻien-lung and re-edited in 1824 at the instigation of the Emperor Tao-kuang, explaining in Manchu transcriptions the foreign names of persons, offices, and localities abounding in the three historical works mentioned and containing important material for the study of the languages of the Khitan and Niüchi, only a few fragments of which have survived.[8] (3) Annals published under the Ming dynasty: the *Shi ki* of Se-ma Tsʻien, the first historiographer of China, printed in 1596, twenty volumes (No.

---

[8] See Laufer, *Sketch of Manchu Literature*, p. 45. Paul Pelliot (*Bulletin de l'École française d'Extrême-Orient*, 1909, p. 71) points to another source of Khitan words in the *Sui shi kuang ki*, embodied in the valuable collection of reprints by Lu Sin-yüan, the *Shi wan küan lou*, a copy of which is in the Newberry Library (No. 974, one hundred twelve volumes).

644), and the *Shi ki p'ing lin* (No. 868), published in 1576 by Ling I-tung in thirty-two volumes, giving the text with critical annotations at the head of the pages; the former edition was the first to print the text together with the commentaries of P'ei Yin, Se-ma Chêng, and Chang Shou-tsieh.[9]   One of the finest Ming printed works is represented by the *Ts'ien Han shu*, the Annals of the Former Han dynasty (B.C. 206–24 A.D.), printed in 1532 in thirty-two volumes, on Korean paper (No. 39).   Further, we have the *Hou Han shu*, the Annals of the Later Han dynasty (25–220 A.D.) of 1596 in thirty volumes (No. 647); the *Han shu p'ing lin* of 1581 in twenty-four volumes (No. 593); the *Nan Ts'i shu* or Books of the Southern Ts'i dynasty (479–501 A.D.) of 1589 in eight volumes (No. 657); the *Ch'ên shu* or Books of the Ch'ên dynasty (556–580 A.D.) of 1588 in six volumes (No. 656); the *Pei Ts'i shu* or Books of the Northern Ts'i dynasty (550–577 A.D.) of 1638 in six volumes (No. 726); the *Wei shu* or Books of the Wei dynasty (386–556 A.D.) of 1596 in sixteen volumes (No. 653); the *Chou shu* or Books of the Chou dynasty (557–580) of 1602 in ten volumes (No. 877); the *Sui shu* or Books of the Sui dynasty (581–617 A.D.) of 1594 in twenty volumes (No. 649); the *T'ang shu* or the Books of the T'ang dynasty (618–906 A.D.) in forty-nine volumes (No. 606); the *Sung shi* or Annals of the Sung dynasty (960–1279 A.D.) of 1480 (some leaves bearing dates 1557 and 1600) in ninety-six volumes (No. 855); the *Liao shi* or Annals of the Liao dynasty (916–1125 A.D.) of 1529 in eight volumes (No. 625), and finally the *Yüan shi*, the History of the Yüan or Mongol dynasty (1206–1367 A.D.), the *editio princeps*

---

[9] Compare E. Chavannes, *Les mémoires historiques de Se-ma T'sien*, Vol. I, p. ccxviii (Paris, 1895).

published under the reign of the first Ming Emperor Hung-wu (1368–1398). This makes a total of thirteen Annals in Ming dynasty editions, which may be considered a very satisfactory result of the search for these works, since complete sets can no longer be obtained. The superiority of the Ming editions has been demonstrated thus far in the case of the *Yüan shi;* but close critical study of the others will presumably reveal similar results. Bretschneider has shown that a learned committee was appointed by the Emperor K'ien-lung to revise the *Yüan shi,* and to change the writing of all foreign personal and geographical names according to an entirely arbitrary system in which the old names can hardly be recognized. The K'ien-lung edition has thus become unserviceable for historical and geographical investigations, and the Ming edition must be made the basis of all serious research. As the same observation holds good for the History of the Liao, the 1529 edition of this work now in the Newberry Library becomes one of fundamental value.

Under the Ming dynasty, three editions of the *Yüan shi* were issued: during the period Hung-wu (1368–1398), Kia-tsing (1522–1567), and Wan-li (1573–1620). The compilation of the Annals began in 1369 and was completed in the middle of 1370. It is curious, however, that our edition, which evidently represents this first original issue of the work, bears on the margin of the first page following the index the imprint "first year of Hung-wu," *i.e.* 1368. There are many leaves in it supplemented from the second Kia-tsing edition, on which dates like 1530, 1531, 1533, etc., and even 1572, are printed. It was a common practice under the Ming to make up books, especially historical works, in this peculiar manner. If single printing-blocks were destroyed or lost, the respective pages were written out and engraved again, and provided with a date-mark on the left margin. A number of our Ming works exhibit this feature, and a well-informed book-expert in Peking told me that this custom was followed in the Government printing-office at Nanking, and that all books of this peculiar make-up come from there.

An indispensable compendium for the study of the Chinese Annals is the *Shi sing yün pien* (No. 1207, twenty-four volumes, 1784), containing an index arranged according to rhymes of all proper names occurring in the Twenty-Four Official Histories. The principle of arrangement is the same as in the *P'ei wên yün fu*. When an historical name is met with in the reading of texts, one is enabled, by consulting this handbook, to refer at once to the chapter in the Annals where the biography of the personage in question may be found.[10]

The Dynastic Histories themselves constitute only a small portion of the historical literature of the Chinese; they form the frame and groundwork on which a lofty structure of investigations, dissertations, and compilations has been built. Next to the Dynastic Histories rank the "Annals" (*pien nien*) the model for which was found in the "Spring and Autumn Annals" (*Ch'un Ts'iu*) of Confucius, a chronicle of events in strict chronological sequence. The work of this class claiming the greatest antiquity is represented by the "Annals Written on Bamboo Tablets" (*Chu shu ki nien*, No. 875) extending to 299 B.C. The most celebrated production of this kind is the *Tse chi t'ung kien* by Se-ma Kuang (1009–1089), completed in 1084 after nineteen years' labor. It is a general history of China from the beginning

[10] Compare F. Hirth in *T'oung Pao*, Vol. VI, p. 319.

of the fourth century B.C. down to the beginning of the tenth century A.D.   About a century later, this work was revised and condensed by the famous philosopher Chu Hi (1130–1200) into fifty-nine chapters.   It was first published in 1172 under the title *T'ung kien kang mu* with an introduction by Chu Hi, and it is a complete copy of this *editio princeps* which the Newberry Library now possesses.   It is a rare and fine specimen of Sung printing and perhaps the most extensive work of that period now known.   This work is still regarded as the standard history of China, and innumerable subsequent editions of it have been published.[11] The fact that this edition is really that of the Sung period is proved by the description of it given by Mo Yu-chi in his valuable bibliographical work (*Lü t'ing chi kien ch'uan pên shu mu*, Ch. 4, p. 14, ed. by Tanaka Keitaro, Peking, 1909). He says that the printing-blocks were engraved in 1172, that the printing was done on pure paper, that each page has eight lines with seventeen characters for each line.   This agrees with our edition, while the reprint of the Yüan period exhibits on each page ten lines with sixteen large characters on each, or twenty-four, if small characters are employed. It is probable that the copy in the Newberry collection is identical with that described by Mo Yu-chi, as a number of other books inspected and attested by this scholar were obtained by me.

The Manchu translation of this history is represented by a Palace edition beautifully printed in 1681 under the generous patronage of the Emperor K'ang-hi and issued in ninety-six large volumes (No. 573).   No library in Europe seems to possess a perfect copy of it; the University Library

[11] A modern reprint dated 1886 is in the John Crerar Library (No. 808, two hundred and forty volumes).

in Kasan owns a fragment in twelve volumes. To the same group of histories belong the *T'ung kien ts'üan pien* (No. 608), compiled under the Ming dynasty in 1559, in twelve volumes; the *Kang kien hui pien* (No. 742), of the same period, giving a history down to 1355; and the *Sung Yüan t'ung kien* (No. 648), a history of the Sung and Yüan dynasties, dated 1566, in twenty-four volumes.

The third class of historical writings is represented by the "Complete Records" (*ki shi pên mo*) in which the authors free themselves from the restraints of the traditional method and treat the whole subject thoroughly from a broad point of view. The most important of these, the *Sung shi ki shi pên mo*, the *Yüan shi ki shi pên mo* of 1606, and *Ming kien ki shi pên mo* (No. 547, twenty volumes, 1648), are all in the Newberry collection; also the *Yi shi* by Ma Su (No. 948, forty-eight volumes), of 1670. In works relating to the history of the Yüan and Ming dynasties, the Library is especially rich. I may mention the *Yüan shi sin pien*, a newly discovered history of the Mongol dynasty published in 1905 in thirty-two volumes; the Code of the Yüan, first printed 1908 in twenty-four volumes; the *Ming shi kao* (No. 607), a valuable history of the Ming dynasty in eighty volumes, inspired by the untiring activity of the Emperor K'ang-hi in 1697, written by Wang Hung-sü and printed 1710; the *K'in ting ming kien* (No. 631), twenty-four volumes, and another record of the same house, compiled by a commission under K'ien-lung; the *Ming ki tsi lio* (No. 856), likewise an account of the history of the Ming dynasty, issued in 1765 in sixteen volumes.

Of the works falling under the category of *chêng shu, i.e.* handbooks of information on the constitution, official administration, and many subjects of national economy; the collection includes the so-called *San t'ung* (No. 920), edition of 1859 in three hundred and twenty volumes, embracing the *T'ung tien* of Tu Yu, who died 812 A.D.; the *Huang ch'ao t'ung tien*, referring to the Manchu dynasty and compiled by order of K'ien-lung, first published about 1790, and the *Wên hien t'ung k'ao* by Ma Tuan-lin, first printed in 1319; further, the *Kiu t'ung lei tien* (No. 973), in which nine works of this class are worked up systematically, in the Shanghai edition by Yü Yüeh in sixty volumes.

A group of historical works not mentioned by Wylie is represented by comprehensive histories of the emperors which seem to have been in vogue under the Ming dynasty. I secured a *Yü ch'i li tai kun lan* (No. 543), a history of the lives of the emperors beginning with the mythical culture-hero Fu-hi and ending with Shun-ti, the last emperor of the Mongols. It was composed by the Ming Emperor Tai-tsung and is in a beautifully printed Ming Palace edition of the year 1453 in five quarto-volumes, probably unique. Of other Ming publications treating of historical subjects may be mentioned the *Ts'in Han shu su* (No. 891) of the year 1558, containing memorials to the throne by eminent statesmen under the Ts'in and Han dynasties.

There are many special records dealing with certain periods and events in the history of the late reigning house. The *Huang Ts'ing k'ai kuo fang lio* (No. 555), Palace edition of 1786 in sixteen volumes of quarto size of fine print, relates the history of the Manchu conquest of China. *Tung hua lu* is the designation of a number of works treating the reigns of the various emperors.[12] We have the *Tung hua lu* by Tsiang Liang-k'i, a summary of events from the origin of

[12] The name means Records of the *Tung hua*, a gate in the east wall of the Palace of Peking, near which there is the *Kuo shi kuan*, the Office of the State Archives.

the dynasty down to the year 1735, printed in 1765 (No. 739, twelve volumes); the *Tung hua se lu* (No. 744) in forty-nine volumes, being the continuation of the former work and treating the history of the long rule of the Emperor Kᶜien-lung (1736–1795); further, the *Tung hua tsᶜüan lu* (No. 988), published in 1884 in one hundred and thirty-four volumes, containing a complete history of the dynasty up to 1874; finally the *Kuang-sü tung hua lu* (No. 972, sixty-four volumes), being the history of the period Kuang-sü (1875–1908).

Wars and rebellions have been frequent during the last two centuries and the official documents relating to most of them have been printed. The formidable war which the Emperor Kᶜang-hi waged against the Kalmuk chief Galdan at the end of the seventeenth century is treated in a Manchu work under the title, "Subjugation of the Regions of the North and West" (Nos. 560, 710), in thirty-five volumes, of which only twenty-three could be secured, no complete copies having survived. So far as I can ascertain, no European library is in possession of this work. Another book of great rarity is the *Pᶜing ting kiao ki lio* (No. 736), "Account of the Pacification of the Sectarian Rebels," published by order of the Emperor Kia-kᶜing and relating to the rebellion of a secret society under the leadership of Li and Lin Tsᶜing, who plotted against the life and throne of the monarch.

The imperial printing-office, which occupied a series of buildings situated to the southwest of the Palace City (called *Tsao pan chᶜu*), was established by a decree of the Emperor Kᶜang-hi in 1680. The superb editions issued from this press by imperial sanction under the reigns of Kᶜang-hi and Kᶜien-lung are known under the name of

Palace editions (*tien pan*). The buildings together with
their entire stock of printing-blocks and types were destroyed
by an accidental conflagration in July, 1869. Palace
editions have therefore become rare and eagerly sought-for
treasures. That disaster resulted in the establishment of a
new press arranged on the plan of movable types which was
connected with the Tsung-li Ya-mên, the former Office of
Foreign Affairs. Fonts of movable lead type were procured
from the supply introduced by Mr. Gamble, superintendent
of the American Presbyterian Mission Press at Shanghai,
and were employed in the production of several official
publications of great bulk and historical importance. The
most noteworthy of these are the *Kiao ping Yüe-fei fang lio*
(No. 654) in four hundred and twenty volumes, and the
Kiao pʻing Nien-fei fang lio (No. 655) in three hundred and
twenty volumes. The former gives the official record of the
Government proceedings in the great Tʻai-pʻing insurrec-
tion, all operations and despatches being given; the latter
work contains a similar record of the great Mohammedan
rebellion. Both publications were issued simultaneously in
1872, and magnificently printed in uniform style. They
form one of the most extensive collections of documents
relating to a particular event ever published by any government, and they deserve the careful attention of the historian interested in these two unique movements; they have
not yet been utilized by any foreign scholar.

Among works relating to the history of modern times, the
following are deserving of special mention: The Collected
Reports and Decisions of the statesman Li Hung-chang
(No. 708) published in thirty-two volumes, 1866, by
Chang Hung-kün and Wu Ju-lun; and the Diary of the
great statesman Tsêng Kuo-fan (1811–1872) published

in the facsimile of his own handwriting (forty volumes, No. 1215).

The *Shi ch'ao shêng sün* (No. 970, one hundred volumes) is a collection of all the decrees issued by the emperors of the Manchu dynasty up to 1874, the year of the death of T'ung-chi; those of the Emperor Kuang-sü have not yet been published in book-form.

Many critical treatises on special historical subjects, many works on biography, memoirs, and local history (several, *e.g.* on the history of Sze-ch'uan Province) are also included in the collection. Taken collectively, these materials provide the means for the detailed investigation of almost any historical problem relating to Eastern Asia.

The cyclopaedic tendency of the Chinese has become almost proverbial. Hardly any nation can boast of such a large number of cyclopaedias. They resemble, on the whole, our own attempts in this direction, except that the method of arrangement is different. The Chinese works of this kind are arranged methodically according to subject-matter, extracts or quotations from older works on the particular subject being given under each heading. The compilers refrain from recording investigations or even opinions of their own, but observe a strictly objective and impartial method in placing only the material itself before the reader. One soon becomes familiar with the mode of arrangement, and finds in a few moments any special subject desired, when accustomed to the system of classification.

Of the more important cyclopaedias in The Newberry Library, the following are deserving of particular mention: the *T'ai p'ing yü lan* (No. 32), edited 1812 in sixty-four volumes by the scholar and statesman Juan Yüan (1764–1849). This is a compilation coming down from the Sung

dynasty and completed by Li Fang and others in 983 A.D.
It is divided into fifty-five sections comprising a thousand
chapters in all.   Extracts are given from 1,690 works all of
which are listed in the introduction.   As scarcely two- or
three-tenths of these are now preserved and a large number
were already lost when the work was compiled, so that the
quotations had to be copied from former cyclopaedias, this
thesaurus is especially valuable since it thus includes a great
deal of information not to be found in other sources.   One of
the most practical works of this class is the *Yen kien lei han*
of which we have the beautiful Palace edition issued under
the patronage of the Emperor Kᶜang-hi in 1710, in one
hundred and forty volumes (No. 36).   F. W. Mayers, who
has traced the literary history of this work (*China Review,*
Vol. VI, p. 287), calls it the most accessible and perhaps the
most generally useful of the imperial compilations of the
Kᶜang-hi period.   An earlier production of the Ming
dynasty, the *Tᶜang lei han* by Yü Ngan-kᶜi,[13] served as
model and foundation of the *Yen kien lei han*.   Arranged in
four hundred and fifty chapters, it amounts to twice the
bulk of the *Tᶜai pᶜing yü lan*, as the chapters are more
voluminous and the types are cut on a smaller scale.   The
cyclopaedia *Yü hai* (*lit.* "Sea of Jade") was compiled by
Wang Ying-lin (1223–1296) in the second part of the
thirteenth century.[14]   It was first printed in 1337–1340.   In
the first half of the sixteenth century, revised and aug-
mented editions began to appear.   The one in the Newberry
collection (No. 33) is the Palace edition of the Kᶜien-lung
period, published in 1738 in one hundred and twenty

[13] The original edition of this work is in the John Crerar Library Collection
(No. 211).

[14] Compare Pelliot, *Bulletin de l'École française d'Extrême-Orient*, Vol. II,
1902, p. 336.

volumes. It is divided into twenty-one sections comprising upward of two hundred and forty articles and is generally prized by scholars, according to Wylie, as one of the best works of its class, although it must be used with discrimination. The latter remark is not restricted, however, to this particular work, but holds good for all cyclopaedias, the quotations of which are sometimes inexact, incomplete, or impaired by misprints, and should be verified in important cases from the originals, if these are available, which is certainly not always the case.

Of the *Ts^cien kio lei shu* (No. 161) the Newberry collection has the original edition of 1632 in forty-four volumes. It is divided into thirteen sections containing upward of fourteen hundred articles. Wylie states regarding this work that in the eleventh book which treats of the bordering countries, and in the fourteenth book on foreign nations, the author speaks with an unguarded freedom respecting the Manchu. This caused the work to be placed on the *Index Expurgatorius*, and these two books were ordered to be suppressed. In our edition, however, they are fortunately retained in full, and it may be a timely task to investigate on what grounds the charge of anti-Manchuism is based.

Of the cyclopaedia *Tse shi tsing hua* (No. 160), the Palace edition executed under the reign of the Emperor Yung-chêng, 1727, in thirty-six volumes, was the one secured for the Newberry Library. This is a voluminous collection, in one hundred and sixty chapters, of extracts from historical and philosophical writers, primarily intended as a convenient manual to aid in the composition of literary essays.

One of the treasures of the Newberry collection is the *Ts^cê fu yüan kuei* (No. 231), edition of 1642 in three hundred and twenty volumes. This is now exceedingly rare and a work of great intrinsic value. It is an historical compendium drawn up in the form of an encyclopaedia with full details of all state matters from the beginnings of history down to the Sung dynasty; it was compiled by a commission of fifteen at the request of Chên-tsung, the third emperor of the Sung, and completed in 1013 A.D., each section being revised by the Emperor in person. The importance of this work rests on the fact that it allows of the comparative study of all existing sources relative to the same events, and that it imparts a great deal of new material not to be found in the official Annals, especially for the history of China under the T῾ang dynasty.

The *Fa yüan chu lin* (No. 38, twenty-four volumes) is a convenient reference work dealing with Buddhistic subjects and affording a comprehensive view of the entire system of Buddhism in cyclopaedic arrangement. It was first issued in 668 A.D. by the monk Tao-shi.

As might be expected from their philosophical trend of mind, philosophy occupies the largest place in the life of the Chinese and in their literary achievements. Of the so-called classical, but more correctly, canonical literature, the Newberry collection contains many Palace editions of the Ming period and of the eighteenth century. The former are nearly all facsimile reprints of the earlier Sung editions; *e.g.* Chu Hi's work on the *Yi king* (No. 661) is a Ming reproduction of the *editio princeps* of 1099. A notable acquisition is the *Huang ts῾ing king tsieh* (No. 623, three hundred sixty-one volumes, edition of 1890) containing one hundred and eighty works of the Manchu dynasty commenting on the Confucian Canon and edited by the famous statesman Juan Yüan (1764–1849). Of philosophical works, the original edition of the *Sing li ta ts῾üan* of 1415 (No. 672), a collection of the writings of the Sung philosophers made at the instigation of the Emperor Yung-lo, and the collected works of the philosopher Chu Hi (1130–1200), Palace edition of 1713, are especially noteworthy. The latter gives a dogmatic interpretation of the ancient canonical books and exercised an almost despotic influence on the subsequent thought and literature of China.

In lexicography the collection is strong. Among the early works in this class may be mentioned the original edition of the *Hung-wu chêng yün* (No. 545, five volumes), a dictionary arranged according to rhymes and composed by the first emperor of the Ming dynasty, Hung-wu (1368–1398), and the *Wu yin pien yün* (No. 887, five volumes), a dictionary of 1467.

Fiction is considered by Chinese scholars an inferior branch of literature and is not grouped with literature proper. It covers a wide field, nevertheless, and is immensely popular. No endeavor at completeness was made, but only the more important novels and those having a certain value as illustrating the history of culture were procured. Poetry, however, has always been viewed as one of the liberal arts and elegant pastimes of a gentleman, and the Chinese have cultivated it to an extraordinary extent. Its study is valuable to us for its high aesthetic merits, but at a future date it will surely fulfill a still greater mission and furnish the fundamental material for the most difficult of all subjects connected with China—the psychology of the Chinese. Here, their sentiments have crystallized, and he who wants to get the spirit of Chinese feeling and thinking must turn to their poetry, which is also the basis for the understanding of their painting and music. This department of literature was made as full and representative as possible, and all poets of distinction are represented. All dynastic collections embracing the poems of certain periods, such as the works of the Han, Leu-ch'ao, T'ang, and Sung, and many individual editions of poets as well as critical investigations of their works were acquired, together with the Palace editions of the collected poems of the Emperors K'ang-hi and K'ien-lung. Among the early poetical works special mention may be made of the *T'ang shi p'in hui*, "Researches into the Poetry of the T'ang Period" (No. 1208), by Kao Sin-ning, printed in 1395 in eighteen volumes; a Collection of Poetry in eight volumes printed under the Yüan dynasty (No. 1151), the fourth volume of which contains the poetical works of the celebrated poet and painter Wang Wei; a Ming edition of the two foremost poets Tu Fu and Li T'ai-po (No. 869, twenty volumes); a Ming edition of the Sung poets of 1504 (No. 1169, twenty-four volumes); and the *Li Sao* of 1586 (No. 871, four volumes). The copy of the *Shuang-ki Hang kung shit tsi* (No. 916), *i.e.* "Collection of the Poems of Hang Huai, or Hang Shuang-ki," a poet of the Ming dynasty, is the only one at present known to be in existence. It was printed in 1559 in the Kia-tsing period, and was formerly in the Lü-t'sing Library. In the Catalogue of this Collection already referred to, it is remarked (Ch. 15, p. 12) that at the end of the last volume a hand-written entry consisting of two lines has been made to the effect that "on the nineteenth day of the month, of the year *hing-se* of the period K'ang-hi (1701), the old man Chu-to has perused this book." This same inscription is found written in red ink at the end of the Newberry copy which consequently must be identical with the one examined by Mo Yu-chi, who died in 1871. Chu-to is the title of Chu I-tsun (1629–1709), a devoted student of

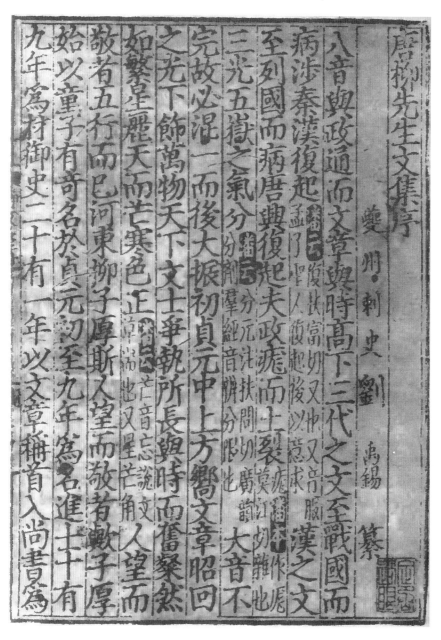

TWO PAGES FROM T'ANG LIU SIEN SHÊNG WÊN TSI,

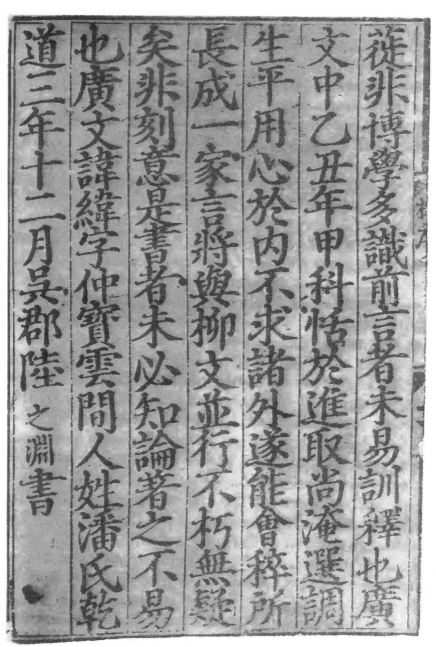

誕非博學多識前言著未易訓釋也廣

文中乙丑年甲科帖於進取尚淹選調

生平用心於內不求諸外遂能會稡所

長成一家言將與柳文並行不朽無疑

矣非刻意是書者未必知論著之不易

也廣文諱緯字仲寶雲間人姓潘氏乾

道三年十二月吳郡陸之淵書

PUBLISHED IN 1167, SUNG PERIOD.

ancient literature and archaeology. We can thus clearly trace the ownership of this copy to two famous scholars.

A work of great importance, and at the same time the earliest printed book in the Newberry Library, is the *Tʻang Liu sien shêng wên tsi* (No. 1174), dated 1167, in twelve volumes, containing the collected poems and essays of Liu Tsung-yüan (773–819 A.D.), one of the most celebrated poets and essayists of the Tʻang dynasty. This edition, in forty-three chapters, has been fully described in the Catalogue of Lü tʻing (Ch. 12, p. 16); it is provided with a commentary by Shi Yin-pien. The margins of the pages show the peculiar black ornament of the Sung period (called "black mouth," *hei kʻou*); there are twenty-six lines of twenty-three characters on each page.

Born bibliophiles and philologists, the Chinese have always devoted the greatest attention to the bibliography of their literature. In the official annals of the various dynasties, there is a section in which the books extant or issued during that particular period are carefully enumerated; this thus becomes an indispensable source for the tracing of the history of books. The best known general catalogue is the *Sze kʻu tsʻüan shu* giving a detailed critical description of the library of the Emperor Kʻien-lung who caused an extensive search for ancient books and manuscripts to be made throughout the empire. The Fan family in Ning-po, which possessed one of the greatest private libraries, distinguished itself in this enterprise and rendered a substantial service to the book-loving monarch by sending up six hundred and ninety-six works not owned by him. The Catalogue of this Library (No. 939) was published in 1808 in ten volumes, by the eminent scholar and statesman Juan Yüan (1764–1849), and registers the titles of 4,094 works. As a characteristic sidelight on Chinese private libraries, the fact may be mentioned that the Fan Library is, or was, guarded with great jealousy. It is the common property of the whole clan, and each member of the clan keeps a key to his own lock, so that the place can be opened only with the consent of all, and it is the strict rule that it shall be opened only in the presence of all. In the Catalogue mentioned, a list of the books presented to the Emperor is drawn up.

The province of Chê-kiang in which Ning-po is situated, and the provinces of Kiang-su and An-hui always excelled in an abundance of books and in a great number of book-lovers and collectors. The total of works despatched from Chê-kiang to Peking amounted to 4,600, of which 2,000 were retained by the imperial bibliographers as deserving of being copied. Critical notes on all these books offered to the throne by that province were edited by Shên Chʿu under the title *Chê-kiang tsʿai tsi yi shu mu* (No. 940), in 1772, eleven volumes. The Newberry Collection includes several other such private catalogues, among which the *Pi sung lou tsʿang shu chi* (No. 943, thirty-two volumes, 1882) deserves special mention. This is a description of the rare books gathered by the famous scholar Lu Sin-yüan,[15] whose library was purchased in 1907 by the Japanese banker Iwasaki for 100,000 Yen ($50,000.00).

A catalogue of special value is the *Hui kʿo shu mu* (No. 953, ten volumes, 1870), a list of two hundred and sixty-nine so-called *Tsʿung-shu* or Repositories.[16] It was first published in 1799 by Ku Siu. Many ancient and most interest-

[15] Compare P. Pelliot, L'œuvre de Lou Sin-yüan (*Bulletin de l'École française d'Extrême-Orient*, Vol. IX, 1909, pp. 211–249).

[16] Compare A. Wylie, *Notes on Chinese Literature*, pp. 76, 255; F. Hirth, *Tʿoung Pao*, Vol. VI, 1895, p. 321.

ing writings have been preserved only in these repositories, a class of publications corresponding to our "Series" or "Library" and usually containing the first printed editions of ancient manuscripts. In some cases these collections are of a heterogeneous nature since they include only such rare books as chanced to fall into the hands of an individual or a publishing house. In other cases they are arranged according to a plan well mapped out before hand, comprising the writers of certain periods or limited to certain classes of literature as philosophy, poetry, geography, or medicine. Thus, the well-known *Han Wei ts͏ung shu* is a collection of authors who lived during the Han and Wei dynasties; the *T͏ang Sung ts͏ung shu* is exclusively devoted to productions of the T͏ang and Sung periods; the *Chêng i t͏ang ts͏ung shu* (No. 753, one hundred and forty-eight volumes, 1709–1710) comprises collections of the treatises of the philosophers of the Sung dynasty. Wylie gives the contents of thirteen such *Ts͏ung-shu*, merely registering the titles of the works embodied in them. Paul Pelliot has seriously taken up this subject and given a detailed critical and bibliographical analysis of several *Ts͏ung-shu* with a stupendous amount of erudition. His high standard should be adopted as the ideal model for all future research in this direction. It is evident that the material incorporated in these enormous collections can be made available for fruitful investigation only by carefully cataloguing and indexing all the single works. It was made a special point to hunt up as many of these Collectanea as possible on account of their intrinsic value. It was a difficult task to trace and find them, owing to the fact that many of them were issued privately for subscribers and no copies in excess of the subscription were struck off. A great many were brought out in the period

Tao-kuang (1821–1850); these are now very difficult to procure. Wylie's remark that "the complete series is issued at once as an indivisible whole" does not hold good for all cases; I know of at least half a dozen Ts‘ung-shu now in process of publication on the subscription plan, the single issues being furnished to subscribers regularly as they come out. The Newberry Library has thirty-two of these works and the John Crerar Library about the same number; altogether they are the equivalent of several thousand useful books.

The most extensive of these publications is the *Wu ying tien tsü chên pan ts‘ung shu* (No. 538, reprint of 1868) in seven hundred and ninety-three volumes, containing one hundred and forty-eight different works, the titles of which have been listed by A. Wylie. *Wu ying* is the name of a building in the Imperial Palace of Peking where a printing-office was established; *tsü chên*, "assembled pearls," is an allusion to the set of wooden movable types cut in 1774 for the printing of the works amassed in the Imperial Library,[17] and constituting the bulk of the works published in this imperial collection. Next in literary importance is the *Chi pu tsu chai ts‘ung shu* (No. 921, two hundred and forty volumes), which means "the Library of the Discontented" *i.e.* those who are not satisfied with the ordinary books published, but who are desirous of delving deeper in branches of literature not easily accessible. Indeed, this series includes a great number of works of the first order for cultural studies, most of which are not obtainable in separate editions. It contains important books on antiquities and inscriptions, and the works of some of the oldest writers on

[17] The history of this event is described by F. W. Mayers in the *China Review*, Vol. VI, 1878, p. 294.

agriculture, mineralogy, and mathematics. Among others are the *Ling wai tai ta*, one of the best sources for our knowledge of mediaeval trade and intercourse of the Chinese with the peoples of western Asia; the *Süan ho fêng shi Kao-li t'u king*, containing a most interesting description of the country, customs, and institutions of Corea, written by Lu Yün-ti in 1167 on the basis of personal experience and observation, and deserving of a complete translation; the *Mêng liang lu* by Wu Tze-mu, giving a vivid account of the culture and social life of the city of Hang-chou during the Middle Ages and being a primary source for the history of games, pastimes, and theatricals.

A remarkable work of modern Chinese scholarship is the *Shi wan küan lou* (No. 974), edited in 1879 by Lu Sin-yüan, in one hundred and twelve volumes. He was a man of vast erudition, wide reading, extensive bibliographical knowledge, and an indefatigable collector of rare ancient manuscripts, part of which have been edited by him under the above title. All students are greatly indebted to the thorough and scholarly analysis which Professor Paul Pelliot[18] has devoted to this important work, and which enables one to put its valuable contents to immediate use. The collection comprises fifty individual works, while No. 43 consists again of twenty different treatises.

The most recent effort in the editing of important monuments of the past is represented by the *Kuo suei ts'ung shu* (No. 983), now being published by the *Kuo hio pao ts'un hui*, a learned society founded at Shanghai in 1906 for the preservation and study of ancient literature and art. This association founded a library and a museum, and seems to

[18] Notes de bibliographie chinoise. III. L'œuvre de Lou Sin-yüan (*Bulletin de l'École française*, Vol. IX, 1909, pp. 211–249, 425–469).

be in the possession of valuable ancient manuscripts which are being printed in the above-mentioned collection. Pelliot[19] remarks that the third section, containing historical works bearing chiefly on the epochs when the peace of the empire was troubled (end of the Sung, Yüan, Ming, Tʿai-pʿing resurrection), is the richest in historical material of immediate interest.

Islam has obtained a strong footing in China and numbers about twenty millions of adherents. A not inconsiderable literature in Arabic and Chinese has been brought into existence by Chinese Mohammedans, of which there are twenty-one works in the Newberry Library, obtained from the mosques in Chʿeng-tu, the capital of Sze-chʿuan, and those in Si-ngan, the capital of Shensi Province. The following books deserve special mention: *Tʿien fang tien li*, Laws and Customs of the Mohammedan Religion, six volumes, Nanking, 1871; *Tʿien fang sing li*, Mohammedan Philosophy, six volumes, Nanking, 1871; *Tʿien fang li yüan*, Origin of the Mohammedan Calendar, in Chinese and Arabic, one volume, 1876; *Tʿien fang wei chên yao lio*, the Islamic Taboos on Food, one volume, 1892; *Tʿien fang huan yü shu yao*, Geography of the Mohammedan World, one volume, 1892, with illustrations of a compass, eclipses, etc.; *Tʿien fang tse mu kieh i*, Explanation of the Arabic Alphabet, one volume, 1894; *Si lai Tsung pʿu*, Life of the Prophet, one volume, 1899; *Tʿien fang jên i pao chên se tse king*, Mohammedan Schoolbook in adaptation of the Chinese Four-Character Primer, one volume, 1897; the Three-Character Primer (*San tse king*), 1838, and the *Great Learning* (*Ta hio*), 1794, in the form of Islamic instruction;

[19] In his article, Les nouvelles revues d'art et d'archéologie en Chine (*Bulletin de l'École française*, Vol. IX, 1909, pp. 573–582).

several works on the history, institutions, and theology of the religion, and some prayer-books. A curiosity is a manuscript Arabic grammar in three volumes written by a Chinese Mollah of the great Mosque of Chᶜeng-tu about forty years ago.[20]

Art, archaeology, and epigraphy are represented by a number of standard treatises. The *Kin shi tsᶜung shu* is a repository of famous works on inscriptions published in 1888 in forty volumes (No. 620). In no department of their philological activity are the Chinese more deserving of praise and admiration than in epigraphy. With true zeal and industry, they have collected the many thousands of ancient stone records of their long past, published them in facsimiles, and displayed a great amount of critical acumen in the identification and interpretation of the old forms of characters. Most of these works are so well known to archaeologists that a detailed description of them need not be given. Suffice it to say that all the necessary material for successful investigations into Chinese antiquities is here, as, *e.g.* the extensive collection of inscriptions entitled *Kin shi tsui pien* (No. 40), sixty-four volumes, 1805; the *Kin shi so* (No. 158), by Fêng Yün-pᶜeng and Fêng Yün-yüan, twelve volumes, in the original quarto edition of 1821; the *Po ku tᶜu* (No. 162), twenty volumes, 1752, the standard work on ancient bronzes with their inscriptions, being the catalogue of bronzes in the Museum of the Sung Emperor Hui-tsung, published by Wang Fu in 1107 A.D.; the *Kin shi chᶜi* (No. 917), a very interesting work on various kinds of antiquities, first edited in 1778, re-edited in 1896, five

[20] A bibliography of Chinese Mohammedan literature is given by A. Vissière, *Études sino-mahométanes*, pp. 106-135 (Paris, 1911), and by the same author in the work of D'Ollone, *Recherches sur les Musulmans chinois*, pp. 389-419 (Paris, 1911).

volumes; *Tao chai ki kin lu* (No. 989), eight volumes, 1908, an illustrated catalogue of the famous collection of the former Viceroy Tuan Fang which is (or was) intended to form the foundation of a Chinese National Museum; *Liang lei hien yi kᶜi tᶜu shi* (No. 942), six volumes, Su-chou, 1873, a finely illustrated description of a valuable collection of ancient bronzes with facsimiles and ingenious explanations of their inscriptions, by Wu Yün, whose work is the best modern contribution to this difficult subject; *Liang lei hien yin kᶜao man tsᶜun* (No. 1152), ten volumes, by the same author, a publication describing the seals of the Han dynasty in his collection, a facsimile rubbing of each seal being given in vermilion, with a transliteration of the ancient script in modern characters and an historical discussion; two other extensive works on seals (No. 1156 and 596) of 1749 and 1904; *Tᶜieh yün tsᶜang kuei* (No. 938), 1904, Ancient Inscribed Tortoise Shells (used for divination), by Tᶜieh yün; the *Kin shi yün fu* (No. 630), an interesting dictionary of the ancient characters as found in bronze and stone inscriptions, printed in red; and many others. For the study of jade, there is the *Ku yü tᶜu pᶜu* (No. 35), the Catalogue of Ancient Jades compiled in the period Shun-hi (1174-1189 A.D.) and printed in 1779; and the *Ku yü tᶜu kᶜao* (No. 863) of 1889, the ingenious work of Wu Ta-chᶜêng. A number of works have reference to the history of painting and the biography of painters; others are collections of drawings, black and white or colored prints. As one of the finest specimens of xylographic art, the *Nan sün shêng tien* (No. 729), in forty-eight volumes, the Palace edition of 1771, deserves especial mention. It contains a description of the travels of the Emperor Kᶜien-lung through the midland provinces, inspection tours with political ends in view, and is sumptuously illustrated with plans and views of scenery encountered along the imperial route. The work is one of our best sources for the study and understanding of the architecture of central China in the eighteenth century.

During the last decade there has been a remarkable renaissance movement in Chinese literature, resulting in an enormous output of books which still seems to be increasing. I do not here refer to the mass of foreign literature made accessible to the Chinese in the form of translations made by missionaries or other foreign teachers engaged by native universities; nor to the awakening of the people at large with respect to political and educational reforms which has resulted in the production of a vast literature on the law, administration, history, and sciences of foreign nations. It is gratifying to observe that despite this reform movement, activity in the domain of native erudition has not been neglected and shows quite unexpected fruits and results. The advocates of the degeneration theory, who diagnosed the whole of Chinese culture as stagnation and decay and were guided rather by hasty impressions and opinions than by a careful scrutiny of actual facts, surely were bad prophets. But a man like Alexander Wylie, gifted with an insight into real conditions, did justice to the literary activity of modern China when he remarked as far back as 1867:

"Apart from the works issued by authority, the publications of private authors under the Manchu rule have been very considerable, and some of them indicate talent of no mean order. Although we have not the dashing flights of the Sung dynasty celebrities, yet we find a deep vein of thought running through the works of some modern authors; and for critical acumen the present age will stand a very fair comparison with most of its predecessors. The views of bygone ages are being freely canvassed; scholars are less under the mental domination of authority; and expositions of the classics which have long been held infallible, are anew submitted to the test of criticism. History, Geography, and Language have each received important accessions, and Mathematical works exhibit an evident tendency to advance."

Whoever takes the trouble to watch the literary activity of the present time will see this sound judgment fully confirmed and will be struck by the variety of topics and the breadth and depth of spirit in which they are treated. The excellent *Bulletin de l'École française d'Extrême-Orient*, edited in Hanoi, gives careful bibliographical notices of the Chinese book-market and also affords to the non-Chinese reader an opportunity of forming an idea of the scope and general contents of modern literature.

Even in poetry and fiction, the old glory has not entirely vanished, and there are promising signs of a new and flourishing era in this department. The modern novel, exhibiting the problems and conflicts of social life, has found an echo in the country and brought forth some remarkable productions. Journalism, which is now fully developed all over the country and certainly does not err in the direction of being too tame or reserved in the expression of opinion, has stamped a far-reaching influence on and given a new stimulus to belles-lettres. Magazines, valuable both for their contents and for artistic features, are an important factor in the culture of modern China and have a large reading public.

Reference has been made to the philological and editorial activity of modern scholars in the example of Lu Sin-yüan.

There are many other examples of this kind. The works of the most fertile contemporary author, Yü Yüeh, from the province of Chê-kiang, have reached one hundred and sixty good-sized volumes (*Ch^cun tsai t^cang ts^cüan shu*, 1902; No. 1182).

There has recently been a notable development of interest in archaeology. Three journals and several large serial publications devoted to this subject are now appearing in Shanghai. One of these, the *Chung kuo ming hua tsi* (No. 997, eleven numbers), *i.e.* Collection of Famous Paintings of China, although its reproductions do not equal similar work done by the Japanese, nevertheless makes most valuable material accessible to the student of Chinese art.[21] The publishing house *Yu chêng shu kü*, in Shanghai, is bringing out a fine series of albums (*Chung kuo ming hua tsi wai ts^cê*) in which both the single and the collected works of an artist are illustrated. Thirty-six numbers had appeared at the end of 1911 (No. 998). This firm has likewise issued a large number of facsimile reproductions of ancient rubbings and manuscripts, the scholarly utilization of which will place sinology on a new and solid basis equal in strength to that of classical philology. I secured for The Newberry Library a complete set of these facsimiles, numbering one hundred and sixty-three works, and relating to the Han, Sui, T^cang, Sung, and Yüan dynasties. They are all got up in tasteful editions to suit the requirements of book-lovers. The achievements of Chinese typography must not be judged from the cheap and flimsy productions thrown broadcast on the market to meet the small purses of the masses. In thorough, elegant, and graceful book-making, they are still

[21] The contents of the first five numbers have been analyzed by E. Chavanens (*T^coung Pao*, 1909, p. 515).

unsurpassed masters, and there is much in the style and technique of their books worth imitating even by us.

In view of the pulsating life animating the production of Chinese literature in all its branches at the present time, I cannot join in the pessimistic outcry with which W. Grube concludes his "Geschichte der chinesischen Litteratur." I see life and progress everywhere and trust in the future of China. I believe that her literature will bring forth new facts and new thoughts, and that the time will come when it will arrest the attention of the world at large. It is hoped that the near future may see many American scholars taking a real interest in this literature, and when that time comes they will have at hand here in Chicago ample foundation material for their studies and investigations.

094

东方绿松石考释

附：书评二则

PORTRAIT OF THE QUEEN OF SIKKIM.

PLATE I

The reigning queen (Mahārānī) of Sikkim, after an oil-portrait by Damodar Dutt, a Bengali artist, in the collection of the Field Museum (Cat. No. 117815, acquired by the writer at Darjeeling in 1908). The queen is a full-blooded Tibetan princess born at Lhasa in 1864 and was married to the present king of Sikkim as his second wife in 1882. Both were taken prisoners by the British in 1893 and held in captivity at Darjeeling. During that time she used to sit to the Bengali painter for the portrait in question which was completed in 1908. The writer had an audience with her in her palace at Gangtak, the capital of Sikkim, at which time she was dressed in the same state-attire and with the same jewelry as in this painting. Her crown, the peculiar headdress adopted by the queens of Sikkim, is composed of broad bandeaux made up of pearls, interspersed with turquoises and corals alternating. Her gold earrings are inlaid with a mosaic of turquoises in concentric rings. The necklace consists of coral beads and large yellow amber balls, and has a charmbox (gau) attached to it, set with rubies, lapis lazuli and turquois. She wears a bracelet of corals and two gold rings set with turquois and coral.

Mr. J. Claude White, the British Political Officer of Sikkim (in his book "Sikhim and Bhutan," p. 22, London, 1909) characterizes her as a striking personality, extremely bright, intelligent, and well educated; her disposition, he says, is a masterful one, and her bearing always dignified; she is always interesting, either to look at or to listen to, and had she been born within the sphere of European politics she would most certainly have made her mark, for there is no doubt she is a born intriguer and diplomat.

Another oil-portrait in the Field Museum from the hands of the same artist represents the Mongol Lama Shes-rab rgya-mts'o ("The Ocean of Wisdom") born in 1821 at Kükā-khota ("Blue City," in Shansi Province, China), teacher to the Pan-ch'en Lama, then interpreter in the Anglo-Indian Civil Service; he was distinguished for his Tibetan scholarship, particularly in the department of astronomy and astrology; he did useful work in assisting English translations of Tibetan books, and died at Darjeeling in 1902, at the age of eighty-one. Damodar Dutt received for this portrait a medal from the Bombay Art Exhibition.

Dimensions of above portrait: 1.74 x 1.06 m.

FIELD MUSEUM OF NATURAL HISTORY.

PUBLICATION 169.

ANTHROPOLOGICAL SERIES.   VOL. XIII, No. 1.

NOTES ON TURQUOIS IN THE EAST

BY

BERTHOLD LAUFER
Associate Curator of Asiatic Ethnology

CHICAGO, U. S. A.

July, 1913.

PREFACE

In April, 1911, Dr. Joseph E. Pogue, mineralogist in the U. S. National Museum of Washington, requested my co-operation in bringing out an extensive monograph which he contemplated on the turquois from the mineralogical, historical, and ethnological points of view. It was originally intended that the following notes should be embodied in the form of an appendix in Dr. Pogue's proposed work which I understand is now complete in manuscript. As adverse circumstances beyond the control of the author have unfortunately delayed for the last two years the publication of his study, and as a recent official journey to Alaska will prevent for some time longer active operations on his part, my contribution to his work is herewith issued in a separate form. It should be understood that only the exhaustive monograph of Dr. Pogue, which it is sincerely hoped will come out in the near future, will lend these notes their proper background and perspective. As at one time a plea was made by me for the co-operation of naturalists and orientalists (*Science*, 1907, p. 894), it is gratifying to note that we have advanced a step farther in this direction, and it will be seen on the following pages that oriental research can also bring to light new and not unimportant facts as yet unknown to our natural science. The occurrence of the turquois in Tibet and China, and to a higher degree, its history and cultural position in those countries, present a chapter of knowledge with which our mineralogists have been hitherto unacquainted. But only concerted action and sympathetic co-operation can lead us to positive and enduring results. The orientalist needs the naturalist as much as the latter, when his inclinations carry him to Asia, may profit from the stimulus of the former, in that he can suggest and encourage problems, the solution of which will turn out to be of vital significance to archæology. Our mineralogical knowledge of Eastern and Central Asia is in a very unsatisfactory condition, and it is desirable that the horizon of our mineralogists should no longer be bordered by the Panama and Suez canals. There is a great and promising field open between the two, and a plan which a mineralogist should follow in aiding the cause of archæology in Asia is briefly indicated on p. 54.

For various information I am under obligation to Dr. Friedrich

i

Hirth, professor of Chinese at Columbia University, of New York; to my friend Paul Pelliot, professor at the Collège de France in Paris; to Prof. Georg Jacob at the University of Kiel; and to Dr. Julius Ruska at the University of Heidelberg. The contributions courteously made by these gentlemen are clearly acknowledged as such in each particular instance. BERTHOLD LAUFER.

CONTENTS

iii

NOTES ON TURQUOIS IN THE EAST

By BERTHOLD LAUFER

I. TURQUOIS IN INDIA

The peoples of ancient India do not seem to have been acquainted with the turquois,[1] nor do they possess an indigenous word for it. The Sanskrit term *peroja* (also *perojā, pīroja*) or *perojana* is a comparatively recent loan word of mediæval times derived from New-Persian *ferozah* (older form *fīrūzag*), from which also the Russian word *biruza* and Armenian *piroza* come;[2] and the Sanskrit designation *haritāçma* is a compound with the meaning "greenish stone." The older Sanskrit treatises on precious stones do not make mention of it. Neither Buddhabhaṭṭa, a Buddhist monk who wrote the *Ratnaparīkshā*, that is, the "Appreciation of Gems,"[3] presumably before the sixth century

---

[1] Our name *turquois* (from French *turquoise*, Old French also *tourques*) means Turkish stone (there is also the word Turkey-stone, formerly turky-stone), not because the stone is found in Turkey, but because the most reputed kind, coming from Persia, first reached Europe by way of Turkey; the Venetians seem to be the first to have imported it (Italian *turchese*), and also to have made of it imitations in glass. The Latinized names were *torcois, turcosa, turchina,* or *turchesia,* and A. BOETIUS DE BOOT, court physician to Emperor Rudolf II (Gemmarum et lapidum historia, ed. A. TOLL, p. 265, Lugduni Batavorum, 1636; the first edition of this interesting work appeared at Hanover in 1609) states: "Omnibus nationibus eo nomine notissima, quod a Turcis ad nos feratur." Others hold that the allusion to Turkey in the stone implies no distinct geographical notion but vaguely means "coming from the Orient" (O. SCHRADER, Reallexikon der indogermanischen Altertumskunde, p. 153, Strassburg, 1901); indeed, Turkey was for a long time a term of uncertain value, almost having the meaning of "strange," and was even connected in Europe with two American products,— our North American bird, and maize (sometimes known as Turkish wheat). At any rate, the Turks were acquainted with the turquois, in particular with that of Persia, calling it by the Persian name *firuze.* According to a kind communication of Prof. GEORG JACOB, the turquois is described in a Turkish work on mineralogy written in 1511–12 A. D. by Jahjà Ibn Muhammad al-Gaffārī (manuscript in Leipzig, Catalogue of FLEISCHER, p. 508, No. 265). Five principal kinds are distinguished: Nishapuri, Gaznewi, Ilaqi, Kermani, and Kharezmi; the first, the well-known turquois from Nīshāpūr in Persia, is regarded as the most valuable, being hard, and fine, and permanent in color; the various sorts are described and followed by reports of celebrated turquoises in the history of Islam.— Shakespeare (*The Merchant of Venice,* Act III, Scene I, in the folio edition) has the spelling *Turkies;* the poet Tennyson adheres to the old form *turkis* (Middle English *turkeis,* on a par with German *türkis,* Middle High German *turkoys, türkīs;* in the seventeenth century, *turkes*). The usual English spelling, in accordance with French, is *turquoise* (formerly it was written also *turcois* and *turkois*); in scientific writings in this country the spelling *turquois* is now generally adopted.

[2] The Persian turquois is not discussed here, as it will be treated by Dr. Pogue in his monograph. The course of our investigation, however, necessitates touching also upon this subject, and some brief notes bearing on it will be found on p. 38.

[3] Edited and translated by LOUIS FINOT, Les lapidaires indiens (Paris, 1896)

I

A. D., nor *Varāhamihira* (505–587 A. D.) in his work *Bṛihatsaṃhitā* allude to the turquois. Agasti, in his versified treatise on gems, the *Agastimata*, and a very late work, the *Ratnasaṃgraha*, each devote a stanza to the turquois.[1] The date of the former work is not satisfactorily established. Inward evidence leads one to think that it is posterior to the sixth century A. D., and that a work under this name possibly existed in the thirteenth century, while in its present shape it is, in all likelihood, of much later date. Of greater importance is the little mineralogical treatise *Rājanighaṇṭu* written by Narahari, a physician from Kashmir, not earlier than the beginning of the fifteenth century. According to Narahari, the two words as given above are used to distinguish two varieties of the stone, as the hue is either ashcolored or greenish. He remarks that it is astringent and sweet to the taste, and an excellent means to provoke appetite; every poison, whether vegetable or mineral, or a mixture of both, is rapidly neutralized by turquois; it also relieves the pain caused by demoniacal and other obnoxious influences.[2] As, in all likelihood, the acquaintance

---

[1] L. FINOT, *l. c.*, pp. 138, 197.

[2] Compare R. GARBE, Die indischen Mineralien, p. 91 (Leipzig, 1882). In the introduction the date of Narahari's work is calculated at between 1235–1250; Prof. Garbe has been good enough to inform me that he has now arrived at the conclusion that the work cannot be earlier than the fifteenth century. The turquois, accordingly, appears on Indian soil very late during the middle ages, in the Mohammedan period. The evidence gathered from mineralogical literature is corroborated by the records of Indian medicine. The famous Bower Manuscript assigned to the year 450 A. D., the brilliant edition and translation of which has just been completed by Dr. A. F. R. HOERNLE (Calcutta, 1893–1912), does not make any mention of turquois, nor do the ancient physicians of India. (For this reason, J. JOLLY, Indische Medicin, does not note the stone.)— It is asserted on the authority of the *Periplus Maris Erythræi* (Ch. 39), a Greek work of an unknown author from the latter part of the first century (probably written between 80–89 A. D., roughly about 85; see the recent discussion of the date by J. F. FLEET, *Journal Royal Asiatic Society*, 1912, pp. 784–7) that turquois was exported from the Indian port Barbaricon. Mr. W. H. SCHOFF, in his new translation of the work (The Periplus of the Erythræan Sea, pp. 38, 170, London, 1912), feels very positive on this point, and explains that "the text has *callean* stone, which seems the same as Pliny's *callaina* (xxxvii, 33), a stone that came from 'the countries lying back of India,' or more definitely, Khorassan; his description of the stone itself identifies it with our turquois, etc." This opinion, however, is more than hypothetical. First of all, as already pointed out by LASSEN (Indische Altertumskunde, Vol. III, p. 14, Leipzig, 1858), it is doubtful whether the *kalleanos* of the *Periplus* is identical with the *callaina* of Pliny, because the localities where, according to the latter, the stone is found are too remote from India to make it possible for it to have been exported from the port of Barbaricon at the mouth of the Indus. Secondly, the supposed identification of Pliny's *callaina* or *callais* with the turquois is no more than a weak guess, and one that is highly improbable; and a mere guess, even though it may be repeated by a dozen or more authors; will never become a fact. It is said in DAREMBERG and SAGLIO (Dictionnaire des antiquités grecs et romains, Vol. II, p. 1463): "On suppose que c'est la turquoise;" and H. BLÜMNER (Technologie und Terminologie der Gewerbe und Künste bei Griechen und Römern, Vol. III, p. 249, Leipzig, 1884) justly concludes that the evidence does not by any means seem to be sufficient to establish the identification of the turquois with the *callais* of the ancients as a positive fact. The vague "description" given by Pliny (XXXVII, 110–2, 151) of the stone bears out no striking reference to the

of the Indians with turquois was conveyed to them by way of Persia, it seems highly probable also that their beliefs in the medicinal properties of the stone, were at least partially derived from Mohammedan lore.[1]

From an interesting text of the Arabic traveler al-Berūnī (973–1048) translated by E. Wiedemann [2] we now see that Persian turquoises were indeed exported from Persia into India, for the Arabic author remarks in his notes on the *firūzag* (turquois) that the people of Irāg prefer the

turquois, its main characteristics not being at all set forth, and may suit many other stones as well; the pale green color (*e viridi pallens*) and the attributes *fistulosa ac sordium plena* by no means fit the Persian turquois which owes its reputation to its deep-blue tinge and its purity, nor has turquois the color of the emerald; the localities pointed out by him (nascitur post aversa Indiæ, apud incolas Caucasi, montis Hyrcanos, Sacas, Dahas) rather militate against the turquois. Mr. Skoff's hint at Khorāsān (not given by Pliny, who only alludes to Carmania) is a somewhat arbitrary opinion prompted by the desire to suit the convenience of his case. The principal point at issue, however, is that there is no evidence for the alleged mining of turquois on Persian soil in the first century A. D. (see p. 40) merely presumed but not proved by Mr. Skoff and his predecessors. If Pliny had known about the quarrying of turquois in Khorāsān, he would have plainly stated the fact with an undisguised reference to Persia or that particular province; but there is not one classical author with a knowledge of Persian turquois, nor is there any evidence proving that turquois was traded from Persia into Greece and Rome. The tradition of India incontrovertibly shows that the Persian turquois, both in its name and as a matter, appeared in India only as late as the Mohammedan period, and the negative evidence of archæology lends further support to this conclusion. Enough archæological work has been carried out in India to prompt us to the positive statement that, despite the numerous precious stones discovered in ancient graves, no find has ever yielded a single turquois. The jewels, for example, in the burial-place of Buddha at Piprāvā discovered by W. C. Peppé (*Journal Royal Asiatic Society*, 1898, p. 573; Rhys Davids, *ibid.*, 1901, p. 397; G. Oppert, *Globus*, Vol. LXXXIII, 1903, p. 225) were carnelian, conch, amethyst, topaz, garnet, coral, and crystal. A similar state of affairs in regard to the Persian turquois, as will be seen on p. 56, obtains in China where the turquois of Nīshāpūr and Kermān became known very late in the middle ages, during the Mongol period of the fourteenth century. It is thus plainly indicated by these two coincidences in India and China which cannot be merely accidental that it was only the Arabs, and after the conquest of Persia in 642 A. D., who imported the turquois from Persia into India and China; and the fact is quite certain that only in this late period the Persian turquois began to conquer the market of the world. There is, accordingly, no reason whatever to interpret the stone in question mentioned by the *Periplus* as the Persian, or any other turquois. The best supposition would be to recognize in the word of the Greek text, as in so many others of the *Periplus*, the transcription of an Indian word (compare J. Bloch in *Mélanges Sylvain Lévi*, p. 3, Paris, 1911), presumably Sanskrit *kalyāṇa*, "good, fine, excellent," which is one of the attributes of gold (R. Garbe, Die indischen Mineralien, p. 33), or in the form *kalyāṇaka* is used with reference to medicines (compare also *kalyāṇī* and *kalyāṇikā*, "red arsenic"). See also p. 41, note 6.

[1] On the other hand it should not be overlooked that certain notions entertained regarding turquois among the Arabs and persisting later in Europe are absent in India and Tibet. Among these are the employment of the stone as an eye-remedy and against the stings of the scorpion, the latter idea first appearing in the Greek physician Dioscorides of the first century (compare L. Leclerc, Traité des simples par Ibn el-Beithar [1197–1248], Vol. III, p. 50, *Notices et extraits des manuscrits de la Bibliothèque Nationale*, Vol. XXVI, Paris, 1883, J. Ruska, Das Steinbuch des Aristoteles, p. 152 [Heidelberg, 1912], and Boetius de Boot, *l. c.*, p. 270).

[2] Ueber den Wert von Edelsteinen bei den Muslimen (*Der Islam*, Vol. II, 1911, p. 352).

smooth ones, those of India like the round ones with convex surface.[1] This is so far also the earliest testimony for the presence of the turquois in India.

The fact that turquois is absent from India is confirmed by the negative testimony of the great merchant traveler JEAN BAPTISTE TAVERNIER (1605–1689), who, as a dealer and expert in precious stones, repeatedly traveled in India and became thoroughly familiar with the customs of that country.   He writes in chapter nineteen of his Travels:[2]

"Turquoise is only found in Persia, and is obtained in two mines.   The one which is called 'the old rock' is three days' journey from Meshed towards the north-west and near to a large town called Nichabourg (Nīshāpūr); the other, which is called 'the new' is five days' journey from it," etc.

Tavernier would have certainly known about the existence of turquoises in India if they ever occurred there *in situ*.   The various reports of modern travelers that turquoises are imported from India into Tibet are therefore to be interpreted in the sense that these Indian turquoises have been imported from Persia.

Also MAX BAUER [3] states that turquois is not found in India, Burma and Ceylon.   But the same author does not note its occurrence in Tibet and China.

Abul Fazl Allami [4] (1551–1602), in his history of Akbar, enumerates among the precious stones in the treasury of the emperor rubies, diamonds, emeralds, and pearls, but not turquois.   Turquois seems to have been everywhere an ornament of the people, but not one of royal personages.[5]

In the modern jewelry of India the turquois is utilized to some

[1] The opinion formerly prevailed in Europe that the turquois was found in India because it was exported from there.   This was the view of FRANCISCUS RUËUS (De gemmis aliquot, p. 54 b, Tiguri, 1565); but the turquoises exported from India were in fact derived from Persia.

[2] Ed. of V. BALL, Vol. II, p. 103 (London, 1889).

[3] Precious Stones, p. 397 (London, 1904).— G. WATT (A Dictionary of the Economic Products of India, Vol. VI, p. 204, London, 1893) says after the Manual of Geology of India: "The existence of the true turquois in India is doubtful.   From the presence of blue streaks in the copper ores of Ajmir, Mr. Prinsep suggested the possibility of the stone being found there.   Subsequently Dr. Irvine reported its existence in these measures, but, according to Ball, the so-called turquoises of Ajmir are only blue copper ore."

[4] The Ain I Akbari, translated from the Persian by H. BLOCHMANN, Vol. I, p. 15 (Calcutta, 1873).   The original was published in 1597.

[5] Compare p. 30, note 3.   In the Arabic account of Abu Zeid of the ninth century (translated by M. REINAUD, Relation des voyages faits par les Arabes et les Persans dans l'Inde et à la Chine, Vol. I, p. 151, Paris, 1845) it is said: "The kings of India are in the habit of wearing ear-pendants consisting of precious stones mounted on gold; they wear necklaces of the highest price composed of red and green stones of the first quality.   But it is the pearls on which they place a greater esteem, and which are eagerly coveted by them; these now form the treasure of the sovereigns, their principal wealth."

extent in connection with pearl, ruby, diamond, sapphire, topaz and emerald, set in silver or gold.[1]

II. Turquois in Tibet

As jade is the recognized jewel of the Chinese, so turquois is the standard gem of the Tibetans.   In the eyes of the Chinese jade is not a stone, but forms a distinct class *sui generis*, as is shown by such constant phrases uttered from the lips of stone dealers: *shi yü pu shi shi-t'ou*, "it is jade, it is not a stone." [2]   To call a turquois a stone means an offense to the Tibetan, and he will exclaim indignantly, *di yü re, dô ma re*, "this is a turquois, and not a stone."   The Tibetan word for turquois, *gyu* (pronounced *yu*, without sounding the prefix g, which, however, appears in the Mongol loan word *ughiu*) [3] is indigenous property, being derived neither from Sanskrit nor Chinese; it shows that turquois must have been known to the Tibetans since remote times.   There are, doubtless, also many ancient turquoises still in their possession as they are inherited from mother to daughter for generations, and thus kept as heirlooms in the same family for centuries; being constantly exposed to the open air, they readily change color and often assume a pale green shade, more or less tainted with black spots.

Two special sorts of turquoises are called *drug-dkar* and *drug-dmar*, that is, white *drug* and red *drug*; the word *drug* designates the number 6, and the two terms are explained to designate very fine kinds of turquoises supposed to be one-sixth part white or red in tint, respectively. Desgodins, in the Tibetan Dictionary published by the French missionaries, translates the two by white and red sapphire, but also reminds

[1] G. C. M. Birdwood, The Industrial Arts of India, Vol. II, p. 25.   In the Higinbotham collection of jewelry in the Field Museum there are several fine specimens of Indian jewelry in which turquois is employed, collected in India by Mr. Lucknow de Forrest of New York.   G. Watt (*l. c.*) remarks that the turquois is largely used by the natives of India in jewelry but that imitations are perhaps more generally employed than the true stone.   While I do not deny that such imitations may occur, I do not believe that they are very generally in use.—Aside from the mineralogical treatises quoted above, as far as I know, the word for turquois has not yet been pointed out in any other work of Sanskrit literature.   The Sanskrit romance Vāsavadattā by Subandhu of the seventh century (translated by L. H. Gray, pp. 85, 109, *Col. Un. Indo-Iranian Series*, Vol. VIII, New York, 1913) mentions a necklace of pearls and sapphires, further emeralds and rubies, diamonds and other stones, but not turquois, which, as shown also by such passages, was a late intruder in such combinations as stated above, and alien to the artistic taste of India.

[2] The agate (*ma-nao*) is, in the eyes of the Chinese, "neither a stone nor a jade," but a thing for itself.

[3] In more ancient texts the word is written also *rgyu*, thus showing that in the ancient pronunciation also the g was sounded.   A singular word for turquois is the Mongol *kiris* which has thus far been pointed out but once in literature (Laufer, *T'oung Pao*, 1908, p. 431), and which presumably represents the ancient Mongol word for the turquois in times before the introduction of the Tibetan loan word.

us of the fact that the native dictionaries interpret them as *gyu* "turquois." While I have encountered a great number of turquoises with white and black veins and streaks, I have never seen any with a red tinge.[1] This classification of turquoises is contained in the ancient Tibetan medical work known under the abbreviated title "The Four Tantra"[2] (*rgyud bži*, Peking edition, Vol. II, fol. 145 b). The literary history of this interesting work remains to be made out.[3] Originally based on a standard Sanskrit work translated into Tibetan in the middle of the eighth century, it passed through the hands of several distinguished Tibetan physicians who revised and increased the work considerably. It contains not only information on Indian anatomy, pharmacology and therapeutics, but also valuable material with respect to the natural products of Tibet and Mongolia. The manifold subsequent interpolations render the utilization of these notes for historical research exceedingly difficult when the question arises as to the time to which they must be referred. A literal translation of the notice regarding turquois runs as follows:

"The turquois, in general, represents one species with two varieties, — that of best quality and the common one. Of the former there are two kinds, the one blue and white, of great lustre, called *drug dkar;* the other blue and red, of great lustre and polished, called *drug dmar.* There is, thirdly, a turquois of superior quality excelling the others in splendor, known as the turquois *sbyad,* 'beauty.' The common ones are 'the intermediate turquois resembling the *drug dmar,*' and 'the blue turquois resembling the *drug dkar.*' There are, further, the Indian turquoises originating abroad, and others. They entirely remove poison and heat of the liver. Substances belonging to the class of turquois and rock-crystal are so-called elements not fusible."[4]

If it could be proved with certainty that this note in the present tenor was already contained in the Sanskrit text or in the Tibetan version of the eighth century, it would be of a certain value in showing that at that comparatively early date turquoises were known in India, perhaps also traded from India into Tibet and then played a rôle in the pharmacopœa of both countries. But such evidence could be established only on the ground of an ancient edition preserving the original status

---

[1] Dr. Joseph E. Pogue informs me that the iron-oxide matrix in turquois from a number of localities is reddish.

[2] See HEINRICH LAUFER, Beiträge zur Kenntnis der tibetischen Medicin, p. 12 (Berlin, 1899).

[3] The brief notes given by Mr. WALSH (*Journal Royal Asiatic Society,* 1910, p. 1218) are not yet satisfactory and far from being exhaustive.

[4] In opposition to the four metals, gold, silver, copper and iron, enumerated shortly before this passage, which are designated as "fusible elements."

of the work. From what has been said above regarding the history of the turquois in India it is not very probable that the passage existed in the Sanskrit original, and if we assume on the basis of the available evidence that the Persian turquois spread in India between the tenth and the fifteenth century, the clause of the Tibetan text relative to Indian turquoises must be regarded as an interpolation, not older perhaps than the sixteenth century. The one feature, however, is conspicuous that the Tibetan terminology of the turquois varieties is not borrowed from India, but created in Tibet; it distinctly refers to the native stones, in opposition to those of India named last, and may well claim a certain age. Altogether six kinds are enumerated, and in the plates illustrating all objects of the *materia medica* described in the text, six kinds of turquois are really pictured.

The plates here referred to are twelve large scrolls or charts preserved in the Great Lama Temple (*Yung ho kung*) of Peking, exact copies of which I had made by an experienced Lama painter; the anatomy and physiology of the human body, and all medicinal substances derived from the three kingdoms are there figured in colors, labeled with their Tibetan names and accompanied by references to the chapters of the Four Tantra where they are described. The turquoises are represented, like the other substances, as being placed in rectangular trays supported by a standard or provided with three feet. The first two kinds, *drug dkar* and *drug dmar*, are painted in a deep-blue color and of oblong shape, no noticeable difference between the two being visible; the edge is marked by a blue line in gold (apparently to express the high quality of the stone) bordered by a line of black ink. It is curious to observe that in each case six stones have been outlined, and it is therefore evident that the draughtsman was guided by a literal interpretation of the two terms *drug dkar* and *drug dmar*, "White Six" and "Red Six." It is hardly plausible that a set of six stones should have been the fixed requirement in ancient times and resulted in this peculiar nomenclature, and I am also inclined to think that the modern Tibetan explanation as given above,— a stone containing one-sixth of white or red tinge,— is a makeshift or an afterthought. It would seem more reasonable to assume that *drug* in this case has no connection with the numeral six, but is an ancient noun signifying this particular variety of turquois. The third variety *sbyad*, also a group of six stones, is painted light-blue; they are pear-shaped, almost globular, surmounted by a curved tip. The two common kinds are each figured as one large stone, the one light-blue, the other grayish-blue, both of curious and fantastic outlines which it is hard to describe. On the second of the two, cloud-patterns in Chinese style of drawing are delineated, probably intended to indicate a "clouded" stone,[1] while the first is decorated with horizontal rows of small black rings presumably expressing veins in the stone. The Indian turquoises, again six in number, are, in distinction from all the preceding ones, light-green in color with fine black veins, and pointed or triangular in shape. It certainly remains an open question as to how far these drawings are faithfully preserved, but despite their imperfection we may learn from them that the appreciation of turquoises by the ancient Tibetans was graduated as follows: Deep-blue, lustrous stones without flaw took the foremost rank;[2] white and red strips or layers were not considered a blemish, but rather a special beauty; the lighter the blue, and the more approaching a gray and green, the more it sank in estimation; stones with black veins and streaks and with cloudy strata were looked upon as common, also those of greenish hues. It is interesting to note that this scale of valuation doubtless going back to ancient times holds good also for the present age.

The small turquoises not larger than a lentil and used for the setting in rings, are designated *pra*.

As famous swords, daggers, saddles and coats-of-mail received in Tibet individual names, so also celebrated turquoises were given special designations. Thus, we read in the History of the Kings of Ladakh that among fifteen turquoises brought from Gu-ge in West Tibet, the best were two, namely, the *Lha gyu od-ldan*, "the resplendent turquois of the gods," and the *Lha gyu dkar-po*, "the white turquois of the gods." [3] Thus, there are also celebrated historical turquoises, as it is recorded in regard to King Du-srong mang-po (beginning of the eighth century) that he found the largest turquois then known in the world, on the top of Mount Tag-tse, a few miles north of Lhasa.[4]

The name of an ancient well-known family of Tibet is *gYu-t'og* (that is, Turquois-Roof). The most celebrated member of it was a physician and author of medical works, who flourished in the eighth century and three times visited India to study medicine at the University of Nālanda. His biography, a very interesting work, is still in existence where it is narrated that he was once visited by gods and demons, who presented

---

[1] A term used also in India (R. GARBE, Die indischen Mineralien, p. 72, note 2) and in our mineralogy (with respect to veins or spots of lighter or darker color than the area surrounding them).

[2] This was the case likewise among the Arabs. The best sort of turquois was considered the one "of a complete purity of color, of a perfect polish, and of a hue entirely uniform" (L. LECLERC, Traité des simples par Ibn El-Beithar [1197-1248], Vol. III, p. 51, *Notices et extraits des manuscrits de la Bibliothèque Nationale*, Vol. XXVI, Paris, 1883). In a similar manner, al-Bērūnī expresses his opinion (WIEDEMANN, Der Islam, Vol. II, 1911, p. 352).

[3] *Journal Asiatic Society of Bengal*, Vol. LX, pt. 1, 1891, p. 123.

[4] *Journal Asiatic Society of Bengal*, Vol. L, pt. 1, 1881, p. 223.

him with an immense quantity of turquoises and other precious stones, heaping them on the roof of his house, hence the origin of his name. The mansion of this family still stands in Lhasa near a bridge called "Turquois-Roof Bridge." A Chinese author, writing in 1792, mentions this bridge and records the following tradition: "In the transparent waters of the river are turquois, colored rocks whose bluish tinge seems on the point of dissolving into water; the tops of the stones are bowl-shaped; if once dug away from the mud around them, they would look as big as elephants. One cannot take pebbles out of this river as an amusement as easily as in other streams."[1] It is not known whether this tradition is founded on fact, or whether the tradition connected with Doctor gYu-t'og and his name gave rise to the notion of turquoises existing in the river whose blue tinge may have lent a support to such a view; for in another Chinese source, according to ROCKHILL, it is said: "At the foot of Marpori (the mountain on which the palace of the Dalai Lamas rises) meanders the Kyi-ch'u, whose azure bends encircle the hill with a network green as the dark green bamboo; it is so lovely that it drives all cares away from the beholder."

In 641 A. D., the powerful Tibetan king Srong-btsan sgam-po married a Chinese princess, the daughter of the Emperor T'ai-tsung of the T'ang dynasty. The story of his wooing of the princess has been made by the Tibetans into a poetical romance in which we find such well-known and world-wide motives of popular tradition as the difficult tasks to be solved by the prospective son-in-law.[2] The candidates for the hand of the princess were many, so the emperor decided that he should obtain her who could best stand a number of tests. One of these was that he laid before the assembled delegates a buckler constructed of a coil of turquois arranged in concentric circles so that one end of it just formed the center; he required that a silk thread should be passed through the apertures of the turquoises from one end of the coil to the other. Nobody could solve the puzzle except the astute Tibetan minister *Gar* who caught a queen-ant and fed it well with milk until it grew bigger. Then he tied a silk thread to its waist, fastening the end of the thread to a silk band which he held in his hand, and placed the ant in the perforation of the first turquois, gently blowing into the hole, till to the amazement of the lookers-on the ant came out at the other end of the coil dragging the thread along.[3]

[1] According to the translation of W. W. ROCKHILL, Tibet from Chinese Sources (*Journal Royal Asiatic Society*, Vol. XXIII, 1891, p. 76).

[2] Compare R. H. LOWIE, The Test-theme in North American Mythology (*Journal of American Folk-lore*, Vol. XXI, 1908, pp. 97–148).

[3] Narrated in the Tibetan Annals of the Kings of Tibet (*rgyal rabs*, manuscript in the writer's possession), Chapter 13, fol. 45a.

The word "turquois" (*gyu*) has become a favorite attribute to designate a sky-blue color; "turquois-lake" (*gyu mts'o*) may be called poetically any blue-glittering lake, but is also the constant epithet of wells and certain favorite lakes, as, for example, for the sacred Manasarovara Lake or the Lake of Yar-brog (Yamdog).[1] Also flowers, the manes of horses, and even bees and tadpoles are described in the same manner; the hair of goddesses and the eyebrows of children born in a supernatural way are called turquois-blue; also the beauty of the body of such beings is compared to the turquois. In Spiti the forget-me-not is called *yu-žung men-tog*, that is, the flower whose essence or main substance is turquois.[2] In ancient mythology "thirteen turquois heavens" are mentioned, and as we speak of the Blue of Heaven, or the sky, the Tibetans say poetically "the turquois of Heaven." In a Tibetan legend, a poetical description of the country is given as follows:[3]

"At the foot of the giant mountains (the Himālaya) supporting the sky, lakes and flowing streams gather, forming plains of the appearance of turquois, and glittering pyramids of snow-clad crystal rise. This mountain range spreading like a thousand lotus flowers is white and like crystal during the three winter-months; during the three months of the summer it is azure-blue like turquois; during the three months of the autumn it is yellow like gold, and in the moons of the spring, striped like the skin of the tiger. This chain of mountains, excellent in color and form, and of perfect harmony, is inexhaustible in auspicious omens."

This passage is very interesting as revealing the innate nature-love of the Tibetan people and showing the connection of the colors of their favorite gems with the general colors of nature in the course of the seasons.[4] With the majority of the people, turquois is favorite, coral

[1] Also in the ancient Egyptian texts, the word turquois is used as a designation for the color of water. "Praises shall be offered unto thee in thy boat, thou shalt be hymned in the Ātet boat, thou shalt behold Rā within his shrine, thou shalt sit together with his disk day by day, thou shalt see the *Ant* fish when it springeth into being in the *waters of turquoises*, and thou shalt see the *Abtu* fish in his hour."— Hymn to the God Rā, in the Book of the Dead, by E. A. WALLIS BUDGE, Vol. I, 1901, p. 78. Interesting studies pertaining to the color of Tibetan lakes and rivers have been made by HERMANN V. SCHLAGINTWEIT, Untersuchungen über die Salzseen im westlichen Tibet, pp. 71 *et seq.* (*Abhandlungen der bayerischen Akademie*, München, 1871).

[2] In this case the word *žung* is to be written *gžung*. A. H. FRANCKE (Ladakhi Songs, p. 13. Reprinted from *Indian Antiquary*, 1902) has proposed to adopt the spelling *žung* in the sense of *chung* "small," so that the name would mean "flower of small turquoises."

[3] Compare I. J. SCHMIDT, Geschichte der Ost-Mongolen, p. 465 (St. Petersburg, 1829). In another passage (p. 439) it is said: "On the plain where diamond rocks glitter is a lake with a mirror like turquois and gold." See also p. 484.

[4] In a Tibetan poem depicting the labors of husbandry (*So-nam bya ts'ul-gyi leu*, published in the Tibetan School Series, No. II, Calcutta, 1890), the awakening of the spring is described, and the first buds on the uppermost branches of the trees are compared with the glimmer of emeralds; the flowers with antlers appear as vomiting sapphires; the great earth is teeming with sap, and resembles the malachite in its medley of blue and green colors.

and amber rank next. The blue, green, and blue-green; the red, rose, and pink; the yellow and brown of these three substances are indeed those tinges which most frequently occur among the flora of the Tibetan plateaus. During the summer, large patches of blue, red and yellow flowers abound on the fine pasture lands, and at this sight I could never suppress the thought that the enthusiasm of the Tibetans for turquois, coral and amber must have been suggested and strengthened by these beautiful shades of their flowers which their women as readily use for ornament as stones; indeed, it seems to me, as if owing to its permanency, the stone were only a substitute for the perishable material of the vegetable kingdom.

Turquoises, usually in connection with gold, belong to the most ancient propitiatory offerings to the gods and demons;[1] in the enumeration, gold always precedes turquois as the more valuable gift. They also figure among the presents bestowed on saints and Lamas by kings and wealthy laymen. The thrones on which kings and Lamas take their place are usually described as adorned with gold and turquoises, and they wear cloaks ornamented with these stones. It may be inferred from traditions and epic stories that in ancient times arrowheads were made not only of common flint, but also occasionally of turquois to which a high value was attached. · A powerful saint, by touching the bow and arrow of a blacksmith, transforms the bow into gold, and the arrowhead into turquois.[2] The hero Gesar owns thirty arrows with notches of turquois.[3]

In the popular medicine of the present time turquois is, as far as I know, not employed; but it is officially registered as a medicament in several medical standard works derived from or modeled after Sanskrit books. There we meet the typical series of ten substances: gold, silver, copper, iron; turquois, pearl, mother-o'-pearl, conch, coral, lapis lazuli.[4] Turquois is credited, as we saw above, with removing poison, and heat in the liver. It seems almost certain that this notion is taken from Indian lore; we remember the words of Narahari that every poison is rapidly neutralized by it, and that it relieves pain caused by demons. Also in the list of 365 drugs published in Tibetan and Chi-

---

[1] Laufer, Ein Sühngedicht der Bonpo (*Denkschriften der Wiener Akademie*, 1900, No. 7, p. 35); Schlagintweit, Die Könige von Tibet, p. 837; A. H. Francke, *Journal Asiatic Society of Bengal*, N. S., Vol. VI, 1910, p. 408.

[2] Laufer, Roman einer tibetischen Königin, p. 153 (Leipzig, 1911).

[3] I. J. Schmidt, Die Taten Bogda Gesser Chan's, p. 283 (St. Petersburg, 1839).

[4] This series occurs also in the Compendium of Tibetan Medicine translated from the Mongol into Russian by A. Pozdnejev, Vol. I, p. 247 (St. Petersburg, 1908).

nese by the Peking apothecary *Wan I*,[1] turquois is listed as a medicament, in the same series as given above.

A curious utilization of turquois is mentioned in the Biography of Padmasambhava (Ch. 53) who is said to have availed himself of gold, silver, copper, iron, lapis lazuli, turquois and minium inks for writing on light-blue paper of the palmyra palm and on smoothed birchbark.[2] Whether it is technically possible to use turquois for the coloring of ink I am not prepared to say; perhaps "turquois" is merely a designation for the blue or green color of the ink.

It seems doubtful whether in ancient times the turquois was considered a precious stone by the Tibetans. There is an old enumeration of jewels in the Annals of the Tibetan Kings (*rgyal rabs*, fol. 7) where the two classes, jewels of the gods and jewels of men, are distinguished, each class forming a series of five. The former comprises: 1. *indranīla*, 2. *indragopi*, 3. *mt'on-ka*, 4. *mt'on-ka ch'en-po*, and 5. *skong-mdzes*. The first two are Sanskrit words; No. 1 is the sapphire; No. 2 a kind of ruby; the word under 3 denotes the color of indigo and corresponds to Sanskrit *nīla* which is a general designation of the sapphire; also the next under No. 4 meaning "the great blue one" = Sanskrit *mahānīla*, denotes a superior quality of sapphire;[3] the signification of the stone No. 5 is unknown. The five jewels of men are gold, silver, pearls, lapis lazuli (*mu-men*), and coral. The turquois does not occur in this group, presumably for the reason that it was not classed among precious stones. It has never been, even in times of old, a stone of any exaggerated value. Among the presents made by the ancient kings of Tibet to the emperors of China we find stones like lapis lazuli and rubies (*padmarāga*), but no mention of turquois; likewise, in the lists of tribute

---

[1] Regarding this work compare BRETSCHNEIDER, Botanicon· Sinicum, pt. 1, p. 104 (Shanghai, 1882). There are several editions of this interesting small work, in Chinese and Tibetan style.

[2] LAUFER, Roman, p. 249. Also in the History of the Kings of Ladakh (A. H. FRANCKE's translation in *Journal Asiatic Society of Bengal*, N. S., Vol. VI, 1910, p. 405) writings in gold and turquois are attributed to five wise men in mythical times.

[3] Buddhabhaṭṭa (FINOT, Les lapidaires indiens, p. 41) explains *indranīla* as a sapphire the interior of which has the lustre of the rainbow colors, and which is rare and highly priced, and *mahānīla* as a sapphire with a color so intense that, thrown into milk of a volume a hundred times larger, it colors it like indigo.— "Sapphires of various colors occur in India. Thus, there is the blue or true sapphire of popular language, the color of which may be any shade of blue, from the palest to a deep indigo, the most esteemed tint being that of the blue cornflower. Violet sapphires (oriental amethysts) are also found in the same localities as those in which the true sapphire is met with. The most valuable sapphire found in the East Indies is the yellow sapphire or oriental topaz. A green gem, called by the Europeans in India an emerald, is often seen. It is, however, a green sapphire, and is much harder than the true emerald, which is a green beryl" (G. WATT, A Dictionary of the Economic Products of India, Vol. VI, p. 474).

sent by the Dalai Lamas to the emperors of China such gifts figure as silk scarfs, bronze images, relics, coral, amber, pearls, incense and woolen stuffs, but turquois does not appear.

In the religious service turquoises are employed, strung in the shape of beads, for rosaries, 108 beads being the usual number. The complexion of the god or goddess to be worshipped sometimes determines the selection in the color of the rosary-beads. Thus a turquois rosary is occasionally used in the worship of the popular goddess Tārā of whom there are two principal forms, one of these being conceived as of a bluish-green complexion.[1]

Turquoises are, further, offered on the altars of the gods, and their brass or copper images are adorned with them. Buddhist images, thus treated, may readily be recognized as Lamaist deities, as the Chinese never adopt this method. The number of stones set in an image varies according to its dimensions, and may reach from a half dozen up to a hundred and more. In any case, however, this is not intended as a mere ornamental addition, but the turquoises are to signify the actual jewelry with which the deities are adorned, and which form part of their essential attributes. One of the finest monuments in Tibet is the sarcophagus of the first Pan-ch'en Lama in the monastery of Tashi-lhunpo near Shigatse. It is of gold, covered with beautiful designs of ornamental work, and studded with turquoises and precious stones. The turquoises, says Captain RAWLING,[2] who has photographed this gorgeous monument, appear to be all picked stones, arranged in patterns, and in such profusion as to cover every available spot, including the polished concrete of the floor. In the oldest temple founded in Tibet about the middle of the eighth century, bSam-yas, which is described at full length in the Annals of the Tibetan Kings, there was a shrine in which the beams are said to have been of turquois; figures of galloping horses of gold were affixed to them, while there were other beams of gold with dragons of turquois attached.[3] This is the earliest Tibetan record regarding carvings from this stone; if the beams of turquois are not merely a metaphor of speech, it may be realized that the turquoises were inlaid in a kind of mosaic.

In the pictorial art of Lamaism jewels take a prominent place. On the first scroll in a set of twelve pictures (in the collections of the Field Museum, Nos. 121,371–382) representing the Eighteen Sthavira or Arhat and the portraits of the Dalai Lamas, we see as the central figure

[1] Compare L. A. WADDELL, The Buddhism of Tibet, p. 209 (London, 1895).

[2] The Great Plateau, p. 184 (London, 1905).

[3] T'oung Pao, 1908, p. 33.

Buddha Çākyamuni holding the alms-bowl of lapis-lazuli color.[1]   On the altar in front is depicted a golden bowl containing rubies, lapis lazuli, white conch-shell and turquois.   In the foreground is a lotus-pond with three flowers widely unfolded; on the central one three gems of oblong form are figured,— lapis lazuli, turquois, and ruby, emblematic of the well-known prayer formula *Om māṇi padme hūm* ("Oh, the jewel in the lotus!") and of the three precious objects (Sanskrit *triratna*), which are Buddha, his doctrine, and the clergy.   In the upper portion of the same painting, two of the Arhat are represented, Aṅgaja and Vakula, the latter holding and stroking an ichneumon which has the ability of spitting jewels; they are gradually dropping into a plate.[2] A tribute-bearer of grotesque racial type is offering to the saint gems in a bowl containing an ivory tusk, a coral-branch, and precious stones of blue, green, rose and pink colors.   This is not the only Arhat to whom

[1] The alms-bowl (*pātra*) of the historical Buddha was a plain pot; the miraculous relics of later times which were passed off as Buddha's alms-bowl form an interesting subject for the historical mineralogy of the East.   The general history of the bowl or bowls has been traced by H. KERN (Manual of Indian Buddhism, p. 90) and H. YULE (The Book of Ser Marco Polo, Vol. II, pp. 328–330).   Here, only the different materials should be pointed out.   Fa Hien who started for India in 399 saw the bowl in Peshāwur (Purushapura) and describes it as being "of various colors, black pre-dominating, with the seams that show its fourfold composition distinctly marked" (LEGGE, Record of Buddhistic Kingdoms, p. 35).   The latter clause in Legge's rendering does not seem to be quite correct; but however this may be, Fa Hien's account, it seems to me, bears out the fact that the bowl seen by him was carved from onyx in various layers in the style of cameo-work (compare G. WATT, *l. c.*, Vol. II, p. 174).   Hüan Tsang (ST. JULIEN, Mémoires sur les contrées occidentales, Vol. I, p. 106; S. BEAL, Buddhist Records of the Western World, Vol. I, p. 99) speaks of the *pātra*, but does not furnish a description of it.   Li Shi of the T'ang period (not of the twelfth century, as WYLIE, Notes on Chinese Literature, p. 192, says), in his *Sü po wu chi* (Ch. 10, p. 2; ed. of *Hu-pei tsung wên shu chü*), makes the statement that Buddha's alms-bowl in Peshāwur was of blue (or green) jade (*ts'ing yü*), or in the opinion of others of blue (or green) stone (*ts'ing shi*); then the text of Fa Hien is repro-duced.   In view of the ultramarine color in which the Buddhist alms-bowls appear on paintings in China and Tibet, it is permissible to think in this case of lapis lazuli; indeed, the word *ts'ing shi*, in this sense, is used in the *Wei lio* (HIRTH, China and the Roman Orient, p. 72).   A still earlier reference to Buddha's alms-bowl in the coun-try of the Ta Yüe-chi, already pointed out by F. HIRTH (Chinesische Studien, p. 251) is contained in the commentary to the *Shui king* written by Li Tao-yüan who died in 527 (his biography in *Pei shi*, Ch. 27) where likewise the term *ts'ing shi* is employed, and I concur with Hirth in the opinion that it should be translated in this case by lapis lazuli.   In Tibetan portrait-statues of bronze, the alms-bowl is often actually represented and carved from lapis lazuli (A. GRÜNWEDEL, Mythologie des Buddhis-mus, p. 79), as the outcome of the tradition that the mendicant's platter brought from Nepal to Tibet by the princess K'ri-btsun in the seventh century and working many miracles was made of lapis lazuli (S. CHANDRA DAS, Narrative of a Journey round Lake Yamdo, p. 79, Calcutta, 1887).   According to MARCO POLO (YULE's edition, Vol. II, p. 320) the dish of Buddha brought to China for Emperor Kubilai from Ceylon was "of a very beautiful green porphyry," while YULE quotes a Chinese account written in 1350 to the effect that the sacred bowl in front of the image of Buddha in Ceylon was neither made of jade, nor copper, nor iron, but that it was of a purple color, glossy, and when struck sounding like glass.

[2] The same attribute of the jewel-spitting ichneumon (Sanskrit *nakula*) appears in the hands of Kubera, the God of Wealth, guarding the northern side of the world mountain Sumeru.

precious stones are offered, but it is the case also with many others.   It is interesting that these tribute-bearers are usually people from Central Asia with unmistakable racial features and appropriate costume, or even turbaned Mohammedans.   We find the same figures also on the corresponding Arhat paintings of the Chinese and Japanese, and they are doubtless intended to express the important rôle which Iranians, Turks and Arabs have played in transmitting to the East the precious stones of western Asia.

In the marriage ceremony when the bridal party has arrived at the gate of the bridegroom's house, the officiating priest recites a few benedictory verses, describing the house of the bridegroom: "May there be happiness to all living beings!   The lintel of this door is yellow, being made of gold.   The door posts are cut out of blocks of turquois. The sill is made of silver.   The door frame is made of lapis lazuli. Opening this auspicious door, you find in it the repository of five kinds of precious things.   Blessed are they who live in such a house."[1]   This is certainly an ideal or poetical description.   In a more ancient marital ceremony described in the Tibetan dramatic play *Nang-sa*, "the turquois sparkling in rainbow tints" is tied to the end of an arrow adorned with streamers of five-colored silk which is fastened to the back of the bride to fix the marriage tie.[2]   In Ladakh, the bride generally receives, on her wedding day, many of the turquoises which her mother had worn.[3]

To describe all objects in which turquois is employed would mean to survey the whole range of Tibetan ethnography, which is certainly beyond the scope of these notes.[4]   But reference should be made to the beautiful Tibetan swords in which the hilts and sheaths worked in repoussé gold or silver are inlaid with large turquois and coral beads. This is an ancient technique practised also by the Turks of Central Asia and the Persians.[5]

So little is known about the localities in Tibet where turquois is found that there have even been authors who doubted its indigenous occurrence.

[1] S. CHANDRA DAS, Marriage Customs of Tibet, p. 12.

[2] L. A. WADDELL, Buddism of Tibet, p. 557 (London, 1895).

[3] A. H. FRANCKE, Ladakhi Songs, p. 13.

[4] For illustrations see Plates I–V.   The Field Museum possesses a rich collection of Tibetan, Nepalese and Chinese jewelry which will give occasion at some future date for a study in decorative and industrial art.   The Tibetan process of covering a gold or silver foundation with a mosaic of turquois agrees with the similar technique practised in Siberia during the bronze age, and therefore becomes an historical factor of great importance.

[5] Compare the Sassanian sword reconstructed by J. DE MORGAN (Mission scientifique en Perse, Vol. IV, p. 321, Paris, 1897) the shape of which is strikingly identical with the Tibetan sword.

A. Campbell, in his "Notes on Eastern Tibet," [1] has the following remarks in regard to turquoises:

"A great merchant of Tibet named Chongpo, who traded ages ago with India, and once crossed the seas beyond India, brought the finest real turquois to his native country. From that time the stone has been known there, and like coined money, it continues to circulate in the country as a medium of exchange. The imitations brought from China are made of common earthen-colored or other compositions. They are easily detected. Those imported via Cashmere are real stones, but not valuable. The only test of a real stone is to make a fowl swallow it; if real, it will pass through unchanged."

This tradition, if at all correct and not rather founded on a misunderstanding, carries little weight. The word *Chongpo* is not a Tibetan proper name, but simply denotes "a dealer, a trader." There is no evidence of the occurrence of turquoises in India proper; the people of India became acquainted with them from Persia only late in the middle ages through Mohammedan influence, and as shown above, they are first mentioned in Sanskrit literature in the beginning of the fifteenth, possibly also in the thirteenth, century. Thus, there is little or no plausibility in the assumption that India could have given the impetus to the introduction of the turquois into a country where almost every individual is in possession of these stones, and where a general national passion for them is·developed among all people high and low, which can have been but cultivated for many centuries and ages. This is corroborated by the facts of language and history, and further by the evidence of localities in Tibet where, in fact, turquois occurs *in situ*.

Marco Polo,[2] speaking of the province of Caindu, which is identical with the western part of the present Chinese province of Sze-ch'uan, a territory largely inhabited by Tibetan tribes, mentions besides a lake in which are found pearls, also a mountain in that country "wherein they find a kind of stone called turquoise, in great abundance, and it is a very beautiful stone. These also [in the same way as the fishing of the pearls] the Emperor does not allow to be extracted without his special order." Yule remarks that Chinese authorities quoted by Ritter mention mother-o'-pearl as a product of Lithang, and speak of turquoises as found in Djaya (or Draya) to the west of Bathang. This latter notice is quite correct and furnished by several Chinese authors who have visited Tibet and written on the subject.[3] They further mention Ch'amdo, that is, not only the small town in Eastern Tibet so

[1] *The Phoenix*, ed. by J. Summers, Vol. I, p. 143 (London, 1870).

[2] The Book of Ser Marco Polo, ed. by Yule and Cordier, 3d ed., Vol. II, p. 53 (London, 1903).

[3] W. W. Rockhill, Tibet from Chinese Sources (*Journal Royal Asiatic Society*, Vol. XXIII, 1891, p. 272).

called, but the whole district in which it is situated, and the territory of the capital Lhasa as places for the production of turquois; this locality seems to be particularly rich in this respect, and we have seen that the largest turquois of his time was discovered in the beginning of the eighth century by a Tibetan king on a hill north of Lhasa.

I have searched through the Chinese Annals of the Mongol or Yüan Dynasty (*Yüan shi*) for a confirmation of Marco Polo's report regarding the imperial turquois monopoly. Though my efforts have not as yet been crowned with success, I do not give up the hope that such an account will be discovered in the future either in this or in some of the other Chinese works treating of the history of the Mongol period. The turquois, however, is repeatedly alluded to in the *Yüan shi*, as we shall note hereafter.[1]

The first European author to report the indigenous occurrence of turquois in Tibet proper, as far as I know, is the Capuchin Friar Francesco ORAZIO DELLA PENNA DI BILLI in his "Breve Notizia del Regno del Thibet" written in 1730.[2]

According to SARAT CHANDRA DAS,[3] the finest turquoises are obtained from a mine of the Gangs-chan mountains of Ngari-Khorsum

---

[1] For the rest, there can be no doubt of the correctness of Marco Polo's statement. The turquois monopoly was the outcome and a part of all other exclusive prerogatives of the emperor extending to all precious metals and stones (compare in particular MARCO POLO, Vol. I, p. 424). This monopoly of the Mongols forms a counterpart to the turquois monopoly of the Persian Shāhs related by J. B. TAVERNIER (ed. V. BALL, Vol. II, p. 104): "For many years the king of Persia has prohibited mining in the 'old rock' for any one but himself, because having no gold workers in the country besides those who work in thread, who are ignorant of the art of enamelling on gold, and without knowledge of design and engraving, he uses for the decoration of swords, daggers, and other work, these turquoises of the old rock instead of enamel, which are cut and arranged in patterns like flowers and other figures which the jewelers make. This catches the eye and passes as a laborious work, but it is wanting in design." According to the opinion of the Persian General C. HOUTUM SCHINDLER who about 1880 was for some time governor of the mining district and acting manager of the mines, operations were probably carried on by the Persian Government up to 1725 (M. BAUER, Precious Stones, p. 394). On Schindler's work see p. 42.

[2] First edited by J. KLAPROTH in the *Nouveau Journal asiatique*, 1835 (the passage referred to on p. 32 of the separate issue: "pietre turchine"). English translation in C. R. MARKHAM, Narratives of the Mission of George Bogle to Tibet etc., p. 317, (London, 1876). I may be allowed to point out that the word "cobalt" in the English version, preceding the turquois stones, is based on a mistranslation of Orazio's *azurro* (present spelling *azzurro*) which is lapis lazuli. Indeed, this Italian word is traced to the Persian and Arabic names of lapis lazuli, *lazvard* and *lāzuward*. We know that this mineral is found in several localities of eastern Tibet (Lho-rong Dzong and Kung-pu Chiang-ta) and in the district of Lhasa (ROCKHILL, *Journal Royal Asiatic Society*, Vol. XXIII, 1891, pp. 272–4, and TIMKOWSKI, Reise nach China durch die Mongolei, Vol. II, pp. 188, 189, Leipzig, 1826), but it may be doubted that cobalt occurs in Tibet (though it may be found in Sikkim, as stated by J. C. WHITE, Sikhim and Bhutan, p. 322, London, 1909).

[3] Tibetan-English Dictionary, p. 1152.

(West Tibet).[1] This is also corroborated by the historical fact that the kings of Ladakh received a tribute of turquoises from Guge.[2]

From my own experience I may say that according to information received in Tibet turquois occurs in several mountains of the great State of Derge in eastern Tibet, though my Tibetan informants were unable to state the exact localities (or, which is more probable, did not want to state them). At any rate, the fact cannot be called into doubt, for in Derge, celebrated for the high development of art-industries and its clever craftsmen, also fine carvings of turquois are turned out, of which several specimens were secured by me that exhibit a peculiar, very pleasing, soft apple-green tinge differing from any other kind met in Tibet and China, and seemingly coming nearer to the Mexican variety. It seems also that in the mountains to the north of Ta-tsien-lu in western Sze-ch'uan a turquois of inferior quality and a sickly green is obtained; it is, however, so poor and insignificant that the Chinese traders there accustomed to the brilliant blue of their home product look down upon it as spurious. A great many of these greenish stones are utilized in a large collection of Tibetan silver jewelry brought together by me in that town. I was first inclined to accept the opinion of the Chinese consulted by me, and to regard these stones as imitations, but DR. JOSEPH E. POGUE to whom I sent three specimens for examination convinced me that this opinion was unfounded. He writes as follows:

"The three specimens give the following specific gravities (theoretical for turquois is 2.6 — 2.83):

1. Small dark-green specimen.................... 2.71
2. Small light-green specimen.................... 2.81
3. Larger perforated green specimen.............. 2.68

All three specimens are phosphates, giving good tests. Washed with strong ammonia, they did not lose their color, as most artificially colored turquoises will do when so treated. The specimens reacted characteristically when heated; and when viewed under the microscope, one contains a little granular quartz attached to its edge."

Captain C. G. RAWLING [3] gives the following summary as the result of his inquiry about the occurrence of turquois in Tibet:

"The rough stones are bought at the fairs held in the country and conveyed by the Indian merchants to Amritsar and Delhi, where they are mounted in gold and silver, and afterwards reimported. Practically every matrix originally comes from Tibet, but though inquiries were made at all the more important places, no information could be obtained as to the situation of the mines. The Phari people obtain their

[1] That is, the three districts of Ngari comprising the territories of Rutok, Guge, and Purang. *Gangs-chan* means the glacier-mountains.

[2] SCHLAGINTWEIT, Die Könige von Tibet, p. 862.

[3] The Great Plateau, p. 294 (London, 1905).

supply from Calcutta, Shigatse from Lhasa, whilst at many other places the people merely said that they did not know where the stones came from, that they had had theirs for years, and that none were to be found in their district or anywhere near. Despite these unsatisfactory answers, the consensus of opinion leads one to believe that they exist in the greatest numbers in the country situated between Lhasa and the western border of China."

I am somewhat doubtful in regard to Rawling's point that Tibetan turquoises are worked up in India and find their way back into Tibet. I am rather under the impression that the reverse is the case, as already stated by George C. M. Birdwood [1] that a good deal of Tibetan jewelry is imported into India through Bhutan, Sikkim, Nepal, and Cashmir, chiefly in silver — ornamented with large, crude turquoises, and sometimes with coral — in the shape of armlets, and necklaces, consisting of amulet boxes, strung on twisted red cloth, or a silver chain, and in various other forms, such as bracelets, anklets, etc., hammered, cut, and filigrained.

I have carefully gone over four volumes of the Trade Statistics of the Government of Bengal.[2] Turquois is not specified in these columns; it cannot, therefore, claim a big share in the trade between Bengal and Tibet. There is, however, a general item: Jewelry, and Precious Stones and Pearls. Jewelry was imported into Bengal from Tibet in 1906–7 at the value of 56 Rupees, precious stones and pearls, unset, at the value of 2,923 Rupees. The export of the latter from Bengal into Tibet for the same year amounted to 27,329 Rupees, in the preceding year, 1905–6, to 32,112 Rupees, in 1904–5, to only 12,460 Rupees (probably owing to Younghusband's expedition). I do not know how large a share is due to turquois in these figures.[3]

Osvaldo Roero [4] gives a list of merchandise imported into Ladakh from the official register kept by the customs of Leh, the capital of Ladakh. Among the products there enumerated he enlists turquoises as coming from Persia by way of Bokhāra, the best and most valuable coming from Seistān. The same view that turquoises are imported into Ladakh from Persia through Bokhāra had previously been upheld by Alexander Cunningham.[5] H. Ramsay [6] enumerates three classes

[1] The Industrial Arts of India, Vol. II, p. 28.

[2] The Trade of Bengal with Nepal, Tibet, Sikkim, and Bhutan. Last volume published, Calcutta, 1907.

[3] There is a pretty lively trade in turquoises on the part of Tibetans in Darjeeling; the stones sold there come from Tibet and China (via Tibet). In most cases it is possible to discriminate between turquoises of Tibetan and Chinese origin.

[4] Ricordi dei viaggi al Cashemir, Piccolo e Medio Tibet e Turkestan, Vol. III, p. 72 (Torino, 1881).

[5] Ladak, p. 242 (London, 1854). Also in Gilgit the turquois is employed (J. Biddulph, Tribes of the Hindoo Kush, p. 74, Calcutta, 1880).

[6] Western Tibet: A Practical Dictionary, p. 162 (Lahore, 1890).

of good turquoises which are free from flaws and with very little green, while inferior kinds are known as "Tibetan" and "Chinese turquoises," which come to Ladakh from Lhasa or China; they are full of flaws and generally very green.   The latter remark holds good only for Tibetan stones, as the Chinese are usually azure-blue.   "The best turquoises," concludes Ramsay, "come up from India.   Ladakhis object to flaws, but they like a little green, as they consider it a sort of guarantee that the turquois has not been manufactured."

In the following notes on China it will be seen that large quantities of turquoises cut into stones or beads and worked into carved objects are imported nowadays from China into Tibet; they are largely used by Chinese traders for purposes of barter with the Tibetans.

III. Turquois in China

The turquois, though found at present in central China *in situ* and commercially exploited by Chinese traders for export trade into Tibet and Mongolia, is not generally known to the Chinese people, for the apparent reason that it is but little employed by them and plays no significant part in their life.[1]   Outside of Peking and Si-ngan fu, where the trade is monopolized by a few of the initiated, the stone is hardly familiar to the people at large, nor to the educated classes; in Shanghai, Hankow, and Canton, it is entirely unknown.   This is glaringly evidenced by the fact that the Chinese commission engaged in working up the "English and Chinese Standard Dictionary," published by the Shanghai Commercial Press, in 1908, is not even acquainted with their own Chinese name for the stone, and speaks of it as a substance entirely foreign to their country; their definition of turquois (Vol. II, p. 2442) is "a Persian gem of a greenish-blue color, etc., first known to Europe through Turkey," and the same translated literally into Chinese, without giving the proper Chinese term for the stone.   Traders who have come in contact with Tibetans or Mongols or even settled among these peoples are certainly acquainted with it, and may even be induced to wear a turquois button, but a "barbarous" odor is always attached to it, and it seldom enters the ornaments of a self-respecting Chinese woman.

Besides the Hon. W. W. Rockhill, S. Wells Williams [2] seems to be the only author to mention turquoises as known to the Chinese. It is somewhat hard to understand how other careful observers could

[1] It follows therefrom that the knowledge of the turquois in China cannot be very old, and this conclusion will be confirmed by our historical inquiry.

[2] The Middle Kingdom, Vol. I, p. 310 (New York, 1901).

have overlooked its presence. F. v. RICHTHOFEN,[1] who gives a fairly complete summary of the commerce of Si-ngan fu does not mention it, nor does he notice it in his enumeration of goods traded from China to Tibet (p. 133). As far as I am aware, no handbook on mineralogy or precious stones makes any reference to the Chinese turquois; it is not noted either by F. DE MÉLY in his otherwise very complete work "Les lapidaires chinois."

The present Chinese name for turquois is *lü sung shi*, that is, "green fir-tree stone," or *sung êrh shi* [2] (also *sung-tse shi*) that is, fir-cone stone. This name must not be confounded with the designation *sung shi*, "fir-tree stone," which is not a stone, but by which petrified pieces of the fir-tree are understood; these are also called *sung hua shi*, "fir-tree transmutation stones," but their very color description as being yellow or purple shows sufficiently that they are entirely distinct from turquois. It will, however, be useful to consider briefly what Chinese authors have to say in regard to these petrefacts, because from these statements we shall gain a clue to the understanding of their name for turquois.

The earliest trustworthy mention of such petrefacts of vegetal origin is made in the "Annals of the T'ang Dynasty" (618–906 A. D.; *T'ang shu*, Ch. 217 B, p. 5) compiled from the records of the dynasty by Ngou-yang Siu (1007–1072) and Sung K'i (998–1061) [3] and completed in 1060. This notice embodied in the chapter on the Uigur (*Hui-hu*) relates to Central Asia, more particularly to the region inhabited by the tribe Bayirku (*Pa-ye-ku*),[4] and runs as follows:

"The country is grassy and produces noble horses and fine iron. There is a river called K'ang-kan. The people cut up fir-trees and throw the pieces into the water. In the course of three years these alter into blue-colored stone, in which the marks of

---

[1] Letters, p. 108.

[2] In the Cantonese dialect *luk ts'ung shek* and *ts'ung i shek*, respectively. The words with this meaning will be found in the Chinese-English Dictionaries of EITEL and GILES, and in the Chinese-Russian Dictionary of PALLADIUS; COUVREUR and others do not give them. The translation by turquois is confirmed by the Great Imperial Dictionary in Four Languages, which has the series: Chinese *lü sung shi*, corresponding to Manchu *uyu*, Tibetan *gyu*, and Mongol *ugyu*, all of which refer to the turquois. In a description of Tibet (*Wei ts'ang t'u chi* by Lu Hua-chu, published in 1785) occurs also the expression *sung jui* (No. 5723) *shi*, "stone of fir-tree buds." The German-Chinese Dictionary published by the Catholic Missionaries of South-Shantung (p. 916, Yen-chou fu, 1906) gives for turquois the word *lü sê shi*, "green-colored stone." G. SCHLEGEL (Nederlandsch-chineesch woordenboek, Vol. IV, p. 232, Leiden, 1890), besides the common *sung êrh shi*, registers for "turkoois" the word *ts'ing yü*, that is, blue or green jade. This must be an artificial modern formation, or rather an error, as the Chinese have never ranged turquois among jade but solely among ordinary stone, on which more will be said farther on.

[3] GILES, Biographical Dictionary, pp. 606, 698.

[4] CHAVANNES, Documents sur les Tou-kiue (Turcs) occidentaux, p. 88 (St. Petersburg, 1903).

the wood are still preserved in delicate outlines. It is generally called K'ang-kan stone."[1]

In 767 A. D. the painter Pi Hung is said to have executed a wall-painting on which fossil fir-trees were depicted, evoking poetical eulogies on the part of admirers.[2]

The Taoists, with their interest in the beauties and wonders of nature, could not fail to seize this attractive subject, and to interpret the phenomenon. The *Lu i ki* (*Ko chi king yüan*, Ch. 7, p. 6), a fabulous book by the Taoist monk Tu Kuang-t'ing of the tenth century,[3] reports:

"In a pavilion on a mountain in Yung-k'ang hien in Wu chou (the modern Kin-hua fu in Chê-kiang Province) there are rotten fir-trees. If you break a piece off, you will find that it is not decayed in the water but a substance altered into stone which previously was not yet transformed in that manner. On examining the pieces in the water, they turn out to be transformations of the same character. These metamorphoses do not differ from fir-trees as to branches and bark; only they are very hard."[4]

[1] Compare D'HERBELOT, Bibliothèque orientale, Vol. IV, p. 165 (La Haye, 1779). The Chinese cyclopædias quote this passage very inaccurately and with arbitrary changes. *Ko chi king yüan* (Ch. 7, p. 6), for example, writes the name of the river K'ang-tse, omits a whole sentence and adds at the end: "The stone has the designs of a fir-tree."

[2] *P'ei wên yün fu*, Ch. 100 A, p. 21 b.

[3] Compare WYLIE (Notes on Chinese Literature, p. 200) who dates this author in the tenth century (likewise p. 221). The *Lu i ki* has been adopted into the Taoist Canon (L. WIEGER, Le canon taoiste, p. 111, No. 586); Dr. WIEGER, however, places the work and the author in the ninth century. M. PAUL PELLIOT (*Journal asiatique*, 1912, *Juillet-Août*, p. 149) fortunately sheds light on the matter by informing us that Tu Kuang-t'ing lived toward the close of the T'ang dynasty, and that all his works come down from the beginning of the tenth century. BRETSCHNEIDER (*Botanicon Sinicum*, pt. 1, p. 172, No. 492) states that a work with the title *Lu i ki* must have been extant in the sixth century, as it is quoted in a book of that time; but it seems not to be known whether the work there referred to is really identical with the *Lu i ki* of Tu Kuang-t'ing. When WYLIE points out that the productions of this author have forfeited all claim to authenticity, this is certainly correct as regards their historical value. He must not be judged, however, in this light, but should be appreciated as a Taoist recluse and dreamer who reveals to us interesting phases of Taoist psychology by describing visions of dragons, tigers, tortoises, serpents and fishes, or relates extraordinary dreams and strange phenomena happening near the graveyards, who now records the principal hills and lakes of the empire famous as retreats of Taoist devotees, now tells the story of the Wu-i Mountain of Fu-kien renowned for its plantations of tea.

[4] The *Po wu chi*, a work by Chang Hua (232–300 A. D.) says: "The root of the fir-tree partakes of the nature of stone; stones, when cracked, are dissolved into sand and produce a fir-tree; and a fir-tree, when reaching three thousand years, again alters into stone." The *Po wu chi* was lost during the Sung period and compiled at a later date from extracts embodied in other publications (WYLIE, Notes, p. 192); there is, consequently, no guaranty that any text of this work, as preserved in the present editions, really goes back to the third century.—The above subject has also an interesting bearing on the Chinese knowledge of fossils, which should be treated some day in a coherent essay. There is a great deal of information on dragon bones and teeth originating from fossil hipparion and rhinoceros, petrified fishes, crabs, and swallows, all procurable in the Chinese drug-stores. There are similar accounts among the Arabs (M. REINAUD, Relation des voyages faits par les Arabes, Vol. I, p. 21; P. A. VAN DER LITH, Livre des merveilles de l'Inde, p. 171), and the palæontologi-

In regard to these petrefacts of Yung-k'ang, another interesting note is given by Tu Wan or Tu Ki-yang in his Treatise on Stones, entitled *Yün lin shi p'u* (Ch. B, p. 3) published in the year 1133 (Sung period), the oldest Chinese *lapidarium* extant.[1] This author speaks of a poet of the T'ang dynasty, Lu Kuei-mêng,[2] who had obtained a pillow and a lute of stone, and left two poems on these objects. In the introduction to the poems, he mentions the fir-trees of Yung-k'ang which from old age had turned into stones, and that one evening, as the effect of a big rainstorm, a whole fir-tree grove on the mountains suddenly changed into stone, and fell to the ground, smashed into pieces from two to three feet in diameter, and these are still there; the natives of the place, then, carried such pieces away and worked them up into footstools, some as small as a fist, or into low tables by breaking the larger pieces.

Another author, Chang Lu-i, states that "there are two varieties of these stones produced by transformation of fir-trees, one of yellow, and one of purple color, of very fine substance and shape, with water marks on the surface, some also with marks of the tree bark, others with marks of the tree knots, such as occur on the T'ien-t'ai mountains (in T'ai-chou fu, Chê-kiang). There are those the transmutation of which is not complete, but which still bear the fir-tree substance; these are useful as medicine. If those perfectly transformed are taken as medicine, they have the effect upon man that he forgets passion and stops longing; this medicine cures love-sickness; if men or women who are unhappily in love partake of it, they will intercept their thoughts and not remember again." This is certainly a sympathetic remedy; in the same manner as the tree has lost its life and changed into a lifeless mass of stone, so it has the effect on the human heart to make it forget, and to render it cold and old like stone.

cal knowledge of the ancients has been treated by E. v. Lasaulx, Die Geologie der Griechen und Römer, pp. 6–16 (*Abhandlungen der bayerischen Akademie*, München, 1851).

[1] It is reprinted in the enormous collection *Chi pu tsu tsai ts'ung shu*, Section 28; also in *T'ang Sung ts'ung shu*. This work is widely different from the class of books styled *pên ts'ao*, in which the therapeutic value of the substances occurring in nature forms the principal point of view. The book of Tu Wan is written from the standpoint of economic geography. The minerals are all named for the localities from which they originate, and the author is chiefly interested in their industrial utilization. This feature lends his notes a practical value, and a complete translation of them, aside from the purely scientific interest, might yield also results for the study of economic mineralogy in China.

[2] He is known as the author of the *Siao ming lu* (Wylie, Notes on Chinese Literature, p. 182) and of a small treatise on the plough (*ibid.*, p. 93, and O. Franke, Kêng Tschi T'u, p. 45, Hamburg, 1913). Bretschneider (*l. c.*, p. 172, No. 493) mentions a work Poems of Lu Kuei-mêng as cited in *T'ang shu*, Ch. 196. The Collection of his Poems (*shi tsi*) is quoted in *Kao chai man lu* (Ch. 1, p. 1; *Shou shan ko ts'ung shu*, Vol. 91).

Also the *Pên ts'ao kang mu* (Section on Stones, Ch. 9, p. 14) of Li Shi-chên, the Chinese standard work on materia medica and natural history completed in 1578 after 26 years' labor,[1] mentions the 'fir-tree stone' (*sung shi*) after Su Sung, an author of the Sung period, as being produced in Ch'u-chou fu (Chê-kiang Province) and being like the trunk of a fir-tree but solid stone.   According to the opinion of some, it is fir-tree which has changed into stone after a long time; it is gathered a great deal on the mountains, and is made into pillows.[2]

It seems to me that similar notions have been active in inducing the Chinese to confer on the turquois the name "green fir-tree stone," because they looked upon it as a transformation from the fir-tree. This may be inferred as a plausible explanation, for as far as I know, there are no definitions of the name in Chinese literature; the word *lü sung shi* can be traced only to the eighteenth century (see p. 60).

A modern author, Chung Kia-fu, in his collected works (*Ch'un ts'ao t'ang ts'ung shu,* 1845, Ch. 29, p. 19) has developed a peculiar view on the origin of turquois which he places in the same category as amber:

"When the moss growing on rock after many years consolidates and assumes color, turquoises arise, those of a deep hue being called *lü sung,* those light in color *sung êrh* ('fir-tree ears').   This is the same process as takes place with respect to fir-tree resin which after many years consolidates and develops into amber, that of a deep shade being called *hu-p'o,* that light in color being called bees'-wax (*mi-la*).[3]

---

[1] The literary history of this interesting work, first printed in 1596, has been traced by BRETSCHNEIDER, Botanicon Sinicum, pt. 1, p. 55.   Despite many efforts I have not succeeded in procuring the original edition which seems to be entirely lost and not now to exist in any Chinese library.   Bretschneider states that the earliest edition extant seems to be that of 1658; but a print of 1645 in 16 vols., edited by Ni Tun-yü of Hang-chou, was secured by me in Tōkyō, now in the John Crerar Library of Chicago, which, besides, has an edition of 1826 in 39 volumes, and one issued in 1885 in 40 volumes, the best print in existence.   An excellent photo-lithographic reprint was published in 1908 by the firm *Tsi ch'êng t'u shu* of Shanghai after an edition of 1657 by Chang Ch'ao-lin.   The text in the Shun-chi editions is more accurate than in the K'ien-lung and Tao-kuang editions.   Prof. HIRTH (*Journal China Branch Royal As. Soc.,* Vol. XXI, 1886, p. 324) mentions a Ming edition printed in 1603, possibly the second edition published.

[2] A. WYLIE (in his treatise Asbestos in China: Chinese Researches III, p. 152, Shanghai, 1897) quotes from the *T'u king:* "Among the hills at Ch'u chou (in Chê-kiang Province) a species of pine stone is produced, resembling the trunk of the pine, but in reality a stone; some say that the pine in the course of time becomes changed into stone.   Many people take it to decorate their mountain lodges, and also shape it into pillows."   This passage is evidently taken from the *Pên ts'ao kang mu,* the abbreviated title *T'u king* being identical with the *T'u king pên ts'ao* of Su Sung. Compare also F. DE MÉLY (Les lapidaires chinois, p. 86, Paris, 1896) where the translation "pour représenter des tranches d'arbres" should read "to represent pillows." On p. 208 DE MÉLY cites an interesting note from DE ROSNY, according to which a fossil pine-tree was found in Japan in 1806.

[3] *Mi-la* is the designation for a light-yellow kind of amber in which presumably also copal and artificial productions occur.   The Imperial Geography of the Manchu Dynasty (*Ta Ts'ing i t'ung chi,* Ch. 274) ascribes its production to Shi-nan fu in Hu-pei Province, but in another passage connects its introduction with the Hollanders.   Other Chinese authors derive the origin of *mi-la* from Yün-nan Province

I once received a water-receptacle to wash writing-brushes in, made from turquois, of the size of a dish, in the shape of lotus-leaves, and onion-green and kingfisher-blue in color."

In the Annals of the T'ang Dynasty (*T'ang shu*), there is a curious word *sê-sê* (No. 9599) occurring in several passages and assumed by HIRTH and CHAVANNES to have the meaning of *turquois*. The one is met with in Ch. 221 B, p. 2 b, in an account of Sogdiana, but relating to the region of Ferghana, where it is said:

"North-east from the capital (modern Tashkend), there are the Western Turks, north-west P'o-la; 200 *li* south one comes to Khojend, 500 *li* south-west to K'ang (that is, Sogdiana, the region of Samarkand). In the south-west is the river Yao-sha (the Yaxartes), and in the south-east are big mountains producing *sê-sê* (or *sö-sö*)."[1]

Another passage containing this word will be found in Ch. 256 of the *T'ang shu*, in the account of Tibet.

and Tibet (the Tibetan name is *ko-shel;* in Mongol: *tabarkhai shel;* in Manchu: *meisile*, an artificial hybrid from Chinese *mi* and Tibetan *shel* 'crystal'). In Ch'êng-tu fu, the capital of Sze-ch'uan Province, a number of small girdle-pendants carved from this substance were obtained by me (Yün-nan being given as the place of production) which have not yet been examined as to their composition.

[1] See F. HIRTH, Nachworte zur Inschrift des Tonjukuk, p. 81 (in W. RADLOFF, Die alttürkischen Inschriften der Mongolei, Vol. II, St. Petersburg, 1899). E. CHAVANNES, Documents sur les Tou-Kiue (Turcs) occidentaux, p. 140 (St. Petersburg, 1903) translating the same passage accepts the rendering of Hirth. Also GILES, in the second edition of his Chinese-English Dictionary, sides with this translation. PALLADIUS, who transcribes the word *she-she*, was not of this opinion, for in his excellent Chinese-Russian Dictionary (Vol. II, p. 569) he gives the definition "azure-colored, transparent precious stone." He has likewise another word *she-she* (written with the character No. 9600 in the Dictionary of Giles) with the meaning of "emerald." COUVREUR (Dictionnaire classique de la langue chinoise, p. 584) explains *sê-sê:* "nom d'une belle pierre et d'une espèce de verre." In his Dictionnaire chinois-français (p. 13), the same author gives the interpretation: "pierre bleue et transparente," and for the plain *sê:* "limpidité d'une pierre précieuse; pur, net." It would be very interesting to have the Chinese source pointed out to which the statements of Palladius and Couvreur in regard to the transparency of the stone go back; in the Chinese records at my disposal I regret I can find nothing to this effect. In view of the mineralogical properties of turquois it is evident that this is a point of importance, for non-transparency is one of the prominent characteristics of turquois. As we can but presume that both Palladius and Couvreur must have founded their definition on some Chinese document, this would present another of the objections which must be raised to the weak hypothesis of identifying *sê-sê* with the turquois. E. H. PARKER (*China Review*, Vol. XVIII, 1890, p. 221) defines *sê-sê* as a sort of jade much used for arrowheads and other purposes by the Tibetans, Tungusians, and even Ta Yüe-chi (Indoscythians) who after their conversion to Buddhism had a sacred *pâtra* or alms-bowl made of the same material (*ts'ing shi*); in his opinion, *sê-sê* is identical with the latter term, which means green or blue (but possibly also dark-colored) stone. This point of view is hardly correct. The arrowheads of the Tungusian tribes, as corroborated by archæological finds made in the Amur region, were of nothing but common flint. The *ts'ing shi* of which the alms-bowl of the Indoscythians was made in all probability was lapis lazuli and would accordingly mean in this case 'blue stone'; on Buddhist pictures alms-bowls are usually painted an ultramarine or lapis-lazuli blue color (see above p. 14). There are several other instances where the word *ts'ing shi* has the same meaning (HIRTH, China and the Roman Orient, p. 72, and Chinesische Studien, p. 250). There is no Chinese text saying that *sê-sê* was a kind of jade, that it was a *ts'ing shi*, or ever used for arrowheads.

"The officers in full costume wear as ornaments — those of the highest rank
sê-sê, the next gold, then gilded silver, then silver, and the lowest copper — which
hang in large and small strings from the shoulder, and distinguish the rank of the
wearer."[1]

BUSHELL comments that *sê-sê* is a kind of precious stone found in the
high mountains north-east of Tashkend. At the outset, it does not
seem very likely that in the latter passage the word has the significance
of turquois, for it outranks gold (compare above p. 11) and however
much appreciated in Tibet, a turquois could never outshine gold nor
have any value equivalent to it, as was and is the case everywhere else;
and as shown above, it was not even looked upon as a precious stone by
the ancient Tibetans. There was still less reason for the Tibetans to
import their turquoises from Tashkend — if *sê-sê* should denote especially the turquoises of that locality — as they found them in great
abundance in their own country. Nor was the turquois apt to serve
for the distinction of the first official rank in Tibet, as it has always been
there part and parcel of the adornment of all classes of people and particularly the ornament of women who are loaded with it. The *sê-sê*
of the Tibetan officials must, therefore, have been something else, a
much scarcer and more valuable gem. An idea of its value is afforded
by a notice in the Annals of the Five Dynasties (*Sin Wu tai shi*, Ch. 74,
p. 4 b) where it is said that the women of the T'u-po (Tibetans) wear beads
of *sê-sê* in the plaited tresses of their hair, and that, as regards the best
quality of these beads, a single one is bartered for, or has the exchange
value of, a noble horse.[2] This seems to me to be sufficient evidence
militant against the identification of *sê-sê* with the turquois, as far as
Tibet is concerned, for a single turquois, whose value in Tibet may range
from a few cents up to a dollar or so, could never have had nor has a
valuation equivalent to a good horse.[3]

[1] See S. W. BUSHELL, The Early History of Tibet, p. 8 (*Journal Royal Asiatic
Society*, 1880). The *T'ang shu* (K'ien-lung edition, Ch. 216 A, p. 1 b) has instead of
sê-sê the reading *k'in-sê*, a frequent compound meaning "lute and harp" (Giles's
Dictionary, No. 2109). It is evident that this way of writing is erroneous, and was
perhaps suggested to a copyist who did not understand the unusual word *sê-sê*. The
passage is not contained in the Old History of the T'ang dynasty (*Kiu T'ang shu*),
but only in the New History (*Sin T'ang shu*).

[2] This passage occurs in the report of the embassy of Kao Kiü-hui of 938 A. D.
ABEL-RÉMUSAT (Histoire de la ville de Khotan, p. 77, Paris, 1820), who has translated
this account, rendered the word *sê-sê* by "pearls."

[3] In the History of the Kingdom of Nan-chao (*Nan-chao ye shi*, published in 1550),
a tribute of *sê-sê* is mentioned for the year 794 as being sent from Nan-chao, comprising the territory of the present province of Yün-nan, to the court of China (C.
SAINSON, Histoire particulière du Nan-Tchao, p. 54, Paris, 1904). At first sight, the
sê-sê in this instance might be regarded as turquoises. R. PUMPELLY, as will be noted
below, has referred to Yün-nan as a locality producing a mineral similar to turquois,
though this report requires confirmation. There is further evidence in the Annals
of the Yüan Dynasty (*Yüan shi*, Ch. 16, p. 10 b) that in 1290 turquoises (*pi tien-tse*)

In the Old History of the T'ang Dynasty (*Kiu T'ang shu*, Ch. 198, p. 11 b), a description of the country Fu-lin (Syria) is given, whose great wealth in precious stones is emphasized.   In the palaces, it is said there, the pillars are made of *sê-sê*.[1]   It is difficult to see, if *sê-sê* should have to be identified with the turquois, how pillars could be made of this material.   The Chinese text does not say that the pillars were adorned or inlaid with this stone but produced from it.

A fourth passage in the *T'ang shu* (Ch. 221 A, p. 10 b) referring to *sê-sê* is contained in an account of Khotan (*Yü t'ien*).   Emperor Tê-tsung (780–805) despatched an emissary, Chu Ju-yü by name, to Khotan on the search for jade, and he obtained there a hundred pounds (catties) of *sê-sê*.[2]   This notice is of great interest in showing that the precious stones of this name were really imported into China, and that the mart for them was Khotan.

There are, however, still earlier references to the jewel *sê-sê*.   It is for the first time mentioned in the *Pei shi* (Ch. 97, pp. 7 b, 12 a) and in the "Annals of the Sui Dynasty"[3] (*Sui shu*, Ch. 83, containing a record of the foreign countries then known to the Chinese).   Both histories

were gathered in the circuit of Hui-ch'uan in Yün-nan Province; my friend Prof. Paul Pelliot was good enough to draw my attention to this passage.   Another passage alludes to a gift of a thousand turquoises sent from Hui-ch'uan in 1284 (*Kin-t'ing se wên hien t'ung k'ao*, Ch. 23, p. 7).   But it seems likely from what will be stated farther on in regard to the first acquaintance of the Chinese with the turquois in the Mongol period that the turquois mines of Yün-nan were opened only shortly before this time.   At any rate I am not inclined to transfer this account without reserve to the date 794, nor to believe in the identity of the different terms *sê-sê* and *pi tien-tse*. While I should merely admit the possibility of such an identification, another historical explanation of the case may be pointed out.   In the eighth century, the T'ai or Shan, the stock of peoples forming the kingdom of Nan-chao, were in close political alliance with the Tibetans who had then reached the zenith of their power.   It would therefore be justifiable to conclude that the *sê-sê* of Nan-chao were derived from Tibet and are to be identified with the ancient Tibetan *sê-sê*, which, as will be shown hereafter, may be the emerald.   In the *T'ang shu* (Ch. 222 A, p. 2 a), the women of the Southern *Man*, the large stock of aboriginal tribes formerly spread over the whole of southern China, are said to fasten in their hair beads, shells, *sê-sê*, and amber.   In this case it is rather tempting at first sight to interpret *sê-sê* as turquois, because this combination of turquois and amber, as pointed out before, occurs indeed among the Tibetan group of tribes.   But the *Man* do not belong to the Tibetan family, and another difficulty is presented by the fact that there are no records either of ancient or modern times pointing to the employment of the turquois among any tribe of the *Man*, so that it is safer to assume that the turquois is not understood in the above text.

[1] HIRTH, China and the Roman Orient, p. 53.   At that time (1885) Hirth had not advanced any identification of this term.

[2] He embezzled the jade objects destined for the emperor, was sentenced, and died in exile (CHAVANNES, Documents, p. 128, note 2).

[3] The *Pei shi*, "Northern Annals," was written by Li Yen-shou (GILES, Biographical Dictionary, p. 474) and completed about the year 644; it comprises the history of the dynasties of the north ruling from 386 to 618.   The Sui dynasty ruled from 589 to 618.   The *Sui shu* was composed by Wei Chêng (581–643; GILES, *l. c.*, p. 856) under the T'ang dynasty and completed in 636.

mention the jewel in two passages,— first, as a product of the country of Sogdiana (*K'ang*) corresponding to the region of Samarkand, and secondly as a product of Persia (*Po-se*, from *Pārs*).[1] The text of the *Pei shi*, with the same indications, is found also in the "Annals of the Wei Dynasty" (*Wei shu*, Ch. 102, pp. 5a and 9b).[2] But this passage

---

[1] It is noticeable that *sê-sê* as products of Persia are mentioned in *Pei shi* and *Sui shu*, but not in the two *T'ang shu*. The *Kiu T'ang shu* (Ch. 198, p. 11) enumerates as precious objects of Persia coral-trees, *ch'ê-k'ü*, agate, and "fire-pearls" (*huo chu*). The *T'ang shu* mentions only coral as a product of Persia and the gift to China of a couch of agate (CHAVANNES, Documents, pp. 171, 174). The exact history of the term *ch'ê-k'ü* which in general denotes a large white conch (Tibetan *dung*, Sanskrit *çankha*, Arabic *shenek*: M. REINAUD, Relation des voyages faits par les Arabes, Vol. I, p. 6), and sometimes seems to refer to a precious stone remains to be ascertained (compare HIRTH and ROCKHILL, Chau Ju-kua, p. 231; PELLIOT, *T'oung Pao*, 1912, p. 481). The "fire-pearls" were lenses of rock-crystal, alleged to have been used for producing fire (F. DE MÉLY, Les lapidaires chinois, p. 60; CHAVANNES, Documents, p. 166; PELLIOT, *Bulletin de l'Ecole française d'Extrême-Orient*, Vol. III, 1903, p. 270; *Pên ts'ao kang mu*, Ch. 8, p. 18 a). In the *Sui shu*, *sê-sê* are enumerated together with genuine pearls, glass, amber, coral, lapis lazuli, agate, rock-crystal, *huo ts'i*, and diamond. The name *huo ts'i* (the alleged identity with *huo chu* remains to be proved) has not yet been properly identified. In the *Nan shi* (Ch. 78, p. 7) these stones are mentioned as products of central India and described as having the appearance of *yün-mu* and the color of violet gold (PELLIOT, *l. c.*); the difficulty is that also the word *yün-mu* which according to PELLIOT seems to designate mica and mother-o'-pearl is not yet determined beyond doubt. Possibly, *huo-ts'i* designates the garnet. The word *sê-sê* is, in the text of the *Sui shu*, followed by the words *hu lo kie lü t'êng*. At first I was inclined to take the verb *hu* in its literal sense "called, designated," and to believe that the words following it represent a gloss, being the Persian or Arabic name of the stone in Chinese transcription. Reconstructing the ancient sounds of those Chinese characters we would arrive at the reading *lok* (or *rok*)-*ket-li-dang;* but there is no word in Persian or Arabic to be identified with such a form. M. PAUL PELLIOT, to whom I submitted this difficult point, has been good enough to write me that this passage had already attracted his attention, and that he does not regard the incriminated words as a gloss; he thinks that the word *hu* is also part of the transcription, and that two further products are enumerated in their Persian names. The passage, accordingly, should be understood in the sense that Persia produces *sê-sê*, *hu-lo(k)*, and *ket-li-dang*. The two latter names, however, are as yet unidentified, but with M. PELLIOT's very plausible point of view, a better attempt at identification might be pursued. Indeed, Prof. A. V. WILLIAMS JACKSON had called my attention to the fact that *katlidang* may be a compound of the word *qatlān*, "link" or "scale," used alike in Arabic, Turkish and Persian, and the Persian word *tan* "body," the content of the term implying scale or chain armor. This is very suggestive, as indeed Persia was the country which supplied China with chainmail (ancient specimens in the Field Museum). The *T'ang shu*, in the account of Samarkand (*K'ang*) states that in the beginning of the period *K'ai-yüan* (713–741) Samarkand sent as tribute to China chain-mail (*so-tse k'ai*). This question will be shortly discussed by me in another place. *Liang shu* (Ch. 54, p. 14 b) attributes to Persia coral-trees one to two feet high, amber, agate, genuine pearls, and *mei-hui*. HIRTH and ROCKHILL (Chau Ju-kua, p. 16, St. Petersburg, 1912), treating the products of Persia after *Wei shu* and *Sui shu*, entirely omit the *sê-sê* (and several others). It seems doubtful if, as stated so positively by the two authors, "most of these products came, of course, from India, or from countries of south-eastern Asia, only a few being products of Arabia, or countries bordering on the Persian Gulf" (and again on p. 7). This is true only to a certain extent; the *sê-sê*, at any rate, are not mentioned by the Chinese as products of India or south-eastern Asia, but exclusively as products of Persia and Sogdiana, to which, later in the T'ang period, Fu-lin, Tashkend, Tibet, and the *Man* are joined.

[2] The Wei dynasty ruled from 386 to 556; the *Wei shu* was written by Wei Shou (506–572; GILES, *l. c.*, p. 867) and presented to the throne in 554.

has no independent value, because Ch. 102 of this work treating of the countries of the west, as well demonstrated by Chavannes,[1] has been merely reproduced from Ch. 97 of the *Pei shi* by a committee of scholars of the Sung period headed by Fan Tsu-yü (1041–1098).

It is thus evident that *sê-sê* were known to the Chinese prior to the age of the T'ang dynasty as occurring in the territory of Persia and Sogdiana, to wit, within the Iranian culture-area. It is noteworthy also that any particular region or mountain producing the stone is not alluded to in these earlier texts as subsequently in the *T'ang shu*, and that *Pei shi* and *Sui shu*, while locating *sê-sê* in Sogdiana, do not allude to it in their notices of Tashkend (*Shi kuo*).

As I did not know on what evidence Prof. Hirth had based his identification of *sê-sê* with the turquois, I consulted him regarding this point, and he was good enough to furnish the following note which is here reproduced with his kind permission.

"The word *sö-sö* (in Cantonese *shat-shat*, *sit-sit*, or *sok-sok*) has, besides others, the meaning of a precious stone, 'a greenish or bluish bead' (*pi chu*), as quoted in *P'ei wên yün fu*, Ch. 93 B, p. 85. The *Pên ts'ao kang mu* (Ch. 8, p. 55) says that the people of the T'ang dynasty called green (or blue) precious stones by the name *sö-sö*. The Japanese sources as quoted in Geerts, Les Produits de la nature japonaise et chinoise, p. 481, do not apparently refer to *sö-sö*, but the *T'u shu tsi ch'êng* (section 27, National Economy, Ch. 335) contains an extract from the *T'ien kung k'ai wu* in which *sö-sö* is classed with greenish precious stones. The *T'ang kuo shih p'u* (*ibid.*) relates the story of a big *sö-sö* which the author thinks was not a genuine one; the same story is told in *Yen fan lu*, Ch. 15, p. 11.

"Bretschneider (*Chinese Recorder*, Vol. VI, p. 6) was, as far as I know, the first to find out that *sê-sê* was not a musical instrument as Pauthier had assumed, but a precious stone.[2] In his translation of a passage regarding precious stones found in the *Cho keng lu* (reproduced in his Mediæval Researches, Vol. I, pp. 173–6), he refers to 'stones called *tien-tze*' which occur in Nishapur and Kirman. Bretschneider says of these: 'I have little doubt that the Chinese author understood by it the turquoise, the Persian name of which is *firuzé*. Both Nishapur and Kirman produced turquoise. So did the hills of Ferghana referred to in *Nachworte*, etc., p. 81, for the territory of Ferghana furnished turquoises, according to von Kremer, Kulturgeschichte des Orients, Vol. I, p. 329. These are the reasons which had induced me to render *sö-sö* by 'Türkis'."[3]

[1] Documents, p. 99.

[2] The word *sê-sê* in the sense of a jewel, as will be seen below, is the Chinese transcription of a foreign word. The single word *sê* denotes a stringed musical instrument, a kind of lute, described *e. g.* by J. A. van Aalst (Chinese Music, p. 62, Shanghai, 1884). But there is a passage (in the *San kuo chi*, *Wei chi*, commentary to the Biography of Ch'ên Se-wang, quoted in *P'ei wên yün fu*, Ch. 93 B, p. 85) where also the compound *sê-sê* seems to have the meaning of a musical instrument. In the *Tsin shu* (Ch. 97, p. 2) it is said in regard to the Shen Han, a Korean tribe, that they are skilled in playing the *sê-sê* which in shape is like a five-stringed lute (*chu*, No. 2575).

[3] It should be added that it is Bretschneider himself (Mediæval Researches, Vol. I, p. 140) who first proposed the translation of *sê-sê* as turquois, but with the restriction of a "probably."

It seems to me that the color designation insisted on by Hirth is hardly conclusive; the color name *pi*, 'bluish-green' (originally a kind of jade) is quite indistinct, and aside from the fact that there are many other green or blue stones like emerald, lapis lazuli, malachite, sapphire, etc.,[1] this attribute with reference to *sê-sê* does not appear in contemporaneous records of the Sui or T'ang periods, but only in later authors who were not personally familiar with the stone. It was known in China under that name only to a limited extent, during that time, and to the later generations the newly coined word simply became a poetical name with no other meaning than that of a rare, precious stone. The one fact stands out clearly, that *sê-sê* was looked upon as a precious stone, a fact for which more testimony will be given, and this is evidence that it can hardly be the turquois. It is always essential to ascertain to what category an object in the views of the Chinese belongs; these categories are always fixed and stable, and suggest an inference as to the nature of the object in question. No Chinese has ever considered turquois a precious stone, but just a common stone good enough for barbarous ornaments.[2] It is worth 5 Taels (about \$3.50) a catty (1⅓ pounds) in Si-ngan fu where it is sold by weight, and if the famous *sê-sê* were nothing more than that, the Chinese authors would not have expressed any enthusiasm about them.[3] Hirth's quotation from Bret-

---

[1] That the definition *pi chu* means little is illustrated by the fact that other jewels are also defined by this term, as, for example, the pearl called *mu-nan* (HIRTH, China and the Roman Orient, p. 59) which is even described as yellow in other texts (*Ko chi king yüan*, Ch. 32, p. 7 b). Compare also *P'ei wên yün fu*, Ch. 7 A, p. 101 b (*pi chu*). The color argument should therefore be disregarded.—The comparative tables of the colors given by W. TASSIN (Descriptive Catalogue of the Collections of Gems in the U. S. National Museum, *Report of National Museum*, 1900, pp. 541, 542) enumerate the green stones as follows: zircon, sapphire, garnet (demantoid and ouvarovite), chrysoberyl (alexandrite), spinel, topaz, diamond, olivine (peridot), tourmaline, beryl (emerald and aquamarine), quartz (chrysoprase, plasma, prase, and jasper), turquois. The blue stones are: sapphire, spinel, topaz, diamond, tourmaline (indicolite), beryl (aquamarine), iolite (water sapphire, dichrolite). It should not be overlooked either that, as shown by the modern word *lü sung shi*, the color of turquois is described by the Chinese with the word *lü*, not *pi*.

[2] A. J. C. GEERTS (Les produits de la nature japonaise et chinoise, p. 202, Yokohama, 1878) was cautious enough to pay due attention to the distinction made between *pierres ordinaires* and *pierres précieuses* in the *Pên ts'ao kang mu*.

[3] In the *Lapidarium* of Pseudo-Aristotle (JULIUS RUSKA, Das Steinbuch des Aristoteles, p. 152, Heidelberg, 1912) it is said in regard to the turquois: "Its color delights those afflicted with sorrow, but it is not employed for the costume of the kings, because it detracts from their majesty." A similar remark is made by Ibn al-Baiṭār: "It is soft and a bit fragile, and is not used for the ornaments of the sovereigns" (L. LECLERC, Traité des simples, *l. c.*, p. 50). In an Arabic work of 1175 it is said: "Many kings hardly have the desire to wear a turquois, because the vulgar frequently utilizes it as sigillum and wears finger rings which are imitations of its best kind" (WIEDEMANN, Beiträge zur Geschichte der Naturwissenschaften, XXX. Zur Mineralogie im Islam, p. 234, Erlangen, 1912). Also in Europe turquoises were low in price. "Admodum magno pretio non venditur, quia magna illarum ex Oriente adfertur copia," says A. BOETIUS DE BOOT (Gemmarum et lapidum historia, p. 271, ed. of A. TOLL, Lugduni Batavorum, 1636). The general rule may be set down that there is a large consensus of opinion as to the value of precious metals and

schneider is hardly in favor of his view that *sê-sê* was the turquois; BRETSCHNEIDER's statement merely shows that at the end of the Mongol dynasty — the *Cho keng lu* was published in 1366 — the Persian turquois became known to the Chinese.

From three practical examples it may be demonstrated that *sê-sê*, as known during the Sung period, cannot be construed to mean turquois.

In the *K'ao ku t'u* (Ch. 10, p. 22 b), a book on ancient bronzes by Lü Ta-lin, completed in 1092, a girdle-clasp is figured and described as being made of *sê-sê*; it is a highly ornamented piece, engraved in fine lines and ending in a curve shaped into a dragon's head. This whole technique would be impossible if the material were turquois, which results only in straight, stiff, angular lines (compare Plates VI–VIII).

The *Ku yü t'u p'u*, "Illustrated Description of Ancient Jades," compiled in 1176 and printed in 1779, describes several jade specimens adorned with the stone *sê-sê*,— a sword possessed by the Sung Emperor T'ai-tsu (968–976), having a hilt ornamented with amber, *sê-sê*, and genuine pearls (Ch. 28, p. 10). The Chinese would hardly display such bad taste as to unite a cheap stone like turquois with genuine pearls. In Ch. 97, p. 10, of the same work a jade lantern of the Sung palace is figured and described, the eight sides of which are adorned with coral, amber, *sê-sê* and such like jewels (*pao*). In this case turquois is again out of the question, as it is not considered by the Chinese a precious stone or a jewel, but just an ordinary stone.[1]

The two works here quoted come down from the Sung period, and it can be shown from another source of the same epoch that the word *sê-sê* designated at that time a stone capable of carving found on the very soil of China, and that, consequently, the *sê-sê* in the age of the Sung dynasty are affairs different from those mentioned in the *Pei shi*, *Sui shu* and *T'ang shu* for Persia, Sogdiana, Ferghana, and Tibet. Kao Se-sun, a poet and essayist who lived in the latter part of the twelfth century,[2] is the author of an interesting work on miscellaneous minerals among peoples of all times, and that the changes which have affected the appreciation of precious stones from the days of antiquity until now are but very slight, chiefly due to the operations of fashion and variations in the sources of supply. Thus it is not very likely that a stone looked upon as non-precious at present by general agreement of opinion was ever prized as a jewel in earlier periods of history.

[1] Such carvings of *sê-sê* are referred to also by other authors of the Sung period. Chou Mi in his interesting work *Yün yen kuo yen lu* (Ch. B, p. 31 b), a review of ancient bronzes, paintings and jades which had come to the notice of the author during his lifetime, mentions the carving of "a crane moaning in the autumn" entirely made from this material. This very subject savors of the impressionism of the Sung artists, and in this case turquois is inconceivable, not only for technical but also, and even more so, for artistic reasons. The work quoted is embodied in the collection *Shi wan küan lou* edited by Lu Sin-yüan and thoroughly analyzed by PAUL PELLIOT (*Bulletin de l'Ecole française d'Extrême-Orient*, Vol. IX, 1909, p. 246).

[2] GILES, Biographical Dictionary, p. 368; WYLIE, Notes on Chinese Literature, p. 161.

subjects, entitled *Wei-lio*.[1] In Ch. 5, p. 3, he has gathered several notes concerning *sê-sê*. He quotes the *Huan yü ki*[2] to the effect that *sê-sê* are mined in Shan-chou [3] and P'ing-lu.[4] Neither of these localities is known as having ever produced turquois. We shall see farther on that turquois became known and was mined in China only under the Yüan dynasty following the Sung, so that we may justly conclude that the Chinese of the Sung period were not yet acquainted with it. Besides, there is the technical evidence that turquois, according to its natural properties, could not have entered such objects as are reported to have been made of *sê-sê*. The *Wei lio* furnishes us with additional evidence on this point, which goes to show that, if these reports are trustworthy, a substance *sê-sê* of Chinese production was utilized as early as the T'ang period. It is related in regard to an official of that time, who presided over the bureau of the salt and iron monopoly in the province of Fu-kien, that he owned a pillow made of *sê-sê* placed on a golden bedstead.[5] Emperor Hien-tsung (806–820) tried to estimate its value, but arrived at the conclusion that it was a priceless treasure, while others said that this pillow was made from a beautiful stone, but not from *sê-sê*. The author of the *Wei lio* adds: "What is circulating among our contemporaries under the name *sê-sê*, I believe is made from molten stone." [6] So it seems that at the Sung period the *sê-sê* may have been, at least partially, artificial productions. It is self-evident that the pillow referred to cannot have been made of turquois. The rectangular shapes of Chinese pillows with convex surface are well known, and it is impossible to carve turquois which is quarried in long slabs [7] into such a form.

[1] Reprinted in *Shou shan ko ts'ung shu*, Vol. 74.

[2] A general, mainly geographical, description of China published by Yo Shi during the period *T'ai-p'ing hing-kuo* (976–981) of the Sung dynasty.

[3] In the province of Ho-nan (PLAYFAIR, Cities and Towns of China, No. 6157).

[4] District in Shan-si Province (PLAYFAIR, No. 5812).

[5] The *Wei lio* quotes this story from the *Yen fan lu* of Ch'êng Ta-ch'ang, written in 1175. The *P'ei wên yün fu* (*l. c.*) gives it after the biography of Lu Kien-tz'e in the *T'ang shu*, so that there is no doubt that it relates to the period of the T'ang dynasty.

[6] There are different versions of this story handed down, the details of which are not of interest in this connection. According to the *T'ang kuo shi p'u*, containing records from 723 to 821 by Li Chao of the T'ang period (as quoted in the *T'u shu tsi ch'êng*), the said official was discharged on account of defraudations; the pillow which was half the size of a peck was confiscated after a judicial trial and sent up to Emperor Hien-tsung who called some shop-keepers as experts to determine its value. Their opinions were divided, the one calling it a priceless treasure, the others a beautiful stone, but not a genuine *sê-sê*.

[7] Several such specimens showing turquois in the matrix, obtained in Si-ngan fu, are in the Field Museum.

The *Wei lio*, further, refers to two stories taken from the *Ming-huang tsa lu*.[1]

Emperor Ming-huang is said to have erected in the palace Hua-ts'ing a bathing establishment consisting of ten rooms where he had a boat built of silver and steel, varnished, and adorned with pearls and jade; moreover, he piled up *sê-sê* in the bathing pool. The author of the *Wei lio* thinks that the use of the word *lei* (No. 6833) "to pile up" in this connection indicates that the *sê-sê* in question were beads, and not stones. But this supposition is hardly correct, for it leaves entirely unexplained what these beads (or pearls) had to do in the bath. In the actual text of the *Ming-huang tsa lu*[2] the story, however, is related in the form that the *sê-sê* were utilized to build up the well-known Three Isles of the Blessed of mythological fame, and this account sounds more plausible. In this case, *sê-sê* seems to have been a kind of building stone.[3] The other story in the *Ming-huang tsa lu* relates to Dame Kuo-kuo, a sister of the celebrated beauty and imperial concubine Yang Kuei-fei,[4] who built a house and rewarded the workmen with two gold cups and three pecks of *sê-sê;* one peck of these, according to the opinion expressed by the author of the *Wei lio*, had the value of a pearl. He further tells after the *Wu lei siang kan chi*, a work of the poet Su Shi (1036–1101), that Emperor I-tsung (860–873) of the T'ang dynasty presented a princess with a screen of *sê-sê* adorned with genuine pearls strung on blue and green silk, whence our author Kao Se-sun infers that in this case *sê-sê* was a kind of pearls of brilliant quality. This discourse leads us to think that the Sung writers did not know any longer what the *sê-sê* of the T'ang dynasty were, that the *sê-sê* peculiar to that age were entirely lost in the Sung period, that substitutes were then in vogue, merely designated by that name and ascribed to two localities, Shan-chou and P'ing-lu, and that even the belief prevailed that the *sê-sê* passed off under this name at that time were artificial productions due to some smelting process.[5]

[1] That is, Miscellaneous Records regarding Emperor Ming-huang (712–754), a work by Chêng Ch'u-hui of the T'ang period.

[2] Printed in the collection *Shou shan ko ts'ung shu*, Vol. 84, Ch. B, p. 4.

[3] This seems to be the case also in a poem of Po Kü-i (772–846) when he speaks of "a piece (or slab) of *sê-sê* stone" (*i p'ien sê-sê shi; P'ei wên yün fu*, Ch. 100 A, p. 47). Neither the addition "stone" nor the word *p'ien* would be used here, if the domestic *sê-sê* had been a precious stone or gem.

[4] Giles, Biographical Dictionary, p. 908.

[5] Under the Yüan dynasty the *sê-sê* are mentioned by Ch'ang Tê, a Chinese envoy who visited Bagdad in 1259, as precious stones in the palace of the Caliph, together with pearls, lapis lazuli and diamonds (Bretschneider, *Chinese Recorder*, Vol. V, p. 5). Bretschneider does not make in this passage an attempt at identifying the stone. When he says that, according to K'ang-hi's Dictionary, it is a kind of pearl, it should be remembered that the Chinese word *chu* means only a bead, regard-

If as early as the Sung period the Chinese had lost all correct notions of the *sê-sê* of the Leu-ch'ao and T'ang periods, there is no reason to wonder that the confusion becomes complete among the later authors who are simply content to repeat the older statements. Characteristic of this state of affairs is the explanation given in the *T'ung ya*, a miscellany written by Fang Mi-chi at the close of the Ming period: "The *sê-sê* are looked upon by some as precious stones, while the *Wei lio* considers them as pearls. Ch'êng T'ai-chi says: The *sê-sê* circulating at our time are all made from burnt stone. There are, however, three kinds of *sê-sê*:— precious stones like pearls are the genuine ones; those passing into blue and changing their color are the burnt ones, which are round and bright; Chinese beads of colored glass and baked clay are also called *sê-sê* by a mere transfer of the name." [1] There is assuredly not one Chinese author to venture the identification of *sê-sê* with turquois; neither under the Yüan nor under the Manchu dynasty when turquois was perfectly known in China did anybody assert that it was identical with the *sê-sê* in vogue during the T'ang dynasty.[2]

less of the material, whereas a pearl is always *chên chu*, a true or genuine pearl. In his Mediæval Researches (Vol. I, p. 140), he says, however, that *sê-sê* is *probably* the turquois. In the Annals of the Yüan Dynasty (*Yüan shi*, Ch. 21, p. 7 b, reign of Emperor Ch'êng-tsung, 1295–1307) there is another reference to *sê-sê*, two thousand five hundred catties of which are reported to have been palmed off on officials in lieu of money; but this transaction was soon stopped by the emperor. The turquois cannot be understood in this case, because, as will be seen below, this stone was known in the Mongol period under the name *pi tien* and is always so designated in the *Yüan shi*. Another text allows the inference that what was known as *sê-sê* in the Yüan epoch was a stone coming from Manchuria. The *Chêng tse t'ung* written in the beginning of the seventeenth century is quoted in K'ang-hi's Dictionary as saying that at the time of Emperor Jên-tsung (1312–20) of the Yüan dynasty it was reported that the subprefecture Kin-chou (in Fêng-t'ien fu, Shêng-king, Manchuria) offered *sê-sê* which had been gathered in a cave. The passage occurs in the *Yüan shi*, Ch. 24, p. 2 b; the Emperor was requested to send an envoy to the place who should gather the stones, but declined, as he regarded them as useless. Consequently the *sê-sê* of the Mongol period, as far as they relate to Chinese territory, cannot have been turquoises, as it was the very turquois which was highly appreciated by the Mongol rulers.

[1] Also Fang I-chi who lived in the first part of the seventeenth century, in his work *Wu li siao shi* (Ch. 8, p. 23 b; edition of Ning tsing t'ang, 1884) states that colored glass beads are designated *sê-sê*. This is the most recent author in whom I have been able to trace this word.

[2] BRETSCHNEIDER (Mediæval Researches, Vol. I, p. 175) is entirely erroneous in his assertion that it is stated in the *Pên ts'ao kang mu* that the stone called *tien-tse* was known under the name of *sê-sê* at the time of the T'ang dynasty. The *Pên ts'ao* does not contain a word to this effect. Its author, Li Shi-chên, states in the beginning of his essay on precious stones that blue ones are called *tien-tse* (No. 11,199); this is not the word *tien-tse* (No. 11,180) used in the *Cho keng lu* and *Yüan shi* (see below). Then follow ten sentences which have nothing at all to do with this subject, whereupon he proceeds to say, as stated above, that blue-green ones were called *sê-sê* by the T'ang people. It is therefore evident that these two statements separated from each other by several lines are not mutually connected, and that, on the contrary, in the mind of Li Shi-chên *tien-tse* and *sê-sê* are entirely distinct affairs; neither in the case of *tien-tse* does he refer to *sê-sê*, nor in the case of *sê-sê* to *tien-tse*; and he says nowhere that the one is identical with the other. Even did he say so, his

Li Shi-chên, the author of the great work on natural history, *Pên ts'ao kang mu,* makes one brief allusion to it (Section on Mineralogy, *kin shi,* Ch. 8, p. 17 b) in the chapter on Precious Stones (*pao shi*). Enumerating the different kinds of jewels mentioned in earlier texts, he says: "As regards the blue-green (*pi*) ones, the people of the T'ang dynasty called them *sê-sê*; as regards the red ones, the Sung people called them *mo-ho*; nowadays the general term is simply precious stones which are used for inlaying head-ornaments and utensils." This passage shows that Li Shi-chên considers the *sê-sê* as a gem peculiar to the T'ang period, and that he regards it as a precious stone, not as an ordinary stone.[1] The lack of any description on his part bears out the fact that he did not know the stone from personal acquaintance, and that he merely speaks of it on the ground of meagre traditions.

It is thus manifest that at various periods and with reference to different localities the Chinese have linked different ideas with the word *sê-sê,* that the later accounts are of no value in its determination as regards the earlier periods of the Leu-ch'ao and T'ang, and that even for the T'ang epoch a clear distinction must be made between the *sê-sê* of the countries outside of China and those within the Chinese dominion.

The various texts of the *Pei shi, Sui shu* and *T'ang shu* relating to foreign countries go to prove that the *sê-sê* of those times were valuable jewels, and that for this reason the word can hardly denote the turquois. It is not known to me on what authority VON KREMER's statement of turquois mines in Ferghana rests (his book is unfortunately not accessible to me), but I should think that he could not be regarded as an

assertion would be valueless, as he simply reproduces literary reminiscences, but does not show any actual knowledge of the stones of which he is speaking. We might well make bold to say that Li Shi-chên (as most of his countrymen during the Ming period) had never seen a turquois. In the official Statutes of the Ming Dynasty (*Ta Ming hui tien*) jade, agate, coral, amber, pearls, and ivory are frequently mentioned in connection with state paraphernalia and court costume, but turquois is conspicuous by its absence.

[1] Also in the great cyclopædia *T'u shu tsi ch'êng, sê-sê* are classified among precious stones (*pao shi*), likewise in the *T'ien kung k'ai wu,* a work on technology by Sung Ying-sing, of 1637 (Ch. 18, p. 58 b). Prof. HIRTH, in his above note, alludes to this book after an extract in the *T'u shu tsi ch'êng.* As an edition of this very scarce and valuable work printed 1771 in Japan is in my possession (despite diligent search I could not find any in China), I may say that it contains nothing to elucidate the subject; it simply says that of green stones there are *sê-sê* beads, emeralds (*tsie-mu-lü*), rubies (*ya-ku*) and the various kinds of *k'ung ts'ing* (on the latter see F. DE MÉLY, Les lapidaires chinois, p. 112, Paris, 1896). The author, accordingly, repeats bookish reminiscences but had no actual knowledge of, or experience with these stones which are to him mere names. It is certainly essential to determine in investigations of this kind, whether a Chinese author speaks of an object from direct knowledge of it, or merely reproduces the statements of his predecessors. In other words, we must adopt sound and critical philological methods before venturing any conclusions. It is manifest that the statements of the Ming and Ts'ing authors concerning *sê-sê* are of a purely bookish character and weak echoes of the past, but have no value whatever for the study of the question as to what the *sê-sê* of the past really were.

authority on mineralogical matters; presumably he refers to the Arabic authors alluded to by MAX BAUER (see below). I do not doubt that, as stated by Hirth, turquois is found in modern times in the region of Ferghana, although the evidence which I am able to find is rather slight.[1] On the other hand it is asserted by BAUER also that turquois occurs *in situ* in the region of Samarkand.[2] BAUER does not state his source, and I have no means of tracing it; the "unknown time" when the turquois mines were operated there is a rather unsatisfactory feature, and it would certainly remain to be proved that turquois was quarried in that region as early as the T'ang period (618–906). But granting the benefit of the doubt to those arguing on the opposite side, the possibility should be admitted that in the one passage of the *T'ang shu* indicated by Hirth and Chavannes the word *sê-sê* could denote the

[1] The opinion that turquois occurs in Ferghana is largely based on a remark of H. LANSDELL (Russian Central Asia, p. 515, London, 1885) who says that turquois is found at Mount Karumagar, 24 miles N. E. of Khojend; but Lansdell was an amateur traveler of journalistic tendencies in whose observations little confidence can be placed. The Armenian *lapidarium* translated into Russian by K. P. PATKANOV (Precious Stones, their Names and Properties according to the Notions of the Armenians, p. 48, St. Petersburg, 1873) mentions Khojend as a source for turquois. On the other hand it should not be passed over with silence that one of the best explorers of Ferghana who has given a detailed description of the region, CH. E. DE UJFALVY (Le Kohistan, le Ferghanah et Kouldja, p. 51, Paris, 1878) remarks on its mineral resources as follows: "In the mountains of Ferghana are found iron, lead, charcoal, quartz, kali, amethyst crystals, rock-crystal, silver, mica schist, sulphur, etc. (a cave near Aravān has stalactites and stalagmites). In the district of Andidjān there are rich sources of naptha of excellent quality, and also sulphurous sources at 38° Celsius." He does not mention turquois. Turquoises are still utilized by the Sart in and around Tashkend for the decoration of silver necklaces, bridles, girdleclasps, etc. (see, for example, H. MOSER, A travers l'Asie centrale, pp. 104–7, Paris, 1885). My colleague Dr. KARUTZ at the Museum of Lübeck, who has traveled extensively in Russian Turkistan, writes me that he encountered two areas in which turquois is diffused, among the Tatars of Russia and among the Sart of Turkistan, but that he did not find it among the Turkmen and the Kirghiz; he therefore concludes that it occurs only in the town population, but not among the nomads of the steppe; he learned nothing about indigenous sources of the stone, but is convinced that it is imported from Afghanistan. There is, he says, a rumor to the effect that turquois is found in the Kirghiz steppe, but he doubts the fact, as it is not employed by the Kirghiz in their ornaments.

[2] MAX BAUER (Precious Stones, p. 396) has the following note on turquois in this region: "It is stated that there are turquois mines, yielding mostly green stones, further to the north-west, beyond the Persian frontier between Herat and western Turkistan. According to the statements of ancient [?] Arabic writers, the precious stone was found at Khojend, from whence came also the green *callais* (*callaina*) of Pliny, now considered to be identical with turquois [this is extremely doubtful: note of the writer. On p. 392 BAUER says: "Whether the ancients were acquainted with turquois is doubtful."]. Other localities in the same region have also been recorded; for example, in 1887 in the mountain range Kara-Tube, fifty kilometres from Samarkand. The turquois occurs here in limonite and quartzose slate, and the place was, at some unknown time, the scene of mining operations. Finds of turquois have been made in the same region in our own time; for example, in the Syr Darya country in the Kuraminsk district (in the Kara Mazar mountains), and also in the Karkaralinsk district in the Kirghiz Steppes (Semipalatinsk territory of Siberia). These and other occurrences in the same region have no commercial importance and need no further consideration."

turquois, if the mineralogical condition of the present time, provided the fact is correct, will be admitted as evidence.[1] It cannot, however, be admitted as already demonstrated, that in other early passages the turquois is disguised under the word *sê-sê*; there is no forcible argument in favor of such a guess (supposition it can hardly be called); on the contrary, the valuation and utilization of the stone speak strongly against it.

Another moot question is the historical position of the turquois in those regions which are covered by the word *sê-sê*; there is no great antiquity and, accordingly, no archæology of the turquois in western Asia. It does not appear in Assyria, Babylonia or ancient Persia; it hardly plays any rôle in Greek and Roman antiquity.[2] Egypt[3] is the only country in the Old World which may lay claim to a great antiquity in the utilization of the turquois mined in the Sinai Mountains, and some objects inlaid with turquois mosaic and assigned to the Siberian bronze period, though neither their locality nor their time is exactly ascertained, may be of considerable age (see p. 58). So far as I know, no really

[1] Personally I am not convinced. It will be seen below that the first actual knowledge of the turquois dawns upon the Chinese as late as the Mongol period when a newly coined word for it appears, and when the word *sê-sê* continues with quite a different meaning. It is inconceivable to me that the knowledge of an object, when it is once acquired (and particularly of an object so striking to the eye as a turquois), can ever become lost. The tradition of the Mongol period is entirely cut off from that of the T'ang, the two not being interrelated. If the *sê-sê* of Ferghana were the turquois, why are the *sê-sê* objects occurring in China in the T'ang period not so described that a plain conclusion as to this material can be drawn? But they were evidently made of some building-stone. For this reason the question may be justly raised whether the account of the *T'ang shu* in regard to the mountain near Tashkend where *sê-sê* is produced really possesses great importance; it is somewhat vague, the name of the mountain not even being given; the report is evidently reproduced from hearsay. The earlier accounts of the *Pei shi* and *Sui shu*, which ascribe these jewels to an adjacent region without making reference to a definite locality, seem to me to be more to the point. All that can be safely laid down therefore is that *sê-sê* occurred in the territory of Ferghana and Sogdiana during the time from the fifth to the seventh century. In view of the other texts quoted above which must be equally taken into account in a consideration of this question, there is no reason to place all emphasis on this one statement; *sê-sê* occurred in Persia and Syria, and were traded in Khotan. Thus, they held the territory of western Asia and the dominion of the Western Turks. And as will be seen farther on, they were brought over to China by Manicheans or Nestorians.

[2] H. Blümner, Technologie und Terminologie der Gewerbe und Künste bei Griechen und Römern, Vol. III, p. 248 (Leipzig, 1884). It is still more doubtful to me than to Blümner if the *callaina* or *callais* of Pliny, as has been supposed, refers to the turquois; the evidence favoring this theory is extremely weak; Pliny's statement that the stone is produced in Farther India or beyond India and in the Caucasus (nascitur post aversa Indiæ, apud incolas Caucasi) where we positively know that no turquois is found proves that he does not speak of the turquois. Compare p. 2, note 2. "Turquois, hardly ever used by the Greeks, was rarely employed by Græco-Roman artists" (D. Osborne, Engraved Gems, p. 284, New York, 1912).

[3] The ancient Egyptian turquois-mines in Wadi Maghara and Wadi Sidreh in the Sinaitic Peninsula were first discovered in 1849 by Major C. Macdonald, then visited by H. Brugsch (Wanderung nach den Türkisminen und der Sinai-Halbinsel, Leipzig, 1866), and examined anew in 1905 by W. M. Flinders Petrie.

ancient carved object of turquois has as yet come to light in Persia or Turkistan,[1] while a great variety of gems appears in the Persian intaglios, particularly in those of the Sassanian epoch (226–642 A. D.) among which turquois is strikingly absent.[2] It is no less because of this lack of archæological evidence that I hesitate to believe in the proposed identification of *sê-sê* with turquois, as regards the older accounts of the *Pei shi* and *Sui shu*.

Furthermore, the important question arises,— what is the antiquity of the turquois in Persia? When were the turquois mines of Persia first operated, at what time did turquois begin to play an active rôle in the culture and life of the Persian people? It is evident that all this is a matter of consequence for our *sê-sê* problem. I am certainly not competent to decide this question, the final solution of which must come to us from one of our co-workers in the Arabic or Persian field; but even to an outsider who has merely a scant knowledge of this subject some observations spontaneously present themselves which render him very cautious or rather skeptic in assuming, as has so often been done without any substantial evidence, a considerable antiquity for the acquaintance of the Persians with the turquois. There is, first of all, no ancient Iranian word for the turquois. Avestan literature, as far as I know, makes no allusion to it. A great authority, W. GEIGER,[3] emphasizes the fact that the minerals characteristic of Irān, as turquois, ruby, lapis lazuli, are not even mentioned in the Avesta. The lack of an Iranian word for it, with the additional absence of an ancient Sanskrit word, renders the supposition highly probable that neither the Aryans nor the Iranians had any knowledge of turquois. The word *ferozah* is New Persian, consequently not older than the ninth century; in Middle Persian or Pahlavī, the language of the Arsacids and Sassanians, no word for the turquois seems to be preserved, unless it is represented by

[1] In a collection of ancient intaglios found in the environment of Khotan and described by A. F. R. HOERNLE (A Report on the British Collection of Antiquities from Central Asia, pt. I, p. 38, Calcutta, 1899) objects of spinel and lapis lazuli occur, but none of turquois. F. GRENARD (Mission scientifique dans la Haute Asie, Vol. III, p. 143, Paris, 1898) found in a cave near Khotan a wooden image with eyes formed by rubies. In the famous treasure discovered in 1877 on the northern bank of the Oxus described by A. CUNNINGHAM (Relics from Ancient Persia, *Journal Asiatic Society of Bengal*, 1881, pp. 151–186) and O. M. DALTON (The Treasure of the Oxus, London, 1905), despite a great number of ornaments, no turquois has been traced. Also in the works on Persian art (M. DIEULAFOY, L'art antique de la Perse; PERROT and CHIPIEZ, History of Art in Persia, London, 1892) no reference is made to turquois.

[2] Compare the Sassanian precious stones as enumerated, for instance, in ED. BAUMANN, Allgemeine Geschichte der bildenden Künste, Vol. I, pt. 2, p. 538, G. STEINDORFF's Description of Sassanian Gems in *Mitteilungen aus den Orientalischen Sammlungen des Berliner Museums*, No. 4, and J. MENANT, Cachets orientaux, Intailles sassanides (*Cat. Coll. de Clercq*, Vol. II, pt. I, Paris, 1890). In none of these publications is turquois pointed out.

[3] Ostiranische Kultur im Altertum, p. 147 (Erlangen, 1887).

the older form *fīrūzag* handed down by the *lapidarium* of Pseudo-Aristotle [1] and al-Bērūnī.   In questioning the archæologists, we meet a slight piece of evidence.   In the kurgans of Anau [2] beads of turquois together with those of carnelian and lapis lazuli have been discovered, and as it is asserted, "in the earliest culture strata."   They were used as burial gifts with the skeleton of a child, and it is concluded in the publication referred to that they must have come from Persia where turquois is known both to the south of Anau and farther eastward on the plateau.   But the chronology of these antiquities of Anau is somewhat uncertain, and by no means seems to me to be settled beyond doubt; aside from this, the deduction that the Anau turquoises, granted that they are what they are presented to be, *must* be of Persian origin is not at all forceful, and not proved.   They *may* have come as well from Siberia where turquois was employed during the bronze age (p. 58), though the locality where the ancient Siberian turquois was mined is not yet known, or (why not?) from Tibet, or from some forgotten mine in Turkistan.   Reverting to Persia and glancing over the pages of the history of the Sassanians (226–642 A. D.) [3] we look in vain for any testimony that turquois formed an essential constituent of the culture of the period.   The only item I am able to trace is a statement made by A. CHRISTENSEN [4] to the effect that King Khosrau II (590 A. D.) possessed a game of backgammon (*nard*), the men of which were carved from coral and turquoises.   With respect to these turquois carvings some doubts may be entertained, particularly for the reason that the Persian mineralogist Muhammed Ibn Mansūr who wrote about 1300 in the translation of Gen. SCHINDLER [5] says that in the environment of

[1] J. RUSKA, Das Steinbuch des Aristoteles, p. 43 (Heidelberg, 1912).

[2] R. PUMPELLY, Explorations in Turkestan, Vol. I, pp. 60, 64, 199 (Washington, 1908).

[3] TH. NÖLDEKE, Geschichte der Perser und Araber zur Zeit der Sasaniden (Leiden, 1879); NÖLDEKE, Aufsätze zur persischen Geschichte (Leipzig, 1887); M. K. PATKANIAN, Essai d'une histoire de la dynastie des Sassanides (*Journal asiatique*, 1866, pp. 101–238); J. MARQUART, Untersuchungen zur Geschichte von Eran (Göttingen und Leipzig, 1896, 1905); K. A. INOSTRANTSEV, Sassanian Studies (in Russian, St. Petersburg, 1909).   There is certainly no doubt that the ancient Persian kings and subsequently the Sassanians possessed quantities of precious stones in their treasuries and graves, but all indications are lacking as to what they were.   ARRIAN (*Anabasis* VI, 29) mentions gold earrings set with precious stones as part of the treasures hoarded in the tomb of Cyrus at Pasargadæ.   Compare further M. DIEULAFOY, L'art antique de la Perse, Vol. V, p. 137, and O. M. DALTON, The Treasure of the Oxus, p. 9.

[4] L'empire des Sassanides, le peuple, l'état, la cour, p. 105 (Copenhague, 1907). The source for this statement is H. ZOTENBERG, Histoire des rois des Perses, p. 700 (Paris, 1900); but the Arabic work (edited and translated by Zotenberg) written by al-Ta'ālibī (961–1038) can hardly claim any historical authenticity; it is purely legendary in character and a counterpart to Firdausī's Shāh-nāmeh.

[5] *Jahrbuch der k. k. Geologischen Reichsanstalt*, Vol. XXXVI, Wien, 1886, p. 310.

Nīshāpūr is found a stone *similar* to turquois from which chess-men are made, but that its color soon disappears.   Thus, also the backgammon men of Khosrau, if at all the report may lay claim to historical authenticity, which is doubtful, may have been worked from this stone material which merely had an outward resemblance to turquois.[1]   We move on safer ground in coming down to the Arabic authors of the middle ages; they indeed are the first to bring to our notice the mining of turquois in Nīshāpūr.[2]   Al-Kindī, who lived in the latter part of the ninth century, as quoted by Ibn al-Baiṭār,[3] briefly mentions the turquois without alluding to Persia, nor does the oldest source for Arabic mineralogy, the *lapidarium*, wrongly connected with the name of Aristotle,[4] which according to Ruska was composed before the middle of the ninth century.   The fact that Persia is not alluded to by these two authors is not decisive; on the contrary, it is highly probable that they had the Persian turquois in their minds, for they designate it by the older Persian form *fīrūzag*, and as pointed out by Ruska,[5] it is noticeable that in the text translated by him Persia, Khorāsān, India and China are most frequently cited among the localities for the minerals described in it.

The earliest allusion to the turquois-mines of Nīshāpūr which I am able to find is made by Ibn Haukal (978 A. D.), who based his account on Iṣṭakhri (951).   He reports as follows: [6]

"The villages and the towns in the plain around Nīshāpūr are numerous and well populated.   In the mountains of Nīshāpūr and Tus are mines, in which are found brass, iron, turquoises, santalum, and the precious stone called malachite; they are said to contain also gold and beryl."

Al-Bērūnī (973–1048) seems to be the second weighty authority with a distinct reference to Nīshāpūr by stating that the turquois is brought from the mountain Ansār, one of the mountains of Rīwand near

---

[1] J. DE MORGAN (Mission scientifique en Perse, Vol. IV, p. 320, Paris, 1897) figures a bas-relief of Takht-i Bostān representing Khosrau II Parwēz (591–628) in full armor on horseback and interprets the medial row of stones inlaid in the sheath of the sword as turquoises.   There is no color displayed on this stone bas-relief, and this view seems wholly arbitrary; it is rejected as fantastic by F. SARRE and S. HERZFELD, Iranische Felsreliefs, p. 203 (Berlin, 1910).

[2] Name of a city and province in northern Khorāsān.   The city was founded by Shāpūr II (309–379) whose name forms the second element in the name of the city. The Old Persian name is Nēw-Shāpūr, the word *nēw* meaning good.   The New Persian form is *Nēshāpūr*, at present *Nīshāpūr*, Arabic *Naisābūr* (compare NÖLDEKE, Geschichte der Perser und Araber zur Zeit der Sasaniden, pp. 59, 67; an interesting sketch of the history of the city is given by A. V. W. JACKSON, From Constantinople to the Home of Omar Khayyam, pp. 246–260, New York, 1911).

[3] L. LECLERC, *l. c.*, Vol. III, p. 51.

[4] J. RUSKA, Das Steinbuch des Aristoteles, p. 151.

[5] *L. c.*, p. 43.

[6] A. V. W. JACKSON, *l. c.*, p. 254.

that city.[1] His contemporary al-Taʻālibī (961–1038) expands likewise
on the turquois of Nīshāpūr.[2] Then we have the testimony of Tīfāshī
whose work on precious stones was written toward the middle of the
thirteenth century (the author died in 1253) who says anent the turquois
that it originates from a mine situated in a mountain of Nīshāpūr
whence it is exported into all countries.[3] This seems to be the first
clear statement of the fact that the Persian turquois of Nīshāpūr had
entered into the commerce of the world. Can it be mere chance now
that we find the first record of the turquois of this place in China in
1366 (p. 56), and that the Persian turquois makes its début in India
only during the Mohammedan epoch? Finally we come to the Persian
mineralogy of Muhammed Ibn Mansūr above referred to which according to SCHINDLER was written about 1300, according to RUSKA [4] in
the thirteenth century; still later is al-Akfānī who died in 1347–48, and
who in his treatise on precious stones deals also with the turquois.[5]

There is also the evidence furnished by MARCO POLO [6] who passed

---

[1] E. WIEDEMANN, *Der Islam*, Vol. II, 1911, p. 352.

[2] WIEDEMANN, Zur Mineralogie im Islam, p. 242 (Erlangen, 1912).

[3] L. LECLERQ, *l. c.*

[4] *L. c.*, p. 31. J. V. HAMMER has translated an extract from this work in *Fund-gruben des Orients*, Vol. VI, pp. 126–142, Wien, 1818; the text has not yet been edited. Compare WIEDEMANN, *l. c.*, p. 208.

[5] Edited by P. L. CHEIKHO (*Al-Machriq*, Vol. XI, 1908, pp. 751–765). Com-pare WIEDEMANN, *Mitt. d. deutschen Ges. für Geschichte d. Med. und Nat.*, Vol. VIII, pp. 509–511. Translation by WIEDEMANN, Zur Mineralogie im Islam, p. 225 (Er-langen, 1912). BRETSCHNEIDER (*China Review*, Vol. V, 1876, p. 124) identifies the Persian vase "reflecting what is going on in the world" (mentioned in the Annals of the Ming Dynasty) with the vase of Djemshid frequently spoken of by the Persian poets and said by Rashid-eddin (1247–1318) to have been made of turquois. This identification can hardly be correct, if the tradition holds good that the Persian vase had "the property of reflecting light in such a way that all affairs of the world could be seen." Turquois is dense, opaque, not at all transparent (in composition a hy-drous phosphate of aluminum containing water, 20.6 per cent, alumina, 46.8 per cent, and phosphorous oxide, 32.6 per cent), and thus in composition as well as opacity, differs from most other gems (O. C. FARRINGTON, Gems and Gem Minerals, p. 170, Chicago, 1903).

[6] YULE and CORDIER, The Book of Ser Marco Polo, Vol. I, p. 90. Marco Polo's itinerary in southern Persia has been elucidated by Gen. A. HOUTUM SCHINDLER (*Journal Royal Asiatic Society*, 1881, pp. 1–8, and 1898, pp. 43–46), further by G. LE STRANGE (The Cities of Kirmān, *ibid.*, 1901, pp. 281–290). On p. 2 of the first of these papers, SCHINDLER has devoted a note to the turquois-mines of the province of Kermān, stating the various localities where they are found; at a place, twelve miles from Shehr-i-Bābek, are seven old shafts, now for a long period not worked, the stones of these mines being of a very pale blue, and having no great value. The inferiority of the Kermān turquois is emphasized also by the Chinese author Tʻao Tsung-i in 1366 (see p. 57). And then, one will make us believe that the turquois should be recognized in PLINY's callais "the best sort of which occurs in Carmania," as if there were no other stones to be found in the big country Carmania, and as if it had been proved that turquois was mined there in the first century. But Marco Polo evidently is the first authority with such a report, and from Pliny to Marco Polo there is a far cry. The Arabic author al-Taʻālibī (961–1038) expressly states that turquois is found only near Nīshāpūr (WIEDEMANN, Zur Mineralogie im Islam, p. 242). Major P. M. SYKES (Historical Notes on South-East Persia, *Journal Royal Asiatic*

through Persia in 1294 and says of the kingdom of Kermān that the stones called turquoises are produced there in great abundance; they are found in the mountains, where they are extracted from the rocks. Gen. A. HOUTUM SCHINDLER [1] to whom we owe an excellent description from the geological viewpoint of the Persian turquois-mines has not solved the problem as to the antiquity of the mining operations; the late report of Ibn Mansūr is the only document quoted by him.[2] In one passage (p. 307) we read the general remark: "Seit Jahrtausenden ist in diesen Gruben gearbeitet worden," which is no more than a personal impression. We live in a skeptic age and are not willing to believe so easily in millenniums, if no evidence of hard and cold facts is advanced. Every human activity is defined by time; language and history seem to militate against such an unfounded surmise.[3] These observations on the history of turquois in Persia form another reason why I am not at all sanguine in accepting the explanation of *sê-sê* by turquois when such early texts as *Pei shi, Sui shu,* and *T'ang shu* come into question and refer to a time when it must be doubted, at least for the present, that turquois was known in Persia or had any significance in her culture. The Chinese accounts plainly refer to Sassanian Persia, while all references to turquois in Persia, at least in the present state of our knowledge, are post-Sassanian.

In our attempts to identify the names of stones mentioned in ancient

Society, 1902, p. 942) who has studied the archæology of the Kermān region reports on the tombs: "In each tomb were a yellow jar of pottery, round bowls of three sizes, a pair of bracelets, two pins, and some arrow and spear heads, all of which were of bronze except the vessels. In addition, two or three carnelian gems were found, and some small silver earrings and bracelets. The custom of placing a carnelian in a dead man's mouth, with the names of the twelve *Imām* engraved on it, is one that obtains nowadays." There is no report of a find of turquoises in a grave of the Kermān region.

[1] Die Gegend zwischen Sabzwār und Meschhed in Persien (*Jahrbuch der k. k. geologischen Reichsanstalt,* Vol. XXXVI, pp. 303–314, Wien, 1886).

[2] When Ibn Mansūr says that the best mine of those at Nīshāpūr is the one discovered by Isaac, the father of Israel, and hence called Isaac's mine, this is certainly a legend without historical value. The account of Ibn Mansūr seems to be pieced together from different sources; the *lapidarium* of Pseudo-Aristotle is evidently utilized (for example, in the statement that turquois is light and brilliant in clear weather, but dim and dull when the sky is clouded). According to RUSKA (*l. c.,* p. 35), matter and arrangement of his work largely depends on Tīfāshī. Regarding the Chinese, Ibn Mansūr remarks that they like the *tarmaleh* (a word queried by Schindler), turquoises intersected by other stone, and employ these for the adornment of their idols and women. This is apparently an error and should read "Tibetans" instead of Chinese (compare p. 13). Travelers who visited the mines are quoted by A. V. W. JACKSON (*l. c.,* p. 259).

[3] The last (eleventh) edition of the *Encyclopædia Britannica* gives two items of information of a contradictory character. In Vol. XIX (p. 710) mention is made of Mādan, 32 miles N. W. of the city of Nīshāpūr, "where the famous mines are which have supplied the world with turquoises for at least 2,000 years." A more moderate attitude is observed in Vol. XXVII (p. 483) where it is said: "In Persia the turquois mines have been worked for at least eight centuries."

records we must never lose sight of the plain facts of archæology, history, and mineralogy. Taking a broader view of the subject we find that ruby and lapis lazuli have been the most prominent jewels of Irān since ancient times, and that both are well attested by the presence of ancient authentic specimens [1] and traceable to a well defined locality. The great jewel-producing district within equal proximity of Sogdiana, Persia and Khotan was the region of Badakshān (in Chinese *P'a-to-shan*), north of the Hindu Kush mountains, well known to the Chinese during the T'ang period, and to every modern mineralogist as a center for the production of two precious stones — lapis lazuli and the balas ruby or spinel.[2] The former stone entered the horizon of the Chinese

---

[1] Compare p. 38, note 1.

[2] Max Bauer, Precious Stones, pp. 278, *et seq.* (German original, Edelsteinkunde, 2nd ed., p. 374); O. C. Farrington, Gems and Gem Minerals, pp. 96, 202. T. Wada (*Beiträge zur Mineralogie von Japan*, No. 1, p. 20, Tōkyō, 1905) describes spinels originating from China, and R. Pumpelly (Geological Researches, p. 118, *Smithsonian Contributions to Knowledge*, Vol. XV, Washington, 1867) seems to have encountered spinels in Yün-nan. Marco Polo (ed. Yule and Cordier, Vol. I, p. 157) has described the ruby and the lapis lazuli mines. They are mentioned by the Arab geographers Iṣṭakri and Ibn Haukal in the tenth century (O. M. Dalton, The Treasure of the Oxus, p. 9, London, 1905). Ibn Haukal's passage has been translated by Wiedemann (Zur Mineralogie im Islam, p. 236, Erlangen, 1912). The Arabic geographer Yāqūt (1179–1229) and the historian Maqrīzī (1365–1442) impart notes on the balas ruby of Badakshān (Wiedemann, *ibid.*, pp. 235–6); al-Ta'ālibī (961–1038) mentions it (*ibid.*, p. 243). The ancients were familiar with the spinel as evidenced by antique intaglios, but its designation in classical times is not known according to H. Blümner (Technologie und Terminologie der Gewerbe und Künste bei Griechen und Römern, Vol. III, p. 236, Leipzig, 1884); but H. O. Lenz (Mineralogie der alten Griechen und Römer, p. 17, Gotha, 1861) includes the spinel under the Greek word *anthrax* (likewise Daremberg and Saglio, Dictionnaire des antiquités grecs et romains, Vol. II, p. 1462, and Pauly's Realenzyklopädie, Vol. XIII, col. 1108). Arabic and Armenian authors relate a legendary tradition that at the time of the dynasty of the Abbassides a terrific earthquake shattered a mountain in Badakshān, in which the spinels appeared (K. P. Patkanov, *l. c.*, pp. 19–20). The great antiquity of the mining operations in Badakshān is illustrated by the wide diffusion in early times of lapis lazuli. In the words of Marco Polo, that of Badakshān is the finest in the world; Yule (Vol. I, p. 162) comments that the mines of *Lājwurd* (whence *l'Azur* and *Lazuli*) have been, like the ruby mines, celebrated for ages. Max Bauer (Precious Stones, p. 442) states that the material which is not sent to Bokhāra (whence it is traded to Russia) goes, together with rubies of the same region, to China and to Persia, and that the lapis lazuli said to occur in these countries, as well as in Little Bokharia and Tibet, has probably been imported from Badakshān. Moreover, according to this author, the material sold in other parts of Asia, for example, in Afghanistan, Beluchistan, and India, and stated by travelers to occur in those regions, in all probability is imported from the locality in the proximity of the Upper Oxus; the lapis lazuli from which the ancient Egyptian scarabs were cut, as Bauer says, presumably came from Badakshān, as did also the material much used elsewhere in ancient times. The early use of lapis lazuli in ancient Babylonia is well attested by numerous finds (P. S. P. Handcock, Mesopotamian Archæology, pp. 76, 102, 315, even from the earliest Sumerian period, p. 340) and the mineralogical analyses of Heinrich Fischer (H. Fischer and A. Wiedemann, Ueber babylonische Talismane, p. 4, Stuttgart, 1881). It is interesting to note among the beads of Sumerian necklaces coral, lapis lazuli, mother-o'-pearl, and agate — all favorite objects of the Tibetans, to the exclusion of turquois, which evidently belongs to a much more recent stratum of culture in Asia. The origin of Babylonian lapis lazuli seems not yet to have been satisfactorily established. Fischer suggested Bokharia; H. Blüm-

at the T'ang period, and likewise from the regions of the Western Turks.[1]
As it has a name of its own (*kin tsing*),[2] and as besides it no other

NER (*l. c.*, Vol. III, p. 275) basing his opinion on Pliny's statement that the best kind
occurs in Media (*apud Medos*) is inclined to think of Tibet where it is found at present.
This fact is certainly correct (see p. 17, note 2), but Tibet cannot come into question
in the times of antiquity, and it seems preferable, at least for the present, to join MAX
BAUER in the opinion that the mines of Badakshān are responsible also for Babylonian
lapis lazuli. Prehistoric occurrence of lapis lazuli in Beluchistan is mentioned by
NOETLING (*Zeitschrift für Ethnologie*, Vol. XXX, 1898, *Verhandlungen*, p. 470); in
Armenia by BELCK and LEHMANN (*ibid*, p. 590). R. LEPSIUS (Les métaux dans les
inscriptions égyptiennes, p. 31, Paris, 1877) derives the lapis lazuli used by the
ancient Egyptians from Badakshān. As to India, the case may be more precisely
made out. There, the stone does not seem to be indigenous. G. WATT (A Dictionary of the Economic Products of India, Vol. IV, p. 587) says: "Though not known
with certainty to occur in India, it is imported into the country, where it is employed
for several purposes." The Sanskrit word *rājavarta* or *lājavarta* (Hindustani *lājward*,
Behar *lājburud*, Guzerati *rājāvaral*) is plainly derived from Persian *lāzuward*
(L. FINOT, Les lapidaires indiens, p. XVIII, connects it with Arabic *lāzurd*), and the
five names enumerated for the stone in the *Rājanighaṇṭu* (R. GARBE, Die indischen
Mineralien, p. 90), though all couched in a Sanskrit form (with the meaning "suitable
for a king's forehead, forehead-jewel") are re-interpretations based on that foreign
word (the Petersburg Sanskrit Dictionary, smaller edition, has still another composite
name *suvarṇābha*). TAVERNIER (ed. by V. BALL, Vol. II, p. 156, London, 1889) who
wrote in 1676 makes a somewhat vague statement: "Towards Tibet, which is
identical with the Caucasus of the Ancients, in the territories of a *Rāja* beyond the
Kingdom of Kashmir, there are three mountains close to one another, one of which
produces gold of excellent quality, another *grenat*, and another *lapis*." The editor
BALL is inclined to think that the *lapis* mine here referred to is near Firgāmu in
Badakshān. Lapis lazuli occupies a place in Indian antiquity, particularly among
the Buddhists, but this subject is still in need of special investigation.

[1] CHAVANNES, Documents, p. 159, and *T'oung Pao*, 1904, p. 66. But lapis lazuli
was perhaps known to the Chinese to a certain extent from the second century A. D.
(compare F. HIRTH, *Zeitschrift für Ethnologie*, Vol. XXI, 1889, *Verhandlungen*, p. 500,
or Chinesische Studien, p. 250). HIRTH refers to the "gold girdles set with blue
stones from Hai-si" presented to the Chinese Court in 134 A. D. by the king and
minister of Kashgar, further to a definition in the glossary *T'ung su wên* from the
end of the second century where the expression "to paint the eyebrows" is explained
as a cosmetic yielded from blue stone (*ts'ing shi*) where ultramarine, a pigment obtained from lapis lazuli, is evidently in question. This is well confirmed by the
report of the *Sui shu* on the country of Ts'ao identified by HIRTH with Badakshān or
the plateau of the Pamir (an identification overlooked by CHAVANNES, Documents,
p. 130), where it is said that Ts'ao produced among other articles *ts'ing tai*, that is,
ultramarine for cosmetic purposes. HIRTH does not state the fact that Badakshān
is the old classical land of lapis lazuli; but just this lends force to his conclusion that
the ancient cosmetic used by the Chinese was of mineral, and not, as later Chinese
authors believed, of vegetal origin.

[2] As I expect to show on another occasion, there is, besides *kin tsing*, an ancient
term *kin sing shi*, "stone with golden stars," for the designation of lapis lazuli. This
is mentioned as a product of Tibet (*T'u-po*) in *Kiu Wu tai shi* (Ch. 138, p. 1 b), as a
product of Khotan in the Geography of the Ming Dynasty (*Ta Ming i t'ung chi*,
edition of 1461, Ch. 89, fol. 25 a), and as a product of Sze-chou fu (in the province of
Kuei-chou) in the Geography of the Ts'ing Dynasty (*Ta Ts'ing i t'ung chi*, Ch. 398,
p. 3 b); compare further *Pên ts'ao kang mu*, Ch. 10, p. 10 a. The word *kin sing*
reflects the same notion as connected by the ancients with the same stone (called
sappheiros, sapphirus, a word of Semitic origin: O. SCHRADER, Reallexikon, p. 152),
described by them as a blue stone with brilliant dots of gold (the small quantity
of sodium sulphide present in the stone being taken for gold), and likened to the
starry sky (compare BLÜMNER, *l. c.*). The modern Chinese name for lapis lazuli is
ts'ing kin shi, that is, "dark-blue gold stone." In the Dictionary of Four Languages
by the Emperor K'ien-lung, this word is rendered into Manchu by *nomin*, Tibetan

important jewel is found within this dominion than the balas ruby or spinel, no other alternative can be seen at present than that, generally speaking, the *sê-sê* is in all probability to be identified with the latter; while in some cases where carvings and building material are mentioned, as will be seen, the onyx might be conjectured.

I have further arrived at the conclusion that, as far as precious stones are concerned, two different species should be understood by the name *sê-sê*,— as far as the Iranian regions are involved, the balas ruby of Badakshān; and, as far as ancient Tibet comes into question, in all likelihood the emerald.

Further evidence may first be adduced for the proposed identification with the balas ruby. The balas ruby is now called in Chinese *pi-ya-se*,[1] correctly translated as early as 1820 by Abel-Rémusat[2] with "le rubis balais"; the Chinese word, according to this author, is derived from *balash* or *badaksh*, whence, as he says, the name of the country Badakshān is derived; but more probably, the name of the jewel is derived from the name of the locality.[3]

The recent valuable paper of E. Wiedemann,[4] allows us to trace the etymology of this Chinese word. Discussing the balas ruby of Badakshān (in Arabic *al balachsh*), al-Berūnī maintains that the best

by *mu-men*, Mongol by *nomin* or *momin*, all of which have the same meaning. Abel Rémusat (Histoire de la ville de Khotan, p. 168) adds the Uigur word *nachiver* which he says is derived from Persian *ladjiver* (*lazvard*). The English and Chinese Standard Dictionary (Vol. I, p. 1308, Shanghai, 1910) translates lapis lazuli by *lan liu-li*. In the Mongol period the Chinese name for lapis lazuli was *lan ch'i* (Bretschneider, Mediæval Researches, Vol. I, p. 151, and *Chinese Recorder*, Vol. VI, 1875, p. 16) doubtless derived from the Persian or Arabic word. It seems to me that the character *lan* "orchid" (No. 6721) used by Ch'ang Tê in writing this word is an error for *lan* "indigo, blue" (No. 6732); the name of the capital of Badakshān where lapis lazuli was mined was *Lan shi*, "Blue Market" (Chavannes, *T'oung Pao*, 1907, p. 188), a designation which apparently refers to the blue color of lapis lazuli.

[1] Giles, Dictionary, No. 9009, who translates "a kind of cornelian", which is not correct.

[2] Histoire de la ville de Khotan, p. 168. The identification is based on the Dictionary in Four Languages (see below p. 48, note).

[3] Yule and Burnell (Hobson-Jobson, p. 52) state in regard to the word *balas*: "It is a corruption of *Balakhshī*, a popular form of *Badakhshī*, because these rubies came from the famous mines on the Upper Oxus, in one of the districts subject to Badakshān," and quote also Ibn Baṭūṭa as saying that the mountains of Badakshān have given their name to the Badakshī ruby.— Eitel (Handbook of Chinese Buddhism, p. 131), I believe, is quite right in recognizing in *rohitaka* or *lohitaka* the Sanskrit word for the balas ruby, for the other Sanskrit words employed for the ruby, as shown by L. Finot (Les lapidaires indiens, p. XXXIX), refer to such species as are found in India. We are much in need of a careful and critical study of all the names of precious stones to be found in Buddhist Sanskrit and Pāli literatures, and these should certainly be examined in connection with the corresponding Chinese and Tibetan renderings; the Buddhist nomenclature, in many cases, deviates from that of the Indian mineralogists.

[4] Ueber den Wert von Edelsteinen bei den Muslimen (*Der Islam*, Vol. II, 1911, p. 349).

sort is the *pijāzakī*, that is, the one coming from the district of Pijāzak. I am inclined to think that we may look upon this word as the source of the Chinese transcription *pi-ya-se*, whereby its meaning is moreover confirmed.   It is true the Chinese word is not traceable earlier than the eighteenth century, but it is doubtless a much more ancient word of the colloquial language which for this reason was not registered in the standard dictionaries; it is, however, entered, as numerous other colloquial words, in the "Dictionary in Four Languages" published by the Emperor K'ien-lung.   Mention is made of *pi-ya-se* in the "Statutes of the Manchu Dynasty" (*Ta Ts'ing hui tien t'u*, Ch. 43, p. 5) where they are granted as a privilege to all imperial court-ladies to be worn on their sable caps.[1]   There is, however, an older trace of the word in a source of the Mongol period where, in my opinion, it has been misjudged by Bretschneider.   The Chinese traveler Ch'ang Tê, who was despatched in 1259 by the Mongol Emperor Mangu as envoy to his brother Hulagu, king of Persia, and whose diary, under the title *Si shi ki*, was edited in 1263 by Liu Yu, reports that a precious stone by the name *ya-se* of five different colors and of very high price is found on the rocks of the mountains in the south-western countries. BRETSCHNEIDER [2] is inclined to identify this word with the Arabic *yashm* or *yashb*, our word jasper, which seems to me very improbable. The stone *ya-se* is mentioned by Ch'ang Tê together with lapis lazuli

---

[1] The character *pi* in the word *pi-ya-se* may explain also why the *sê-sê* are designated by some authors as *pi* "green," which may simply be due to a reminiscence of the word *pi-ya-se* where the *pi* enters as an attempt at reproducing a foreign sound.   Besides, al-Bērūnī, in the passage quoted by WIEDEMANN, speaks of four color variations in the balas ruby,— red, violet, *green* and yellow.   The Armenian *lapidarium* (K. P. PATKANOV, *l. c.*, p. 19) ascribes to spinels a red color, the colors of the garnet, of fire, of vinegar, of wine, of the scorpion, and of peas.   A. BOETIUS DE BOOT (*l. c.*, p. 149) states in regard to the color of *Rubinus Balassius:* "Habet iste Rubinus laccæ florentinæ, aut cremesinum colorem, ita ut parum cærulei coloris vero rubro admixtum videatur, rosei coloris rubentis instar."   R. MIETHE (in KRÄMER, Der Mensch und die Erde, Vol. V, p. 377) describes spinels as black or brown-black, frequently brownish, rarely green and blue.   O. C. FARRINGTON (Gems and Gem Minerals, p. 96), besides red, gives also the colors orange, green, blue, indigo, white and black.   MAX BAUER (Precious Stones, p. 297) has the following: "Spinels of a rose-red or light shade of color inclined to blue or violet are referred to as 'balas rubies.'   They not infrequently combine with this character a peculiarly milky sheen which considerably detracts from their value.   Stones the color of which is more decidedly blue or violet resemble, although much paler, some almandines, and are known as 'almandine spinels.'   Violet spinels, which are not too pale in color, often resemble both the true amethyst and the 'oriental amethyst,' and indeed have sometimes been put on the market under the latter name."   Blue and black spinels are discussed by the same author on p. 299.   Red and purple balas rubies are distinguished also by Yang Shên (*Ko chi king yüan*, Ch. 33, p. 1) who, according to MAYERS (Chinese Reader's Manual, p. 270) lived from 1488 to 1559 (GILES, Biographical Dictionary, p. 912, sets the date of his death at 1529; but WYLIE, Notes on Chin. Lit., p. 154, assigns 1544 to one of his works).

[2] Mediæval Researches, Vol. I, p. 151, and *Chinese Recorder*, Vol. VI, 1875, p. 16 (where the Chinese characters are given). .

(*lan-ch'i*) as occurring in the same locality, and since Badakshān is *the* locality producing lapis and balas ruby, the greater probability is that Ch'ang Tê's *ya-se* is identical with *pi-ya-se*, the balas ruby of Badakshān. The fact that Ch'ang Tê, in writing the name, employs other characters than those in use at later times is certainly not in the way of this identification.   On the other hand, as stated above (p. 33, note 5), Ch'ang Tê notices in the palace of the Caliph *sê-sê* together with lapis lazuli, and hence it may be concluded that his *sê-sê* is identical with *ya-sê*, *i.e.* that *sê-sê* is the balas ruby of Badakshān.   And if it is permissible to interpret the word *liu-li*, occurring in the account on Persia in the *Sui shu* (Ch. 83, p. 11 b) and having its place between coral, agate and crystal, in the sense of lapis lazuli, we are there confronted with an analogous case.   The great antiquity of lapis lazuli in Egypt and Western Asia, corresponding to its relatively early appearance in China, leads one to the inevitable conclusion that also balas ruby, originating from the same mines, must be of proportionately equal age.

From the few accounts we have in regard to the *sê-sê* — there is no contemporaneous description of them — we cannot surely be too positive on the subject of identification.   But the spinel or balas ruby tentatively proposed suits the situation far better than the turquois, for it is a precious stone, it was found (and still is found) in the heart of those regions participating in the property of *sê-sê*, and is well authenticated historically and archæologically.   The word *sê-sê* is evidently not Chinese, but derived from a foreign language; it may be either a Chinese attempt at transcribing a Turkish or Persian designation of the stone, or the name of some locality, mountain or river.[1]

---

[1] Marco Polo's designation of the mountain where the balas ruby is mined, *Syghinan = Shignān*, is very suggestive as a possible foundation of the word *sê-sê* (Cantonese: *sok-sok*).   It is also significant that the ancient Chinese name for Shighnāṅ is recorded in the Annals of the T'ang Dynasty in the form *Sê-ni* or *Sê-k'i-ni*, and that this syllable *sê* is written with the same character as used in the jewel *sê-sê* (see CHAVANNES, Documents sur les Tou-kiue occidentaux, pp. 162, 322).   It is therefore possible, after all, that *sê-sê* derives its name from this locality and means "stone of Shighnān," — a case from a philological point of view analogous to the above mentioned *pijāzakī* and *badakshī*.   During the Mongol period the balas ruby is designated *la* in the *Cho keng lu* (Ch. 7, p. 5 b, edition of 1469), — a word which is traced by BRETSCHNEIDER (Mediæval Researches, Vol. I, p. 173) to Persian *lāl*.   The Chinese author T'ao Tsung-i — a fact not mentioned by BRETSCHNEIDER — states that this word is only dialectic.   The adoption of this foreign word indicates a change in the commercial conditions; in the T'ang period the balas rubies were traded to China from the country of the Western Turks (Khotan), in the Mongol period from Persia.   An error of Bretschneider here deserves correction.   The *Cho keng lu* enumerates four red stones and expressly says that they come from the same mine; since the balas ruby is mined in Badakshān, the three others, *viz.*, *pi-che-ta*, *si-la-ni*, and *ku-mu-lan*, must be derived from the same locality, and it is impossible to conjecture with Bretschneider that *si-la-ni* probably means "from Ceylon" — a view untenable also for philological reasons.   The *pi-che-ta* corresponds to Arabic *bigādī*, "garnet" (WIEDEMANN, *Der Islam*, Vol. II, p. 352, and Zur Mineralogie im Islam, pp. 217, 236, Erlangen, 1912).   If Bretschneider adds that nowadays the

We noticed from the statements of the *Pei shi* and *Sui shu* that *sê-sê* are attributed to the Persians, and from the *T'ang shu* that these jewels were known in Syria.   In the epoch of the T'ang dynasty (618–906) the three great religions of Western Asia, Mazdeism, Nestorianism and Manicheism reached a high degree of expansion and spread over

name for ruby in China is *hung pao shi* ("red precious stone"), it should be understood that this is not the balas ruby, but the Burmese ruby (see also G. E. Gerini, Researches on Ptolemy's Geography, pp. 39, 741, London, 1909). Russian manuscripts of the seventeenth century mention expressly "Chinese *lāl*" (K. P. Patkanov, *l. c.*, p. 21) which goes to show that spinels really existed in China and were traded from there to Russia.   Julius Ruska (Das Steinbuch des Aristoteles, p. 32) doubts the correctness of the identification of the Persian word *lāl* with the spinel, and is inclined to regard it as tourmalin, as it is stated that the colors of the stone are red, yellow, violet and green (which, however, is no conclusive argument), that the same stone is often half red and half green, that it is found in a matrix of white stone, and smaller stones frequently lie around a bigger one.   I have no judgment on this matter, but wish to point out on this occasion that it is impossible to reach any certain results in this line from the Chinese field of research, before our colleagues in the Persian and Arabic quarters have satisfactorily settled their questions and furnished us with the material to build our conclusions.   From a purely philological point of view, however, it does not appear that Ruska's opinion can be upheld.   There can be no doubt that the word *pijāzakī* denotes the balas ruby, and that the Chinese word *pi-ya-se* phonetically corresponds to it.   The Imperial Dictionary in Four Languages renders this Chinese term by the Tibetan and Mongol words *nal* and Manchu *langca* (Laufer, Jade, p. 109, note 3); these two forms are nothing but variations of the Persian word *lāl*, and consequently Persian *lāl* must designate the balas ruby, so it does also in Osmanli and some other Turkish dialects which have adopted this word.   Likewise in Russian the spinel was known under the name *lāl* in the seventeenth century (K. P. Patkanov, *l. c.*, p. 20).   Dr. Ruska, to whom I submitted this observation, was good enough to write me that he does not mean to reject in principle the interpretation of *lāl* by spinel, but mainly wishes to point out the difficulties of the case arising from the confused descriptions of the stone.   He thinks the green color is so rare in this mineral that it is impossible that the same stone should be often half green and half red, a feature which, however, not seldom occurs in tourmalin (especially with neutral, colorless intermediary zone). He further states that he has no doubt of the correctness of the equation *pi-ya-se* = *pijāzakī* and *nal* = *lāl*, and obligingly refers to the fact that the word *pijāzakī* is contained also in Vullers' Lexicon Persico-Latinum with the additional form *piyāzī* which comes still nearer to the Chinese word.   According to Vullers, *pijāzek* is the name of a district where are the mines of the *lāl*.   As the word *pijāz* means "onion," it was wrongly translated into Arabic as *baṣalī* "onion-like *lāl*," and even *pijāzī* was perhaps understood in this sense.   Dr. Ruska refers to a study on Qazwīnī now in print, where he has commented on this subject.   Meanwhile E. Wiedemann (Zur Mineralogie im Islam, p. 216) has commented on Dr. Ruska's opinion in a translation of al-Akfānī's mineralogy, who identifies *al-balachsh* with Persian *lāl*.   Wiedemann, on the ground of the specific gravities given by al-Khāzinī, holds that in this case *lāl* should be translated by ruby-spinel, but admits that in other cases it could signify tourmalin.   With obliging courtesy Dr. Ruska has recently sent me a proof of the note previously referred to (the above remarks were jotted down a year ago) where he says that the identification of the *lāl* with the *balachsh*, the stone of Badakshān, and with the *rubis balais* is confirmed by al-Akfānī; but he adds: "This does not exclude a freer usage of the word for all possible red gems, for who might say that the only then known means of their distinction, the determination of specific gravity, has always been employed?"   This result is entirely satisfactory, and we return to the former conclusion that *lāl*, in general, designates the balas ruby, with the restrictions made by Dr. Ruska.   In no case can it certainly be expected that any Oriental names of minerals, plants, or animals will exactly coincide with our scientific species; it is always necessary to grant the former a certain latitude.   But for purposes of translation we have to adhere to the one principal notion connected with the object.

Central Asia, also into China.   A curious document allows us to establish the fact that it was representatives of these religions, in all likelihood Manicheans, who brought these jewels to China.   This text occurs in the *Shu tien*, a collection of interesting notes on the province of Sze-ch'uan, compiled in 1818 by Chang Chu-pien (4 vols., reprinted in 1876). The passage (in Ch. 8, p. 5) [1] is derived from the *Hua yang ki*, a work containing records relative to Sze-ch'uan, whose time and authorship is not known to me. [2]   The text runs as follows: "The family K'ai ming [3] erected a several-storied building of seven precious objects; there were screens composed of connected genuine pearls.   At the time of Emperor Wu (B. C. 140–87) of the Han dynasty, a conflagration in the district of Shu (Sze-ch'uan) destroyed several thousand houses, and even several-storied houses were consumed by the flames.   At the present time people constantly find genuine pearls preserved in the sandy soil.—Chao Pien, [4] in his work *Shu tu ku shi* ("Ancient Affairs of the Capital of Sze-ch'uan"), says: "The Monoliths are outside of the west gate of the Yamen, two shafts being extant.   This is the site of the building of the genuine pearls.   Formerly people of Central Asia (*Hu jên*) erected at this spot a Temple of Ta Ts'in (*Ta Ts'in sze*) with gates and storeys consisting of ten rooms.   By means of genuine pearls and bluish jade (*ts'ui pi*) which were strung, they made screens.   Later on, this building was destroyed and fell into ruins.   Even now whenever a big rain has fallen, people pick up at this place genuine pearls, *sê-sê*, gold, blue jade, and strange things.   The poet Tu Fu (712–770), in his 'Poem on the Monoliths' has the verse: 'During a rainfall, they constantly obtain *sê-sê*,' which is an allusion to this affair." [5]

[1] It is quoted with exactly the same readings also in *Ko chi king yüan*, Ch. 32, p. 7.   The same story is narrated in the *Nêng kai chai man lu* (edited in *Shou shan ko ts'ung shu*, Vols. 70, 71; Ch. 7, p. 22 b) by Wu Tsêng of the twelfth century (WYLIE, Notes on Chinese Literature, p. 160), which goes to prove that the story was known in the Sung period.   Also Wu Tsêng connects this tradition with the verse of Tu Fu (712–770) relative to the finds of *sê-sê* at a rainfall; this is the heading of his essay, and the story is given in explanation of the poem which in our text, as translated above, follows at the end.   If this interpretation is correct, the event of the destruction of the temple of Ta Ts'in must have happened contemporaneously with, or prior to the age of, Tu Fu.   Thus, the Ta Ts'in temple here in question may have been founded toward the end of the seventh or in the beginning of the eighth century.

[2] It is not identical with the *Hua yang kuo chi*, ancient records of Sze-ch'uan by Ch'ang K'ü of the Tsin dynasty.

[3] According to an information received by M. PELLIOT, *K'ai ming* is the name of the later Emperor Ts'ung; he is identical with the personage called Pie Ling in GILES (Biographical Dictionary, No. 2071).   His record is contained in *Hua yang kuo chi* (ch. 3, p. 2).

[4] An official of the Sung period (994–1070), celebrated for his integrity and benevolence, popularly known as "the Censor with the Iron Face," acted as governor of Sze-ch'uan (GILES, Biographical Dictionary, p. 73).   The *Nêng kai chai man lu* designates him by his posthumous name Chao Ts'ing-hien.

[5] Wu Tsêng, after giving the text of this tradition as above, winds up with a comment on the precious stones and pearls of the country of Ta Ts'in (the Roman

Ta Ts'in is the peculiar name hitherto unexplained in its origin by which the Roman Orient was known to the Chinese, a subject ably and thoroughly discussed by Prof. HIRTH in his book "China and the Roman Orient" on the basis of all available documents. The "Ta Ts'in temples or churches" are first mentioned in Chinese records for the year 631 when the magus Ho-lu arrived from Persia at Si-ngan fu, and an imperial edict ordered to establish in the capital a temple of Ta Ts'in.[1]

In the year 745 an edict was issued that the Manichean churches heretofore called "Persian temples" should throughout the empire change this designation into "temples of Ta Ts'in."[2]   The tenor of this edict leaves no doubt that the Manicheans are understood: for the purpose of the imperial order is to do justice to the true name of their religion; their places of worship had heretofore been called "Persian" for the mere reason that they had hailed from Persia, but the foundation of their religion was Christian and had originated in Ta Ts'in, in Syria. I am therefore inclined to think that also in the above text the "Temple of Ta Ts'in" should be identified with a Manichean church. If this document can be looked upon as authentic we here have the interesting fact, which I believe was not known before, that the Manicheans, probably in the first part of the eighth century, had extended their settlements to far-off Sze-ch'uan, and the point at issue in this connection is that they must have brought over to Sze-ch'uan a large quantity of precious stones, among these sê-sê or balas ruby. In this case it should rather be positively asserted that the turquois is out of the question. The mere idea that the Manicheans should have employed turquois, and especially in combination with genuine pearls and precious jades, for the decoration of their churches, seems absurd. Precious stones played a significant part in the religious system and symbolism of the Manicheans, and as their religious notions centered around the

Orient) and concludes that the founders of the temple have therefore been indeed men from the country of Ta Ts'in.   It will be seen from the *P'ei wên yün fu* (Ch. 93 B, p. 85) that under the word *sê-sê* a passage from the History of the Liao Dynasty (*Liao shi*, chapter on Rites) is quoted to the effect that "Jo-han selected an auspicious day to practice the ceremony of *sê-sê*, in order to pray for rain." This word *sê-sê* has nothing to do with the jewel in question, but is a word of the Tungusic language of the Khitan, *seseli* (explained as such in *Liao shi*, Ch. 116, p. 4 b), meaning a ceremony of rain-prayer, in which no stone at all is employed but a willow at which a lance is thrown by the emperor, the princes and ministers.   The ceremony is fully described in the second chapter of the *Liao shi* (see H. C. v. D. GABELENTZ, Geschichte der grossen Liao, p. 31; further *Liao shi*, Chs. 27, p. 3; 55, p. 1 b; 56, p. 1 b).   The K'ien-lung scholars identify the Khitan word with Manchu *sekseri* (*K'in ting Liao shi yü kiai*, Ch. 10, p. 1).

[1] CHAVANNES, Le Nestorianisme (*Journal asiatique*, 1897, p. 61).

[2] This edict has been translated by CHAVANNES (*l. c.*, p. 66) and PAUL PELLIOT (*Bulletin de l'Ecole française d'Extrême-Orient*, Vol. III, 1903, p. 670).

dualistic principle of light and darkness, it is evident that brilliant jewels were conceived by them as emblems of light and for this reason were employed in their churches.[1]

Several examples of this kind may be gleaned from the newly discovered Manichean treatise brilliantly translated by M. Chavannes and M. Pelliot in collaboration.[2] Compassion is likened there to "the precious pearl called the bright moon which is the first among all jewels" (p. 67). The Messenger of Light is compared with "the perfumed mountain, vast and grand, of all jewels," and with "the precious diamond pillar supporting the multitude of the beings" (p. 90). "Our heart has received the majestic splendor of the pearl granting every wish," it is said at the end of this treatise (p. 92). Such like thoughts may explain the utilization of pearls in the Manichean church of Szech'uan. We know also that the adherents of Mani were fond of flowers, perfumes and ornaments, and in the same book (p. 61), there is even a legend in regard to the origin of jewels which seems to be connected with their beliefs of resurrection. The dying Manichean was adorned with rich ornaments (apparently symbolic of light) to be prepared for admission into the luminous regions. The gods approach the dead with ornaments which have the effect of putting the present devils to flight.[3] It does not seem to be known what symbolism the Manicheans attached to the balas ruby.[4] But as the lost literature of

---

[1] The basis of this symbolism certainly is to be traced to the writings of the New Testament, especially Revelation XXI, 18–21, and presents the counterpart to the mystic and moralizing ideas associated by mediæval Christian writers with the twelve precious stones in the breastplate of the Jewish High Priest, the twelve jewels forming the foundations of the wall of Heavenly Jerusalem just referred to, and the jewels in the crown of the Virgin. Among the latter, the balas ruby appears in the fifteenth century in the story of the visions of Sainte Françoise where four other stones not figuring among the twelve of the Bible are listed,— the diamond, garnet, carnelian and turquois (L. Pannier, Les lapidaires français du moyen âge, p. 225, Paris, 1882; see ibid., pp. 280–2, on the mediæval beliefs regarding balas ruby). Dante (Paradiso IX, 67) extols the lustre of the balascio: L'altra letizia, che m'era già nota Preclara cosa, mi si fece in vista Qual fin balascio in che lo sol percota. The ruby, in general, was emblematic of glory, and with predilection, chosen for the rings of the bishops (H. Clifford Smith, Jewellery, p. 148, New York, 1908, and D. Rock, Church of our Fathers, Vol. II, p. 171, London, 1849, where a gold pontifical ring with a sapphire surrounded by four balas rubies is mentioned).

[2] Un traité manichéen retrouvé en Chine. Extrait du Journal asiatique, Paris, 1912.

[3] G. Flügel, Mani, seine Lehre und seine Schriften, pp. 268, 339 (Leipzig, 1862).

[4] Some of the symbolism associated with the spinel in western Asia may be gleaned from the Armenian lapidarium translated into Russian by K. P. Patkanov (p. 19): "The spinel shares with the ruby in the quality that it quenches thirst, as soon as it is placed in the mouth. When pounded and mixed with a medicinal extract, it gladdens man and removes from him grief and sorrow. Mixed with an unguent and administered to the eyes, it strengthens their vision and renders man farsighted. Its nature is warm and dry. The sages say that the wearing of a spinel protects one from all diseases, from pain in the loins; it safeguards man from bad dreams and devils. The wearer of a spinel becomes agreeable to people."

this religion is gradually being rediscovered thanks to the scholarship of F. W. K. Müller, Le Coq, Pelliot and Chavannes, there is hope that the future will reveal this fact, and that also the puzzling word *sê-sê* will occur in the writings of the Manicheans.[1]

I have no definite opinion as to the indigenous *sê-sê* mentioned in the Sung period and in the fanciful stories of the T'ang dynasty. It is evident that neither the spinel nor the turquois is here involved, but that it is the question of some Chinese stone of fine appearance which is beyond the possibility of positive identification, as the accounts are too vague and elastic. It is manifest that *sê-sê* was a favorite word in the age of the T'ang, perhaps owing to its pleasing rhythm, that the far-off countries where the jewel was first discovered lent it a nimbus of romance, and that the name could easily be transferred to other similar stones. If I am allowed to express a personal opinion, I may say that this kind of *sê-sê* possibly refers to onyx. We see from Pseudo-Aristotle's *Lapidarium*[2] that China was known to the Arabs as a place of production for onyx, and it might even be conjectured that the Arabic word for onyx *djaza* (Persian *djiza*) which has penetrated into Sanskrit in the form *çesha*[3] and into Tibetan in the form *zé*[4] may have been instrumental in the shaping of the Chinese word *sê-sê* of this meaning. The existence of onyx in ancient China has not been heretofore recognized, because the indigenous word for it, on traditional convention, was accepted to have the meaning of jade, nobody knowing what kind of jade was understood. This is the compound *pi yü* (Giles's Dictionary, No. 9009) usually translated "greenish or bluish jade." A. FORKE[5] was the first to express his doubts of the correctness of this translation, and to point out that there are Chinese authors who distinguish *pi yü* from jade. Now we find in the English and Chinese

---

[1] We could perhaps even go so far as to connect the importation of rubies into China with the Manicheans. According to the Arabic author Qazwīnī (1203–83) there were several kinds of precious stones like rubies and others, and plenty of gold in Sandābil, identical with Kan-chou, the capital of the kingdom of the Tangutans (Chinese *Si-hia*, 1004–1226), and according to him, Manicheans lived there at the same time and enjoyed there perfect liberty (compare J. MARQUART, Osteuropäische und ostasiatische Streifzüge, pp. 87–88, Leipzig, 1903).

[2] JULIUS RUSKA, Das Steinbuch des Aristoteles, p. 145, Heidelberg, 1912. Also Ibn al-Baiṭār, 1197–1248 (L. LECLERC, Traité des simples, Vol. I, p. 354, Paris, 1877), makes the statement that onyx is found in Yemen and in China. Regarding onyx in Persia compare G. P. MERRILL, The Onyx Marbles, pp. 577–9 (*Report of U. S. National Museum*, 1893). See further E. WIEDEMANN, Zur Mineralogie im Islam, pp. 245–9 (Erlangen, 1912).

[3] L. FINOT, Les lapidaires indiens, p. XVII (Paris, 1896).

[4] Mr. ROCKHILL (The Ethnology of Tibet, p. 692) tells us that he has seen in certain portions of Tibet (Miri, near Shobando, for instance) the men wearing necklaces of coral beads and a substance which he believes is onyx, and which is called by them *zé*.

[5] *Mitteilungen des Seminars für Orientalische Sprachen*, Vol. VII, 1904, p. 147.

Standard Dictionary published in 1908 by the Commercial Press of Shanghai (Vol. II, p. 1561) the word onyx translated by this very term *pi yü* [1] (also by *tai wên ma-nao*, "streaky agate"), and I have no doubt that also in ancient texts the word *pi yü* designates the onyx. Thus, for example, in the account of Ta Ts'in given in the *Wei lio* [2] where the five-colored (that is, variegated) *pi*, in my opinion, is onyx; likewise the pillars in the country of Fu-lin, as reported in the *Kiu T'ang shu* (see above p. 27) were of onyx or *sê-sê*. In the older account of the *Wei lio* compiled prior to the year 429 A. D., the Chinese designation is still retained, while in the epoch of the T'ang a preference was manifested for the West-Asiatic name which was then transferred also to the home product. In this manner, a plausible explanation may be found for the occurrence of *sê-sê* on Chinese soil, for the use of this word with reference to the Roman Orient, and particularly for the carvings described in the traditions of the T'ang period and in the archæological works of the Sung dynasty, which could have indeed been made of onyx, a stone material ranking next to jade in Chinese eyes. [3] The difficulty of research in this line is enhanced by our lack of knowledge of the mineralogy of China, so that we are still deprived of a solid scientific foundation for our studies.

[1] This is likewise the case in the German-Chinese Dictionary published by the Catholic Missionaries of South-Shantung, p. 613 (Yen-chou fu, 1906).

[2] Hirth, China and the Roman Orient, pp. 73, 113.

[3] There are two references pertaining to Herat and Samarkand in a text of the fifteenth century where the word *sê-sê*, without any doubt, signifies a building-stone. In 1415 Ch'ên Ch'êng returned to China from a journey through Central Asia which had taken him through seventeen different countries. He published the information gathered by him in a book entitled *Shi si yü ki*, "Record of an Embassy to the Western Regions" (compare *Ming shi*, Ch. 332, and Bretschneider, *China Review*, Vol. V, 1876, p. 314). The original seems to be lost, but extracts from it are quoted in the Imperial Geography of the Ming Dynasty (*Ta Ming i t'ung chi*, edition of 1461). Under the heading of Herat (*Ho-lie*), Ch'ên Ch'êng is cited as saying (Ch. 89, fol. 23 b): "They are fond of clean clothing which is white in color, and which is exchanged for dark in case of mourning. The windows and walls of the palace in which the ruler of this country lives are adorned with gold, silver, and *sê-sê*." With reference to Samarkand the same author reports (*ibid.*, fol. 22 b): "There are many workmen there skilful in all handicrafts and clever in erecting palaces, buildings, gates, and pillars, with carvings in open work, and with windows connected by *sê-sê*." Both turquois and balas ruby are out of the question in these two cases; it is a building-stone, and most probably the onyx, which is here referred to.—As our knowledge of ancient Chinese sculpture advances, we may hope to obtain several exact definitions for the ancient names of stones, as in many of the votive inscriptions engraved in the monuments the name of the stone is expressly stated (though most frequently only the designation "stone image" is employed). The term *yü shi* (*lit.* jade stone) mentioned by Pelliot (*T'oung Pao*, 1912, p. 435) is also well known to me as occurring on Buddhist statuary of the Wei and T'ang periods, and it had never been assumed by me that it has the meaning of jade; *yü shi*, as also Pelliot says, means a jade-like stone, probably only a highly prized or valuable stone. The term *yü Fu* (*lit.* jade Buddha) may have well been employed in the figurative sense of "precious Buddha." The word *sê-sê* I have not yet traced in the inscription of any sculpture, but it is possible that it will turn up some day or other.

The problem as to the precious stones utilized by the T'ang dynasty could very well be solved in the Imperial Treasury (Shōsōin) of Nara, Japan, if an experienced mineralogist might be admitted there to make a close investigation of the numerous· precious stones lavished on Chinese objects of that period. In the *Tōyei Shuko* published by the Imperial Household where these treasures are splendidly illustrated, but inadequately described, the importance of this subject is overlooked. We read, for example, on p. 5 of Vol. I of discs used in playing games, 35 of crystal, 35 of amber, 20 of yellow lapis lazuli, 20 of azure lapis lazuli, 15 of slightly green lapis lazuli, 15 of green lapis lazuli; in the description of swords, green lapis lazuli is repeatedly mentioned. Needless to say there is no yellow or green lapis lazuli, and that these definitions rest on guesswork, not on investigation. But the same remark holds good for most of our archæological collections. A competent examination of the intaglios discovered in Turkistan, especially Khotan, and of the engraved gems of the Sassanian period of Persia would likewise yield new results for this interesting branch of research.

At a future date, precious stones will occupy a prominent place also in Chinese archæology, and the practical utility of studies like the present one will then become manifest. Already now the fact is apparent that precious stones are found in Chinese graves, and there is a certain number of them (especially lapis lazuli, carnelian, agate, and others as yet undefined) in the collections of the Field Museum. But for lack of evidence this subject is difficult to treat at present. Turquois, as far as I can judge, and as far as I know from Chinese experts, has not yet been discovered in any Chinese grave. The day will not be far when also Chinese archæology will be based on the actual evidence of the finds, and — *qui vivra verra*.

In regard to Tibet a plausible interpretation may be offered, and the Chinese transcription *sê-sê* referring to a jewel greatly prized by the ancient Tibetans seems to be traceable to a Tibetan word. There is a Tibetan word *zé* (or *zé-ba*, *ba* being only a suffix), a different word from the one mentioned before, which in the Tibetan-Sanskrit dictionaries is translated by Sanskrit *açmagarbha*; the latter is, according to the *Rājanighaṇṭu* (ed. GARBE, p. 77), an epithet of the emerald, and Tibetan *zé* (= Chinese *sê-sê*) would accordingly designate the emerald.[1] This identification is quite in keeping with what Chinese authors report

[1] Compare the Chinese word *she-she*, "emerald," cited by PALLADIUS (see above p. 25, note 1). It is noteworthy that the Chinese accounts of *sê-sê*, in a measure, present a curious analogy with the notices of the emerald on the part of the ancients, in that the latter have mingled with the genuine emerald other statements which cannot relate to the latter, for example, fabulous reports on Egyptian emeralds four cubits long and three cubits wide, and on obelisks of emerald (H. BLÜMNER, *l. c.*, Vol. III, p. 239, and LESSING, Briefe antiquarischen Inhalts, XXV).

regarding the high value of *sê-sê* in Tibet. Also the Tibetan word *mar-gad*, derived from Sanskrit *marakata*, sufficiently shows that the Tibetans were acquainted with the emerald. Capt. A. GERARD, speaking of the people of Spiti in the extreme western part of Tibet, remarks that they have beads of coral and other precious stones which resemble rubies, *emeralds* and topazes.[1] The surface of a mausoleum in Yamdo Samding is studded over with large turquoises, coral beads, rubies, *emeralds* and pearls.[2] SAMUEL TURNER,[3] in "a list of the usual articles of commerce between Tibet and the surrounding countries," has registered emeralds exported from Bengal to Tibet.

ABEL-RÉMUSAT[4] has wrongly ascribed the meaning of emerald to the word *lü-sung shi*, meaning "turquois"; on the other hand he erroneously translates by "chrysolith, or perhaps turquois" the Chinese word *tsie-mu-lu*, corresponding to Manchu *niowarimbu wehe* (that is, greenish stone), which is the emerald. This is proved by the Imperial Dictionary in Four Languages, where this Chinese and Manchu term corresponds to Tibetan *mar-gad* (written also *ma-rgad*, Tāranātha 173, 19) and Mongol *markat*, both derived from Sanskrit *marakata* which itself is a loan word from Greek *zmaragdos* or *maragdos*.[5] The Chinese word

[1] See ROCKHILL, The Ethnology of Tibet, p. 694.

[2] S. CHANDRA DAS, Journey to Lhasa, p. 183 (London, 1904).

[3] An Account of an Embassy to the Court of the Teshoo Lama, in Tibet, p. 383 (London, 1800). On p. 261 he tells that he saw the rosaries owned by the deceased Pan ch'en rin-po-ch'e, made of pearls, emeralds, rubies, and sapphires; and on p. 336, he describes the necklace in the possession of a Lhasa lady of high rank, in which were employed balas rubies, lapis lazuli, amber, and coral in numerous wreaths, and in her hair she wore pearls, rubies, emeralds, and coral.

[4] Histoire de la ville de Khotan, p. 168 (Paris, 1820).

[5] A. WEBER, Die Griechen in Indien (*Sitzungsberichte der Berliner Akademie*, 1890, p. 912). The oldest reference given in the Petersburg Sanskrit Dictionary as to the occurrence of the word *marakata* is the *Rājanighaṇṭu*. It is found, however, in the Sanskrit romances, for example, in the *Vāsavadattā* (edition and translation of L. H. GRAY, p. 109, *Col. Un. Indo-Iranian Series*, Vol. VIII, New York, 1913) of the seventh century. Also the Tibetan derivate *mar-gad*, appearing as equivalent of *marakata* in the Sanskrit Buddhist dictionary Mahāvyutpatti (ed. of MINAYEV and MIRONOV, p. 77, St. Petersburg, 1911), allows us to infer that the Sanskrit word is older than the seventh century. It occurs likewise in Buddhabhaṭṭa (L. FINOT, Les lapidaires indiens, p. XLIV) who probably wrote before the sixth century A. D. For the first part of the sixth century we have the testimony of Cosmas Indicopleustes (J: W. McCRINDLE, Ancient India as described in Classical Literature, p. 164, Westminster, 1901) who states that the White Huns living farther north than India highly prize the emerald, and wear it when set in a crown, for the Ethiopians, who traffic with the Blemmyes in Ethiopia, carry this same stone into India, and with the price they obtain make purchases of the most beautiful articles. The tradition of the *Agastimata*, as pointed out by FINOT (p. XLIV), seems to allude likewise to Egypt as to the derivation of the emerald. The Egyptian emerald has been studied by O. SCHNEIDER and A. ARZRUNI (*Zeitschrift für Ethnologie*, Vol. XXIV, 1892, pp. 41–100). The Greeks seem to have obtained their emeralds from Egypt; the Greek word is connected with Semitic *baraqt* or *bāreqet* (DAREMBERG and SAGLIO, Vol. II, p. 1467, and O. SCHRADER, Reallexikon, p. 153), but possibly also with Egyptian *mafek-ma* or *mafek-en-mā* (R. LEPSIUS, Les métaux dans les inscriptions égyptiennes, p. 43, Paris, 1877). The reports of the Arabic geographers on the Egyptian emerald-mines are translated by WIEDEMANN (Zur Mineralogie im Islam, p. 239, Erlangen, 1912).

tsie-mu-lu seems to go back directly to the Persian word *zumurrud*,[1] and it seems quite plausible that the Chinese obtained emeralds in their considerable trade with Persia. Possibly the Chinese have made their first acquaintance with emeralds at the end of the Mongol period.[2]

Let us now revert to the history of the turquois in China. T'ao Tsung-i, the author of the interesting work *Cho keng lu* (first published in 1366) replete with valuable information concerning the Mongol period, has embodied in it a brief enumeration of the precious stones of the Mohammedans, which were traded to China in his time (Ch. 7, pp. 5 b–7 b, edition of 1469). The last group of these stones is designated *tien-tse* (Giles's Dictionary, No. 11,180, but not noted with this meaning).[3] Three kinds of these are distinguished: first, *Ni-she-pu-ti*, that is, stones from Nīshāpūr in Persia, called the Mohammedan *tien-tse* whose veins are fine; secondly, *Ki-li-ma-ni*, that is, stones from Kermān in Persia, called *Ho-si tien-tse* (that is, *tien-tse* used in the country of

[1] Horn, Neupersische Schriftsprache, p. 6 (*Grundriss der iranischen Philologie* I, 2) and F. Justi, Kurdische Grammatik, p. XVI (St. Petersburg, 1880). In the Taoist novel *Fêng shên yen i* emeralds are mentioned as composing the umbrella of Virūpāksha. In W. Grube's posthumous work (Metamorphosen der Götter, p. 512, Leiden, 1912) the name *tsu-mu-lu* has not been recognized as a foreign word and is literally translated from the meaning of the Chinese characters "Grandmother green," while the editor H. Mueller in the index compiled by him (p. 651) explains it as "pearls"; also the variant *tsu-mu-pi* there employed means "emeralds." The same manner of writing the word (*tsu-mu-lu*) is employed also by Yang Shên (*Ko chi king yüan*, Ch. 33, p. 1) and by Ku Ying-t'ai in his *Po wu yao lan* (*ibid.*) written between 1621 and 1627. A curious error occurs in R. Pumpelly, Geological Researches, p. 118 (*Smithsonian Contributions to Knowledge*, Vol. XV, Washington, 1867) who in a discussion of the mineral production of Yün-nan remarks: "Emeralds are very rare, and although the Chinese name is lieupaoshī [*i. e. lü pao shi*] (green precious stone), they are known among lapidaries as Sz'mulu, the name of Sumatra, whence they are probably obtained." The Chinese word in question is not a designation of Sumatra which was known to the Chinese under the names Shi-li-fo-shi and San-fo-ts'i (see Hirth and Rockhill, Chau Ju-kua, p. 63), nor is emerald known to be found on Sumatra.

[2] Bretschneider, Mediæval Researches, Vol. I, p. 174. But at an earlier date they heard of emeralds in the translations of Buddhist Sanskrit works. In one of the series of the so-called Seven Jewels (*saptaratna*) the emerald appears in the second place following the diamond, and is transcribed in the form *mo-lo-kia-t'o* rendered into Chinese "green-colored bead" (*lü sê chu*) identical with *harinmani*, one of the Sanskrit synonyms of the emerald. Compare *Kiao ch'êng fa shu* (Ch. 7, p. 3, Hang-chou, 1878), a Buddhist dictionary of numerical categories written by Yüan Tsing in 1431 (Wylie, Notes, p. 211).

[3] In fact, none of our dictionaries contains the word *tien-tse* with the meaning of turquois, nor even does K'ang-hi's Dictionary. The origin and significance of the word is somewhat embarrassing, as it cannot be explained from any meaning assigned to the character *tien* (No. 11,180). In my opinion, a confusion of characters has been in operation in writing the word. It was intended for the character *tien* (No. 11,179) whose meaning is "to inlay objects with stones, inlaid or incrustated work." From this verbal noun, the new word *tien-tse* was derived with the sense "stones for inlaying," one of the main purposes for which turquoises are employed, and hence quite an appropriate designation. In the Ming period (*Ko ku yao lun* and *Pên ts'ao kang mu*, see below) we find a new mode of writing the word *tien-tse*, with the character *tien* (No. 11,199) meaning "indigo," which might have been prompted by an association of the color of the stone with that of indigo.

the Tangutans) whose veins are coarse; thirdly, stones of King-chou, called *tien-tse* of Siang-yang whose color is changeable. It is easy to see, and BRETSCHNEIDER [1] has already pointed it out, that the turquois of Nīshāpūr and Kermān is understood here,[2] and it hence follows that *Siang-yang tien-tse* must have the meaning "turquois of Siang-yang." Siang-yang is a city and prefecture of Hu-pei Province, and King-chou is the name of an ancient province comprising parts of the present provinces of Hu-nan and Hu-pei. The changing of color, indeed, fits the turquois, since its blue shades often fade to a pale green on long exposure to the light.[3] If this conclusion is correct, this would be the oldest Chinese reference to a turquois-producing locality in China proper in the latter part of the fourteenth century, and the first authentic use of a word for turquois in the Chinese language.[4] It is interesting that Tu Wan in his *lapidarium* of 1133 (Ch. A, p. 11; see p. 23) devotes a notice to stones of Siang-yang employed for building purposes, but has no allusion to turquois of this or any other locality. It is therefore obvious that, while quarries existed in that place during the Sung period, the turquois had then not yet made its appearance, and the fact is confirmed that the turquois mines were not operated before

---

[1] Mediæval Researches, Vol. I, p. 175.

[2] See above pp. 40-42.

[3] This peculiar property of the turquois is well known. Ibn al-Baiṭār says after al-Kindī who (according to WIEDEMANN, Zur Alchemie bei den Arabern, *Journal für praktische Chemie*, 1907, p. 73) died shortly after 870 A. D., that "the turquois changes its color on contact with an oily substance; also perspiration affects it [this is mentioned likewise by BOETIUS DE BOOT, *l. c.*, p. 269 and MAX BAUER, Edelsteinkunde, p. 488] and entirely deprives it of its color; contact with musk has a similar effect and destroys its value; Aristotle holds the opinion that a stone thus changing color has no value for its wearer" (L. LECLERC, *l. c.*, Vol. III, p. 51). In Pseudo-Aristotle (J. RUSKA, *l. c.*, p. 152) the purity of color in the stone is ascribed to the purity of the atmosphere which, when the latter becomes impure, causes the stone to become dim; when it comes in contact with molten gold, its beauty disappears. The latter clause is dubious. The sentence imputed to Aristotle is not traceable to him; neither Aristotle nor Theophrast make mention of the turquois. The alteration of color gave rise to the belief in the west that the stone foretold misfortune, or that the stone, when its owner sickens, will grow pale, and at his death lose color entirely. BEN JONSON (1574-1637), the dramatist, in his *Sejanus*, has the verse: "And true as Turkise in the deare lord's ring, Looke well or ill with him." FENTON (Secret Wonders of Nature, 1569) says: "The Turkeys doth move when there is any perill prepared to him that weareth it."

[4] The quarrying of turquois in Hui-ch'uan, Yün-nan Province, mentioned for the year 1290 in the *Yüan shi* (see p. 26, Note 3) is, of course, older in fact. The *Cho keng lu*, however, is our starting-point in unraveling the mystery, as it affords the means of determining in an unobjectionable manner the significance of the word *tien-tse*. With this authentic evidence in our hands we can hope to attack successfully the passages in the *Yüan shi* where this word is employed, while it is not there explained. Besides, we have the important testimony of MARCO POLO, as pointed out before, which enables us to establish with certainty the fact that turquois was known and mined in China during the Mongol period. Marco Polo was familiar with the turquois, as shown by his remarks on the turquois-quarries in the province of Kermān in Persia, so that his turquois in the province of Caindu cannot be called into doubt.

the Mongol period. It is noteworthy, as we shall see hereafter, that at present turquoises are still mined in Hu-pei. If we now recall the account of Marco Polo (p. 16), it seems we are justified in saying that the Chinese became acquainted with the turquois not earlier than in the Yüan or Mongol period, that is to say, the thirteenth and fourteenth centuries. This early mining in Hu-pei cannot have been of great importance, as it is not alluded to in later sources. We further observe that the Persian turquois became known to the Chinese in the fourteenth century, and was considered as superior to the domestic stone.[1]

This identification enables us to recognize the turquois also in the *Yüan shi*, the Chinese Annals of the Mongol Dynasty, where it is called *pi tien*, or *pi tien-tse*, "green or blue *tien*" (the word *tien* being identical with the above-mentioned word of the *Cho keng lu*). It entered the robe of the emperor and courtiers, and gold beads and turquoises are especially mentioned as used for earrings (*Yüan shi*, Ch. 78, p. 13 b). The Mongols were doubtless acquainted with the turquois long before their occupation of China, either through the Tibetans or Turkish tribes or through both. We know that among the antiquities of the bronze age of Siberia gold plaques incrusted with turquois and emeralds have been found [2] and, aside from Egypt, this ancient Siberian technique possibly represents the oldest employment of the turquois in the world. To the Chinese it was an alien substance which never became a national factor in their jewelry. They made its acquaintance through Turks, Persians, Tibetans, and Mongols, and I am under the impression that the Mongol rulers were the first to introduce it into China, and that their utilization of the stone gave impetus to the discovery of turquois mines on Chinese soil, and led to the turquois monopoly related by Marco Polo which has been mentioned above (p. 16) in the notes on Tibet. Also the reports of turquoises sent from the circuit of Hui-ch'uan in Yün-nan Province in the years 1284 and 1290 related in the *Yüan shi* (compare above p. 26, note 3) may be set in causal connection with the craving of the Mongol sovereigns for this stone.

At the rise of the Yüan Dynasty, the rule was established that products like gold, silver, pearls, jade, copper, iron, mercury, cinnabar, and

---

[1] At this point we should stop to reflect again whether, after all, the mining of turquois in Persia on an extensive scale is not earlier than the Mohammedan period. Compare above p. 40.

[2] KONDAKOFF, TOLSTOI and REINACH, Antiquités de la Russie méridionale, pp. 404, 405; S. REINACH, La représentation du galop dans l'art ancien et moderne, p. 66 (Paris, 1901). As far as I know, these turquoises have never been examined by a competent mineralogist, nor have they been traced to their place of origin. BAUER has nothing to say concerning the occurrence of turquois in Siberia. The Armenian *lapidarium* of the seventeenth century (translated by K. P. PATKANOV, *l. c.*, p. 48) gives Siberia as the fourth source for the turquois, and adds that this kind does not command any price.

turquoises offered to the throne by the inhabitants of the regions where the said products were found should be charged to the annual taxes due to them.[1]   The localities for the production of turquoises, on this occasion, are given as Ho-lin and Hui-ch'uan. The former place is mentioned again under the Mongols in the year 1271 to the effect that a certain Umala collected turquoises (*pi tien-tse*) at Ho-lin.[2]

Ho-lin is the Chinese designation for Karakorum,[3] the famous residence of the first successors to Chinggis Khan, Ogotai, Kuyuk and Mangu, and a large industrial and commercial centre in the Mongol period. It is not known whether turquoises occurred or are now found in the environments of that ancient capital, that is, in the basin of the Orkhon river, and it may very well be that Karakorum, during the thirteenth century, was merely a staple-place for them, whence they were traded to the Mongol and Tungusian tribes. In *Yüan-shi* (Ch. 94, § 2, p. 1) turquoises are enumerated among the natural products of the empire together with gold, silver, pearls, jade, copper, iron, mercury, vermilion, lead, tin, alum, saltpetre, and carbonate of natron.

It is thus evident that in the Mongol period at least three turquois-mines were in operation, in Hu-pei, Yün-nan, and Sze-ch'uan (Marco Polo's Caindu).

Also from Tibet turquoises were imported into China during that period. This may be inferred from a remark of Ts'ao Chao who published in 1387 the *Ko ku yao lun*, a collection of notes on art and antiquities. This was in the beginning of the Ming dynasty which rose in 1368, so that the author must have lived through the last years of the Yüan. He avails himself for the turquois of the peculiar Yüan expression *pi tien-tse*, but using a different character to write *tien*, identical with the word for 'indigo' (No. 11,199), so that the name would mean 'blue indigo sons.' He gives as localities for these stones the regions of the southern and western Tibetans (*Nan-fan*, *Si-fan*[4]) and describes them as of blue and green color, adding that good ones come somewhat near to the price of a horse, a statement evidently copied from the *Wu Tai shi* in regard to the *sê-sê* of the Tibetan women, referred to above.[5] He further says that they are of the class of beads, and that

[1] *Kin-t'ing se wên hien t'ung k'ao*, Ch. 23, p. 3.

[2] *Ibid.*, p. 6 b. Hui-chou, as written in this work, is a mistake for Hui-ch'uan, as proved by the parallel passage in *Yüan shi*, Ch. 94, §2, pp. 1 b, 2 a.

[3] Bretschneider, Mediæval Researches, Vol. I, p. 122; Vol. II, p. 162.

[4] From our standpoint eastern Tibetans; they border on the west of China and live partially on territory belonging to the political administration of China.

[5] For the rest, this is merely phraseology which cannot be taken very seriously. In China as elsewhere stock-phrases are formed in the way of literary allusions and *bons mots* for stylistic purposes, and it would be preposterous to see in these a foundation of real fact. Thus the phrase, "bead or pearl of the value of a horse," is one of the

there are also those of black and green hues which are low in price. In another work of the Ming period, the *Po wu yao lan* (published between 1621 and 1627), the precious stones of Tibet are enumerated, the series being closed by a "blue precious stone, light-blue in color like the hue of the sky." As the word *pao shi*, 'precious stone' is used here, I am not certain whether the turquois is meant.

In Chinese pottery occurs a deep-blue glaze well known to collectors under the name of turquois glaze. It has sometimes been supposed that this glaze was intended to imitate the color of turquois, as is, *e. g.*, stated in the "Catalogue of the Morgan Collection of Chinese Porcelains." [1] This view, however, is erroneous; "turquois glaze" is merely a designation of foreign origin, whereas, in the minds of the Chinese, the glaze has no relation to the turquois. This glaze is produced by means of a silicate of copper known to the Chinese as *fei ts'ui* from its resemblance to the color of the plumes of the kingfisher, or as *k'ung-tsio lü*, "peacock green." [2] This glaze appears for the first time in the pottery of the Sung period (960–1279) [3] and was in full swing during the time of the Ming dynasty, being applied to porcelain as successfully as to faience. During these two periods, the turquois was hardly known to the Chinese, or played no rôle in their life. [4]

The modern word *lü sung shi*, as far as I can see, does not occur earlier than the eighteenth century, [5] and it may be presumed also that

school reminiscences reiterated by several authors. As early as in the *Lapidarium* of Tu Wan (*Yün lin shi p'u*, published in 1133) we find it stated (Ch. 2, p. 7) that on the waste land of the Temple of the White Horse (*Pai ma sze*) east of Ho-nan fu, after a heavy rain, fine stones of a deep purple and green color are found in the ground which belong "to the class of beads having the price of a Tibetan horse; others of these beads are light-green with many veins and speckles, and some are made into carvings of objects and images; deep-green ones are high in price." "These stones," concludes the author, "are produced in foreign countries, and are found also in the soil near the ancient capitals of Si-ngan fu and Lo-yang (Ho-nan)." If it were permissible to regard these stones as imported turquoises, additional negative evidence would be furnished that turquois was not yet mined in China during the Sung period. Also Li Shi-chên (in his *Pên ts'ao kang mu*, Ch. 8, p. 17 b) uses the term 'bead of the value of a horse' (*ma kia chu*) as a designation for kingfisher-blue stones.

[1] Vol. II, p. 78 (New York, 1911).

[2] Compare S. W. BUSHELL, Oriental Ceramic Art, pp. 265, 315, 376 (New York, 1899).

[3] LAUFER, Chinese Pottery, p. 316.

[4] There is a popular tradition in Tibet in regard to blue-glazed Chinese faience tiles with which some temples are roofed that the first king Srong-btsan sgam-po of the seventh century had produced the glaze by melting an immense quantity of turquois for the purpose (S. CHANDRA DAS, Narrative of a Journey round Lake Yamdo, p. 49, Calcutta, 1887).

[5] It is certainly possible, as in the case of the word *pi-ya-se*, that also the word *lü sung shi* belonging to the colloquial language may be of earlier date than we at present suspect; but as the older sources regarding the every day language are very scarce, we can not yet offer any positive evidence.

the exploitation of turquois mines in China was taken up again only at that time, while it was interrupted during the Ming period. In the records of the Ming and Ts'ing dynasties, there is no reference to quarrying turquois. The great work on natural history of the sixteenth century, the *Pên ts'ao kang mu*, has nothing to say regarding this matter. In the K'ien-lung period (1736–95) turquois was occasionally used in the imperial manufacture at Peking, as we may ascertain from several specimens in the Bishop collection in the Metropolitan Museum of New York. It contains, for instance, a scabbard of chiseled repoussé gold decorated with the eight Buddhist emblems (*pa pao*) carved in turquois, apparently intended as a gift for some Mongol prince, and an imperial knife marked with *K'ien-lung's* seal, the handle being studded with lapis lazuli, carnelian and turquois.[1]

It appears that the Manchu emperors with their predilection for Lamaism and their interests in the Mongols and Tibetans derived the application of turquois from these peoples, and followed in this respect the trail of the Mongol emperors. Among the Chinese these stones never became popular.[2] They were occasionally employed for inlaying, but then in connection with other stones to produce certain color effects. Bushell[3] figures a box of carved red lacquer, decorated with floral designs, the fruit, flowers and other details inlaid in green and yellow jade, lapis lazuli, turquois and amethystine quartz. In the Chinese collection in the Field Museum there is a pair of jade trees in pots of cloisonné enamel, the leaves of which are beautifully carved out of turquois.[4]

It was for the first time also in the K'ien-lung period that the stone was officially adopted and its use sanctioned for the imperial cult.

Turquoises enter the imperial robe on some occasions, as recorded

[1] See Bishop, Investigations and Studies in Jade, Vol. II, p. 244.

[2] This lack of popularity is best evidenced by the fact that the turquois does not appear in the Chinese materia medica as it does in India and Tibet, nor are there any superstitious beliefs regarding it. This is remarkable considering among the Chinese the widest utilization for medicinal purposes of all substances occurring in the three kingdoms of nature.

[3] Chinese Art, Vol. I, p. 133.

[4] Figured in Jade, Plates LXVI and LXVII. The model of the Chinese jade trees presumably is to be looked for in the Bodhi trees of India made from precious stones and metals. The Great Chronicle of Ceylon (W. Geiger, The Mahāvaṃsa, p. 203, London, 1912) from about the sixth century A. D. has this report: "In the midst of the relic-chamber the king placed a Bodhi tree made of jewels, splendid in every way. It had a stem eighteen cubits high and five branches; the root, made of coral, rested on sapphire. The stem made of perfectly pure silver was adorned with leaves made of gems, had withered leaves and fruits of gold and young shoots made of coral. The eight auspicious figures [these are, lion, bull, elephant, water-pitcher, fan, standard, conch-shell, lamp] were on the stem and festoons of flowers and beautiful rows of four-footed beasts and rows of geese." Then follows a description of the canopy consisting of pearls and precious stones.

in the "Institutes of the Manchu Dynasty" (*Ta Ts'ing hui tien t'u*, Ch. 42). When the emperor officiates in the Temple of Heaven (*T'ien t'an*), he wears a rosary of lapis lazuli beads; in the Temple of Earth (*Ti t'an*), one of amber beads, yellow being the color of Earth; in the Temple of the Sun (*Ji t'an*), one of corals; and in the Temple of the Moon (*Yüe t'an*) one of turquoises (*lü sung shi*); while the girdle for the service in the latter temple is set with white jade.[1] The ordinary imperial court-girdle consists of yellow silk and is adorned with rubies or sapphires and turquoises. Also in the State Handbook of the Manchu Dynasty (*Huang ch'ao li k'i t'u shi*) turquoises are repeatedly mentioned as entering imperial helmets and sword-sheaths, also as employed for the jewelry of the empress and the court-ladies. They usually were combined with river pearls, corals, and lapis lazuli.[2]

[1] Color symbolism is an ancient and conspicuous feature of Chinese rites, and was originally associated with the four quarters and the cosmic deities who were linked with the latter (compare Jade, p. 120); at a later time it was affiliated also with the five elements and other categories of five (a comparative table of these associations is given by A. FORKE, Lun-hêng, Vol. II, p. 440). The Chinese system has already been compared with those found in North America and Mexico by Mrs. ZELIA NUTTALL, The Fundamental Principles of Old and New World Civilizations (*Arch. and Ethn. Papers, Peabody Museum*, Vol. II, Cambridge, Mass., 1901, pp. 286, 293), with the result that, "whilst the fundamental principle of the system was identical, the mode of carrying it out was different in China and America, a fact which indicates independence and isolation at the period when elements and colors, etc., were chosen and assigned to the directions in space." The whole problem, of course, is not historical but purely psychological.—In the imperial worship of the Manchu dynasty, as shown above, color symbolism was still fully alive. In the Temple of Heaven covered with blue-glazed faience tiles, everything was blue during the ceremonies, the sacrificial utensils being of blue porcelain, the participants in the rites being robed in blue brocades, and Venetian shades made of thin rods of blue glass were hung over the windows, in order to lend also to the atmosphere a tinge of blue. At the Temple of Earth, all was yellow; at the Temple of the Sun, red; and at the Temple of the Moon, everything was brilliant with a moonlight white.

[2] In the Sungari River, a light-green stone of unctuous appearance is found which is utilized for the making of ink-slabs. It is called in Chinese *sung hua yü*, lit., pine-tree flower jade, but it has nothing to do with turquois nor with jade. The Manchu name Sungari means in the Manchu language the Milky Way, and is popularly called in Chinese, with reference to the Manchu sounds, *Sung hua kiang*, "Pine-tree Flower River," while the designations of the Chinese written language are Hun-t'ung Kiang or Hei Shui ("Black Water"). The meaning of the stone *sung hua yü*, accordingly, is "precious stone of the Sungari River." Compare *Man-chou yüan liu k'ao*, Ch. 19, pp. 1–2 (a work on the History of the Manchu published in 1777). When the records of the Manchu dynasty mention turquois in combination with pearls, this is not a contradiction to what has been stated above regarding this point (p. 31). The pearls of the imperial house were cheap river pearls known as eastern pearls (*tung chu*), a product of Manchuria. They were fished in the Sungari and its side-rivers, and are described as brilliant-white nearly half an inch (Chinese) big, even the smallest of the size of the seed of a soy-bean. They were chiefly utilized on the crowns of the caps of royal princes, their number marking differences of rank (*Man-chou yüan liu k'ao*, Ch. 19, p. 1). The shell yielding this pearl has been identified with *Anodonta plicata Sol.* (compare GRUM-GRZHIMAILO, Description of the Amur Province, in Russian, p. 358, St. Petersburg, 1894, where some information regarding the pearl industry of the Amur region is given).—The word for lapis lazuli in the State Handbook is *ts'ing kin shi* (see p. 44). Beads made from this stone were chiefly employed for ornaments of the empress and court-ladies, likewise for the adornment of ceremonial head-dresses.

At the present time there are two distributing centers in China for the trade in turquoises,— Peking commanding the market of Mongolia, and Si-ngan fu controlling the trade with Tibet. In Si-ngan fu there may be a dozen traders engaged in the business. They are all settled in the same street and work up the raw material in their own shops. They produce beads and flat stones (Plate VI, Fig. 1) in any desired dimensions, by grinding and polishing, and drill perforations through them. The latter is an essential operation as Tibetans are averse to accept any others (except the small beads to be set in rings or the plaques for inlaying earrings and charm boxes). The first experiment that a Tibetan will make with a turquois offered is to ascertain the quality of the perforation by blowing or spitting through it, or by boring it with a needle. If the experiment is unsuccessful, he will return it at once. At Si-ngan the stones are sold by weight, prices ranging according to quality from 5–8 Taels (about $3.50 to $5.60) a catty (1⅓ pounds). Exceptionally beautiful stones or very small and carefully polished beads are sold as individual items only. Beads and stones are purchased there by Chinese commercial travelers trading with Tibetans and employed by them as a means of barter. Their example was duly adopted, and a great many specimens were secured by me in Tibet in exchange for turquoises.

Of worked articles of the Chinese the quadrangular, flat stones (Plate VI, Fig. 1) and the large beads for use in rosaries come first. Then there are fanciful carvings formed into the appearance of rocks (Plate VI, Figs. 2–4) or birds (Plate VII, Fig. 1) destined to adorn the table of a Lama and to serve as paper-weights; further, figures of animals like that of a tiger or a fish to be suspended as ornaments from a girdle (Plate VII, Figs. 3 and 4); snuff bottles (Fig. 2) skilfully hollowed out, and buttons (Fig. 5) with double edge cut into the petals of a flower to be sewed on to a cap, or a fillet worn by women. The image of carved turquois on Plate VII, Fig. 6, represents the Dhyānibuddha Amitābha, made in Peking. The twelve animals of the solar zodiac (Plate VIII) constituting a cycle of twelve years, each year being named for one animal, are each carved from turquois, of Peking workmanship; they represent rat, ox, tiger, hare, dragon, serpent, horse, sheep, monkey, cock, dog, pig or boar. Such sets are made for wealthy Mongols to facilitate the counting of years.[1]

[1] The carving of such sets is not a modern idea. It was practised as early as the T'ang period when marble was listed for this purpose. A complete set of the animals does not seem to have survived from that epoch, at least none has come to my notice; but a certain number of single animals belonging to different sets, obtained by me in Si-ngan fu, is in the collections of the Field Museum. A curious set carved from nephrite is in the Bishop collection in New York (see Bishop, Investigations and Studies in Jade, Vol. II, p. 241, No. 730); the representatives of the zodiac have hu-

The Hon. W. W. ROCKHILL,[1] who passed through Si-ngan in 1889, was given the information that turquois is found in Ho-nan. There is no reason to doubt the correctness of this statement for the mere reason that it was not confirmed to me in 1909; for even in China considerable changes are bound to come about within a period of twenty years, and I am inclined to think that it is quite possible that the mines of Ho-nan have since been exhausted.

man bodies but animal heads, they are clothed in costume of Chinese style and hold objects as attributes in their hands. This iconographic composition is also traceable to the T'ang period, as may be evidenced by a tombstone in the collections of the Field Museum; this contains an epitaph (*mu chi*) yielding the date 861 A. D. The twelve animals of the cycle are here arranged in four groups corresponding to the four cardinal points, a group of three facing one of the sides of the quadrangular stone slab, in the same manner as in the one published by M. CHAVANNES (*T'oung Pao*, 1909, p. 74) from a Chinese rubbing. While, however, the illustration of M. Chavannes shows figures of men, that is, Chinese officials in official costume, holding in their arms the respective animal, there are engraved on our tombstone figures of men with human bodies clad in official robes and holding jade insignia of rank in their hands, but each having the head of the particular animal. This is the same principle as in the set of the Bishop collection in which each piece is an independent all-round carving. In the rubbing of M. Chavannes the idea is brought out of the officials presiding over the twelve animals, whilst in the two other series the animals are themselves conceived as officials. The same ideas are expressed in the iconography of the gods of the Twenty-eight Lunar Mansions, which will shortly give me occasion for some remarks with reference to a group of masks in our collection representing this series of deities. It is known that the origin and diffusion of this solar zodiac based on a division into twelve parts of the celestial or ecliptic equator has given rise to many discussions and theories. I was formerly inclined (*T'oung Pao*, 1907, p. 400, and 1909, p. 71) to accept the theory of CHAVANNES (*ibid.*, 1906, pp. 51–122) according to which the cycle of the twelve animals would have originated among Turkish tribes who transmitted it to the Chinese. Having meanwhile studied the work of FRANZ BOLL, "Sphæra" (Leipzig, 1903), and the same author's recent paper, Der ostasiatische Tierzyklus im Hellenismus (*T'oung Pao*, 1912, pp. 699–718), I hold that his arguments in favor of an origin of the cycle within the sphere of Egyptian Hellenism are, in general, convincing to a certain extent, though much would remain to be done in detail to prove the migration of the system from this centre to the Turks and to China. There is, however, an objection to be made to the first piece of evidence offered by BOLL on behalf of the dependence of the Chinese cycle (p. 705): "In the Chinese list sacred Egyptian animals have survived, particularly the monkey which does not occur on the cold plateau of Central Asia." This is merely an old European fable which seems to be inexterminable, and which has already been refuted by me in *T'oung Pao*, 1901, p. 28; it would mean to shoot sparrows with cannon to march up here the whole evidence known to every zoölogist, to the effect that monkeys are propagated from the Himalaya through Tibet into the mountains of Yünnan, Sze-ch'uan and Kukunör region, and throughout central and southern China. Chinese, Tibetan, and all other Indo-Chinese languages possess ancient indigenous words for several species of monkeys, and at the time when the cycle was received by the Chinese, the monkey was very familiar to them and frequently represented in art. Another more serious objection to be advanced to the essay of Boll is that he has paid no attention to the arguments which induced L. DE SAUSSURE (*T'oung Pao*, 1910, pp. 583–648) and A. FORKE (Lun-hêng, Vol. II, pp. 479–494; compare also the additional remarks of P. PELLIOT, *Journal asiatique*, 1912, *Juillet-Août*, p. 163) to defend the indigenous origin of the cycle in China. The Chinese tradition entirely unheeded by Boll can not be so easily run down, and though he has stated the case clearly on its historical side, there remains to be solved the psychological part of the problem which has not yet been touched upon.

[1] The Land of the Lamas, p. 24 (London, 1891).

Bretschneider[1] quotes Pumpelly (Geological Researches in China, Mongolia, Japan, p. 118) as mentioning the existence of *sung ur shi*, a mineral similar to turquois, in the province of Yün-nan; but this statement requires confirmation, as it is not found in other sources relating to Yün-nan (compare above p. 26, note 3).[2]

From one of the turquois dealers in Si-ngan fu the information was given me that the turquoises traded there come from the prefecture of Yün-yang in Hu-pei Province, while another more especially pointed to the district of Chu-shan, situated in the same prefecture, as the place of production. The Imperial Geography (*Ta Ts'ing i t'ung chi*, Ch. 272),[3] in the chapter dealing with Yün-yang fu, contains no allusion to this fact, and mentions in an enumeration of the mountains of the Chu-shan district only one producing stones, the *Fan shi shan*, deriving its name from the *fan shi* or alum formerly produced there. It should be borne in mind that the Imperial Geography, as far as products are involved, does not reflect the present conditions of China based on actual research, but merely gives occasional quotations from older literature going back as far as the T'ang dynasty, so that this feature of the Geography is very incomplete and unsatisfactory. The products of Yün-yang fu, for instance, are all cited from the Geography of the Ming Dynasty (*Ta Ming i t'ung chi*). It is very probable that the turquois production of Yün-yang fu is of recent date, and presumably posterior to the publication of the Geography; it seems to me that the exhaustion of the turquois mines in Ho-nan may have given the impetus to a search for a new locality in Hu-pei. It would be gratifying if these lines would

---

[1] Mediæval Researches, Vol. I, p. 176.

[2] The work of R. Pumpelly is published in *Smithsonian Contributions to Knowledge*, Vol. XV, Article IV, Washington, 1867. I do not understand Pumpelly with Bretschneider as saying that "a mineral similar to turquois" is actually found in Yün-nan. Pumpelly enumerates it in a series of eight other stones of which he says that "they are carved, with great labor and patience, in very intricate forms." He does not point out any locality in Yün-nan, where turquois is obtained, but merely intends to say that he has seen in Yün-nan carvings made from this material which, judging from the Chinese name given by him, doubtless was turquois. But this turquois may have been imported into Yün-nan as well. G. Soulié (La province du Yün-nan, p. 24, Hanoi, 1908) only states: "The south-western part of the province furnishes a certain quantity of precious stones amassed in the beds of torrents or rivers; the west and south-west of the province are renowned for their amethysts, sapphires and rubies."

[3] First printed in 1745, second edition 1764. The modern Shanghai photolithographic reprint is a poor production. The Palace editions of this work are now exceedingly scarce. When at Si-ngan fu in 1902, an official there informed me that the late Empress Dowager, while living as an exile in that city in 1900, was anxious to obtain a copy for personal reading and wired to all Governors General making a requisition for it, but was unable to procure it. Eight years later fate treated me more kindly than the Empress by permitting me to see the *editio princeps* in the hands of a Peking bookseller, but lack of cash (the price demanded was 400 Mexican Dollars) unfortunately barred me from the privilege of acquiring it.

cause a mineralogist or geologist to pay a visit to those Chinese turquois mines, and to give us information on their extent, the working methods employed, and the magnitude of the output and trade in the material.

In Japan the turquois does not occur, and it has been unknown to the Japanese. The Japanese mineralogists, on becoming acquainted with it through our literature, coined the artificial word *turkodama*.

ADDITIONAL NOTES

pp. 1–4. The date of the introduction of the turquois into India may be somewhat more exactly defined by referring to the negative evidence presented by the great Sanskrit-Buddhist dictionary, the Mahāvyutpatti (Th. Zachariae, Die indischen Wörterbücher, p. 39) the Sanskrit text of which, accompanied by a Tibetan translation, is printed in the Tibetan Tanjur (Sūtra, Vol. 123). In Ch. 235 (ed. of Minayev and Mironov, p. 77, St. Petersburg, 1911) giving the names of precious stones, the word for turquois, *peroja*, is not included, quite in accordance with the fact that the turquois is not spoken of in Buddhist literature. We are therefore justified in concluding that at the time when Buddhism was introduced from India into Tibet, in the seventh and eighth centuries, the stone was not yet known in India, whereas at the same time it was widely known and appreciated in Tibet; thus, Tibetan knowledge of the turquois is not due to an impetus received from India. The Sanskrit-Tibetan equation, *peroja=gyu*, which we might expect does not exist in lexicographical literature. The earliest historical testimony for turquois in India, as shown above p. 3, remains that of al-Bērūnī in the post-Buddhistic or Mohammedan period, and even at his time the turquois cannot have been very generally diffused over India, as at that time it had not yet entered the horizon of the Indian mineralogists.

p. 1. The Persian word *ferozah* or *firozah* (*firūza*) for the turquois means "victorious," and is derived from the word *feroz* or *firoz*, "victory, victorious, successful" (see, for example, Johnson and Richardson's Persian-English Dictionary, ed. by Steingass, p. 944). Also the Arabic mineralogist al-Akfānī explains the Persian name of the turquois as signifying "victory"; hence, he says, it is called also "stone of victory" (Wiedemann, Zur Mineralogie im Islam, p. 225); likewise, al-Ta'ālibī (*ibid.*, p. 242) has an allusion to this effect. A similar notion seems to be underlying the first of five turquois varieties established by the Lama Klong-rdol (Chandra Das, Tibetan-English Dictionary, p. 1152), called *zil-gnon gyu spyang*, in which term the first element has the significance "overcoming, vanquishing."

p. 3, note 1. It should not be understood that Dioscorides had any knowledge of turquois; he does not mention it (in the same manner as his contemporary Pliny) nor does he have any name that could be interpreted as such. In Ch. 157 of his *Materia Medica* he speaks of the sapphire (*sappheiros*), *i.e.*, lapis lazuli, and says that those bitten by a scorpion will be relieved by taking this stone as a potion (compare F. de Mély, Les lapidaires grecs. Traduction, p. 24, Paris, 1902). It is only in the mediæval work of Ibn. al-Baiṭār, the Arabic version of Dioscorides, that the same notion is transferred to the turquois.

p. 14, note 1. The statement that Li Shi belongs to the T'ang period is based on the fact that in the editions of the *Sü po wu chi* he is assigned to the T'ang. This, however, seems to be a mere traditional opinion, while in fact the work is said to date from the Sung period (Pelliot, *Journal asiatique, Juillet-Août*, 1912, p. 155).

p. 21. M. Pelliot, who showed me the favor of looking over the galley-proofs of this paper, kindly calls my attention to another interesting text mentioning a fossil tree. This is the *Tu yang tsa pien* (Ch. C, p. 1, edition of *Pai hai*) written by Su Ngo in the latter part of the ninth century (WYLIE, Notes, p. 194).   Under the year 841 A. D. mention is made of a tribute sent to Emperor Wu-tsung (841–846) of the T'ang dynasty by the country of Fu-yü.   The latter were a tribe, presumably belonging to the Koreans, residing in Liao-tung and in the valley of the Sungari, and are first mentioned in the Annals of the Later Han Dynasty (Ch. 115, p. 2).   Their tribute consisted of two objects, three pecks of obsidian (*huo yü, lit.* fire jade, that is, stone of volcanic origin; compare the discussion on obsidian at the end of these notes) and a petrified fir-tree (*sung fêng shi, lit.* fir-tree wind stone) measuring ten (Chinese) feet all round and lustrous like jade.   Inside of the stone substance the outlines of a tree were visible.   As an old fir-tree bends from the action of the wind, so a cold blast came from the branches of that petrified tree.   In the midst of the summer, the emperor ordered the tree to be placed in the rooms of the palace; gradually there arose the sound of the whizzing of the autumn breeze; when the rooms were cooled off, he had the tree brought out again.

p. 23, note 2.   According to a communication of M. PELLIOT, the collected works of the poet Lu Kuei-mêng have been published under the title *Li* (No. 6957) *tsê* (No. 11,666) *ts'ung shu* (4 chapters and an appendix), of which there are several modern editions.   I find a biographical sketch of his embodied in the *Pei mêng so yen* (WYLIE, Notes, p. 194), Ch. 6, p. 10b (edition of *Pai hai*).

p. 25.   The first European author who treated of *sê-sê* was A. PFIZMAIER (Beiträge zur Geschichte der Edelsteine und des Goldes, *Sitzungsberichte der Wiener Akademie*, 1868, p. 210) in translating the two texts relative to the stone in the *Ming huang tsa lu*.   He did not explain it, though he was always ready to translate Chinese names, even those being transcriptions of foreign words which are not·capable of a literal interpretation.

p. 25, note.   M. PELLIOT thinks that the source for the definition of Palladius is K'ang-hi's Dictionary *sub voce sê* (No. 9600) where after the *Yün hui* of the thirteenth century *sê-sê* is defined as a *pi chu*, while the foundation for Couvreur's statement is the commentary to the *Shi king* (K'ang-hi, *sub voce sê*, No. 9599); this, however, refers only to the single word *sê*, not to the later compound *sê-sê* which, as pointed out on p. 47, is the Chinese transcription of a foreign word.   No conclusions, accordingly, can be built on the definitions of Palladius and Couvreur in regard to the nature of *sê-sê*.

p. 26, note 3.   M. PELLIOT remarks that the text of the *Nan-chao ye shi* is derived from the older work *Man shu* of the T'ang period where the passage relative to *sê-sê* occurs in Ch. 10, p. 48.   The *Man shu* is the work of Fan Ch'o and was published about 860; the history of the work is given by PELLIOT (*Bulletin de l'Ecole française d'Extrême-Orient*, Vol. IV, 1904, p. 132).

p. 33.   The word *sê-sê* occurs several times in the *Tu yang tsa pien*.   In Ch. A, p. 3 (edition of *Pai hai*), its author, Su Ngo, speaks of a peculiar kind of silk threads sent as tribute in 765 A. D. by the country Mi-lo in the Eastern Sea.   These threads, of great strength, were knitted into a kind of bag or sheath which on both sides was perfectly translucent like strung *sê-sê*.   It follows from this important passage that the *sê-sê* were bright and lustrous stones, and therefore cannot denote the turquois which is dense and non-transparent.   Further (Ch. A, p. 8), there is described a marvelous screen which originally belonged to Yang Kuo-chung, a cousin of Yang Kuei-fei (above, p. 33), who died in 756 (GILES, Biographical Dictionary, p. 909).   On this screen the figures of the beauties and hetairas of the times of antiquity were engraved, and it was framed with tortoise-shell and rhinoceros-horn.   A fringe was suspended from the lower edge and formed by genuine pearls and *sê-sê*,— the whole of such ingenious workmanship that one could hardly believe it was produced by a human hand.   A screen having the color of *sê-sê*, thirty feet wide and a hundred feet long, is mentioned (Ch. C, p. 9 b) as having been in the possession of Princess T'ung-ch'ang, curtains and screens made from gold, silver, and *sê-sê* (*ibid.*, p. 12 b); Buddhist pennants or streamers composed of coral, agate, genuine pearls and *sê-sê* (*ibid.*, p. 14 b) to be used in a procession when some sacred bones of Buddha were sent from the temple Fa-mên in Fêng-siang to the capital Si-ngan fu to be shown in the palace and in the monasteries of the city (compare *Kiu T'ang shu*, Ch. 15 A and De Groot, *Album Kern*, p. 135). These five passages relate to the age of the T'ang dynasty and show that *sê-sê*, as then employed in China, were precious stones of transparent quality, on a par with genuine pearls and precious metals; they further bear out the fact that *sê-sê* were jewels not bigger than a pearl, otherwise they could not have been strung together with pearls. All this renders the assumption of *sê-sê* being the turquois impossible and confirms my opinion that it was the balas ruby. We insisted above on the popularity of the word *sê-sê* in the T'ang period. This is fully corroborated by the interesting work *Tu yang tsa pien* where it enters into comparisons from which it becomes clear that the word was very familiar and generally understood at that time. In Ch. c, p. 5 are described three marvelous plants which, when eaten, guard man from old age. The first of these is called *shuang lin chi*, "the agaric with the double *lin* (female unicorn)," and is described as a plant with one stalk and two flowers so hidden away that they are scarcely visible, and shaped like a *lin* with head and tail, all complete; and they produce seeds like *sê-sê*.

p. 36, note 1. Turquois-mines in the district of Upper Nasiyā in Ferghana are mentioned by Ibn Haukal (978 a. d., ed. of De Goeje, *Bibl. Geogr. Arab.*, p. 397), as Mr. Guy le Strange, the excellent Persian scholar of Cambridge, England, has been good enough to write me. M. Pelliot refers, as regards turquois-mines of Khojend, to Pavet de Courteille, Baber nameh, Vol. I, p. 7. This work is not accessible to me, but I find in the new English translation of A. S. Beveridge (The Memoirs of Bāber, p. 8) the passage as follows: "To the north of both the town [Khojend] and the river lies a mountain range called Munūghul; people say there are turquois and other mines in it, and there are many snakes." There is, accordingly, no longer any reason to doubt the indigenous occurrence of turquois in the territory of Ferghana, and it will be correct to assume that it was mined there from the latter part of the tenth century.

p. 40. Mr. Guy le Strange has been good enough to refer me to the fact that the text of the passage of Ibn Haukal (Hauqal or Hawqal, as others spell it), the continuer or re-editor of Iṣṭakhri, is found in De Goeje's Bibl. Geogr. Arab., p. 313; the turquois mines, according to him, were near Nūqan, which is Tus to the north of modern Meshed. On p. 362 of the same work, celebrated turquois mines are mentioned in Transoxania near the mountains called Jabal-Buttam.

p. 43, note 2. In regard to the *uk-nu* stone mentioned in the Assyrian inscriptions Mr. Pinches has shown that it denotes lapis lazuli from the Zagros range (*Journal Royal Asiatic Society*, 1898, p. 259, note 1). I have no judgment on this point.

p. 44, note 2, and p. 62. It is assumed by several authors that lapis lazuli is found in China. A. Williamson, who has written an interesting article on the productions of northern China (*Journal China Branch R. As. Soc.*, Vol. IV, 1868, p. 41), asserts on hearsay reports that in Shan-si, and among the hills in the south of Shen-si, precious stones, such as lapis lazuli, ruby, etc., abound, and that he has every reason to believe the report correct. F. Porter Smith (Contributions towards the Materia Medica and Natural History of China, p. 129, Shanghai, 1871) who takes the word *liu-li* in the sense of lapis lazuli says that the blue mineral known by this name is met with in very fine specimens in China and Central Asia. In the "Catalogue spécial des objets exposés dans la section chinoise à l'exposition de Hanoi, 1902" (p. 121) mention is made of the lapis lazuli of the Island of Hainan. In the latter case it is more than probable that the determination is wrong and merely due to a confusion with cobalt which, as well known, is obtained on Hainan (Hirth, Chinesische Studien, p. 251). According to R. Pumpelly (Geological Researches, p. 117, *Smithsonian Contributions to Knowledge*, Vol. XV, Washington, 1867) lapis lazuli is found at Mount Nien in the district of Ch'ang-shan, prefecture of K'ü-chou, Chêkiang Province, and in the district of Lo-ts'ing, prefecture of Wên-chou, of the same province. These statements, however, as the entire list of minerals in which they are contained, are based on the *Ta Ts'ing i t'ung chi* and other Chinese sources examined by "the author's Chinese secretary" (p. 109). But I am at a loss to explain where the Chinese secretary found these statements.   There is nothing to this effect to be met in the *Ta Ts'ing i t'ung chi.*   Among the products of K'ü-chou fu (Ch. 233, p. 9) are mentioned, after the *Ta Ming i t'ung chi,* ink-slabs produced in the two districts of Ch'ang-shan and K'ai-hua, but no other kind of mineral; in the account of Wên-chou fu (Ch. 235, p. 10) no stone is registered.   Perhaps his statement is derived from his "other Chinese sources"; but even then we should like to know the Chinese word translated by him "lapis lazuli," and as he does not give it, his note is rather valueless.   As pointed out on p. 44, note 2, we meet in the *Ta Ts'ing i t'ung chi* the word *kin sing shi* (*lit.* gold star stone) in Ch. 398, p. 3 b, description of Sze-chou fu in Kuei-chou Province, where it is said that this stone occurs east of the city of Sze-chou in the Kia-ch'i Lake, and that, according to the Provincial Gazetteer, stars and spots appear on its surface, that it is hard and glossy and can be worked up into ink-slabs.   GILES (in his Dictionary, p. 252c) explains *kin sing shi* by "iron and copper pyrites," in agreement with F. PORTER SMITH (Contributions towards the Materia Medica and Natural History of China, p. 123, Shanghai, 1871).   I do not wish to push this discussion any further, as the second word cannot be spoken before the first has been said.   Specimens suspected of being lapis lazuli must be procured from the various localities where they are reported to occur, and examined by competent mineralogists.   Others like T. WADA (*Beiträge zur Mineralogie von Japan,* No. 1, p. 21) deny that lapis lazuli is found in China, and are of the opinion that it is imported from Central Asia.

p. 45.   Incidentally I wish to refer here to a now antiquated investigation of T. DE LACOUPERIE, On Yakut Precious Stones from Oman to North China, 400 B. C. (*Babylonian and Oriental Record,* Vol. VI, 1893, pp. 271–4), in which the Chinese *ye kuang chu,* "the bead or pearl shining at night," is set in relation with the Arabic word for the ruby *yakut* (an etymology impossible for philological and historical reasons; in fact, the Chinese term is not a transcription of any foreign word but a thoroughly Chinese formation) and identified with the ruby, "probably from Badakshān, the chief source of these stones at that time."

p. 46, note 1.   The date of Yang Shên, as given by MAYERS, is correct; it is given as the same by CHAVANNES (*T'oung Pao,* 1904, p. 474) who notes his biography after *Ming shi* (Ch. 192).   Yang Shên wrote the *Nan-chao ye shi* in 1550.

p. 49.   For the elucidation of this text I am greatly indebted to M. PELLIOT who will himself take it up in his proposed study of the history of the Nestorians in China. M. PELLIOT says that he has not found the text in the *Hua yang kuo chi,* but on the other hand has not succeeded in discovering a trace of an independent work *Hua yang ki.*   The quotation given from Chao Pien is not contained in the latter work, but is taken from another source.   According to the *Sung shi,* Chao Pien lived in fact from 1006–1084, as will be demonstrated by M. PELLIOT in the publication mentioned, but it is not certain whether he is the author of the *Shu tu ku shi.*   What is rendered above by monolith is in the text *shi sun* (Nos. 9964, 10438) by which M. PELLIOT is inclined to understand megalithic monuments.   These stone ruins are the remainders of an ancient tomb (compare *Ch'êng-tu hien chi,* 1873, Ch. 2, p. 3 b, and *Sze-ch'uan t'ung chi,* Ch. 48, p. 66 b), and have nothing to do with the Temple of Ta Ts'in.   In the last number of the *Journal asiatique* (*Mars-Avril,* 1913, p. 308), CHAVANNES and PELLIOT alluding to the text of *Neng kai chai man lu* incline toward the opinion that the temples of Ta Ts'in are due to the Nestorians.

p. 52.   Also Qazwīnī (1203–83) mentions the onyx of China (Sīn) with the curious addition that the people of Sīn repudiate the quarrying of the onyx mines, which is left only to slaves who cannot otherwise eke out a living and sell the stone in countries outside of Sīn (J. RUSKA, Das Steinbuch aus der Kosmographie des al-Qazwīnī, p. 12).

p. 53, note 3.   The statement that the *Shi si yü ki* seems to be lost is based on BRETSCHNEIDER'S authority; besides the quotation as given there, his Mediæval Researches (Vol. II, p. 268) ought to have been pointed out, where the same is repeated.   But this is contradictory to what Bretschneider says on p. 147 of the same volume that the book in question seems to be still extant and is noticed in the Imperial Catalogue (Ch. 64, p. 5). Also M. PELLIOT thinks that the work is extant, but there are no modern editions of it.

p. 55, note 5. As the emerald is not made mention of in the Bower Manuscript of about 450 A. D., it would be justifiable to conclude that, taking the positive testimonies into consideration, the emerald was introduced into India not earlier than the beginning of the sixth century A. D. The passage of Cosmas regarding the emerald will be found on p. 371 of MAC CRINDLE's translation (Christian Topography, ed. of *Hakluyt Society*, London, 1897).

p. 56, note 3. T. WATTERS (Essays on the Chinese Language, p. 352) believed he recognized the Persian word *fīrūza* in Chinese *pi-liu* (Nos. 9009 and 7245) or *pi-liu shi* (stone) to which he ascribes the meaning of turquois (observation of M. PELLIOT). But the source from which the Chinese word is derived is not given.

The discourse on *sê-sê* has furnished sufficient proof for the fact that the Chinese designation of a stone may refer to different species according to different localities, and that the significance of such a word may undergo changes in course of time. Moreover, we observe that the name of a stone used with reference to a foreign country does not necessarily denote the same species as the same name when applied to the domestic variety. An interesting case of a similar bearing is presented by the account on Japan in the Annals of the Later Han Dynasty (*Hou Han shu*, Ch. 115, p. 5 b) where white pearls (*pai chu*) and (what from a Chinese point of view would be a literal translation of the term) "green jade" (*Ts'ing yü*) are mentioned as products of Japan; indeed, the term has thus been translated, for instance, by E. H. PARKER (*China Review*, Vol. XVIII, p. 219 a). But it is evident that this translation cannot be correct, for we know surely enough that Japan does not produce jade (see Jade, pp. 351-4). It is therefore manifest that the word *ts'ing yü* in the above text relates not to any kind of jade but to a Japanese stone, and that the term must be taken from a Japanese, not a Chinese viewpoint, and it may be inferred also that it must designate a mineral peculiar to Japan and absent in China. The Chinese character *yü* is read in Japanese *tama*, and this Japanese word signifies any gem or precious stone in general, or even more commonly a bead or ball of any stone. The color name *ts'ing* (Japanese *aoi*, Sinico-Japanese *sei*) is of uncertain quality and refers to the general color prevalent in nature, green, blue, black, gray, usually meaning any dark neutral tint. Such a substance playing a large rôle in the antiquity of the Japanese and the Ainu is obsidian. It is unknown in China, but found in several localities of Japan (Bungo, Izu, Kai, Shinano, Tokachi: N. G. MUNRO, Prehistoric Japan, p. 292, Yokohama, 1908). It was largely utilized, as in ancient Mexico, for the manufacture of arrowheads, and abundant flakes scattered around in the sites mentioned testify to its popularity. As elsewhere, it was worked up also into beads and balls to enter into personal adornment.

P. F. v. SIEBOLD (Geogr. and Ethnogr. Elucidations to the Discoveries of M. G. Vries, p. 175, Amsterdam, 1859) reports on obsidian balls received from Yezo, "from two feet to two feet and a half in diameter, coal-black of color, and some small blue pieces of stone, of which probably the so-called *Krafto* (properly *Karafuto*) *tama*, or precious stone of *Krafto* is formed." It could appear from this statement that obsidian and the blue *Karafuto tama* are considered by Siebold as different stones; but, in another passage of the same book (p. 105), he comments on a blue bead chain noticed by Vries in the ears of an Ainu woman of Saghalin that "the most precious are the blue obsidian which they call *Krafto tama*, precious stone from *Krafto;* these blue corals [?] are found among all the peoples of the frigid zone, of the northern hemisphere, from the Great Ocean up to Behring's Straits, where they were found by von Kotzebue in the Sound which bears his name." A. J. C. GEERTS (Les produits de la nature japonaise et chinoise, p. 294, Yokohama, 1878) describes precious stones under the name *ruri-tama* (written with the Chinese characters *liu-li yü*) as of deep-blue color and entering into the necklaces of the ancient Japanese (the *shitogi* of the Ainu). He identifies them with lapis lazuli, and says that these very rare stones have been found on the Kurile Islands, several specimens of which are in the Museum of Tōkyō. Under the name Karafuto-tama, that is, jewels of the Island of Saghalin, the precious stone par excellence of Saghalin and the Kuriles is understood, made into the necklaces called *shitogi*. They are well polished lustrous balls of blue or bluish color, but less dark than the *ruri-tama*. They are the product, adds GEЁRTS, of dark blue obsidian varying much in size; they belong, as the preceding stone, to a period posterior to the *maga-tama* and still serve as ornaments to the natives of the Kuriles. It is more than likely that the *ruri-tama* and *Karafuto-tama* are identical, and that the material in question is obsidian. Obsidian, as well known, is not a mineral proper but a natural glass, a black vitreous volcanic rock being produced where a rapid cooling of certain liquid lavas has taken place and occurring in many parts of the world, the coloration being black, gray, brown, yellow, red, green, sometimes also blue. A peculiar variety is known to our mineralogists from the river Marekanka near Okhotsk in eastern Siberia, hence called *marekanite;* these obsidian balls are partly colored evenly, partly of brown and gray, frequently also of yellow and red hues (MAX BAUER, Edelsteinkunde, p. 551). O. C. FARRINGTON (Gems and Gem Minerals, p. 181) gives for it also the name "mountain mahogany," and says that it makes a pretty stone, which is used for the manufacture of some objects. This material is doubtless the source for the precious beads of Saghalin and the Kuriles. Nothing is known to Japanese or foreign mineralogists of lapis lazuli found on the Kuriles, and the definition of Geerts must be considered an error. In the Ainu collection of the Field Museum there is a necklace (Cat. No. 88037) coming from Hakodate on Yezo, in which are strung six large, black obsidian balls (about 3 cm in diameter), together with many blue, green and white glass beads. J. BATCHE-LOR (The Ainu and Their Folk-lore, p. 154, London, 1901) states that the glass beads of which the Ainu women are extremely fond are of Japanese make, others appear to have come from China; the people believe that the ancients got them from the *Rushikai,* that is, Russians and Manchu. In the Annals of the Later Han Dynasty (*Hou Han shu,* Ch. 115, pp. 5a, 5b) the countries of the Fu-yü (see above) and the Yi-lou who lived over 1000 *li* north-east of the Fu-yü are reported to produce "red jade" (*ch'i yü*). Also in this case, the word *yü* cannot be construed to have the literal meaning of "jade," as no jade is found in those localities which were inhabited by the Fu-yü and Yi-lou, and I am inclined to regard the term *ch'i yü* as having likewise the significance of obsidian. The evidence for this supposition is furnished by the *Tu yang tsa pien* (Ch. C, p. 1) in the passage above alluded to. In the tribute sent by the Fu-yü in 841 A. D. to Emperor Wu-tsung there were three pecks of "volcanic jade" (*huo yü san tou*), which was red (*ch'i*) in color. The pieces were half an inch long, pointed on top, and round below; they emitted their brilliancy at a distance of ten paces. Gathered in a cauldron they could be ignited, and the heat of such a cauldron placed in the house was sufficient to dispense with double quilted garments [which the Chinese use to wear in the winter, heating their bodies instead of their rooms]. The court-ladies of inferior rank availed themselves of this fire to heat a brand of wine called "clear wine" which had been sent as the gift of a foreign country. There can be no doubt of the identification of the term *huo yü* with obsidian; this expression literally means "fire jade," and "fire mountain" (*huo shan*) is the Chinese word for a volcano; *huo yü,* accordingly, is a fine stone of volcanic origin, and such a product of volcanic outflows is obsidian. The account of this substance being utilized as a combustible is quite credible, for obsidian "fuses rather easily before the blowpipe to a porous, gray mass" (O. C. FARRINGTON, *l.c.,* p. 180). This "fire jade" was red in color; accordingly, it was a *ch'i yü,* and this is the very designation which we encounter in the Annals of the Later Han Dynasty. For this reason we may conclude that the term *ch'i yü,* as pointed out in the above passage, serves for the designation of obsidian which itself was unknown in China.

Woman from southern Tibet, to illustrate the mode of wearing jewelry: gold earrings inlaid with a mosaic of turquois; gold amulet box (*gau*), the surface being filled with a network of designs formed of gold filigree and inlaid with seven choice turquoises of first quality; necklace composed of large turquois, amber and coral beads; and silver chatelaine with ornamental halberd, toothpick, ear spoon, tweezers and small brush (lost) for oiling the hair (see Jade, p. 203).

All ornaments worn by this woman were acquired for the Field Museum (exhibited in the Gem Room, with a large collection of other Tibetan and Nepalese jewelry).

TIBETAN WOMAN SHOWING MANNER OF WEARING JEWELRY.

Tibetan woman in festival dress of Chinese silk.   The chaplet is worn over an artificial wig of long flowing hair imported from China.   The turquoises are sewed on to a foundation of stiff red cloth, and bandeaux formed by rows of artificial pearls are laid around the sides.

TIBETAN WOMAN IN FESTIVAL DRESS

PLATE IV

Tibetan woman wearing chaplet set with turquoises and artificial pearls; turquois earrings; copper charm-box inlaid with turquoises suspended from a necklace; a quadrangular silver charm-box attached to the rosary; a silver chatelaine with five utensils; a silver belt with chain (called *digra*) falling down over the apron, caught up and fastened to the belt, and then again to the bodice, where it terminates in the figure of a rooster of silver. It is covered with plaques of gold filigree set with turquoises. She wears silver rings set with turquoises on the middle and fourth fingers of both hands, a white conch-shell as bracelet on her right arm, and a Chinese silver bracelet on her left arm.

TIBETAN WOMAN WITH COMPLETE JEWELRY.

PLATE V

Pair of Tibetan earrings, foundation of gold inlaid with a mosaic of turquois. The requirement of these mosaics is that the stones shall be well matched in color, resulting in an harmonious color arrangement. With this end in view, Tibetan women gather turquoises during many years, till they have the desired colors in the required number of stones. Such earrings belong to the most cherished property of a Tibetan woman and range from a hundred to six hundred rupees and more in price. From collection of Field Museum (exhibited in Gem Room).

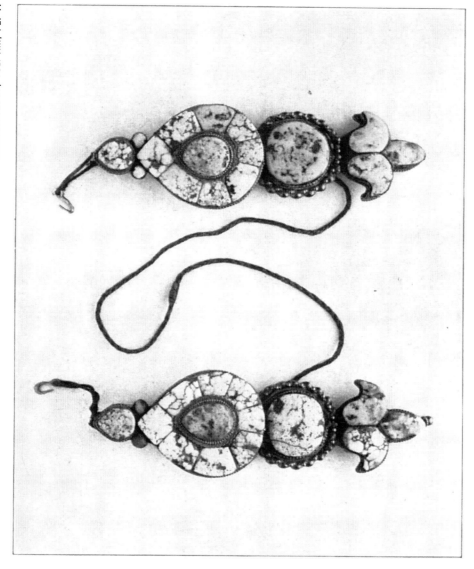

TIBETAN TURQUOIS EARRINGS.

Chinese Turquois Carvings.

Fig. 1. Flat, polished and perforated turquois of dark-blue color with black strata, mined in Hu-pei Province and worked in Si-ngan fu. Cat. No. 116679/3.

Figs. 2–4. Fanciful carvings of turquois, Si-ngan fu, made for Tibetan and Mongol Lamas who use them as decorations on their tables, and also as paperweights, 10, 8.5, and 12 cm high, respectively. Cat. Nos. 116663, 116664, 116666.

1
3
CHINESE TURQUOIS CARVINGS.
2
4

Chinese Turquois Carvings.

Fig. 1.   Bird carved from turquois, serving for decorative purposes, and also as paper-weight, Si-ngan fu.   12.4 cm long, 3.8 cm high.   Cat. No. 116665.

Fig. 2.   Snuff-bottle carved from turquois, Peking.   3.3 cm high.   Cat. No. 116670.

Fig. 3.   Turquois carving of recumbent tiger, used as girdle-pendant, Si-ngan fu.   5.3 x 3 cm; 2.2 cm high.   Cat. No. 116668.

Fig. 4.   Turquois carving of fish, both sides alike, used as girdle-pendant, Si-ngan fu.   5.4 x 3.3 cm.   Cat. No. 116667.

Fig. 5.   Turquois carving of ornamental button in shape of blossom with a double row of petals, worn in front of cap or fillet, Si-ngan fu.   3.5 x 3 cm.   Cat. No. 116669.

Fig. 6.   Image of the Dhyānibuddha Amitābha, Lamaist type, carved from turquois, Peking.   6.7 cm high.   Cat. No. 116673.

Chinese Turquois Carvings.

The Twelve Animals representing the periodical cycle of twelve years. Carved from turquois, Peking. Average dimensions of figures 5 x 3 cm, with a height of 2 cm. Cat. No. 116674.

The animals are:

1. Rat.	2. Ox.	3. Tiger.
4. Hare.	5. Dragon.	6. Serpent.
7. Horse.	8. Ram.	9. Monkey.
10. Rooster.	11. Dog.	12. Boar.

See p. 63.

CHINESE TURQUOIS CARVINGS.

The
Burlington Magazine

for Connoisseurs

Illustrated & Published Monthly

Number CXXXVI   Volume XXV

LONDON
THE BURLINGTON MAGAZINE, LIMITED
17 OLD BURLINGTON STREET, W.

NEW YORK: JAMES B. TOWNSEND, 15-17 EAST FORTIETH STREET
PARIS: BURLINGTON MAGAZINE, LTD., 10 RUE DE FLORENCE, VIII°
BRUSSELS: LEBÈGUE & CIE., 36 RUE NEUVE
AMSTERDAM: J. G. ROBBERS, SINGEL 151-153
LEIPZIG: FR. LUDWIG HERBIG (Wholesale Agent), 20 INSELSTRASSE
KARL W. HIERSEMANN, 29 KONIGSSTRASSE
FLORENCE: B. SEEBER, 20 VIA TORNABUONI
BASLE: B. WEPF & CO.

July 1914

are still people living who consider that architecture died with the Gothic period ; others would have it perish with Michelangelo and Palladio ; but by far the strongest opinion of the present day asserts that it lives perpetually, rising and falling in waves of achievement, and always expressing its age. Can we, from the muddle of our modern street architecture, afford to despise a style so strong, so varied and yet so uniform as baroque was in Italy and Austria in the 17th century ? Mr. Briggs does not attempt to prove it equal to its predecessors ; he merely lays before us the facts and leaves us to judge. " Baroque architecture " he says, " may be limited to an historical period, varying in date in different countries and cities, but in general beginning as the renaissance spirit declined to pedantry, and ending with the return to pedantry in the 18th century. Its buildings may be recognized by the general principles which govern their design rather than by the abundance of their ornament ". Again, baroque and rococo are not interchangeable terms. Baroque was sometimes rococo, as in Germany in the 18th century, but the latter is rather a phase of the former than a distinct style. All this is well explained in the first chapter, with adverse quotations from Ruskin and other authorities. There follows a quite unbiassed description of the religious movements of the day : the counterreformation and the rising power of the Jesuits. Wherever the church regained its hold, there baroque flourished, and the style is traced in its growth from its cradle at Rome in the end of the 16th century to more distant parts of Europe. There are chapters on Bernini, Roman palaces and gardens, Genoa, Venice, Northern Italy, Malta and the kingdom of Naples. These occupy half the book and conclude with a clever analysis of the whole style in Italy. Three chapters are then given to Austria and Germany, in which cities like Vienna and Salzberg are treated in detail, and one each to France, Spain, Portugal and Spanish America, Belgium and Holland, and England. The concluding chapter is the best ; in it the characteristics of the style are summed up admirably. The author admits the artificiality of baroque, but claims with truth that the 17th century witnessed a real advance—especially in planning. The style "taught us many things about the broader aspects of architecture in an age when life was easy and spacious—how to plan and design on a monumental scale, how to make our surroundings less austere, how to glorify the gifts of nature in garden and fountain, how to appreciate the grace of the human form, and lastly, how to beautify our cities".

If a criticism has to be made, one might say that Mr. Briggs has made his subject too comprehensive. The inclusion of Bernini, for instance : his baldacchino and tombs are essentially baroque, but his colonnade can hardly be classed so ; though in a book like this, where a period has to begin and end,

it is difficult to know what to do with a man who lived sometimes in a previous one. Also, the buildings mentioned at La Rochelle (p. 176) belong rather to the period following baroque ; they were built chiefly in the middle of the 18th century— during the "return to pedantry"; and the cathedral was built by Gabriel, and not in 1603. But these are minor criticisms, and cannot mar the general excellence of the book. There are a great many photographs, and several clear and delightful drawings by the author, of which that illustrating Moreglia cathedral in Mexico and several of the Brenta canal may be especially mentioned. At the end of each chapter there is a bibliography, and there are a few plans. More of these would be acceptable, especially of large secular buildings, and perhaps even a few measured details to a large scale would not make the book too technical, nor frighten the general reader. Lastly, a German writer would probably add a chronological chart, showing the activities of the style in different countries : but these are no more than suggestions for the possible enhancement of so valuable a piece of work. Mr. Briggs is to be congratulated ; so is Mr. Unwin.   A. S. G. B.

NOTES ON TURQUOIS IN THE EAST. By B. LAUFER. Chicago (Field Museum of Natural History).

Mr. Laufer seems to be endowed with superhuman energy. Hardly have we absorbed his admirable work on jade when it is followed by an equally admirable monograph on the turquoise. The latter will not appeal to such a wide public as the former, but it is just as thorough and scholarly, an exhaustive work on a relatively obscure subject, and the kind of book which makes us thankful for the existence of such scientific institutions as the Field Museum of Natural History. On the subject of the turquoise it will no doubt remain for a very long time the standard authority, but, needless to add, it will be consulted besides on a hundred and one questions relative to precious stones in general, and it abounds in pithy notes on the lithology of the East. About one-third of the text is devoted to the turquoise in India and Thibet. The stone appeals most strongly to the feminine taste in the latter country, and the turquoise-inlaid jewellery worn by Thibetan women has supplied some very attractive illustrations. The remaining two-thirds treat of the turquoise in China, but of these the greater part is occupied by the identification of a troublesome stone named *sê sê*, which is not, after all, the turquoise, but a member of the ruby family. Having cleared up this question and incidentally put on record the results of much laborious research, Mr. Laufer proceeds to prove that the turquoise is a relatively recent acquisition in China. Its modern name is *lü sung shih* (green fir-tree stone), varied to *sung êrh shih* and *sung tzŭ shih* (fir-cone stone), which, however, is not to be

249

Reviews

confused with *sung shih*, the fossil fir. Apparently the name *lü sung shih* is not older than the 18th century; and though it is true that under the name of *pi tien* or *pi tien tzŭ* the turquoise is mentioned in the Yüan dynasty (1280 to 1367 A.D.), we are expressly told that during the Sung and Ming periods which respectively preceded and followed "the turquois was hardly known to the Chinese or played no *rôle* in their life". It was, in fact, in the Ch'ien Lung period (1736–95) that "the stone was officially adopted and its use sanctioned for the imperial cult", and the illustrations which Mr. Laufer gives of carved turquoise are relatively modern work, made in Hsi-an Fu and Peking, the two chief centres of the turquoise cutting industry to-day. Apparently the stone is unknown in Japan. These are a few of the many important points which Mr. Laufer emphasizes in his book and they are supported by a wealth of quotations and cross references which will make the work invaluable to the student. "At a future date", is Mr. Laufer's creed, "precious stones will occupy a prominent place also in Chinese archæology, and the practical utility of studies like the present one will then become manifest". It is a modest prophecy, for the fulfilment of which we need wait no longer than the time required for obtaining a copy of Mr. Laufer's monograph.   R. L. H.

L'EXPOSITION DE LA MINIATURE À BRUXELLES EN 1912. Brussels (Van Oest).

The exhibition of portrait-miniatures held at Brussels two years ago was notable in many ways. It was, we believe, the first exhibition of the sort to be held on international lines. It also both established the line of succession from the miniaturists of the middle ages to the fashionable portrait-miniatures of the present day, and at the same time pointed a moral as to the limitations of this particular art, and the sincerity, or lack of it, of those who practise it nowadays. Thanks to the surprising energy of Baron H. Kervyn de Lettenhove, who has a real genius for the organization of exhibitions, a very adequate collection was shown in very advantageous and attractive circumstances. The exhibition, though hardly one to attract the average sightseer, was well patronized by royalty and society. It is generally accepted that the English section, which was admirably organized under the presidency of Lord Hothfield, with Mr. M. H. Spielmann as secretary, was the *clou* of the exhibition, due not only to the quality of the portrait-miniatures shown, but to the tasteful installation in rooms specially decorated by Messrs. White, Allom & Co. to suit the purpose of the exhibition. Here, although the great collections at Windsor Castle, Welbeck Abbey, and Montagu House could not be drawn upon, a really interesting and illustrative collection of portrait-miniatures from Hilliard to the present day could be studied. The remainder

of the exhibition was of a less coherent character, and would have been the better for some more stringent sifting. It included, however, the valuable loans from the Queen of the Netherlands, and thus illustrated the art of Holbein, which it had been impossible to supply from England. The Rijks Museum at Amsterdam also sent a valuable contribution. A memorial catalogue has now been issued which will be treasured by all lovers of this beautiful art. Not only does it contain excellent reproductions of the best miniatures in the collections of Earl Beauchamp, Mr. Henry J. Pfungst, and other English collectors, as well as the pick of the continental exhibits, but many of these are reproduced in colours. In an art where delicacy of touch is so essential an ingredient it is not surprising to find that colour-printing, although producing a pleasing result in itself, fails to reproduce the tenderness and refinement of an original miniature. It is, in fact, the coarser and less skilful miniature-paintings which suffer least from the printing in colours. The book, however, is a fine tribute to the industry and artistic skill of Baron Kervyn and his collaborators, and should retain its interest and value.   L. C.

THE ART OF BOTTICELLI. An Essay in pictorial criticism; by LAURENCE BINYON. (Macmillan) £12 12s. net.

Throughout this brilliant and sympathetic essay Mr. Binyon lays stress on his indebtedness, especially as regards the historical side of his subject, to Mr. Horne's classical work on Botticelli, which, as Mr. Binyon truly puts it, "cannot be too much admired as a monument of English scholarship". Such new facts as have come to light between the production of these two books have, however, been conscientiously noted by Mr. Binyon; but the recounting of Botticelli's life and analysis of his artistic development and achievement have not been the sole or principal aims of Mr. Binyon. What he above all has been interested in is the significance of Botticelli for modern art. The opening section of the book is devoted to a discussion of this subject, and the views expressed in it by Mr. Binyon—especially on the value of imagination in art and on the habitual misapplication of evolutionary principles in judging of art—seem so sound in substance and so admirably and convincingly put, that they are bound to be of great helpfulness in these days of keen æsthetic argument. After a brief but vivid sketch of Botticelli's life, Mr. Binyon goes on to discuss first the influences under which he was formed and then his works grouped according to their subjects; and for sympathetic and illuminating interpretation some of these pages must undoubtedly take rank among the finest that have ever been written on Italian art. I should like especially to instance the charming sketch of Fra Filippo Lippi and the chapter on the Dante illustrations by Botticelli. But if it is impossible to speak of Mr. Binyon's share in the volume

T'OUNG PAO

通報

OU

ARCHIVES

CONCERNANT L'HISTOIRE, LES LANGUES,
LA GÉOGRAPHIE ET L'ETHNOGRAPHIE
DE
L'ASIE ORIENTALE

Revue dirigée par

Henri CORDIER
Membre de l'Institut
Professeur à l'Ecole spéciale des Langues orientales vivantes
ET
Edouard CHAVANNES
Membre de l'Institut, Professeur au Collège de France.

VOL. XIV.

LIBRAIRIE ET IMPRIMERIE
CI-DEVANT
E. J. BRILL
LEIDE — 1913.

BULLETIN CRITIQUE.

Berthold LAUFER, *Notes on turquois in the East* (Field Museum of Natural History, Publication 169, Anthropological Series, Vol. XIII, N⁰ 1; in-8 de 71 p. + 8 planches hors texte. Chicago, July 1913).

M. LAUFER a groupé et discuté les textes relatifs à la turquoise en Inde, au Tibet et en Chine; à tous ceux qui ont lu ses excellentes études sur le jade et sur l'ivoire de morse et de narval, il est inutile de recommander ce nouveau travail; dans cette série de recherches, on trouve toujours la même union intime de l'observation directe et de l'érudition philologique, le même sens de la réalité, la même abondance d'information. Notre seul regret est que l'auteur n'ait pu introduire dans son article sur la turquoise les caractères chinois qui eussent été bien nécessaires.

Je ne puis entrer dans la discussion même des faits groupés en grand nombre par M. Laufer; je me bornerai à signaler deux points sur lesquels le doute me paraît permis. En premier lieu (p. 14, n. 1), d'après M. Laufer, une phrase de Fa-hien relative au bol du Buddha à Peshawer signifierait que ce bol était taillé dans différentes couches d'une substance travaillée comme un camée; à mes yeux, le texte de Fa-hien signifie simplement que, sur le bord du pâtra, on avait tracé des rainures de manière à rappeler la légende suivant laquelle ce bol unique était formé de la réunion miraculeuse des quatre bols offerts au Buddha par les quatre rois divins (cf. Foucher, *Les bas-reliefs gréco-bouddhiques du Gandhára*, p. 419—420, et *B.E.F.E.O.*, t. III, p. 433, n. 2). En second lieu, le sö-sö 瑟 瑟 qu'on identifiait jusqu'ici avec la turquoise, serait d'après M. Laufer, le rubis; je crois en effet que l'identification avec la turquoise doit être abandonnée, mais il me semble bien difficile, quelles que soient les variations de couleurs du rubis, de croire que le sö-sö dont la teinte était bleu-verdâtre soit le rubis.   ED. CHAVANNES.

> William COHN, *Einiges über die Bildnerei der Naraperiode* (mit 47 Abbildungen). (Extrait de l'*Ostasiatische Zeitschrift*, Jahrg. I, p. 298—317, p. 403—439; Jahrg. II, p. 199—221).

Dans une série d'articles très remarquables, M. William COHN a traité de la sculpture japonaise de la période Nara; cette période comprend les années 710 à 784, mais l'auteur y ajoute les années 645 à 709, c'est-à-dire les années postérieures à la période Suiko (552—645). L'histoire de la sculpture en Extrême-Orient a été jusqu'ici fort négligée par la science occidentale; aussi M. Cohn a-t-il vraiment fait œuvre de pionier; il a eu cependant cet avantage sur les sinologues de pouvoir se servir des travaux déjà nombreux des savants japonais, tandis que, en Chine, rien n'a été tenté par l'érudition indigène au sujet de la sculpture postérieure aux Han. Une autre différence essentielle entre la sculpture japonaise de la période Nara et la sculpture chinoise de l'époque correspondante provient de la matière employée; tandis que, en Chine, c'est la pierre des grottes de *Long men* et de *Kong hien* qui nous a conservé les plus beaux spécimens de l'art religieux des T'ang, au Japon c'est en bronze, en laque ou en terre que sont faites les statues bouddhiques.

M. W. Cohn montre bien l'influence profonde qu'a exercée l'art

095

阿拉伯和中国的海象和独角鲸牙的贸易

T'OUNG PAO

通報

OU

ARCHIVES

CONCERNANT L'HISTOIRE, LES LANGUES,
LA GÉOGRAPHIE ET L'ETHNOGRAPHIE
DE
L'ASIE ORIENTALE

Revue dirigée par

Henri CORDIER
Membre de l'Institut
Professeur à l'Ecole spéciale des Langues orientales vivantes
ET
Edouard CHAVANNES
Membre de l'Institut, Professeur au Collège de France.

VOL. XIV.

LIBRAIRIE ET IMPRIMERIE
CI-DEVANT
E. J. BRILL
LEIDE — 1913.

ARABIC AND CHINESE TRADE IN WALRUS AND NARWHAL IVORY

BY

BERTHOLD LAUFER.

EILHARD WIEDEMANN, the well-known physicist and Arabist at the University of Erlangen, published two years ago a paper on the value of precious stones among the Moslems [1]) which contains a great deal of material interesting to a student engaged in Chinese research. The bulk of these notes is based on a mineralogical work written by al-Bērūnī (973—1048), the eighth section of which contains the following on a product called *al-chutww* [2]): "It originates from an animal; it is much in demand, and preserved in the treasuries among the Chinese who assert that it is a desirable article because the approach of poison causes it to exsude. It is said to be the bone from the forehead of a bull. Its best quality is the one passing from yellow into green; next comes one like camphor, then the white one, then one colored like the sun, then one passing into dark-gray. If it is curved, its value is a hundred dīnār at a weight of one hundred drams; then it sinks as low as one dīnār, regardless of weight". At the end of another treatise dealing with the volumes of metals

---

1) Über den Wert von Edelsteinen bei den Muslimen. *Der Islam*, Vol. II, 1911 pp. 345—358.

2) *L. c.*, p. 353.

21

and precious stones, al-Bērūnī expands on the fashions to which the latter are subjected, and speaks again on the *chutww*: "It is asserted that it is the frontal bone of a bull living in the country of the Kirgiz who, it is known, belong to the northern Turks. The preference (for the one or other gem) changes with different social strata and peoples. The Bulgar bring from the northern sea teeth (*nāb*) of a fish over a cubit long. White knife-hafts (*nisāb*) are sawed out of them for the cutlers. The middle portion (of the tooth) is distributed among the single hafts, so that every piece of the tooth has a share in them; it can be seen that they are made from the tooth itself, and not from ivory, or from the chips of its edges. The various designs displayed by it give the appearance of wriggling. Some of our countrymen bring it to Mekka where the people regard it as white *chutww*. The Egyptians crave it and purchase it for a price equal to two hundred times its value. Likewise (as in the case of the teeth mentioned before) I conclude from the appearance of the *chutww* that it is the main portion of a tooth or horn. If it were really found among the Kirgiz, it would have certainly not been imported from the 'Irāq into a country nearer to this tribe". In a footnote Prof. WIEDEMANN remarks: "The significance of *al-chutww* is not clear. Perhaps mammoth-teeth are understood. A passage in al-Afkānī's dissertation on precious stones regarding this material runs thus: *Chartūt* is called also *chutww*. Abū'l Raihān al-Bērūnī says: it originates from an animal. It is said to be obtained from the forehead of a bull in the regions of the Turks in the country of the Kirgiz, and it is said also (by others than al-Bērūnī) that it originates from the forehead of a large bird which falls on some of these islands; it is a favorite with the Turks and with the Chinese. Its value comes from the saying that the approach of poisoned food causes it to exude. The Ichwān al-Rāzījūns state that the best is curved, and that it changes from yellow into red, then comes the apricot-colored one, then that passing into a dust-color and down to black (*kahūba*). Formerly there were pieces whose price amounted to from one hundred to one hundred and fifty dīnār. It has been established by experience that together with the vapors of perfume it has an excellent effect in the case of hemorrhoids".

At the end of Wiedemann's paper G. Jacob [1]) imparts information on the subject from a Turkish work on mineralogy written in 917 (1511/12 A.D.) by Jaḥjà Ibn Muḥammad al Gaffārī, who makes the following statement: "On the Ḥutū Tooth. The *ḥutū* is an animal like an ox which occurs among the Berber and is found also in Turkistan. A gem is obtained from it; some say it is its tooth, others, it is its horn. The color is yellow, and the yellow inclines toward red, and designs are displayed in it as in damaskeening. When the *ḥutū* is young, its tooth is good, fresh, and firm; when it has grown older, its tooth also is dark-colored and soft. The padishahs purchase it at a high rate. Likewise in China, in the Magrib, and in other countries it is known and famous. It is told that a merchant from Egypt brought to Mekka a piece and a half of this tooth and sold it on the market of Minā for a thousand gold pieces. Poison has no effect upon one who carries this tooth with him, and poison placed near it will cause it to exude. For this reason it is highly esteemed". G. Jacob [2]) has the further merit of pointing to Bretschneider's Mediaeval Researches (Vol. I, p. 153) where it is said in Ch'ang Tê's travels: "The *gu-du-si* is the horn of a large serpent. It has the property of neutralizing poison". He further refers to Ursu (Die auswärtige Politik des Peter Rares, p. 28) who says that in 1527 envoys from Moldau demanded passage from Poland to Moscow *pro comparandis*

---

1) *L. c.*, p. 357.
2) *Der Islam*, Vol. III, 1912, p. 185.

sobellis et aliis pellibus et similiter dentibus (piscium) [1]), *quibus indiget ad solvendum tributum Turco.*

The oldest Chinese source referred to in the *P'ei wên yün fu* as containing an allusion to *ku-tu-si* is the *Sung mo ki wên* 松漠紀聞 "Historical Memoranda regarding the Kin Dynasty", written by Hung Hao 洪皓 (1090—1155 A.D.) who was sent on an embassy to the Kin where he remained for fifteen years (1129—1143) [2]). His statement runs as follows: "The *ku-tu-si* is not very large. It is veined like ivory, and of yellow color. It is made into sword-hilts (or knife-handles). It is a priceless jewel" [3]).

The report of Hung Hao led me to think that the word *ku-tu-si* might be derived from a Tungusic language, either from that of the Niüchi or the Khitan. Accordingly, I made a search through Ch. 116 of the *Liao shi*, in which the words of the Khitan language are explained, and found (p. 17 a): "*ku-tu-si*: the horn of a thousand years' old snake; there is also the word *tu-na-si*" 榾柮犀、千歲蛇角、又爲篤納犀。[4]) To make sure that these trans-

---

1) It will be seen farther on from a consideration of Russian sources that these 'fish-teeth' were walrus-tusks.

2) A. WYLIE, *Notes on Chinese Literature*, p. 32, who adds: "During his residence in the neighborhood of their capital, he had jotted down a large collection of notes, but these were committed to the flames by the authorities, when he was about to return to his country. The present work consists of a portion of his more extensive manuscript, written from memory after his return, and is of value as a record of the time". The work is reprinted in the collection *Ku kin yi shi*. The life of the author is described by MAYERS (*Chinese Reader's Manual*, p. 64) and GILES (*Biographical Dictionary*, p. 344); compare also CHAVANNES, *Voyageurs chinois chez les Khitan et les Joutchen* (*Journal asiatique*, Mai-Juin, 1898, p. 370).

3) 骨咄犀不甚大、絞如象牙、帶黃色、作刀靶者已爲無價之寶也。Quoted in *P'ei wén yün fu*, Ch. 8, p. 89 b; in the same way in *Pén ts'ao kang mu* (Ch 43, p. 13 b) except that the word *si* 犀 is added after *ku-tu-si*, meaning "*the horn of the ku-tu-si*".

4) PALLADIUS, in his *Chinese-Russian Dictionary* (Vol. I, p. 504) has entered the word *ku-tu-si* (but adopting the orthography of the *Cho keng lu* 骨咄犀) with the meaning "horn of a snake, extraordinarily poisonous, but notwithstanding effectual against poisons". As will be seen below, this definition is based on the *Cho keng lu*. Palladius is the only one of our dictionaries to take notice of the word *ku-tu-si*.

criptions had not been tampered with by the K'ien-lung editors, as
it is well known has been done in the case of the *Yüan shi*,
I looked up the passage in an edition of the *Liao shi* printed in
1529 where it occurs (p. 24) with exactly the same wording and
written with the same characters; the date "eighth year of the
period Kia-tsing" is imprinted on the margin of this very page.
We may therefore be sure of the fact that this passage and the
mode of writing the word *ku-tu-si* were contained in the original
edition of the *Liao shi* and are peculiar to the Khitan period. This
brief text consisting of only twelve words is very valuable: it
shows that the product was known in the period of the Liao
(907—1125), the beginning of which is coeval with the lifetime of
al-Bērūnī, apparently the first Arabic author who had a knowledge
of the same product; it further gives a definition of it, which,
though fanciful, will assist us in recognizing its character, and two
appellations of the product, both of which are clearly characterized
as words of the Khitan language [1]). The second of these words *tu-na-si*
does not seem to occur in any later source.

The glossary of the *Liao shi* is not intended to embrace au

---

1) In the *Sui shi kuang ki* 歲時廣記 by Ch'ên Yüan-tsing 陳元靚
of the Sung period (Ch. 40, p. 11; edited by Lu Sin-yüan in his *Shi wan kuan lou ts'ung
shu*; see PELLIOT, *B.E.F.E.O.*, Vol. IX, 1909, p. 224) occurs the word *ku-tu* 骨 骷.
The character 骷 (not in Giles) is read *k'u* in the tribal name *Yue* (月)-*k'u*, but
otherwise *tu* 咄 (according to *Tsi yun*), and according to the *Yü p'ien* of 543 means
'divination from the voices of birds' (鳥 鳴 豫 知 吉 凶). It is the question
of the customs observed on the last night of the old year (歲除), and one of these
consists in burning *ku-tu* to illumine the hall, and to strengthen the male principle (*i. e.*
to ward off demons, calamities, diseases etc.). The essential condition of this observance is
the bright, open fire which may be effected also by torches and the pods of *Gleditschia
sinensis* (*tsao kio* 皂 角), and there can hardly be any doubt that the above *ku-tu*
represents likewise a combustible substance of vegetable origin (not listed in BRETSCHNEIDER's
Botanicon Sinicum), and has therefore no relation whatever to the *ku-tu-si* of the Liao
and Kin periods.

exhaustive list of Khitan words, but it is its purpose merely to explain such Khitan words as masqueraded in a Chinese garb appear scattered through the Annals. They are consequently arranged in the sequence of the chapters in which they occur. The word *ku-tu-si* is placed under the heading "Biographies" 列傳, so that it is bound to have been used in this section of the Annals. There is an instance of the application of the word in Ch. 96, p. 3 b, where it is written in the manner as above indicated and mentioned as a gift together with jade; but no inference as to the nature of the product can be drawn from this passage [1]).

There are three references to *ku-tu-si* in the Mongol period. But these pertain to the Mohammedan countries of the west, while the *Kin* author distinctly describes a product in the far north of China. The one is indicated by BRETSCHNEIDER [2]) in the *Si shi ki* 西使記 edited by Liu Yü, containing the diary of Ch'ang Tê 常德 who was dispatched by the Mongol Emperor Mangu in 1259 as an envoy to his brother Hulagu, king of Persia. He mentions among the products by the western countries *ku-tu-si* 骨篤犀 as the horn of a large snake which has the property of neutralizing every poison. It is curious that the *Pên ts'ao kang mu* of Li Shi-chên (Ch. 43, p. 13 b) quotes the same passage (the work is called *Shi Si-yü ki* 使西域記 by Liu Yü 劉郁) to the effect that "the *ku-tu*(篤)-*si* is the horn of a large snake produced in *Si-fan* 西番" [3]).

_____

1) Others better read in the *Liao shi* or having more time for reading will probably be able to reveal more passages of this kind. It may be presumed that the word will be found also in the *Kin shi*.

2) *Chinese Recorder*, Vol. VI, 1875, p. 19, or *Mediaeval Researches*, Vol. I, p. 153.

3) In the first edition of his translation of the work which appeared in the *Chinese Recorder* (Vol. VI, p. 19) BRETSCHNEIDER said that the statement of the *Si shi ki* has passed into the *Pên ts'ao kang mu*; in *Mediaeval Researches* (Vol. I, p. 153), this reference is omitted. It is strange that Bretschneider, who had doubtless perused this section of the *Pên ts'ao*, omits to call attention to the fact that *Si-fan* is there given as the place of

T'ao Tsung-i 陶宗儀, the author of the interesting work *Cho keng lu* 輟耕錄, published in 1366, has devoted a brief notice to this subject. The edition referred to is that printed in 1469 (Ch'êng-hua period) which is liable to afford a guarantee for

---

production. Nevertheless it may be that in the editions of the work consulted by Bret-schneider the word *Si-fan* does not occur. He states (p. 110) that many typographical blunders have crept into the different editions, which render it difficult for the reader to understand who has access only to one edition, and that he has compared the texts of four different editions so as to be enabled to reconstruct the complete original. This variant, at all events, should have been noted, for a traditional opinion seems to exist among the Chinese that *ku-tu-si* is also a product of Tibet This view is expressed in the *Wei Ts'ang t'u chi* 衛藏圖識 (Ch. 下, p. 22 b, in the original edition of 1792, where *ku-tu-si* 骨篤犀 is enumerated in a list of the strange products 異産 of Tibet and described as "pale blue-green, and when struck, emitting a clear sound like jade; it is scented and can overcome all poisons". This passage inclusive of the other *mirabilia* mentioned is quoted from a work *Yi shi* 譯史 (not to be confounded with the *Yi shi* 繹史 by Ma Su of 1670 in 48 vols.), a curious small book written in four chapters by Lu Ts'e-yün 陸次雲 (T. Yün-shi 雲士) full of marvelous notes regarding real and imaginary countries. WYLIE (Notes, p. 64) mentions the work under the fuller title *Pa hung yi shi*, and adequately describes its contents (a copy of it is in my library). According to WYLIE (Notes, p. 60), the author who wrote also a miscellany concerning the antiquities on West Lake near Hang-chou lived in the middle of the seventeenth century. It hence follows that the two officials Ma Shao-yün and Shêng Mei-k'i, the authors of the *Wei Ts'ang t'u chi* (see WYLIE, Notes, p. 64, and ROCKHILL, *J. R A. S.*, N. S, Vol. XXIII, pp. 23—26), do not speak of the subject on the ground of a personal experience but of mere bookish knowledge, nor do they assert that they actually encountered the product in Tibet. The *Yi chi* on which they depend is a pure story-book of the wondrous kind, devoid of historical value. Moreover it will be noticed from the text of the *Ko ku yao lun* of the Ming period, given farther on, that the statement of the *Yi shi* is a literal extract modeled after the latter work, and therefore forfeits any claim to consideration as an independent observation; the *Ko ku yao lun*, in its notice on *ku-tu-si*, makes no allusion to Tibet. The author of the *Yi shi*, consequently, links two literary reminiscences into one by combining the text of the *Ko ku yao lun* with the supposed reading *Si-fan* in one of the editions of the *Pen ts'ao kang mu*. His makeshift, not sustained by any palpable evidence, cannot therefore be considered as a contribution to the eventual question as to whether *ku-tu-si* may have existed in Tibet, and which to all appearances will shrink into the clerical error of a copyist. The fancy of the *Yi shi* is copied again in a recent work on Tibet, *Si-ts'ang t'u k'ao* 西藏圖考, by Huang P'ei-k'iao 黃沛翹 of Hu-nan (first published in 1886, reprinted in the geographical collection *Huang ch'ao fan shu yu ti ts'ung shu*, 1903, vols. 1—2; Ch 6, p. 27 b). Here again it is merely a case of repro-duction without the evidence of a personal experience.

representing the text of the original issue. The passage (Ch. 29, p. 7 b) runs as follows: "*Ku-tu-si* is the horn of a large snake, and as it is poisonous by nature, it can counteract all poisons, for poison is treated with poison. For this reason it is called *ku-tu-si* ("*ku*-poison horn") [1]. In the Annals of the T'ang dynasty it is the question of the country of *Ku-tu* 古 都, so that it seems that this place is responsible for this product. It is therefore erroneously that the people of the present time write the word *ku-tu* 骨 咄" [2].

---

[1] The conception that *ku-tu-si* cures *ku-tu* rests on a notion of sympathetic magic elicited by a pun upon the words. The substitution of the word *ku*, it seems to me, has been suggested by the passage regarding rhinoceros-horn in the *Shén-nung pén ts'ao king* (Ch 2, p. 31 a; edition of *Chou-shi hui k'o I hio ts'ung shu*, 1891) where it is said: "The taste of rhinoceros-horn is bitter and cold; it cures all poisons and the *ku* poison" 犀 角 味 苦 寒 主 百 毒 蠱. The nature of the *ku* poison is discussed at some length by S. WILLIAMS (Witchcraft in the Chinese Penal Code, *J. China Branch R. A. S.*, Vol. 38, 1907. pp. 71—74); it has been made the subject of a monograph on the part of A. PFIZMAIER under the somewhat startling title *Das Ereignis des Wurmfrasses der Beschwörer* (*Sitzungsberichte der Wiener Akademie*, 1862, pp. 50—104), which despite the questionable correctness of the translations makes interesting reading. In my opinion the numerous intestinal parasitic worms causing many diseases in China (now fully discussed in the remarkable work by Dr. JAMES L. MAXWELL, *The Diseases of China*, p. 137, London, 1910) form the basic foundation of the *ku* poison, with a later development into an alleged practice of witchcraft; but it seems very doubtful if *ku* has ever the meaning of insanity attributed to it by Giles. Cases of insanity are rare in China, as may be seen from MAXWELL, p. 256. The flesh of the fox which was eaten by the ancient Chinese was formerly considered as a preventive remedy against *ku* poison (SCHLEGEL, *Uranographie chinoise*, p. 167).

[2] 骨 咄 犀 大 蛇 之 角 也、其 性 至 毒 解 諸 毒、盖 以 毒 攻 毒 也、故 曰 蠱 毒 犀、唐 書 有 古 都 國 必 其 地 所 產、今 人 訛 爲 骨 咄 耳。*P'ei wén yun fu* (Ch. 8, p. 89 b) gives only the first clause with the variant 解 蠱 毒 如 犀 角 "it counteracts the *ku* poison like rhinoceros-horn", which is evidently derived from a different edition of the *Cho kêng lu*. This phrase occurs also in the quotation from this work as given in *Pén ts'ao kang mu* (Ch. 43, p. 13 b) under the heading "snake-horn". The last clause is cited there in a different way: 唐 書 有 古 都 國 亦 產 此 則 骨 咄 又 似 古 都 之 訛 也。This seems to mean: "The *T'ang shu* mentions the country of *Ku-tu* as producing this (horn), so that the word *ku-tu* 骨 咄

T'ao Tsung-i, evidently, does not speak from any personal experience with the object which he is discussing, but reflects and philosophizes on it. The definition of the *ku-tu-si* as a snake-horn, is derived, apparently, from Ch'ang Tê, while in the writing of the name with the character *tu* 咄 [1]) the tradition of the Kin period inaugurated by Hung Hao is retained. The opinion that the object in question is poisonous and therefore cures poison is peculiar to the author; it is by no means, however, his original idea, but one transferred from the ancient beliefs in the properties of rhinoceros-horn to the *ku-tu-si*. The Taoist adept and writer Ko Hung who lived in the first part of the fourth century A.D. is the father of the theory that the rhinoceros feeding on brambles devours all sorts of vegetable poisons affecting the horn which, according to the principle that poison cures poison, becomes an efficient antidote [2]).

A country *Ku-tu* 古都 is not known to me; but *T'ang shu*, Ch. 221, contains a notice of the country *Ku-tu* 骨咄 identified

---

seems to be erroneous for *ku-tu* 古都". — Another way of writing is introduced into a work entitled *Liang ch'ao chai yü* 兩鈔摘腴 (quoted in *P'ei wén yün fu*, Ch. 92, p. 18 b) where it is said: "What is now called *ku-tu-si* 骨拙犀 is the horn of a snake; being poisonous by nature, it is capable of neutralizing poisons, and is therefore called *ku tu si* 蠱毒犀". The date of this work is not known to me; but the definition being identical with that of the *Cho keng lu*, it may be concluded that it is posterior to the latter book.

1) The *P'ei wén yün fu* regards this as the standard mode of writing. The transcription 篤 occurs again in the *Ko ku yao lun* (see farther on).

2) *Pên ts'ao kang mu*, Ch. 51 上, p. 6. I do not enter here into a discussion of the rhinoceros and its horn, as I have just completed a lengthy investigation of this subject which it is hoped will be embodied in a publication to come out in the near future. The contention of Prof. GILES (*Adversaria Sinica*, p. 394) that the words *se* 兕 and *si* 犀 originally refer to a bovine animal is not at all justified, and none of the arguments advanced by him in favor of this point of view can be defended. All available evidence philological, historical, archaeological, zoological and palaeontological leads me to the result that the words *se* and *si* very well apply to the rhinoceros, and to this animal exclusively, and that from earliest times two distinct species are understood, the word *se* referring to the single-horned rhinoceros (*Rhinoceros unicornis*), and the word *si* to the two-horned rhinoceros (*Rhinoceros sumatrensis*).

by M. CHAVANNES [1]) with Khottal on the upper Oxus north-east of Tokharestan. There is evidently some confusion in the passage quoted, but however this may be, there is no connection between the product *ku-tu-si* and the country of *Ku-tu*, for the text of the *T'ang shu* as translated by M. CHAVANNES attributes to Khottal excellent horses, red leopards and black salt mined in four mountains, but not snake horn or any other horn. The combination of *ku-tu-si* and *Ku-tu* is therefore arbitrary and suggested only by their phonetic similarity. This confusion may be accounted for by "the snake-horn of Ku-tu" 古都之蛇角 mentioned in the *Shan hai king* and explained by a commentator as a designation for "the blue-green rhinoceros-horn" 碧屖 [2]). This seems to be also the reason why the *Ko ku yao lun* (see below) gives this definition for the *ku-tu-si*.

Nevertheless it is probable that the product in question was known in the age of the T'ang dynasty. At least the *K'in ting Man-chou yüan liu k'ao* [3]) (lithographic reprint of 1904, Ch. 19, p. 15) quotes the following statement from the *T'ang hui yao*: [4]) "In the country of the *Mo-ho* [5]) there is a great number of sable-skins, *ku-tu* horn 骨咄角, white hares, and white falcons". The *T'ang hui yao* is not accessible to me, and I am not inclined to regard this passage as conclusive as to the occurrence of the word *ku-tu* in the T'ang period, unless more substantial evidence will be forthcoming. Yet it will be seen below that the product represented

---

1) *Documents sur les Toukiue (Turcs) occidentaux*, p. 168 (and see Index).

2) Quoted in *P'ei wén yün fu*, Ch. 92, p. 18 b.

3) WYLIE, *Notes on Chinese Literature*, p. 44. The work was published in 1778 (not 1777) by order of the Emperor K'ien-lung.

4) A work relating to state matters of the T'ang dynasty compiled by Wang P'u of the tenth century (WYLIE, p. 69).

5) The *Mo-ho* were settled in the north of Korea and extended east of the Sungari to the ocean; the Shi-wei were their neighbors in the north, the T'u-küe in the west (*Kiu T'ang shu*, Ch 199 下, p. 7 b).

by the word *ku-tu* or *ku-tu-si* was known in that epoch, but under a different name [1]).

Finally there is a brief reference in the *Yüan shi lei pien* (Ch. 42, p. 53 a; edition of 1795) [2]) to the effect that "*ku-tu*(篤)-*si* is originally the horn of a large snake and is capable of neutralizing all poisons". It is listed there among the products of Central Asia (*Si yü*). The passage has no independent value and is doubtless copied from the account of Ch'ang Tê.

A work of the Ming period, the *Ko ku yao lun* 格古要論, a collection of essays in thirteen chapters on objects of art and antiquities by Ts'ao Chao 曹昭, published in 1387 (revised and enlarged edition by Wang Tso 王佐 in 1459) [3]) makes the following allusion to this subject: "*Ku-tu-si* is a blue-green rhinoceros-horn; it is in color like a pale blue-green jade and is also yellow to a small extent. It veins resemble those of a horn; when struck, it emits a clear sound, much more so than jade. When you [rub or scrape and] smell it, you will find it is scented; but when burnt, it is odorless. It is very highly prized, for it can reduce swellings and neutralize poison" [4]).

---

1) The great historical importance of this passage will be discussed below in our attempt to identify this product.

2) In *Skizze der mongolischen Literatur* (*Keleti Szemle*, 1907, p 213) the name of the author, on the authority of Bretschneider, had been given by me as *Kiai-shan*. PAUL PELLIOT, with obliging courtesy, has been good enough to inform me that *Kiai-shan* 戒山 is only his *hao*, and that his real name is Chao Yüan-p'ing 邵遠平. The passage of the *Yüan shi lei pien* is not quoted in *P'ei wén yün fu*.

3) BRETSCHNEIDER, *Botanicon Sinicum*, pt I, p. 162; HIRTH, *Ancient Chinese Porcelain*, p. 141; BUSHELL, *Description of Chinese Pottery and Porcelain*, p. 175.

4) 骨篤犀碧犀也、色如淡碧玉稍有黃色、文 [理] 似角、扣之聲清越如玉、[磨刮] 嗅之有香、燒之不臭、最貴重能消腫 [解] 毒。*P'ei wén yün fu* (Ch. 8, p. 89 b) and *Pén ts'ao kang mu* (Ch. 43, p. 13 b). The characters enclosed in brackets are additions occurring only in the latter work.

Li Shi-chên, the author of the *Pên ts'ao kang mu*[1]), has devoted a full discussion to the *ku-tu-si* (Ch. 43, p. 13 b; edition of *Tsi ch'êng t'u shu*, Shanghai, 1908, reprint of the edition of 1657). He takes note of the two different ways of writing the word and records also the name *pi si* "blue-green rhinoceros-horn" due to the *Ko ku yao lun*. Nevertheless he does not entertain this explanation seriously, for the subject is treated under the heading "snake-horn"

---

1) The literary history of this work, completed after 26 years' labor in 1578 and first printed in 1596, has been traced by BRETSCHNEIDER (*Bot. Sin.*, pt. 1, p. 55), who states that the earliest edition now extant seems to be that of the year 1658. But there are older ones in existence. HIRTH (*J. China Branch R. A. S*, Vol. XXI, 1886, p. 324) refers to a Ming print of 1603, possibly the second edition published. An edition of 1645 in 16 vols, edited by Ni Shun-yü 倪純宇 of Hang-chou, was secured by me in Tōkyō and is now deposited in the John Crerar Library of Chicago, which, besides, has an edition of 1826 in 39 vols, and one issued in 1885 in 40 vols, the best modern reprint. The text of the Shun-chi editions is more accurate than that of the K'ien-lung and Tao-kuang editions. Despite diligent search and many efforts I failed to discover in China the *editio princeps* which seems to be entirely lost, and not to exist any longer in any Chinese library; positively I may say it exists in no private library of Si-ngan fu. The recent reprints are based on the Shun-chi issues. Also Mo Yu-chi 莫友芝, the author of the excellent bibliographical work *Lü t'ing chi kien ch'uan pên shu mu* (Ch 8, p. 11; compare the notice of CHAVANNES, *T'oung Pao*, 1910, p. 146) does not know any earlier edition than that of 1603; he further enumerates re-editions of 1640, of the period Shun-chi (1644—61), of 1684, and 1735. The value of the *Pên ts'ao kang mu* is vitiated by occasional carelessness and defectiveness with which extracts from previous works are quoted, and in important cases it is not safe to rely exclusively upon its text; this feature must have adhered to the original edition, while the misprints of the later editions, of which Bretschneider complains, may be overcome. For a revision of the text, good services are rendered by the *Chêng lei pên ts'ao* (on which a bibliographical notice is given farther on), as will be seen from the chapter on the rhinoceros where the whole text of the *Kang mu* has been restored and supplemented by me on the basis of the *Chêng lei*. It is further necessary to resort to the *Pên ts'ao kang mu shi i* 本艸綱目拾遺 written by Chao Hio-min 趙學敏 of Hang-chou in 1650, reprinted in 1765. This important work, not made use of by Bretschneider, to which I called attention in the Publication of the *Congrès international des Américanistes à Quebec*, Vol. I, 1907, p. 260, in connection with a study of the introduction of maize, ground-nut and other cultivated American plants into Asia, contains in the first chapter a long list of rectifications of Li Shi-chên's errors 正誤, while the nine remaining chapters embrace a most valuable supplement and are chiefly taken up with interesting notes regarding the newly introduced plants and products of the sixteenth and seventeenth centuries.

蛇角 which is arranged in the section on snakes, while it is not dealt with at all in. the essay on the rhinoceros and rhinoceros-horn (Ch. 51 A, pp. 5 a *et seq.*) where the word *ku-tu-si* is not even mentioned. It is thus perfectly evident from the texts of the *Liao*, *Kin* and *Yūan* periods as well as from the view taken by Li Shi-chên in the matter that rhinoceros-horn and *ku-tu-si* are entirely distinct substances in Chinese eyes. It could hardly be expected to be otherwise, as the Chinese were thoroughly familiar with the rhinoceros-horn ages before the *ku-tu-si* entered upon their horizon, and have woven many wondrous legends around the former beginning with the Taoist adept Ko Hung of the fourth century.

Li Shi-chên quotes the *Cho keng lu* discussed above, and then makes reference to the *Ta Ming hui tien*, "the Statutes of the Ming Dynasty" [1]) as saying that "snake-horn is produced in the district of Hami" [2]). But the name *ku-tu-si* is not mentioned here.

After quoting Ch'ang Tê, the *Ko ku yao lun* and *Sung mo ki wên*, Li Shi-chên arrives at the conclusion that *ku-tu-si* is poisonous and capable of reducing swellings and neutralizing all poisons as well as the *ku* poison, as poison is treated with poison. It is evident that he had a good literary knowledge of the subject and knew the principal sources relating to it, except the earliest passage in the

---

1) A copy of this work (edition of 1620) is in my possession, but I cannot find in it, after a cursory search, any allusion to the snake-horn of Hami; it would be difficult to guess in which chapter to look for this information. PAUL PELLIOT (*B.E.F.E.O.*, Vol. IX, 1909, p. 37) has given valuable notes on the literary history of this work. The *Ta Ming i t'ung chi* (edition of 1461, Ch 89, fols 19 a, 21 a) mentions a horn *yin ya kio* 陰 牙 角 as product of Hami and Qarā-khodjo 火 州 (see PELLIOT, *Journal asiatique*, 1912, pp. 579—603), and a *su ho kio* 速 霍 角 as a product of the latter locality only. According to the geographical section of the *T'ang shu* (quoted in *P'ei wên yin fu*, Ch. 92, p. 26) both these products were sent as tribute (no date given) from *Pei t'ing* 北 庭. I cannot explain these names which do not seem to occur elsewhere; even Palladius has not registered them; they are not listed in the Glossary of the *T'ang shu*.

2) 蛇 角 出 哈 密 衛。

Liao shi. But the principal question to be raised is whether he had any personal experience with, or actual knowledge of the object, and this must be flatly denied. In this account no word of his own is uttered which would justify the conclusion that he had ever had a *ku-tu-si* before his eyes. This is in striking contrast with his notes on rhinoceros-horn which furnish ample proof that he had really seen and studied it. Of rhinoceros-horn he states expressly that it is not poisonous (and this is a fact corroborated by a scientific investigation made years ago in London), while in the above case he blindly accepts the purely imaginary assertion of T'ao Tsung-i.

The most recent author in whom I have been able to find the word *ku-tu-si* is Fang I-chi 方以智 in 1640 [1]), in his *Wu li siao shi* 物理小識 (edition of *Ning tsing t'ang*, 1884, Ch. 8, p. 20) who merely states that "*ku-tu-si* is a snake-horn of blue-green color", a sentence embodied in a notice on rhinoceros-horn and apparently the echo of former statements.

In attempting to identify the character of the product *ku-tu-si* it is apparent that the epigone, purely bookish utterances of the Ming authors are devoid of any practical value, and that the earliest accounts of the *Liao shi* and *Sung mo ki wén* must primarily be taken into consideration. Hung Hao, the author of the latter work, had evidently had the product under his eyes on the occasion of his visit to the Khitan country, and reports it in plain and sober language without a gleam of imagination. First of all it becomes evident from his definition that *ku-tu-si* is a kind of ivory, and that for this reason it is utterly impossible to assume that it is anything like rhinoceros-horn, which is most assuredly not "veined like ivory", as Hung Hao expressly states. The definition

---

1) HIRTH, *T'oung Pao*, Vol. VI, p. 428.

of the *Liao shi* "the horn of a thousand years' old snake", moreover,
militates against such an hypothesis, for there would be in all the
world no reason to designate a rhinoceros, or to confound it with,
a snake, especially for a people like the Chinese who were acquainted
with the single-horned and two-horned species of rhinoceros from
the earliest days of antiquity. The *ku-tu-si* was a kind of ivory,
but could have been neither elephant nor mammoth [1]) ivory, for
this was always called and is still called *siang ya* 象牙, and the
Chinese, in the epoch of the Khitan, were surely familiar enough
with the elephant and the mammoth to be sufficiently sophisticated
not to classify these animals with snakes [2]). Besides the elephant

---

1) The mammoth has become known to the Chinese to a certain extent from the stories
of Siberian natives, under the name *yin shu* 隱鼠 'the hidden rodent' (first mentioned
by T'ao Hung-king), as the belief prevailed in Siberia that the mammoth lives and moves
underground, shatters the banks of rivers, and dies as soon as it comes up to the surface
(compare especially S. PATKANOV, *Die Irtysch-Ostjaken und ihre Volkspoesie*, Vol. I,
pp. 123—124, St. Pet., 1897). Li Shi-chên has gathered the principal notes on the subject
in his *Pên ts'ao kang mu* (Ch. 51 下, p. 10). Klaproth, I believe, was the first to resort
to this work for information when he found mammoth-bones in the Chinese drug-stores at
Kiachta and had the name of the animal pointed out to him in that book. F. W. MAYERS
treated the subject in *China Review*, Vol. VI, pp. 273—6, with an additional note in
Vol. VII, p. 136; and J. EDKINS popularized it in a brief essay inserted in his *"Modern
China"*, p. 24 (Shanghai and London, 1891). The subject, though practically finished,
would be capable of a more critical and exact treatment. The curious fact has strangely
been overlooked that the older texts as quoted in the *Pên ts'ao* fail to allude to the
mammoth as the animal furnishing the fossil ivory of Siberia, nor is any reference at all
to the tusks, and the Chinese seem not to have been aware of this fact, until the attention
of the Emperor K'ang-hi was called to it by Russians presenting themselves at his court
in 1721 (see MAYERS, *l. c.*, p. 274). There is, as far as I know, no ancient Chinese reference
to mammoth-ivory and its importation from Siberia, and the evidence for such a trade
mainly rests on Russian-Siberian reports, one of the oldest of which is contained in the
learned book of the Swedish Captain PH. J. v. STRAHLENBERG (*Das nord- und östliche
Theil von Europa und Asia*, p. 393, Stockholm, 1730). It should be understood, of course,
that the mammoth and its ivory tusks were known to the natives of Siberia ages before
it came to the notice of our scientists.

2) There is also a logic of imagination, inherent even to the wildest fairy-tales. The
building of a snow-hut in an equatorial region, the handling of a palmleaf fan near the
North Pole, the assigning to an animal a rôle which in accordance with its natural
qualification it could not represent would offend the imaginative faculties of a child's

and the mammoth there are only two other creatures on this globe furnishing ivory, and these are the narwhal and the walrus, and for this reason our first conclusion is that *ku-tu-si* is nothing but ivory obtained from walrus and narwhal [1]). Ample historical evidence

----

susceptible mind and be immediately rejected. The former conception of whale, seal and walrus as fish was perfectly logical and compatible with the mental working of a primitive mind which first clings to some exterior trait in observing a new phenomenon and links with the new an old familiar experience; it thus arrives at a series of classifications or a system of associated notions widely differing from ours, and here is the germ of the fundamental diversity in the intellectual make-up of the various nations. The Chinese, in agreement with the peoples of Siberia, have affiliated the mammoth with the ox, the water-buffalo, the pig, the mole; all this is perfectly logical and consistent with their imaginative traits. Yet an association of the mammoth with a serpent has never entered their minds, and such a conception flatly contradictory to any law of the logic of imagination would be utterly impossible in any human society. On the other hand when referred to the narwhal and walrus, the simile with the snake becomes a logical transcript of what the emotional flight of primitive imagination has suddenly and swiftly perceived at the sight of a novel object.

1) In the zoological system the walrus belongs to the order Pinnipedia which consists of the three families *Otariidae* (eared seals), *Trichecidae* (walrus), and *Phocidae* (seals); the genus *Trichecus* consists of the two species *rosmarus* occurring on the coast of Labrador northward to the Arctic Ocean, along the shores of Greenland, and in the polar areas of the eastern hemisphere to western Asia, and *obesus* occurring on the north-west coast of America, in the Arctic Sea and Bering Strait as well as along the north-eastern coast of Asia. The most striking characteristic of the animal is the pair of tusks corresponding to the canine teeth of other mammals and descending almost directly downward from the upper jaw, sometimes attaining a length of twenty inches or more. Some information on the various names of the walrus is given farther on. — In the zoological system the narwhal belongs to the order *Cetacea*, family III *Delphinidae*, sub-family I *Delphinapterinae*, genus *Monodon*, species *monoceras*, or *monoceros*. The animal frequents the icy circumpolar seas, and is rarely seen south of 65° N. lat. It resembles the white whale in shape and in the lack of a dorsal fin. Its peculiar feature is the absence of all teeth, except two in the upper jaw arranged horizontally side by side. In the male, usually the left tooth, and occasionally both teeth, are strongly developed into spirally twisted straight tusks passing through the upper lip and projecting like horns in front. They often reach a length of half, and even more, that of the entire animal which in the state of maturity may attain to fifteen feet. Its life-history is unfortunately little explored, and the biological function of the tusk or tusks is more conjectured than accurately ascertained (weapon of defense, for breaking ice in order to breathe, and for killing fish). — "The ivory of the narwhal is esteemed superior to that of the elephant, and far surpasses it in all its qualities; it possesses extreme density and hardness, has a dazzling whiteness, which does not pass into yellow, and easily receives a very high polish" (W. JARDINE, *The Natural History of the*

will be furnished for the fact that an ancient trade in the ivory of these two arctic sea-mammals existed, in Russia at least from the ninth century, also that the Chinese received this article probably over two commercial routes and still obtained it in recent times at least as far down as the middle of last century, and presumably even at present, and further that the Japanese cultivated this product obtained by them in the channel of trade.

First, to return to our earliest definitions of *ku-tu-si*, — they most excellently fit the proposed identification, for it is the very designation of 'horn' under which narwhal and walrus ivory was at all times current all over the northern hemisphere, as may be learned from the pieces of evidence brought together in the footnote [1]). The report that the narwhal was described as a snake is

---

Ordinary Cetacea or Whales, p. 190, Edinburgh, 1837) In regard to walrus ivory J. A. ALLEN (*History of North American Pinnipeds*, p. 133, Washington, 1880) remarks: "The ivory afforded by the tusks, though inferior in quality to elephant ivory, is used for nearly the same purposes. It is said, however, to sooner become yellow by exposure, to be of coarser texture, and hence to have less commercial value".

1) The narwhal tusks were always designated "horn" in Europe, hence the term *monoceros* and the "unicorn of the sea", the name being even retained in our natural history. "The two tusks, long and pointed, are usually called horns", says Sir William Jardine (*The Natural History of the Ordinary Cetacea or Whales*, p. 182, Edinburgh, 1837). "The creature grows to a length of about fifteen feet; such an individual would have a 'horn' of some seven feet" (F. E. BEDDARD, *A Book of Whales*, p. 247, New York, 1900). ANSELMUS BOETIUS DE BOOT, court-physician to the Emperor Rudolf II (*Gemmarum et lapidum historia*, ed. A. TOLL, p. 434, Lugduni Batavorum, 1636; the first edition of this interesting work had appeared at Hanover in 1609) describes a walrus-tusk (*rosmari dens*) which he had seen at the end of the sixteenth century in the possession of a druggist at Venice (*simplicista rerum exoticarum studiosissimus*) and expressly states that during and before his time these tusks were confounded with, and sold in the place of, rhinoceros-horn, the basest substitute of which, however, was cervine antlers; all of these, according to the experience of many, were believed to have no small properties against poison (*cornu multorum experientia non exiguas adversus venena habet vires*) — Also in the Eskimo story of the origin of the walrus and the caribou, according to which the walrus at first had the caribou's antlers, and the caribou the tusks of the walrus, till an exchange was effected by a woman magician, an idea of relationship between tusks and antlers seems to be at the root (compare BOAS, *The Eskimo of Baffin Land and Hudson Bay*, p. 167, *Bull. Am. Mus. Nat. Hist*, Vol. XV, 1901). The Yakut indiscriminately designate mam-

22

perfectly believable and has nothing surprising for him who has studied the interesting story of the gradual development of our knowledge of narwhal and walrus which has become somewhat accurate only during the last decenniums, while it has been an unbroken chain of myth and fable ever since the days of Albertus Magnus and Olaus Magnus. The "thousand years' old snake" is nothing but the fossil narwhal occurring on the northern shores of Siberia, especially in the valley of Kolyma River, on which v. DITMAR and v. NORDENSKIÖLD (see footnote) have reported. Stress should be laid on the continuity of Chinese tradition: the snake-horn of the Liao period appears again persistently in the age of the Mongols and is finally endorsed by Li Shi-chên. There is

---

moth and walrus ivory as *muos* 'horn' (PEKARSKI, *Short Russian Yakut-Vocabulary*, pp. 37, 108, Irkutsk, 1905). The mammoth tusk is regarded by the native tribes of Siberia as a horn, the Yukaghir word *xolhutonmun* signifying 'the horn of the mammoth' (JOCHELSON, *Sketch of the Animal Industry and Fur Trade in the Kolyma District*, in Russian, p. 107, St. Pet., 1898). 'Horn' has thus developed in Siberia into a commercial term which may comprise mammoth, walrus, and narwhal tusks, and certainly also fossil rhinoceros-horn. This point of view is easy to understand when we consider that mammoth and rhinoceros occurring there only in fossil remains are utterly unknown to most people as animals, and that tusks and horn are often enough found scattered and detached from any bodily parts; further, that narwhal and walrus are familiar to a minority of maritime people only and again unknown to the inland tribes, and that along the northern shores of Siberia stretches of land occur where immense masses of mammoth and rhinoceros bones are accumulated together with those of stranded walruses and fossil tusks of the narwhal (compare A. E. F. v. NORDENSKIÖLD, *Die Umsegelung Asiens und Europas auf der Vega*, Vol. I, p. 378, and K. v. DITMAR, *Reisen und Aufenthalt in Kamtschatka*, p. 37, St. Pet., 1890). It further remains to be noted that in many cases it is not the complete horn or tusk which is traded by the Siberian and Russian ivory hunters, but merely a fragment; hollow and rotten portions are cut off as useless, as soon as the best preserved pieces have been picked out, and the remainders which are still of a considerable size are again sawn into parts of smaller dimensions to be rendered fit for transportation on the pack-horses. Hence perhaps the statement of Hung Hao that the *ku-tu-si* is not very large. The dealer who buys up this material, and the final consumer remote from the place of production, therefore, have little or no occasion to obtain a clear idea of the origin of the product, still less of the character of the animal from which it may have come. The door was thus open for fabulous speculations of all sorts, and part of the lore which the Chinese and Arabs coined in regard to the 'horn', may have reached them directly from Siberia.

no confusion whatever in the early Chinese authors (as it has crept into the accounts of the Arabs) with any other animal than the one indicated; the association of the tusk with rhinoceros-horn is a subsequent development nourished by the similar medicinal employment of both substances and arising only in popular belief, but not proving in fact that both were alike [1]). Another argument in favor of our identification is the yellow color emphasized by Hung Hao, which is peculiar to walrus ivory after long exposure to air and moisture (see below), and another proof is presented by the statement of Hung Hao that *ku-tu-si* is made into sword-hilts or knife-handles, and there is the interesting coincidence in the report of al-Bērūnī that the Bulgar cut the same implement out of "fish-teeth brought from the northern sea." This northern sea is the sea of the northern coast of Russia, and from the Russian accounts to follow it will be seen that the "fish-teeth" of the old Russian documents, as proved long ago by the famous historian KARAMSIN, were walrus tusks,

The earliest reference to such sword-hilts is contained in GAIUS JULIUS SOLINUS, who lived in the first half of the third century A. D., author of *Collectanea rerum memorabilium*, revised in the sixth century under the title of *Polyhistor*. In Chap. XXXV he has a report regarding sword-hilts made by the inhabitants of ancient

---

1) BRETSCHNEIDER's (*Mediaeval Researches*, Vol. I, p. 153) contributions to the elucidation of *ku-tu-si* are now, of course, without any value, as the *Liao* and *Kin* texts were unknown to him; these do not refer to Africa with its horned adders nor to any locality where the rhinoceros occurs, but to the extreme north-east of Asia where neither exists, and only walrus and narwhal come into question. We shall see that, besides the inward evidence yielded by the early Chinese definitions of the name, there are convincing geographical and ethnological reasons upheld and corroborated by recent trade relations which explode any speculations connecting *ku-tu-si* with rhinoceros, mammoth, or suchlike, and which raise the identification with walrus and narwhal to a well assured fact.

Ireland from the teeth of a marine animal[1]). K. E. v. BAER[2]) is inclined to derive this ivory from the narwhal rather than from the walrus which does not occur at all in the British seas, while the narwhals sometimes descend far southward; in the eighteenth century a narwhal was seen stranded at the mouth of the Elbe, and another at the mouth of the Weser, while no similar example exists in the case of the walrus.

According to L. v. SCHRENCK who traveled in the Amur region from 1854 to 1856, the walrus was known to the Gilyak at that time only by name from its teeth which they received through the medium of the northern neighboring tribes in times prior to the Russian colonization on the Amur. Since 1853 they have traded them from the Russian-American Compagnie at Nikolayevsk, for the purpose of bringing them to the Chinese on the Sungari, and exchanging them with profit for other objects[4]). A long-enduring familiarity with the work of L. v. SCHRENCK has accustomed me to place great confidence in the observations of this scholar; while engaged in a study of the ethnology of the Amur region in 1898 — 99, I naturally had his publications in my hands almost daily and had ample occasion to test his observations which, though they can certainly be widened, supplemented, and deepened, I generally found accurate to a high degree. On his authority it may therefore be accepted as a fact that in the nineteenth century the Gilyak were the middlemen in the trade of walrus ivory between the high northeast corner of Asia and the Chinese on the Sungari, and probably so long before that time. It is noteworthy that it was the Sungari region where the distant arctic products coming down the valley

1) Qui student cultui dentibus mari nantium belluarum insigniunt ensium capulos. Candicant enim ad eburneam claritatem.

2) *Anatomische und zoologische Untersuchungen uber das Wallross* (*Mémoires de l'Académie imp. des sciences de St -Pétersbourg*, sixième série, Vol IV, 1838, p. 224).

3) *Reisen und Forschungen im Amur-Lande*, Vol. I, p. 179 (St. Petersburg, 1858).

of the Amur finally reached their destination, for this recent fact gives us a welcome clue as to how the same articles may have found their way into the realm of the Khitan at an earlier period. We know that the Gilyak are very shrewd and energetic tradesmen and have taken an active part in the distribution of commercial goods resulting in long journeys which bring them in contact with Manchu and other Tungusian tribes, as well as Chinese, Ainu, Japanese, Yakut and Kamchadal. The observation which is due to L. v. Schrenck bears out the fact that walrus ivory has really transgressed the boundary of China; thus, this ivory trade is not a purely academic construction based on documentary evidence exclusively.

W. Jochelson[1]) has compiled a list of the goods exported in 1899 from Gishiginsk and Baron Korff's Bay, the territory of the Koryak, to Vladivostok. The quantity of walrus-tusks in that year is figured at 25 *pud* (the equivalent of 900 pounds English) to the value of 620 rubles[2]). I have no information on the further handling of this merchandise at Vladivostok, but am under the impression that it arrives there only in transit bound for other ports. Ivory is not worked there, and it seems plausible to assume that China and Japan will receive a due share in these spoils. It remains open for investigation as to how far walrus and narwhal ivory have been

---

1) *The Koryak* (*Memoirs Am. Mus. Nat. Hist.*, Vol. X, p. 775).

2) This is certainly only a small percentage of the total output of this material "It is stated on the highest authority that for several years preceding 1870 about one hundred thousand pounds of walrus ivory was taken annually, involving a destruction of not less than six thousand walruses. Later statistics show that for many years following this date the catch of walrus in Bering Sea was not far from ten to twelve thousand annually. The wholesale slaughter continued until the herds became so reduced in numbers that their pursuit was commercially unprofitable This destruction was additional to the number usually killed by the natives to supply their domestic needs and for barter" (J A Allen, *Am. Mus. Journal*, 1913, p. 42). In this interesting article Allen sounds a timely warning: the extermination of the walrus will be accomplished in a few years unless steps are immediately taken for effective protection. In some districts the life of the natives, for this

or are still utilized in the ivory carvings of those two countries[1]). F. E. BEDDARD[2]) makes the statement that the tusk of the narwhal was employed in Europe in the past as a drug and is so used in China to-day; I am not prepared to confirm or to refute the latter assertion, but should not wonder if it were correct. And finally it should be mentioned that S. WELLS WILLIAMS[3]) gives the following information: "The teeth of the sperm whale, walrus, lamantine, and other phocine animals, form an article of import in limited quantities under the designation of 'sea-horse teeth'; these tusks weigh from sixteen to forty ounces, their ivory being nearly as compact though not so white as that of the elephant[4])."

We read above in the account of al-Bērūnī that the Bulgar bring from the northern sea teeth of a fish over a cubit long. Now this matter has been made the subject of a profound and ingenious historical research as early as 1835 on the part of K. E. v. BAER[5]) whose work is still considered (and justly) by naturalists as a classical treatise. Had Wiedemann had access to it, he could not have

---

reason, is in a precarious condition. BOGORAS (*The Chukchee, Jesup North Pac. Exp.*, Vol. VII, p. 122) reports that to the south of Anadyr the walrus have greatly diminished in numbers, and that the Kerek on the southern shore of Anadyr Bay, who in former times subsisted on walrus, are now rapidly starving to death.

1) The most interesting account of the ivory industry in the East will be found in A. DE POUVOURVILLE, *L'art indochinois*, pp 183—191 (Paris, no year).

2) *A Book of Whales*, p. 248 (New York, 1900).

3) *The Middle Kingdom*, Vol. II, p. 400 (New York, 1901).

4) The "List of Chinese Medicines" published by order of the Inspector General of Customs (Shanghai, 1889, p. 445) registers *hai ma* 海馬 *Hippocampus* sp. as production of Kuang-tung, but not *hai ma ya*. The term *hai ma* has been adopted as the rendering of walrus by the English and Chinese Standard Dictionary (Vol. II, p. 2605), and 海馬或獨角魚之齒 is the translation of marine ivory (Vol. I, p. 1261).

5) *Anatomische und zoologische Untersuchungen über das Wallross (Trichechus Rosmarus)*, in *Mémoires de l'Académie imp. des sciences de St.-Pétersbourg*, sixième série, Vol. IV, 1838, pp 96—236 (with a map showing the distribution of the walrus). What later authors have written on the historical development of our knowledge of the animal, is nearly all derived from this fundamental investigation. A good deal of it is reproduced by J. A. ALLEN, *History of North American Pinnipeds*, pp 82 *et seq* (Washington, 1880).

doubted for a moment that the *chutww* of the Arabs is the tusk of the walrus (the narwhal, though an entirely distinct animal, must be included, as in commerce hardly any distinction is made between the ivory yielded by the two species).

According to the thorough investigations of the great naturalist K. E. v. Baer the first acquaintance of Europe with the walrus dates from the latter part of the ninth century and is connected with the daring exploits of the Norseman Ohthere from Helgeland in Norway who between 870 and 880 sailed around the North Cape to Biarmia (the modern word Perm) and reported on this enterprise to King Alfred the Great of England[1]). The main purpose of his voyage was to obtain "horsewhales (*horshvael*), which have in their teeth bones of great price and excellencie." It appears that on the coast of the North Polar Sea the chase pursuit of the walrus had been going on for some time, and this is confirmed by Russian accounts. The Anglo-Saxon report (and this makes its historical value on which v. Baer lays great emphasis) bears out the fact that walrus-hunting and trade in walrus-teeth took their starting-point in the ninth century from the northern coast of Russia and long preceded the discovery of Greenland. In the sources of Russian history walrus-teeth are known as fish-teet[2]). The famous Russian historian Karamsin has solved this question by appealing to Herberstain who published in 1549 his work *Rerum Moscoviticarum Commentarii*, a primary source for the history of Russia. This author gives a very plain and reliable account of the walrus, insists on the great value attached by the Russians, Turks and Tatars to the teeth,

---

1) Ohthere's account is preserved in the first chapter of King Alfred's (848—901) edition of the *History of the World* (De miseria mundi) by Paul Orosius. The first chapter contains a geographical introduction to the work, in which the account of another Norseman, Wulfstan, regarding a voyage into the southern part of the Baltic Sea is included.

2) Not only in Europe but also in Asia the naive conception prevailed that the walrus was a fish. The Yakut simply call the animal *balyk*, *i. e.* a fish.

and remarks that they are called fish-teeth. Still earlier in 1517 the learned Pole MATTHIAS MECHOVIUS in his *De Sarmatia Asiana et Europaea*, after giving a correct description of the walrus, says: "Hos illae gentes colligendo dentes eorum satis magnos latos et albos pondere gravissimos capiunt: et Moschovitis pendunt atque vendunt: Moschoviae vero his utuntur: ad Tartariam quoque et Turciam mittunt, ad parandum manubria gladiorum, framearum, cultrorum, quoniam gravitate sui majorem et fortiorem impressionem impingunt." Karamsin observes that the expression 'tooth' (зубъ) was not understood in later times, and was taken for a corruption due to copyists, but that walrus-teeth are evidently involved which were used in Novgorod like marten and squirrel-skins in the manner of monetary values; in old Russian tales, these fish-teeth appear as highly priced objects (*e. g.* a precious chair of fish-teeth), in which case only walrus or narwhal-teeth can be understood. In 1159 the Grand duke Rostislav and the Prince Svätoslav Olgovich made gifts to each other on the occasion of an alliance effected at Morovsk; Rostislav presented sables, ermines, black foxes, polar foxes, white bears, and fish-teeth. During the sway of the Mongols and Tatars frequent demands for this product were made from Asia[1]), and Ivan Vasilyevich received in 1476 a fish-tooth as a gift from a citizen of Novgorod. So far v. BAER[2]). We see that, from the ninth century at least, walrus-tusks formed an important article of trade in the north-east of Europe, that they were known as fish-teeth, and that they were traded to the Turks, and probably reached also inner Asia during the middle ages.

In the period K'ai-yüan (713—742) Hing Kuang, king of Sinra

---

1) This affords an explanation for the *ku-tu-si* described by Ch'ang Tê as a product of the Western Regions. It was in my opinion walrus ivory coming from Russia.

2) On p. 224 he refers to FRÄHN (*Ibn-Fozzlan*, p 229) as saying that in Khiwa mammoth-teeth coming from southern Russia and brought to that place by the Bulgar were worked there in early times. These "mammoth-teeth" doubtless were walrus-teeth.

新羅 (in Korea), sent to China as tribute *kuo hia* ponies[1]), silk textiles called *chao hia*[2]), silk textiles called fish-tusks 魚牙紬,

---

[1]) 果下馬. First mentioned in *Hou Han shu* (Ch. 115, p. 5) as an animal occurring in the country of Wei 濊 in Korea, and explained by the gloss 高三尺乘之可於果樹下行 *i. e.* "It is three feet high; when riding it, one can conveniently pass under the fruit-trees" (on account of its low stature). According to *San kuo chi* (*Wei chi*, Ch 30, p. 8 a), such a horse from Wei was offered to China at the time of the Emperor Huan 桓 (147—167 A.D.) of the Later Han dynasty; the gloss there added is the same, with the addition at the end "hence it is called *kuo hia*" (故謂之果下). According to *Wei shu* (Ch. 100, p. 2 a) and *Pei shi* (Ch. 94, p. 4 a: account of Kokuryö) these three feet horses, called *kuo hia*, were reared also in that Korean kingdom and believed to descend from those which Chu Mung had broken in; Chu Mung 朱蒙 is the legendary founder of the kingdom of Kokuryö (*Kao-kou-li*), the son of the sun-god, who was in his youth the groom of the king of Fu-yü, at which time he made a close study of horses (compare *P'ei wén yün fu*, Ch. 51, p. 9 b, which does not quote the oldest reference in the *Hou Han shu*). A. PFIZMAIER (*Nachrichten von den alten Bewohnern des heutigen Corea, Sitzungsberichte der Wiener Akademie*, 1868, p. 501) did not recognize *kuo-hia* as the name for this breed of horse and translated the phrase 即果下也 in the latter passage: "But they truly are inferior horses". — This dwarf-breed of pony is still a famous production of Korea. "In size when alongside of a Western horse, it looks like a ten-year-old boy accompanying his grandfather, or like an ordinary Japanese walking out with Li Hung-chang", remarks J. S. GALE (*Korean Sketches*, p. 119) who has devoted to the animal an essay accompanied by a photograph of it. H. B HULBERT (*The Passing of Korea*, p. 256) has the following: "History and tradition have much to say about this breed of horse. As far back as ancient Yemak, which flourished at the beginning of our era, we read that the horses were so small that men could ride under the branches of the fruit trees without striking their heads against them. From time immemorial the island of Quelpart has been the famous breeding-place of the hardy pony, and the Mongols established themselves there very strongly in order to breed horses for use in their wars". The reference to Quelpart, pointed out also by A. HAMILTON (*Korea*, p 270, New York, 1904) as the place of production of large numbers of pack-ponies, is very suggestive as to the origin of this equine race; it is well known that insular isolation has a tendency to produce diminutive forms of mammals, and this observation has especially been made in regard to insular stocks of horses, as *e. g.* those of Ireland and Iceland, and the much smaller ones of the Isle of Man, the Hebrids, Orkneys, and Shetlands. The dwarf horses of Corsica and Sardinia are described also as being three feet high (see particularly E. HAHN, *Die Haustiere*, p 188, Leipzig, 1896, and C. KELLER, *Studien über die Haustiere der Mittelmeer-Inseln*, p. 125, Zürich, 1911). The above Chinese data are presumably the oldest on record anent such an insular dwarfish breed.

[2]) 朝霞紬. Compare HIRTH and ROCKHILL, *Chau Ju-kua*, p. 218 (and the correction proposed by GILES, *Adversaria Sinica*, p. 394).

and skins of *Phoca equestris*[1]) (*T'ang shu*, Ch. 220, p. 9b). In the
Ts'ê fu yüan kuei (as quoted in *K'in ting Man-chou yüan liu k'ao*,
Ch. 19, p. 5) an embassy from Sinra is more specifically assigned
to the year 723, the list of products being the same, with the addition
of bezoar 牛黃, ginseng, human hair, steel bells to be tied to
the necks of falcons, gold and silver. In 748 the *Mo-ho* of the
Sungari (*Hei shui* 黑水) sent likewise silk textiles called fish-
tusks and *chao hia* silks, and the same objects are enumerated again
in *Kiu T'ang shu* (Ch. 199 上, p 9a), among the tribute gifts offered
by Sinra in 773. But in the latter text we meet an important
variant reading 獻金銀牛黃魚牙、納朝霞紬等 which
means: "They offered gold, silver, bezoar, and fish-teeth; and received
(in exchange from the Chinese Court) *chao hia* silk and other goods."
Also in *P'ei wén yün fu* (Ch. 21, p. 124b) this passage is quoted
under the catchword *yü ya* 魚牙, though in the text *ch'ou* 紬

---

1) 海豹 *hai pao*, 'marine panther'. In 730 five of these skins were offered by
the P'o-hai and Mo-ho; in 734 sixteen skins were sent by Sinra. The name *hai pao*
(probably identical with the 'speckled fish produced in the ocean' 海出班魚,
mentioned in *Hou Han shu*, Ch. 115, p. 5, and *San kuo chi, Wei chi*, Ch. 30, p. 8)
distinctly refers to the finest of all marine mammals, the ribbon seal, *phoca equestris*
Pallas (also *histriophoca fasciata*), first briefly described by PALLAS (*Zoographia Rosso-
asiatica*, Vol. I, p. 111, St. Pet, 1811), then more accurately by L v. SCHRENCK (*Reisen
und Forschungen im Amur-Lande*, Vol. I, Säugetiere, pp. 182—8, St. Pet., 1858). He
who will look up Plate IX in this work and admire the wonderful design on the skin of
the male, will readily grasp the appropriateness of the name 'sea panther'. I am the
fortunate owner of a skin of this now almost extinct phoca obtained in the northern part
of Sachalin in 1898; the Tungus (tribe Emunkun) there call it *alakú*, the Gilyak *alχ*,
the Ainu *targa*. Its habitat is formed by the Bering Sea, the coasts of Kamtchatka, the
chain of the Kuril Islands, the Okhotsk Sea and Tartar Strait down to the southernmost
part of Sachalin. *Ta Ming i t'ung chi* (Ch. 89, fol. 10b; edition of 1461) lists *hai pao*
skins as products of Korea and the Su-shên country; other Pinnipedia are enumerated there
as the sea-ass, sea-badger, sea-ox, sea-dog, and sea-pig. The latter can safely be identified
with *Otaria ursina* L, the so-called fur seal of commerce, as the Tungusian tribe of the
Mangun calls it also *mu-nyghty*, 'water-boar' (SCHRENCK, *l.c.*, p. 189). SCHLEGEL's explanation of *hai pao* as sea-lion (*T'oung Pao*, Vol. III, 1892, p 505) is not very fortunate;
the skins of sea-lions being without value would not have been sent to China as tribute.

takes the place of *na* 納. It seems to me that the text of the *Kiu T'ang shu* preserves the correct reading, and that it is the question there of fish-tusks. On the other hand, the existence of the term *yü ya ch'ou* cannot be denied in the other passages where the words *yü ya* are followed by the word *ch'ou*, and apparently two kinds of silks are understood. The expression "fish-tusk silk," as far as I know does not occur otherwise, nor is it interpreted to us in this case, and it can only be guessed that it may have been a weaving with a fanciful design somewhat resembling the natural veins occurring in the "fish-tusk." But whatever the relation of the latter to the weaving may have been, it is obvious that a product like "fish-tusk" must have been known to the people of Sinra and the Mo-ho to enable them to draw such a comparison, and the "fish-tusk" surely was nothing but walrus or narwhal tusk[1]), in

---

1) It could not have been whalebone which is known under the name *k'ing ya* 鯨牙 (*P'ei wén yün fu*, Ch. 21, p. 126). The oldest account of the whale (defined by the *Shuo wén* as a 'big sea-fish', by the *Yü p'ien* as 'the king of the fishes'), I believe, is extant in the *Ku kin chu* 古今注 by Ts'uei Pao of the middle of the fourth century (Ch. 2, p. 9 b; edition of *Han Wei ts'ung shu*) where it is said: "The whale is a sea fish. The biggest are a thousand *li* long, while the smallest reach a size of a hundred feet. One individual brings forth numerous young ones. In the fifth or sixth month they are in the habit of going to shore for the purpose of propagation. In the seventh or eight month they return with their young ones into the open ocean where they cause an uproar like thunder in rousing the waves and almost produce rain in spirting water out of their jaws. All the water animals, terror-stricken, take to flight, no one daring to offer resistance. The female is called *i(ni)*; the biggest attain likewise a length of over a *li*, and their eyes make bright-moon pearls". 鯨魚者海魚也、大者長千里小者數十丈、一生數萬予、常以五月六月就岸邊生予、至七八月導從其予還大海中、鼓浪成雷、噴沫成雨、水族驚畏皆逃、匿莫敢當者、其雌曰鯢、大者亦長于里、眼爲明月珠。 It is well known that the pearls bright like the moon are listed by the Chinese among the products attributed to the Roman Orient and are frequently mentioned in the texts relative to Ta Ts'in, as may be ascertained by referring to HIRTH's China and the Roman Orient, and CHAVANNES, *T'oung Pao*, 1907, p. 181. It should not be supposed, however,

the same manner as we hear of fish-tooth in the Slavic regions.
In this connection the account of the *T'ang hui yao* quoted above
may claim great significance, if it can be proved that the passage
already occurred in the edition of the T'ang period. It is an interesting coincidence, as we now observe, that the Mo-ho, on the one
hand, are reported to possess *ku-tu* horn, and on the other hand

that the term *ming yue chu* was coined only at the time of Chinese relations with Ta
Ts'in, but as shown by the quotations given in *P'ei wén yün fu* (Ch. 7 A, p. 97), it occurs twice in the *Shi ki* of Se-ma Ts'ien. In the periods K'ai-yüan (713—742) and T'ien-
pao (742—756) the Mo-ho of the Sungari sent as tribute pupils of the eyes of whales
鯨 睛, sable-skins and white hare-skins (*T'ang shu*, Ch. 219, p 6), and according to
Ts'é fu yüan kuei (quoted in *K'in ting Man-chou yuan liu kao*, Ch. 19, p 5), the Mo-ho
sent the same objects (鯨 鯢 魚 睛) in 719. *Ta Ming i t'ung chi* (Ch. 89, fol 10 b;
edition of 1461) lists whale-pupils among the products of the country of the Su-shén
(compare also SCHLEGEL, *T'oung Pao*, Vol. VI, 1895, p. 41). — The coincidence of Gilyak *keñ* (among the Orochon, a Tungusian tribe on Sachalin Island, who call themselves
Ulča, I noted *kū̀ñia*) 'whale' with Chinese *k'ing* is curious. MÖLLENDORFF, in a somewhat
inconsiderate notice on the Gilyak language (*China Review*, Vol. XXI, p 143) in which
he "proves" to his own satisfaction the relationship of Gilyak with the Ural-Altaic languages,
makes Grube say that *keñ* is a Chinese loanword in Gilyak. GRUBE has never said anything
so foolish as that, but has simply recorded the word, without further comment, from the
notes of Schrenck in his Giljakisches Wörterverzeichnis, p. 56 (Anhang zu Schrenck's
Reisen und Forschungen im Amur-Lande, St. Pet , 1892). It is entirely out of the question
that the Gilyak word is derived from Chinese If there is any people in eastern Asia
thoroughly familiar with the whale, it is certainly the Gilyak; they are the only ones
among the Amur tribes to hunt the whale (*Balaenoptera longimana*) and surely know as
much about the animal and its habits as the Chinese The beach along the east coast of
Sachalin is strewn with skeletons of castaway whales, and whale-bone is amply utilized by
the Gilyak in their industries, e g. for the runners of their sledges. The Orochon word
kū̀ñia, moreover, shows that the word *keñ* or more correctly *kañ* is a specific Sachalin
word, for the Tungusians on the mainland designate the whale as *kā'lym*, the Gold on
the Amur as *kā'lyma*. As the word *kañ* is absent on the mainland, it is most improbable
that the Chinese word can be traced back to it, and the coincidence may be accidental. —
The subject of the whale has also a slight bearing on Chinese art, in that the whale has
sometimes been associated with the dragon. SCHLEGEL (*T'oung Pao*, Vol VI, 1895, p. 42)
has furnished an example af a whale being cast off on the coast of Chê-kiang and regarded
by the people as a dragon. In the *Fang shi mo p'u* of 1588 (Ch. 4 下, p. 52) is
an illustration of a *hüan k'ing pao chu* 玄 鯨 寶 柱 'precious pillar (*ratnastambha*)
of the black whale'. The pillar crowned with five rows of jewels in Buddhist style and
adorned with lotus designs at the base is wound around by a dragon, head downward and
tail upward, with the body and tail of a fish

to be acquainted with fish-teeth. The one, however, must be identical with the other. The Mo-ho were a Tungusian tribe related to the Khitan, and it would be no marvel after all if they had been in possession of that Tungusian word as early as the T'ang period. There can be no doubt of the fact that the trade in the article makes itself felt in that epoch, and that the Mo-ho and the Koreans took an active part in it. This affords the strongest historical evidence for the fact that *ku-tu-si* cannot have been the product of the rhinoceros nor of the mammoth, neither of which occur in the territories of Korea and the Mo-ho, but this ethnographical indication opens the way to the northern Pacific Ocean and brings us in immediate contact with the ivory produced in its waters. The Moho bordered on the ocean along the shore stretching between the Korean peninsula and the mouth of the Amur[1]), and thus were next-door neighbors to that stock of North-east Asiatic tribes which are often designated Palae-Asiatic, but which I prefer to comprise under the term of the North-Pacific culture-area.

As a last resort, Chinese trade in marine ivory leads us back to the culture of those arctic peoples settled along the northern shores of Asia and America who hunt the narwhal and walrus for the sake of their flesh, blubber, and tusks, and whose work in ivory carving forms an essential feature of their cultural achievements. The wide geographical distribution of this industry over vast and scattered tracts of circumpolar land is amenable to the belief that

---

1) A rather clear allusion to walrus-tusk in the same region is made in the Imperial Geography of the Ming Dynasty (*Ta Ming i t'ung chi*, edition of 1461, Ch. 89, fol 10 b) where *shu kio* 殊 角 is recorded as one of the products of the country of the *Su-shên*, a gloss being added to the effect 即 海 象 "this is the sea-elephant". Sea-elephant and latinized *Elephas marinus* was the common name by which the walrus was known in Europe during the sixteenth and seventeenth centuries, and more particularly in those countries of Europe where only the teeth were traded but little accurate accounts of the animal were spread (K. E v. BAER, *l. c.*, pp. 109, 117).

it is very ancient, and not only the art but also the religion and mythology of the Eskimo, in particular their highly organized system of taboos, with which narwhal and walrus are closely interwoven [1]), point to a great antiquity as regards their acquaintance with these animals.

W. Bogoras, our great authority on the Chukchi, has given a vivid description of walrus-hunting as practised by this people [2]). The trade formerly carried on by it in the tusks must have been enormous: official records among the archives of Kolyma reveal the fact that, in 1837, 1563 walrus-tusks were sold at the fair of Anui first established in 1788 [3]).

The Koryak employ for carving, W. Jochelson [4]) informs us, different kinds of wood, the antler of reindeer and the horn of

---

1) Boas, *The Eskimo of Baffin Land and Hudson Bay* (*Bulletin Am. Mus. Nat. Hist.*, Vol. XV, 1901, pp. 122, 123). Narwhal and walrus themselves are carved from ivory by the Eskimo (Boas, *The Central Eskimo*, plates 8 and 9; Murdoch, *The Point Barrow Eskimo, Ninth Annual Report Bureau of Ethnology*, 1892, p. 272); the walrus also by the Chukchi (v. Nordenskiöld, *Umsegelung Asiens und Europas auf der Vega*, Vol. II, p. 129) and the Koryak W. Jochelson, *The Koryak, Memoirs Am. Mus. Nat. Hist.*, Vol. X, p. 662). The objects carved by the Eskimo from ivory are numerous: knife-blades, handles for skin-scrapers, ends for back-scratchers, tops for spinning, dice, combs, needle-cases, snuff-boxes, tobacco-pipes, beads for hair-ornaments, bows, ear-trumpets used by hunters at the seal-hole to hear more readily the noise made by the emerging seal, and animal-carvings.

2) *The Chukchee*, I. Material Culture, *Jesup North Pac. Exped.*, Vol. VII, 1904, pp. 122—123.

3) *Ibid.*, p. 56. American whalers now accept walrus ivory in payment of goods furnished to the Chukchi (p. 63). They carve from ivory beads and buttons used for personal adornment (pp. 259, 260), and large numbers of animal figures, many examples of which are illustrated in the work of v. Nordenskiöld (*l. c.*, Vol. I, p. 463; Vol. II, pp. 128—141).

4) *Material Culture and Social Organization of the Koryak* (*Memoirs Am. Mus. Nat. Hist.*, Vol. X, p. 646). Before their acquaintance with iron the Koryak used stone implements in working bone. The walrus-tusks were split into strips by means of stone chisels and wedges, and the work was continued with the aid of stone knives and awls. At present tusks are sawed with an iron saw, home-made or imported, and the rest of the work is accomplished with a knife (p. 670).

mountain-sheep, bone of whale, teeth of the white whale and the bear, walrus-tusks, and mammoth ivory. Sometimes the horn of the narwhal, brought from the shores of the Arctic Ocean, is also used. The material most suitable, on account of its solidity and fineness of grain, is ivory of the walrus and mammoth, especially the latter, which is as hard as the former, walrus-tusk being used to a greater extent than mammoth-tusk, because the latter is not found so frequently in the Koryak territory as in the more northern regions of the Chukchi. Jochelson points out that both kinds of ivory, when exposed to the air and moisture for a long time, lose their original whiteness and acquire a yellow tobacco color [1]). The sculpture of the Maritime Koryak who carve figures of wrestlers and drummers is most remarkable for the lifelike action and motion of representation and sharply contrasts in this point with similar efforts of the Maritime Chukchi and Eskimo who merely grasp to a certain extent the exterior forms of an animal but represent it in a stiff and motionless manner. Besides artistic carvings, the Koryak further make thimbles, rings, and particularly chains of ivory [2]), the latter carved out of a single piece of bone.

If the length of the preceding notes may seem somewhat unduly out of proportion with the subject proper, I wish to say, by way of apology, that it was necessary to point to the central region from which this peculiar Chinese trade in *ku-tu-si* has radiated, and to insist upon the antiquity and importance which the marine product must have had in the extreme north-east of Asia. The mere lack of historical documents for that culture-area cannot prevent us from

---

1) This chimes in with the yellow color which Hung Hao attributes to *ku-tu si*. I have examined a large number of Eskimo ivory carvings in the collections of the Field Museum, and find that stained pieces range from a light yellow to a deep brown, while others have retained their pure whiteness; but the latter may be supposed to be narwhal ivory.

2) *Ibid.*, pp. 626, 670.

regarding the utilization of ivory there as being of considerable age[1]), and as having given the impetus to a trade in this product moving in a southerly direction and reaching the Mo-ho and Korea before the eighth, and the Khitan before the tenth century. *Vice versa,* the Chinese accounts corroborate the necessary supposition that an ivory industry must have existed in those early days in the far north, and that the peoples living there must have pursued the capture of the sea-mammals yielding the precious material[2]).

---

1) Also v. NORDENSKIÖLD (*l.c.*. p. 137) justly says that long before historical times the walrus has been captured by the polar peoples, and that implements of walrus-bone appear among the grave-finds in the north of Europe.

2) The location of *ku-tu-si* in the country of the Khitan, the two Khitan words *ku-tu-si* and *tu-na-si*, the acquaintance of the Koreans with walrus-tusk, and the modern trade in this article of the Koryak and Gilyak necessarily lead us to the inference that the transportation of walrus and narwhal ivory always moved along the north-east coast of Asia, as an offshoot of the North-Pacific culture-area. It remains to be considered that besides this natural maritime route there may have been an inland commercial high-road from inner Siberia into the region of the Khitan. An indication to this effect may be gleaned from the interesting geographical text inserted in *Wei shu,* Ch. 100, p. 6 (identical with *Pei shi,* Ch. 94, p. 10 b). The Shi-wei, a tribe akin to the Khitan, in-habited under the T'ang a territory bordered in the east by the Mo-ho of the Sungari, in the west by the Tu-küe, in the south by the Khitan, in the north by the sea, the centre of their habitat being formed by the basin of Kerulen River (compare CHAVANNES, B. E. F. E. O., Vol. III, 1903, p. 225). According to the text of the *Wei shu,* more than a thousand *li* west of the Shi-wei was the country of the *Ti-tou-kan* 地豆干, (or *Ti-tou-ya?*), and more than 4500 *li* north of the latter was the country of the *Wu-lo-hou* 烏洛侯. "North-west of this country there is the river *Wan* 完水 which flows in a north-easterly direction and unites with the river Nan 難水, The small streams of this territory all discharge themselves into the Nan which flows in an easterly direction into the sea. After a twenty days' journey toward the north-west is encountered the Great Water *Yü-ki(i?)-ni* 于已尼大水 which is called the Northern Sea". 其國西北有完水、東北流合于難水、其地小水皆注於難東入于海、又西北二十日行有于已尼大水所謂北海也。 WASILYEF (Труды Восмочнаго Омд. Цми. Археолог. Общесмва, Vol. IV, p. 33, St. Pet., 1859) regards this Northern Sea as Lake Baikal and the two rivers as Onon and Selenga, and there is certainly much in favor of such a view. On the other hand, there are grave obstacles in the way of such an interpretation; the Selenga falling into Lake Baikal seems to me

While the North-Pacific world was still unknown and covered by a dense veil, we hear the pulsation of human labor beating there in the Chinese records of Arctic ivory. No part of the world, to our modern way of thinking, stands any longer in rigid isolation; lands and peoples of the farthest Thule draw nearer and nearer and join into the general frame of history. Those who have pursued the epoch-making results of the Jesup North Pacific Expedition — the publications of which are still in progress under the energetic editorship of Franz Boas, its spirited leader, — are now familiar with the fact that Asia and America are overbridged, and that migrations of tribes as well as currents of thought and culture have passed from one continent to the other. With reference to our present subject, another matter of Asiatic-American interest here deserves mention, as briefly as possible. In the Annals of the Three Kingdoms (*San kuo chi*, *Wei chi*, Ch. 4, p. 13 a) it is on record under the third year of the period King-yüan (262 A. D.) that the country of the Su-shên sent a tribute of thirty bows three feet and five inches long, arrows of the wood *hu* 楛 [1]) one foot and

---

quite out of the question as capable of being identified with the river Nan which, as is plainly said in the text, flows into the eastern ocean. The Amur can hardly be intended, being too well known to the Chinese to suppose that a suppression of its usual name might be intended in this context. If the Shi-wei were located in the basin of the Kerulen, the distance of 5500 *li* partly west and partly north of this territory would apparently carry us much farther than the valley of the Onan and probably lead us into the river system of the Witim and Lena. The identification of the *Pei hai* with the Arctic Sea of the Siberian coast, however, would be beset with no small difficulties. Far from pretending to solve the problem, I merely wish to intimate that the text of the *Wei shu* is capable of a different interpretation than the one advanced by Wasilyef. The mode of interpretation has no direct principal issue for the point under consideration. With reference to the pending question the vital point of the argument is that the Khitan (as later the Niüchi), in the west and north-west, were backed by a number of tribes connecting them and their culture with the very heart of Siberia, and were in-fluenced by commercial and mental currents coming from that direction.

1) An unidentified tree, mentioned as early as in the *Yü kung* (compare BRETSCHNEIDER, *Bot. Sin.*, pt. 2, N°. 543). In the place of *hu*, other texts write *jo* 楉, also unexplained

eight inches long, three hundred stone crossbows 石砮, a mixed lot of twenty armors of leather, *bone*, and iron 皮骨鐵雜鎧 二十領, and four hundred sable-skins. Hide armor and bone armor formed the national defensive weapons of the Su-shên, as may be inferred from a passage in the Annals of the Tsin Dynasty (*Tsin shu*, Ch. 97, p. 2 b) where the characteristic weapons of the tribe are enumerated as, "stone crossbows, hide and bone armor 皮骨之甲, bows from the timber of the tree *t'an*[1]) 檀弓, three feet and five inches long, arrows from the wood *hu*, one foot and eight inches long 長尺有咫." The subject revealed by these two memorable passages has a large bearing on American ethnology and the history of plate armor in America and Asia, and has been discussed at full length by me in an address delivered on January 2 of this year before the meeting of the American Anthropological Association at Cleveland under the title "Plate Armor in America, a sinological contribution to an American problem"[2]). Only a few indications can find place here. It is noteworthy that the Chinese do not ascribe bone armor to any other of the numerous tribes with which they came in contact during their long history, and whose culture they have described to us. In all likelihood the term 'bone armor' occurs in their records only in those two passages, and it is not at all ambiguous. There is but one

---

(*ibid.*, N°. 569). The arrowheads of the Su-shên and allied tribes were chipped from flint. The principal passages relating to the flint arrowheads 石鏃 of the ancient Tungusian tribes are *Hou Han shu*, Ch. 115, p. 2 b; *San kuo chi, Wei Chi*, Ch. 30, p. 7 b; *Tsin shu*, Ch. 97, p. 2 b; *Wei shu*, Ch. 100, p. 4 a; *Pei shi*, Ch. 94, p. 7; *T'ang shu*, Ch. 219, p. 5 b (compare *Jade*, pp. 57 *et seq.*). *Hu* arrows and stone crossbows of the Su-shên were sent as tribute from Korea in 458 A. D. (*Nan shi*, Ch. 79, p. 1 b).

1) *Dalbergia hupeana*, yielding the well-known blackwood of commerce from which carvings and furniture are turned out at Canton and Ningpo. In the above case, another species of the Amur Region seems to be meant.

2) A brief abstract of this address has appeared in *Science*, Vol. 37, 1913, p. 342. Its publication in full is hoped for in the near future.

thing that can be understood by it (and my friends working in the field of American ethnology are agreed with me on this point), — the well-known type of bone plate armor, consisting of rows of overlapping plates of ivory, as still occurs among the tribes occupying the northern shores of the Pacific on the American and Asiatic sides, particularly among the Eskimo and Chukchi, and in that region exclusively. The plates in this type of armor are usually carved from walrus ivory, as naturally possessing a greater elasticity than other ordinary kind of bone. The point at issue, then, is the fact that the entry of the Chinese annalist under the year 262 regarding the presentation of bone armor on the part of the Su-shên is the earliest recorded reference in history to plate armor of presumably walrus ivory, and hence the earliest instance of an object wrought from this material. We now recognize also that the Geography of the Ming Dynasty, as previously stated, is quite right in assigning walrus ivory to the country of the Su-shên. In the tracing of this article we are thus carried far beyond the time when the word *ku-tu-si* made its début; we see that, prior to the age of the Khitan, Mo-ho and Koreans, the Su-shên were in possession of walrus ivory, at least earlier than the year 262, and probably worked it themselves into plates for defensive armor. Narwhal and walrus ivory became known likewise to the Japanese. F. W. K. MÜLLER[1]) called attention to the fact that the word 一角 is read in Japanese *unkōru* or *unikōru* (our word *unicorn*), when a commercial product brought to Japan by the Dutch (more correctly perhaps in earlier times by the Portuguese) comes into question, and quotes

---

1) Ikkaku Sennin (reprint from *Bastian-Festschrift*, p. 24). The English and Chinese Standard Dictionary (published by the Commercial Press of Shanghai, 1908, Vol. II, p. 1505) has adopted the old European term 'sea-unicorn' in the translation of the word narwhal by 獨角魚 or 一角魚.

RÉMUSAT[1]) as saying that in this case rhinoceros-horn is hardly understood but rather narwhal-tusks. The walrus is equally entitled to consideration, as the teeth of the two animals are not discriminated in commerce. At the end of the eighteenth century shipwrecked Japanese sailors cast off on the Aleutian Islands acquainted their countrymen with a somewhat romantic but unmistakable sketch of the walrus[2]), and it happens that walrus get astray into Japanese waters. Captain H. J. SNOW[3]) remarks on this point: "The writer has never seen the walrus about the Kurils, or even south of Avatcha Bay, on the Kamchatka coast. A stray one, however, was taken some years ago near Hakodate, in Tsugaru Strait, which must have passed along the Kurils from the north." It seems, however, that prior to the time of Portuguese and Dutch trade narwhal and walrus ivory were known in Japan. At least, A. BROCKHAUS[4]), evidently from a Japanese source, makes the statement that both materials inclusive of elephant ivory (*zōge*) were utilized for the carving of *netsuke*, and remarks that narwhal tooth, alabaster-like, was taken during the middle ages also in Japan for the horn of the unicorn, being regarded as an infallible antidote against poison and paid dearer than gold[5]).

---

1) *Notices et extraits des manuscrits de la Bibliothèque Nationale*, Vol. XI, pt. I, p. 198 (Paris, 1827). Dr. Müller accepted this interpretation of Rémusat without reserve; but whoever will look up the sketch of the horn supplied by the *Wa-Kan San-sai-zu-e* (the Japanese edition of the Chinese *San ts'ai t'u hui*) on which the argument of Rémusat and Müller is based, can not fail to notice that the horn slender and curved as there represented can only be tusk of a walrus, not that of a narwhal which is perfectly straight, pointed, and twisted in grooves. The Japanese illustration is very distinct and true to nature; it strictly excludes any notion of a rhinoceros-horn, and as could be confirmed from actual comparison with the stuffed specimen of a walrus in the Field Museum, refers only to the tusk of this animal.

2) Reproduced by A. E. v. NORDENSKIÖLD, *Die Umsegelung Asiens und Europas*, Vol. I, p. 140, where a detailed account of those Japanese sailors and their diary is given.

3) *Notes on the Kuril Islands*, p 28 (London, 1897).

4) Netsuke, *Versuch einer Geschichte der japanischen Schnitzkunst*, p. 25 (Leipzig, 1905).

5) It would be interesting to know from what Japanese source this information is

Under the Liao and the Kin, *ku-tu-si* does not seem to have entered the pharmacopœa[1]); at least we do not know, Hung Hao mentions only the one practical utilization for knife-hilts. It is

---

derived, and what the Japanese names for narwhal and walrus ivory are. On p. 26 of his interesting and attractive work, Mr. Brockhaus quotes from the Japanese book *Sōken kishō* of 1781 the sentence that "there is a material sold by tricky dealers under the name *ningyo kotsu* (人魚骨), said to be the lower jaw of a shark", and concludes that this might be an error, as the bones of the 'siren' or 'mermaid' of antiquity, the dugong and the whale, were used like ivory. But *ningyo* (Chinese *jên yü*) is an ancient general designation for Pinnipedia which occurs as early as Se-ma Ts'ien's *Shi ki* (CHA-VANNES, Les mémoires historiques de Se-ma Ts'ien, Vol. II, p. 195), and G. SCHLEGEL (*T'oung Pao*, Vol. III, 1892, pp. 506—9) has proved with good arguments that these 'human fish' or 'mermaids' of Chinese lore are nothing but seals. The Chukchi carving of a seal with a human head figured by W. BOGORAS (*Memoirs Am. Mus. Nat. Hist.*, Vol. XI, p. 329) is the offshoot of these beliefs in the human character of seals, which is emphasized also by modern observers (*e. g.* Steller). Moreover, Captain SNOW (*l. c.*, p. 84) observes in regard to the sea-lion that its large canine teeth, some of which are nearly four inches in length, and of the consistency of ivory, are sometimes carved by the Japanese into *netsuke*.

1) This is confirmed by the fact that the work of Li Shi-chên is the first and only *Pên ts'ao* to make mention of *ku-tu-si*, while the article is absent from the *Chêng lei pên ts'ao*, the *materia medica* published in 1208 by the physician T'ang Shên-wei 唐慎微 . BRETSCHNEIDER (*Bot. sin.*, pt. 1, p. 47), while accurately describing this work, confesses that he never came across it, though it is still extant. Two Ming editions in folio were secured by me in Si-ngan fu, those of 1523 and 1587. The following bibliographical references are based on the notes of Mo Yu-chi (Ch. 8, p. 5) quoted above. The *editio princeps* of 1108 known as the Ta-kuan edition was followed by a reprint issued under the Sung in the period 1111—18 and hence designated as the Chêng-ho edition. The latter was republished under the Kin in 1204, with re-editions in 1206 and 1214. A fac-simile of the Sung print saw the light under the Yuan in 1302. From the Ming period no less than six editions are noted by Mo Yu-chi: 1468, reprint of the Kin edition of 1204; 1523, facsimile of the Sung print; 1572; 1577, reprint of the Yüan edition, followed by a new edition in 1579; finally 1598, the last three falling within the period Wan-li. Under the Manchu only one edition was published in 1656 in the reign of Shun-chi, which seems to be the last. It will be recognized that the *Chêng lei* main-tained its place till the appearance of the *Pên ts'ao kang mu* in 1596 supplanted it. The T'ien-lu-lin-lang 天祿琳琅 Library possessed copies of the Sung, Kin, and Yüan editions. The importance of the work rests on the fact that it reflects the tradition of the science of the Sung period and contains many ancient texts excluded from its successor, while other extracts often mutilated by the latter are reproduced in a more complete or more correct form. Also its illustrations are of interest, and there are many not adopted by Li Shi-chên.

apparent from the two words *ku-tu-si* and *tu-na-si* stated as belonging to the Khitan language that the last syllable is part of the Khitan word-stem, and not a Chinese addition to a stem *ku-tu* or *ku-tur*, and *tu-na*. If the character *si* 犀 was chosen to transcribe in Chinese the syllable *si* in the two Khitan words, the reason was, as has been explained, that walrus and narwhal tusk was looked upon as a horn, and *si* 犀 means not only rhinoceros but also rhinoceros-horn [1]). This conception of the tusks as horns and the suggestive writing of the word resulted during the Mongol period in the thought development that *ku-tu-si* was regarded as an effi-

---

1) The Turkish and Arabic forms *ḫutu* and *chutwic* naturally presuppose a Chinese *ku-tu* in which the final *si* was dropped. This hypothetical *ku-tu* I take as the Chinese colloquial word formed after the Khitan word *kutusi*. The peculiar way of writing this word leading to its association with rhinoceros-horn produced among the Chinese the notion appearing during the Mongol period that the word *ku-tu-si*, separated into *ku-tu si*, was a formation by analogy with the numerous varieties of rhinoceros-horn, as there are *t'ung-t'ien si* 通天犀 'the horn communicating with the sky', *pi-han* 辟寒 *si*, 'the cold-dispelling horn', *pi-shu* 辟暑 *si*, 'the heat-dispelling horn, *ye-ming* 夜明 *si*, 'the horn shining at night, *kwan-fên* 蠲忿 *si*, 'the wrath-removing horn', *pi-ch'en* 辟塵 *si*, 'the dust-dispelling horn, and others (which are all discussed in the forthcoming publication previously alluded to). For this reason the process of eliminating the word *si* was easy, and in the same manner as *t'ung-t'ien* was said in lieu of *t'ung-t'ien si*, also *ku-tu si* was by way of analogy abbreviated into *ku-tu*; hence the corresponding Turkish and Arabic forms whose existence renders the supposition of a Chinese *ku-tu* necessary. The above remarks are made without any regard to the word *ku-tu* ascribed to the Mo-ho in the *T'ang hui yao*; as said above, it remains to be seen whether this passage was extant in that work, as it existed in the T'ang period. If this should be the case, my opinion would require a certain modification in that we should have two Tungusian words, a *Mo-ho* word *kutu*, and a Khitan word *kutusi*; this would mean that the latter is a compound in which *kutu* is the name of the animal itself and *si* may have the significance of tooth or horn. Indeed the parallel Khitan word *tunasi* may lead one to the same view. In the present state of our meagre and inaccurate knowledge of Tungusian languages it would be "love's labor lost" to speculate on the origin and meaning of the two Khitan words; in their phonetic make-up they are Tungusian all right, though there is the possibility that they may have been adopted with the goods from a farther North-east-Asiatic tribe. But neither in Gilyak nor in Kamchadal, Koryak, Yukagir or Chukchi can I discover anything that would be comparable with them. We have to wait. The two words do not occur in Manchu.

cient remedy on a par with rhinoceros-horn, and like this one could neutralize every poison[1]). We see how this belief was gradually aggrandised, if we compare the simple statement of Ch'ang Tê with the more elaborate note of T'ao Tsung-i about a century later where *ku-tu* is wittily interpreted as the *ku* poison, and with the fanciful dream of Ts'ao Chao who simply plagiarizes a text relative to rhinoceros-horn heading it with the title *ku-tu-si*. Thus, the final outcome was that *ku-tu-si* was regarded as a substance closely akin to, or identical with, rhinoceros-horn. It is no doubt this peculiar development of beliefs in China which has imparted itself to the Arabs. If the word *chutww* cannot be explained from Arabic, as Prof. Wiedemann says, it would be reasonable to infer that it is derived from Chinese-Khitan *ku-tu-si*, and Turkish *ḫutū* would appear as the intermediary form. If this identification is correct, it is logical to conclude also that the Arabic and Turkish words refer to walrus

---

1) Not only in China and among the Arabs but also in Europe narwhal and walrus ivory was employed medicinally, at least as far down as the seventeenth century, as already shown above by a reference to Boetius de Boot. W. JARDINE (*The Natural History of the Ordinary Cetacea or Whales*, p. 190, Edinburgh, 1837) remarks on the former use of the narwhal-tusk in Europe: "At a time when the origin of the horns of these animals was less known, and when they were more rare than in the present day, they were considered as invaluable, and brought a high price. The physician, and still more the charlatan, employed them, and superstition converted them to its own use; for it is stated that the monks in various convents procured the *true* horn of *the* unicorn, endowed with unheard of powers, and far and near obtained for them the credit of curing the most inveterate diseases". It is well known that the narwhal became the unicorn of European fables and largely figures as such in the mediaeval bestiaries. Dr. E. L. TROUESSART of the Muséum d'Histoire Naturelle of Paris remarks on this subject (*Proceedings of the Zoological Society of London*, 1909, p. 200): "Le Rhinocéros blanc (*Rhinoceros simus cottoni*) est très probablement l'*Unicorne* ou *Licorne* des anciens. Ctsésias (410 av. J. C.) nous apprend que, dès cette époque, on creusait dans la corne de Rhinocéros des coupes qui avaient la réputation de mettre ceux qui s'en servaient pour boire à l'abri de l'effet des poisons. C'est seulement au moyen âge que la défense de *Licorne de mer* ou Narwal (*Monodon monoceros*) fut considérée comme ayant la même propriété, et placée sur le front de la Licorne héraldique qui figure comme *support* dans les armes de la Grande-Bretagne". The passage here referred to in Ctesias will be found in *Indica Opera*, ed. Baehr, p. 254.

and narwhal ivory. And this can implicitly be inferred from the Arabic and Turkish texts: it is true beyond cavil, as shown above, in regard to the fish-teeth traded by the Bulgar and coming from the northern sea. W. REINHARDT[1]) made an emphatic plea on behalf of the *chutww* of the Arabs being nothing but rhinoceros-horn imported from India, and this is "quite indubitable" to him. But the Indian rhinoceros does not occur on any northern sea nor in any of the other localities mentioned in the Arabic and Turkish texts. The Arabs following the example of the Chinese have merely transferred to the walrus-tusks certain popular beliefs entertained regarding rhinoceros-horn. If anything in the case is quite certain, it is that rhinoceros-horn is not understood by *chutww*. Why should the Egyptians have craved it and purchased it for a price two hundred times its value, if *chutww* was rhinoceros-horn which they could have obtained easily and in great abundance from inner Africa? And were the Arabs themselves not familiar with the rhinoceros and its horn, called *kerkeden*? True it is that the bull in the country of the Kirgiz savors of the mammoth[2]). But notwithstanding mammoth ivory is not involved in this case, because the Arabs, I am inclined to believe, in the same manner as the Chinese, would call this material simply ivory, and further, because no such superstitious beliefs as come here into question exist in regard to ivory in Siberia, China, or elsewhere. The bull of the Kirgiz rests on a confusion of notions which may be accounted for in various ways. It seems to me that the Kirgiz were the mediators in the trade of *ku-tu-si* between the Chinese and the Turks, and possibly the Arabs. Naturally the Kirgiz were questioned by their neighbors

---

1) *Der Islam*, Vol. III, 1912, p. 184.

2) According to S. PATKANOV (*Die Irtisch-Ostjaken*, Vol. I, p. 123, St. Pet., 1897) the mammoth is called 'Earth-ox' by the North Ostyak. T'ao Hung-king compares its size with that of the water-buffalo and the taste of its flesh with that of beef; Su Sung says that it resembles the ox (*Pên ts'ao kang mu*, Ch. 51 下, p. 10).

and customers as to the nature and origin of the article and the animal to which it belonged; naturally they knew as much about walrus and narwhal as the Chinese and the Arabs, and any explanation was therefore acceptable. As transpires in such cases, an imported word is easily understood or interpreted with a word of one's own language, and it seems to me that the foreign word *kutu-si* was taken by the Kirgiz or a related Turkish tribe on account of some real or alleged similarity in sound in the sense of a word of their language signifying 'bull'[1]). At all events, while I strictly

---

1) When I incidentally refer to Djagatai *kotas* and Taranchi *kotaz*, 'yak' (RADLOFF's *Versuch eines Wörterbuches der Turk-Dialekte*, Vol. II, col. 608), I simply mean to furnish an example as to how in my opinion the process might have evolved, but I do not mean to say that these actual words have been the agency instrumental in bringing about this end. The fact that the understanding of the word of a foreign language in one's own or pure misunderstanding of it will lead to fabulous speculations in regard to an animal is well proved by the Russian name of the walrus МОРЖЪ. In the Latin account of Matthias Michovius of 1517 (quoted above) it appears as *morss* (hence French and English *morse* first coined by Buffon), and the accidental similarity with Latin *mors* 'death' seems to have contributed much to the West-European notions of the formidable character of the animal, while there is no word to that effect in the Russian accounts K. E. v. BAER, *l. c.*, p. 111). A translation of Herberstain published at Basle in 1567 describes the walrus as an animal of the size of an ox, called by the natives *Mors* or Death. In the historico-topographical work *De gentium septentrionalium conditionibus cet.* (Rom, 1555) the etymological joke is perpetrated to derive the word from *mordere* 'to bite': Norvagium litus maximos ac grandes pisces elephantis magnitudine habet, qui morsi seu rosmari vocantur, forsitan ab asperitate mordendi sic appellati etc. (*ibid.*, p. 112). The latinized *rosmarus* is derived from Skandinavian *rosmhvalr* (= horsewhale; Norwegian *rostungr*, Anglo Saxon *horsewael*, Dutch *wallrus*). Hence the popular names sea-horse (*cheval marin*) and sea-cow (*vache marine*; latinized *bos marinus*; the early French settlers in America used the expression *bête à la grande dent*). — G. SCHLEGEL (*T'oung Pao*, Vol. VI, 1895, p. 24) remarked: "Le Narval est bien connu des Chinois qui l'appellent *Loh-sze-ma* 落斯馬 : un nom que nous n'avons pas pu identifier". Apparently this *lo-se-ma* is a regular transcription of the word *rosmarus*, and the meaning intended is walrus, and not narwhal: this Chinese word is not found in any dictionary, and if this identification is correct, it was evidently formed by a missionary or foreign scholar who translated a European treatise on zoology into Chinese. It is certainly an absurdity to say that the narwhal, an animal restricted to the arctic regions, should be well known to the Chinese (as already stated, little is known about its life even to our modern science and the only people familiar with it is the Eskimo), but it should not be forgotten that

adhere to the conclusion that Arabic *chutww* and Turkish *ḫutū* like Chinese *ku-tu-si* principally denote walrus and narwhal ivory, it must be admitted that a confusion with mammoth ivory was possible, in view of the fact that it seldom was the complete tusk which was the object of trade, but prepared fragments or wrought articles.

The propagation of walrus and narwhal ivory is one of the stories of romance in the history of trade, and if not a page of great importance in the development of culture, yet a picture not devoid of a certain human touch with a grip of fascination upon our minds. The wonders of the Arctic Seas and the indomitable energy of the polar peoples far in the background, then a sudden flash of the daring exploits of the Norsemen, the steel-hard audacity of Siberian adventurers and treasure-seekers, castaway Japanese sailors adrift among the Aleutians, the Mo-ho and Khitan as receivers and distributors of the northern goods, the commerce of the Mongols uniting East and West, and the marvels of the Arctic finally landing at the foot of the Egyptian pyramids, — all this makes a little chapter of human effort and activity furnishing food for some reflection.

Schlegel found the narwhal described in the *Shi i ki* (*ibid.*, pp. 21, 23). Aside from the doubtful authenticity of this work (compare WYLIE, *Notes*, p. 192) in which Schlegel placed absolute confidence unrestricted by any sound criticism, it is questionable whether the narwhal *must* be recognized in this "fish a thousand *chang* long, spotted, having a horn at the end of its nose". The extraordinary length and the further note that it spirts forth water appearing from a distance like colored clouds would rather be suggestive of a species of whale. It is moreover incorrect on the part of Schlegel to assert that the narwhal is generally called by the Japanese *shachihoko*; this word denotes the grampus, and the Japanese were not acquainted either with the narwhal as an animal species.

Additional Notes on Ku-tu-si.

An interesting text relative to *ku-tu-si* occurs in the *Yün yen kuo yen* *lu* 雲烟過眼錄, a work inserted in the *Shi wan küan lou ts'ung shu* of Lu Sin-yüan (compare P. PELLIOT, *B.E.F.E.O.*, Vol. IX, 1909, p. 246). My first knowledge of this passage was intimated by *Ko chi king yüan* (Ch. 33, p. 11 b) where it is quoted *in extenso* and correctly, the word being written in the style of the Yüan period 骨咄犀, whilst the edition of Lu Sin-yüan (Ch. 上, p. 17), in the first paragraph, has twice altered the syllable *ku* into *kuo* 國 but farther on has again the normal 骨; the former way of writing seems to be faulty. When first reading the text in the *Ko chi king yüan*, I was naturally struck by the mention in it of Ye Sen of the Yüan period and the date 1320, for as M. PELLIOT informs us, the *Yün yen kuo yen lu* was written by Chou Mi of the Sung[1]. It was therefore reasonable to expect that we might light upon the passage in the appendix to this work compiled by T'ang Yün-mo of the Yüan, especially as the name of Ye Sen is cited in the postscript. In fact, however, it is not contained therein, but in the first chapter of the main treatise attributed to Chou Mi. May be this author was still alive in 1320; the date of his death is not ascertained. May be, as M. PELLIOT assures us that his work has come down in a somewhat bad condition, an editorial confusion of notes has come into play, a record

---

1) WYLIE (*Notes on Chin. Lit.*) gives three different and contradictory dates for the lifetime of this author; on p. 166: latter half of the thirteenth century; on p. 198: former part of the fourteenth century; on p. 250, he wrote "somewhere about the same date", the one previously mentioned being 1138. BRETSCHNEIDER (*Bot. Sin.*, pt. 1, p. 141, N°. 48) makes him live: latter part of the thirteenth and beginning of the fourteenth century. According to *Se k'u...* (Ch. 141, p. 34), he lived under the Southern Sung in the thirteenth century (HIRTH in the writer's *Chinese Pottery*, p. 5).

of T'ang Yün-mo having been accredited to Chou Mi. However
this may be, the account itself is of great interest and value. The
first paragraph is exactly the same as the statement of the *Cho
keng lu*, as given above (p. 321), and although the date of T'ang
Yün-mo is not known to me, yet the reference to the year 1320
renders it obvious that T'ao Tsung-i writing in 1366 is indebted
to this work for his information on *ku-tu-si*. The second paragraph
runs as follows: "When Ye Sen 葉森 in the summer of the year
keng-shen in the period Yen-yu (1320) paid a visit to his son Pi-
ming 必明, (Pi-ming) brought him two knife-hilts of *ku-tu-si*
骨咄犀刀靶二, the material here under discussion. The natural
designs displayed on it resembled the sugar-cakes now sold in the
markets 其花絞似今市中所賣糖糕. Some have white
spots, which are somewhat like the spots of cakes and pastry candied
with sugar 或有白點或如嵌糖糕點. When you feel it
with your hands, it emits an odor of *yen* cinnamon [1]); when you
rub it, and it remains odorless, it is a counterfeit 以手摸之
作巖桂香若摩之無香者乃偽物也."

We here have, accordingly, a precise chronological indication
for the presence of *ku-tu-si* on Chinese soil in the year 1320, and
we notice that the objects made from it were the same in the Mongol
as in the previous Kin period, — knife-hilts, the same as is chronicled
regarding the fish-teeth of the Bulgar on the Wolga in the West.
On reading this passage I experimented on a walrus-tusk in the
possession of my colleague Dr. Cory, the well-known zoölogist, but
while we are agreed that on being rubbed it emits a certain odor,
we do not feel sufficiently qualified to issue a definite statement as
to the peculiar character of this odor.

----

1) According to BRETSCHNEIDER, *Botanicon Sinicum*, II, p. 384, *yen-kouei* is an old
name of the *Olea fragrans*.

Ko chi king yüan (Ch. 33, p. 11 b) quotes a text from the *Sü* 續 *Sung mo ki wên*, apparently a continuation to the *Sung mo ki wên* by Hung Hao; the date of this appendix is not known to me. This passage is as follows: "The Khitan hold the *ku-tu-si* in esteem. The horn is not big; (it is so rare that) among numerous pieces of rhinoceros-horn there is not one (of this kind). It has never been worked into girdles [as is the case with rhinoceros-horn]. Its designs are like those in ivory, and it is yellow in color. Only knife-hilts are made from it, which are considered as priceless. Emperor T'ien Tsu[1]) hade made from this substance a *t'u-hu* (gloss: called in Chinese: *yao t'iao p'i* 'leather strip for the loins') fastened to and hanging down the head (?)." 契丹重骨咄犀、犀不大、萬株犀無一、不曾作帶、紋如象牙帶黃色、止是作刀把已爲無價、天祚以此作兔鶻 (gloss: 中國謂之腰條皮) 插垂頭者。

The word *t'u-hu* evidently belongs to the property of the Khitan language, but is not listed in the glossary of the *Liao shi*; it is perhaps preserved in the first element of the Gold word *túgbule*, 'girdle-pendant'[2]).

The word *kuo hia ma* 果下馬 figures as a Khitan word in the glossary of the *Liao shi* (Ch. 116, p. 14); the explanation given is the same as the one in the commentary to *Hou Han shu*. It occurs, for example, in *Liao shi*, Ch. 55, p. 3. It would be reasonable to expect that the word is of Korean origin; but I am unable to trace it in the Korean Dictionary published by the French Missionaries or in that of Gale.

---

1) The ninth and last emperor of the Liao dynasty who reigned from 1101 to 1119 and died in 1125 (GILES, *Biographical Dictionary*, p. 932). The history of his reign is recorded in *Liao shi*, Chs. 27—30.

2) W. GRUBE, *Goldisch-deutsches Wörterverzeichnis*, p. 79.

The Manchu equivalent of *hai pao* 海豹 is *huwethi* (*Yü ch'i se t'i ts'ing wén kien*, Ch. 31, p. 18). SACHAROW (Manchu-Russian Dictionary, p. 452) who writes *huwethe* explains the word as a seal with short hair of dark color with a greenish tinge.

The Ostyak word for mammoth *mī-χor* is discussed by Ö. BEKE (*Keleti Szemle*, Vol. XIII, 1912, p. 120) and compared to Wogul *mā-χar* (*mā*, 'earth,' and *χār*, 'reindeer').

On the occasion of a review of a paper by P. L. Cheikho concerning a treatise on precious stones by al-Afkānī who died in 1347/48, E. WIEDEMANN (*Mitt. d. deutschen Ges. f. Geschichte der Medizin und Naturwissenschaften*, Vol. VIII, p. 510) had already drawn attention to *al-chartūt* or *al-chutww* (rendering it by mammoth-teeth followed by an interrogation-mark) by reproducing a statement of al-Bērūnī as embodied in the work of al-Afkānī. Speaking of the fish-teeth wrought into knife-hafts, al-Bērūnī here concludes that *al-chutww* is likewise a tooth or horn; this would mean that he is convinced as to the identity of the two terms "fish-teeth" and *al-chutww*. In this place WIEDEMANN alludes also to FRÄHN's Ibn Fozzlān (St. Pet., 1823, pp. 228—9) where according to the *Sīrat al-Mulūk* ("Chronicle of the Kings") of 1076 by the Vesīr Niẓam al-Mulk Ḥasan are mentioned teeth resembling the tusks of elephants which were obtained in the country of the Bulgar then living on the Wolga, thence exported to Khiwa and there worked up into combs, capsules, etc. (compare above p. 316). It seems that in this case the mammoth cannot come into question, no mammoth having ever been found in the region of the Wolga, and that the trade in these tusks can only be connected with the walrus-teeth captured by the Russians, as shown above.

GEORG JACOB had already confronted Arabic *chutww* and Chinese-Khitan *ku-tu-si* in his treatise "Welche Handelsartikel bezogen die Araber des Mittelalters aus den nordish-baltischen Ländern?", p. 58 (Berlin, 1891) and commented on the term in his "Die Waren beim arabisch-nordischen Verkehr im Mittelalter," p. 9 (Berlin, 1891). With correct instinct he remarks that the word *ku-tu-si* does not seem to be originally Chinese.

In the last number of *Der Islam* (Vol. IV, May, 1913, p. 163) Dr. J. RUSKA contributes a note under the title "Noch einmal al-Chutww." Wrongly assuming that it is now certain that *al-chutww* means rhinoceros-horn, he furnishes very interesting material regarding the latter, chiefly after Qazwīnī of the thirteenth century (1203—83), but without noticing that this account is copied from the report of the merchant Soleiman of 851 translated by M. REINAUD (*Relation des voyages faits par les Arabes*, Vol. I, p. 28) [1]), and that the story of the rhinoceros with jointless legs occurs as early as in the *Physiologus* (Ch. XIX) where the same fable is related in regard to the elephant. This story is of particular interest to us, as a purer and more original version of it is preserved in a Chinese account. Su Sung, author of the *T'u king pên ts'ao* published by imperial order in the Sung period, in his account on the rhinoceros (*Pên ts'ao kang mu*, Ch. 51 上, p. 5 b) has the following story attributed to Wu Shi-kao 吳士皐, a physician of the T'ang period; according to the fuller version of the *Chéng lei pên ts'ao* (Ch. 17, fol. 21 b), this physician served in an official capacity

---

1) Also the passage translated by RUSKA from Damīrī is cited by REINAUD in his notes (Vol. II, p. 69). In regard to Qazwīnī, G. JACOB (*Ein arabischer Berichterstatter aus dem 10. Jahrhundert*, p. 56, Berlin, 1896) observes that he repeatedly copies without quoting.

on the maritime coast of southern China and picked up the fable from a captain whom he encountered there. It is a real captain's story. "The maritime people intent on capturing a rhinoceros proceed by erecting on a mountain path many structures of decayed timber, something like a stable for swine or sheep. The front legs of the rhinoceros being straight without joints, the animal is in the habit of sleeping by leaning against the trunk of a tree. The rotten timber will suddenly break down, and the animal will topple in front without being able for a long time to rise. Then they attack and kill it." 唐醫吳士皋言、海人取犀先於山路多植朽木如豬羊棧、其犀前脚直常依木而息、爛木忽然折倒仆久不能起因格殺之 (*Chéng lei pén ts'ao* adds: 而取其角 "and capture its horn")[1].

The coincidence with the elephant story of the *Physiologus* is obvious. "When the elephant has fallen, he cannot rise, for his knees have no joints. But how does he fall? When he wants to sleep, he leans against a tree, and thus he sleeps. The Indians familiar with this peculiarity of the elephant saw the tree a bit. The elephant comes to lean toward it, and as he draws near to the tree, it falls to the ground, taking him with it. After falling he is not able to rise. He begins to scream. One elephant, and then twelve others arrive to help him, — in vain, until at last the small elephant appears, lays his trunk around him and lifts him"[2].

---

1) GROENEVELDT (*T'oung Pao*, Vol. VII, 1896, p. 131), without stating his source, refers to a similar story told by the natives of Java in regard to a wild cow of diminutive size, said to live in the loneliest recesses of the jungle.

2) Compare P. LAUCHERT, *Geschichte des Physiologus*, p. 43 (Strassburg, 1889); E. PETERS, *Der griechische Physiologus und seine orientalischen Uebersetzungen*, p. 39 (Berlin, 1898); K. AHRENS, *Das Buch der Naturgegenstände*, p. 40 (Kiel, 1892); F. HOMMEL, *Die äthiopische Uebersetzung des Physiologus*, p. 89 (Leipzig, 1877), etc.

The *Physiologus* plainly refers to India as the source of the tradition, but has arbitrarily changed the rhinoceros into the elephant. The Arabic report of Soleiman and our Chinese version go to show that the story was associated in India with the rhinoceros; it would be difficult to understand also that people so intimately familiar with the elephant as those of India should have ever conceived of it with jointless knees. The fundamental value of the Chinese text lies in the fact that it mirrors the primeval form of the Indian story which served as basis to that adopted by the *Physiologus*. The Chinese story is consistent in relating the capture of the rhinoceros in consequence of the human ruse founded on the alleged anatomical quality and life-habit of the animal. The *Physiologus*, however, only tells the operation of the trick, and quite illogically, forgets the hunter waiting in ambush and has the animal rescued in a miraculous manner. This feature is due to the religious tendencies of this book in which all animal stories are subjected to a symbolic Christian interpretation. In the present case the big fallen elephant is Adam, the twelve elephants are the prophets, and the elephant coming to the rescue is Christ. Our Chinese text does not directly allude to India proper, and "the maritime people" is a somewhat vague expression hinting at the inhabitants of the southern sea, as Annam, Cambodja etc.; but the captain repeating the story to the Chinese physician of the T'ang period had doubtless hailed from some southern port within the culture sphere of India, so that we may well assume that the story was diffused at that time over the Archipelago and Farther India. The version of the *Physiologus* proves that it is far older in India proper, and there are indications that it must have spread to the antique world at a time somewhat

24

anterior to the composition of the original *Physiologus*. It is well known that PLINY (*Nat. Hist.* VIII, 39) and CAESAR (*De bello gallico comm.* VI, 27) have similar yarns to tell about the elk whose legs are without joints, wherefore it does not lie down in sleeping, but only leans against a tree which is sawed through to trap the animal[1]. As to Pliny (23—79 A.D.), F. HOMMEL[2]) assumes that among the Greek works ransacked by him there was also the *Physiologus*; it is not known to me whether this opinion is shared or still upheld by classical philologists. As to Caesar (B. C. 100—44), I do not venture to set forth an opinion as to the possible dependence of his story on that of the primeval *Physiologus*, but must leave this question to the decision of those competent to judge. There can be no doubt, however, of the close historical interrelation of the occidental and oriental versions of this fable, and of its localization in India confirmed by Soleiman and our Chinese text which despite its relatively recent record contains the primitive form of the story. While it must be recognized that the Greek *bestiaire* arising during the Alexandrian epoch in that curious medley of Egypto-Hellenic thought is mainly composed of Egyptian and Semitic ideas, it is covered also by a certain stratum of Indian elements deserving careful study.

---

1) Compare O. KELLER, *Die antike Tierwelt*, Vol. I, pp. 282, 283 (Leipzig, 1909).
2) *L. c.*, p. XXXIV.

ADDENDA

PAR

PAUL PELLIOT.

———

C'est à la demande de mon ami B. Laufer que je me permets d'ajouter quelques notes à son article si intéressant sur l'ivoire de morse et de narval. Je crois que M. Laufer a parfaitement établi l'identité du produit *al-chutww* des Arabes et du *kou-tou-si* des Chinois, et mes notes ne visent qu'à préciser quelques points de détail et à faire connaître un ou deux textes nouveaux.

En premier lieu, je relève dans l'article de M. Laufer une expression qui demeure pour moi assez mystérieuse; c'est celle de 碧 犀 *pi-si*, que M. Laufer traduit par «corne de rhinocéros bleu-verte» (p. 324, 325). Littéralement, tel paraît bien être le sens, mais cette expression semble avoir pris d'assez bonne heure une valeur spéciale qu'il reste à déterminer. Un examen rapide ne m'a pas fait retrouver, malheureusement, le passage du commentaire du *Chan hai king* que cite le *P'ei wen yun fou*; il devrait cependant s'agir en principe du commentaire de Kouo P'ouo, ce qui attesterait l'existence de l'expression *pi-si* au moins au début du IVe siècle [1]. Mais ce qui est bien certain, c'est qu'en chinois mandarin moderne, *pi-si* désigne une pierre précieuse et non une corne de rhinocéros. Nous avons tous vu à Pékin cette pierre rose veinée très transparente qu'on appelle *pi-si*, et pour laquelle certains lettrés, faute d'une orthographe absolument consacrée, songent à une forme 碧 璽 *pi-si* à côté de 碧 犀 *pi-si*. Par contre 黃 碧 犀 *houang pi-si*, le

---

[1] Il ne résulte pas de la citation du *P'ei wen yun fou* que «la corne de serpent du Kou-tou» soit mentionnée dans le *Chan hai king* lui-même, comme l'admet M. Laufer.

«*pi-si* jaune», est sans aucun doute la topaze et est donné comme
tel dans le dictionnaire de Giles [1]). Une fois de plus, nous nous
apercevons ici que notre connaissance de la terminologie chinoise
des pierres précieuses est encore très peu satisfaisante, et il faudra
tâcher de retrouver l'expression dans les textes.

En ce qui concerne l'expression même de *kou-tou-si*, M. Laufer
en a cité (p. 320) un exemple dans le chap. 96, fol. 3 v⁰, du
Leao che et a supposé qu'on devait la retrouver dans le *Kin che*.
En effet, au chap. 64, fol. 2 r⁰, du *Kin che*, il est question de
«poignard à [manche de] *kou-tou-si* des anciens Leao» (故遼
骨睹犀佩刀) [2]).

Aux p. 340—341, je ne suis pas d'accord avec l'interprétation
que propose M. Laufer pour le texte du *Kieou t'ang chou*. Le mot
納 *na* ne peut signifier ici «recevoir» et la coupure qui résulte de
cette leçon est très anormale; en réalité 納 *na* doit être une simple
faute d'impression pour 紬 *tch'eou*, et il n'y a, selon moi, rien à
tirer directement de ce passage, où il est question d'une étoffe, pour
attester qu'on ait connu en Chine l'ivoire de morse ou de narval à
l'époque des T'ang.

Le *Siu song mo ki wen*, dont M. Laufer dit ne pas connaître
la date (p. 358—359), est en réalité le second chapitre du *Song mo
ki wen* lui-même; il est dû, lui aussi, à Hong Hao, et fut écrit en 1143
ou très peu après. Cf. à ce sujet le *Catalogue impérial*, chap. 51,
fol. 19—20.

Il n'est pas douteux que le terme de 兔鶻 *t'ou-hou*, que le
Siu song mo ki wen a fourni à M. Laufer, désigne bien une espèce
de ceinture. S'il se présentait isolément, on pourrait hésiter, puisque,
traduit mot-à-mot, *t'ou-hou* signifie «le faucon [qui prend] les lièvres»,

1) Cf. aussi par exemple A. Guérin, *Dialogues chinois*, un album oblong sans lieu ni
date [1911], p. 73, 75.

2) On notera cette orthographe de *kou-tou-si* qui jusqu'ici ne s'est pas rencontrée ailleurs.

et tel est en réalité le nom d'un oiseau de proie qui correspond au *aïtalɣu, Falco sacer*, des Turcs d'Asie centrale [1]). Mais la glose qui accompagne ici le nom montre bien, comme l'a vu M. Laufer, qu'il s'agit de la transcription d'un mot khitan. D'ailleurs, à côté de l'orthographe que nous avons ici, on rencontre plus souvent une autre orthographe 吐鶻 *t'ou-hou* [2]); le mot a désigné une ceinture, ou plutôt un pendant de ceinture, aussi bien au temps des Leao que sous les Kin.

Dans ses notes additionnelles (p. 357—358), M. Laufer a traduit un curieux texte du *Yun yen kouo yen lou* de 周密 Tcheou Mi où il est question du *kou-tou-si*, mais sur la date de ce texte, notre confrère laisse en suspens certaines questions qu'il n'est pas impossible de résoudre. M. Laufer s'étonne en effet, si Tcheou Mi est bien de la fin des Song, qu'on trouve dans son ouvrage la date de 1320. Tcheou Mi est un écrivain abondant et qui a laissé des œuvres d'un grand intérêt historique. Je ne crois donc pas inutile de serrer le problème d'un peu plus près qu'on ne l'a fait jusqu'ici.

Malgré l'importance de son œuvre, Tcheou Mi n'a pas eu les honneurs d'une biographie dans l'histoire officielle des Song. Mais, de nos jours, Lou Sin-yuan a tenté de suppléer à cette lacune en groupant dans son 宋史翼 *Song che yi* (chap. 34, fol. 8 v⁰—9 r⁰) les principaux renseignements qui nous sont parvenus sur ce personnage [3]). Il en résulte que Tcheou Mi dut naître au plus tard vers 1230. En 1253—1258, il était sous-préfet de 義烏 Yi-wou, puis fut secrétaire du préfet de Hang-tcheou en 1261, inspecteur des greniers en 1274. A la chute des Song, il se retira au 癸辛街 Kouei-sin-kiai [4]) de Hang-tcheou et passa le reste de sa vie à s'oc-

---

1) Cf. D Ross, *A polyglott list of birds*, p. 274

2) Par exemple dans *Leao che*, chap. 96, fol. 3 v⁰, dans *Kin che*, chap 64, fol 2 r⁰, etc.

3) Sur le *Song che yi*, cf *B.E.F.E.-O.*, IX, 813.

4) Ainsi s'explique le titre de 癸辛雜識 *Kouei sin tsa che* donné par Tcheou Mi à un de ses principaux ouvrages.

cuper de littérature et d'archéologie. Il est pratiquement certain qu'il était mort en 1320, ou tout au moins que toutes ses œuvres, et en particulier le *Yun yen kouo yen lou*, sont antérieurs à cette date. Tcheou Mi a laissé les œuvres suivantes: 1° 齊東野語 *Ts'i tong ye yu*; 2° 癸辛雜識 *Kouei sin tsa che*; 3° 志雅堂雜鈔 *Tche ya t'ang tsa tch'ao*; 4° 浩然齋雅談 *Hao jan tchai ya t'an*; 5° 浩然齋視聽鈔 *Hao jan tchai che t'ing tch'ao*; 6° 澄懷錄 *Tch'eng houai lou*; 7° 乾淳起居注 *K'ien tch'ouen k'i kiu tchou*; 8° 乾淳歲時記 *K'ien tch'ouen souei che ki*[1]); 9° 武林舊事 *Wou lin kieou che*: 10° 武林市肆記 *Wou lin che sseu ki*; 11° 湖山勝概 *Hou chan cheng kai*; 12° 弁陽客談 *Pien yang k'o t'an*; 13° 雲烟過眼錄 *Yun yen kouo yen lou*; 14° 絕妙好詞 *Tsiue miao hao ts'eu*. Presque toutes ces œuvres nous sont parvenues et il y en a des rééditions modernes. Selon le 元藝文志 *Yuan yi wen tche* de Ts'ien Ta-hin, qui est devenu le chap. 94 (fol. 3 v°) du 元史新編 *Yuan che sin pien* de Wei Yuan, il faudrait encore ajouter le 蠟屐集 *La ki tsi* en 1 chapitre et le 弁山詩集 *Pien chan che tsi*, en 5 chapitres.

Mais comment expliquer alors la mention de la date de 1320? D'une manière très simple: le passage traduit par M. Laufer, ainsi qu'il résulte du texte lui-même, est une de ces additions dues à Ye Sen et dont il est question dans la notice finale. Mais à quoi rime cette addition? C'est ici qu'il faut faire intervenir le paragraphe précédant celui que M. Laufer a traduit et qui seul justifie la glose de Ye Sen. En réalité, Tcheou Mi rapporte plusieurs propos qu'il met sur le compte d'un certain 伯幾 Po-ki. L'identité de ce dernier personnage n'est pas douteuse; Po-ki, plus souvent écrit 伯機 Po-ki, est le surnom d'un calligraphe et poète de la fin du

---

1) C'est là l'ouvrage dont il est question dans *B.E.F.E.-O.*, IV, 288, et *k'ien-tch'ouen* y est bien un *nien-hao* de la fin des Song.

XIIIᵉ siècle, 鮮于樞 Sien-yu Tch'ou ¹). C'est donc Sien-yu Tch'ou qui a tenu à Tcheou Mi le propos relatif au *kou-tou-si* qui est «la corne d'un serpent»; le passage fait bien partie de la rédaction primitive du *Yun yen kouo yen lou*. Quant au deuxième paragraphe traduit par M. Laufer, c'est une note ajoutée par Ye Sen, qui visita «son fils» Pi-ming en 1320 et vit chez lui deux manches de poignard en *kou-tou-si*. Qui est «son fils»? Mais évidemment le fils de Sien-yu Tch'ou; ce fils possédait encore en 1320 les objets dont son père avait parlé à Tcheou Mi quelque trente ans plus tôt. Quant à ce «nom» de 必明 Pi-ming, c'est certainement un surnom. Il doit s'agir en réalité de 鮮于去矜 Sien-yu K'iu-king, qui lui aussi s'acquit quelque réputation comme calligraphe. A vrai dire, le *P'ei wen tchai chou houa p'ou* ²) donne à Sien-yu K'iu-king le surnom de 必仁 Pi-jen et non de 必明 Pi-ming. Mais on sait qu'il y a généralement un rapport entre le nom personnel (*ming*) et le surnom (*tseu*). Or je ne vois pas comment justifier Pi-jen pour un nom personnel K'iu-king. Pi-ming s'explique bien au contraire par allusion à une phrase de Siun-tseu ³). Ainsi, en définitive, le

1) Sien-yu Tch'ou, *tseu* Po-ki, *hao* 困學 K'ouen-hio, avait laissé un 困學齋集 *K'ouen hio tchai tsi* aujourd'hui perdu. Deux morceaux écrits par lui sont incorporés au chap. 4 du 元文類 *Yuan wen lei*; d'autres se trouvent au chap. 4 (fol. 8 v°—9 r°) du 元詩紀事 *Yuan che ki che*; cf. aussi *P'ei wen tchai chou houa p'ou*, chap. 37, fol. 2 v°. Les bibliographes de K'ien-long (*Catalogue impérial*, chap. 122, fol. 1 r° et v°) ont accepté l'attribution à Sien-yu Tch'ou d'un 困學齋雜錄 *K'ouen hio tchai tsa lou*, proposée par 曹溶 Ts'ao Jong dans une notice finale de 1682. D'après ces bibliographes, Ts'ao Jong avait incorporé l'ouvrage à son 學海類編 *Hio hai lei pien* (sur cette collection, cf. *Catalogue impérial*, chap. 134, fol. 21 v°—22 v°); mais le *Hio hai lei pien* est resté longtemps manuscrit, et l'édition en caractères mobiles qui en a été donnée en 1831 ne contient pas le *K'ouen hio tchai tsa lou*; par contre, cet opuscule se trouve dans le *Tche pou tsou tchai ts'ong chou*; un passage me paraît gêner l'attribution à Sien-yu Tch'ou

2) Chap. 37, fol. 2 v°.

3) 有兼聽之明而無舊矜之容。

passage du *Yun yen kouo yen lou* traduit par M. Laufer nous aurait conservé la forme véritable d'un surnom qui a été altéré dans la source du *P'ei wen tchai chou houa p'ou*. Quant au texte essentiel relatif au *kou-tou-si*, celui qui a ensuite passé en 1366 dans le *Tcho keng lou*, il est bien de Tcheou Mi lui-même et se place dans les dernières années du XIIIᵉ siècle.

Dernières additions. P. 355. — Sur *lo-sseu-ma*, cf. encore *T'oung Pao*, V, 1894, p. 370. Il faudrait rechercher si le terme se retrouve réellement dans le 正字通 *Tcheng tseu t'ong*, comme le dit Schlegel; en tout cas, les «deux cornes recourbées» prouvent bien qu'il s'agit du morse et non du narval. [B. L.]

P. 340—341 et p. 359. — Sur *yu-ya-tch'eou* et sur *kouo-hia-ma*, cf. Courant, dans *T'oung Pao*, IX, 1898, p. 15 et 16. [B. L.]

096

关于按指印的历史

ANNUAL REPORT OF THE BOARD OF REGENTS OF

THE SMITHSONIAN INSTITUTION

SHOWING THE

OPERATIONS, EXPENDITURES, AND
CONDITION OF THE INSTITUTION
FOR THE YEAR ENDING JUNE 30

1912

WASHINGTON
GOVERNMENT PRINTING OFFICE
1913

HISTORY OF THE FINGER-PRINT SYSTEM.

By Berthold Laufer.

[With 7 plates.]

On May 2, 1906, the *Evening Post* of New York announced in an article headed "Police Lesson from India" the first successful application in this country of the thumb-print test. A notorious criminal had robbed the wife of a prominent novelist in London of £800, had made his escape to New York, and was captured after committing a robbery in one of the large hotels in that city. The Bertillon Bureau of the Police Department took a print of one of his thumbs, which was mailed without any other particulars to the Convict Supervision Office, New Scotland Yard, London, where he was promptly identified. He was convicted and sentenced to seven years in prison. The system of finger prints is now successfully utilized by the police departments of all large cities of this country, central bureaus of identification having been established in the capitals of the States. The admissibility of finger-print evidence as valid proof of guilt in murder trials was upheld in the case of a colored man executed in Cook County, Ill., on February 16, 1912. He was convicted of murder largely on a showing by the prosecution that the imprint of a finger on the woodwork in the slain man's house corresponded with that in the records of Joliet prison, where an imprint of the accused's fingers had been taken when he was discharged from the penitentiary a short time before the murder. Likewise, in our relations with illiterate people the system has come to the fore. On the approval of the Secretary of the Interior Department, the Commissioner of Indian Affairs instructed officials throughout Oklahoma in 1912 that hereafter every Indian who can not write his name will be required to sign all checks and official papers, and indorse checks and warrants covering Indian money, by making an impression of the ball of his right thumb, such imprint to be witnessed by an employee of the Indian agency or by one of the leading men of the tribe who can write. If an Indian is not living with his tribe, his thumb-mark signature must be witnessed by the postmaster of the place where he resides. Prominent banks

631

of Chicago have adopted finger prints in the case of foreign-born customers who can not sign their names in English, and it is reported that the scheme has worked out to perfect satisfaction. The cashier of one of the large Chicago banks stated in an interview in the *Chicago Tribune* of May 14, 1911:

We have never had a complaint or error from this system. There are absolutely no two thumbs alike, and the thumb-print mark is an absolute identification. We have had complaints over signatures, but never over thumb prints. Men have claimed that they did not sign withdrawal slips, but no one has ever denied his thumb mark.

It is well known that the honor of having developed the system of finger prints and placing it on a scientific basis is due to Sir Francis Galton, explorer and scientist, born at Birmingham, England, February 16, 1822, and who died in London in January, 1911. The results of his studies are contained in two books, Finger Prints (London, 1892) and Finger Print Directories (London, 1895).[1] The system is based on two observations—the widely varying, individual character of the finger marks (in Galton's words: "It is probable that no two finger prints in the whole world are so alike that an expert would fail to distinguish between them") and the persistency of the form of the marks in the same individual from childhood to old age. Galton comments on the latter point as follows:

As there is no sign, except in one case, of change during any of these four intervals which together almost wholly cover the ordinary life of man (boyhood, early manhood, middle age, extreme old age), we are justified in inferring that between birth and death there is absolutely no change in, say, 699 out of 700 of the numerous characteristics of the markings of the fingers of the same person such as can be impressed by him wherever it is desirable to do so. Neither can there be any change after death up to the time when the skin perishes through decomposition: for example, the marks on the fingers of many Egyptian mummies and on the paws of stuffed monkeys still remain legible. Very good evidence and careful inquiry is thus seen to justify the popular idea of the persistence of finger markings. There appear to be no bodily characteristics other than deep scars and tattoo marks comparable in their persistence to these markings; at the same time they are out of all proportion more numerous than any other measureable features. The dimensions of the limbs and body alter in the course of growth and decay; the color, quantity, and quality of the hair, the tint and quality of the skin, the number and set of the teeth, the expression of the features, the gestures, the handwriting, even the eye color, change after many years. There seems no persistence in the visible parts of the body except in these minute and hitherto disregarded ridges.

The permanency of the finger marks certainly refers to the features of the design, especially the character of the ridges, but not to their measurements, which are subject to the same general changes associated with the growth of the body. Galton himself admits his great

---

[1] Of later books on the subject, E. R. Henry, Classification and Uses of Finger Prints, 3d edition, London, 1905, may be specially mentioned.

indebtedness to Sir William J. Herschel,[1] and from him he appears to have received the first impetus for an investigation of this subject. Galton's attention was first drawn to it in 1888 when preparing a lecture on Personal Identification for the Royal Institution, which had for its principal object an account of the anthropometric method of Bertillon. "Wishing to treat the subject generally," he says, "and having a vague knowledge of the value sometimes assigned to finger marks, I made inquiries, and was surprised to find both how much had been done, and how much there remained to do before establishing their theoretical value and practical utility."[2] This confession implies that Galton did not discover the idea himself, but derived it from, and relied solely on, his predecessors, chiefly Herschel, who, moreover, can not claim that the idea was wholly his own.

This method of identification had been suggested to Sir William Herschel by two contracts in Bengali, dated 1858. "It was so difficult to obtain credence to the signatures of the natives that he thought he would use the signatures of the hand itself, chiefly with the intention of frightening the man who made it from afterwards denying his formal act. However, the impression proved so good that Sir William Herschel became convinced that the same method might be further utilized. He finally introduced the use of finger prints in several departments at Hooghly (in Bengal) in 1877, after 17 years' experience of the value of the evidence they afforded. A too brief account of his work was given by him in *Nature*, volume 23, page 23 (Nov. 25, 1880). In 1877 he submitted a report in semiofficial form to the Inspector General of Gaols, asking to be allowed to extend the process; but no result followed." "If the use of finger prints ever becomes of general importance," remarks Galton, "Sir William Herschel must be regarded as the first who devised a feasible method for regular use and afterwards officially adopted it."[3]

It is difficult to believe that Herschel, stationed in India, should have conjured up, entirely from his own resources, a system which had been known and applied in the East ages before his time. Had he designed it in his home study in England, the matter might be looked upon in a different light. But he resided at Calcutta, where a large colony of Chinese had been settled for a long time, and if a European, living in the Orient in close official and private relations with its people, conceives an idea which seems to belong to his very surroundings, it would be proper to credit his environment with its due share in shaping that idea. The man laboring on his "invention" for years may easily forget this first impetus. It matters little

[1] Finger Prints. London, 1892, p. 4. Herschel was born in 1833 and engaged in the Civil Service of India from 1853 to 1878.

[2] Ibid., p. 2.

[3] Compare Galton, Finger Prints, p. 28.

also whether or not he himself is conscious of outward influences; the cool and impartial historian, in the light of observed facts, can reach no other conclusion than that Herschel must have conceived his idea from observations of similar affairs made on the spot. A similar judgment was early rendered by a writer in the *Ninteenth Century* (1894, p. 365) who championed the cause of the Chinese in the priority of the finger-print system. Herschel himself, however, was of a different opinion and indignantly rejected such a point of view.

In a letter addressed to *Nature* (vol. 51, 1894, p. 77) Herschel claimed for himself that "he chanced upon finger prints" in 1858 and followed it up afterwards, and that he placed all his materials at the disposal of Galton. While vindicating the honor of the invention for himself, he at the same time deprecated "as being to the best of his knowledge wholly unproved the assertion that the use of finger marks in this way was originally invented by the Chinese." "I have met no evidence," he continues, "which goes anywhere near substantiating this. As a matter of fact, I exhibited the system to many passengers and officers of the P. and O. steamship *Mongolia* in the Indian Ocean during her outward voyage in February, 1877, and I have the finger prints of her captain, and of all those persons, with their names. It is likely enough that the idea, which caught on rapidly among the passengers, may have found a settlement in some Chinese port by this route, and have there taken a practical form; but whether that be so or not, I must protest against the vague claim made on behalf of the Chinese until satisfactory evidence of antiquity is produced."

The notion here expressed by Herschel that his thought might have spread to some Chinese port is, to say the least, somewhat naïve, and the fact remains that the use of finger prints is well authenticated in China long before his lifetime. The gauntlet brusquely thrown down by him was soon taken up by two scholars—a Japanese, Mr. Kumagusu Minakata,[1] and the always combative Prof. G. Schlegel,[2] of Leiden. Both were actuated by the sincere intention of furnishing proof of the antiquity of the method of finger prints in China and Japan; but both failed in this attempt for lack of proper understanding of what the finger-print system really is. Both confused with the latter the hand stamp; that is, a slight impression taken from the palm. These are entirely different affairs, and in view of the general knowledge now existing in regard to the significance and effects of finger prints it is needless to emphasize the fact that a mere impression of the palm can never lead to the identification of an individual, which is of first importance in finger prints. The entire argument of Schlegel is restricted to two references occurring in his Dutch-Chinese

---

[1] The Antiquity of the "Finger-Print" Method (*Nature*, vol. 51, 1894, pp. 199–200).

[2] *T'oung Pao*, vol. 6, 1895, p. 148.

Dictionary, one pertaining to bills of divorce which are authenticated by a print of the hand of the husband, and the phrase *ta shou yin*, "to produce a hand seal"; that is, to make an impress with the blackened palm.[1]

The Chinese origin of the finger-print system has been upheld by several writers on the subject.[2] The correspondent of the *Evening Post* quoted at the beginning of this paper said: "As a matter of fact, it is one of those cherished western institutions that the Chinese have calmly claimed for their own, and those who doubt this may be convinced by actual history, showing it to have been employed in the police courts of British India for a generation or so back." In 1908 Prof. Giles,[3] the well-known sinologue, wrote: "It should always be remembered that the wonderful system of identification by finger prints was borrowed straight from China, where it has been in vogue for many centuries." But this "straight from China" is the very difficult point in the matter. While the chronological priority of the Chinese in the practice of finger prints may be satisfactorily established, there is no evidence to show that Herschel received a stimulus directly from China, nor that the people of India, from whom Herschel may well have borrowed the idea, were ever influenced in this direction by the Chinese. As a matter of principle it should be stated that it is most unlikely that a complex series of ideas as presented by the finger-print process was several times evolved by different nations

---

[1] It should not be supposed that this is a common Chinese practice. It may be a local custom of which Schlegel heard in Amoy or its vicinity, where he derived his knowledge. The Chinese marriage and divorce laws (comp. P. Hoang, Le mariage chinois au point de vue légal, Shanghai, 1898) make no reference to such procedure. J. Doolittle (Social Life of the Chinese, London, 1868, p. 75) has the following: "It is not necessary for the husband, in giving a bill of divorcement to his wife, to do it in the presence of an officer of the Government as witness in order to make it legal. He does it on his own authority and in his own name. It is often written in the presence of her parents and in their house. Very few divorces occur in China." In a recent work (Dr. L. Wieger's Moral Tenets and Customs in China. Texts in Chinese, translated and annotated by L. Davrout, Ho-kien-fu, 1913, on plate opposite p. 193) is illustrated a divorce bill stamped with the hand and foot of the husband in black ink. It is remarked in the text that the impress of a finger is sometimes used as a seal, that the paper would be invalid without such a stamp, and that in case of contestation the document thus stamped proves the divorce.

[2] In the second chapter of his "Finger Prints," which treats of the previous use of them, Galton refers also to many impressions of fingers found on ancient pottery, as on Roman tiles. These nail marks, used ornamentally by potters, especially in prehistoric pottery, are well known to every archeologist, but they move on a line in psychological and technical regard entirely different from the finger-print system and can not by any means be connected with its history, as Galton inclines to establish. Thus also the coin of the T'ang dynasty, "bearing a nail mark of the Empress Wen-te in relief" and figured by Galton, does not belong at all to this category. The Chinese works on numismatics (e. g., *K'in-ting ts'ien lu*, ch. 11, p. 2, ch. 16, p. 14) explain this mark occurring on many issues of the T'ang and Sung dynasties—apparently the mark of a mint—as a picture of the crescent of the moon. Handcock (Mesopotamian Archeology, London, 1912, p. 83) has an allusion to "finger-marked bricks" of the Sargon period. This vague hint, from which no inference whatever as to the use of these marks for identification can be drawn, has led astray a well-known egyptologist into proclaiming the origin of the invention of finger prints in Babylonia, but as this statement appeared only in sensational newspaper reports, I refrain from discussing it. Finger marks may naturally arise anywhere where potters handle bricks or jars, but every expert in finger prints will agree with me that these are so superficial as to render them useless for identification. A clear and useful finger impression in clay presupposes a willful and energetic action, while the potter touches the clay but slightly. However this may be, we are not willing to admit as evidence for a finger-print system any finger marks of whatever kind occurring in pottery of any part of the world, unless strict proof can be furnished that such marks have actually served for the purpose of identification.

[3] Adversaria Sinica, No. 6, p. 183.

independently. If there is one thing that we know surely, it is the fact of the scarcity of original ideas among mankind, which may stand in relation to reproduced ideas as 1:100. The fact remains that, however simple and self-evident the system may now look to us, the most advanced civilized nations have never hit upon it, that no trace of it can be discovered among Egyptians or Babylonians, Greeks or Romans, and that its so very recent adoption into our culture, after prolonged contact with east-Asiatic nations, is in itself suspicious of a derivation from a foreign source. The hypothesis, therefore, seems to be justified that Chinese immigrants into India may have carried the idea over, or that the long religious and commercial intercourse between the two countries may be responsible for the transmission. It is out of the question to assume the reverse course of events, for the application of finger prints in China is of great antiquity, even greater than ever suspected heretofore, while nothing of the kind can be proved for its antiquity in India.

At all events it seems certain that finger impressions were known in India prior to the time of Herschel.   George A. Grierson,[1] one of the best connoisseurs of modern Hindu life, in describing the ceremonies at the birth of a child, mentions the fact that the midwife, using red lead, makes a finger print on the wall, with the intention of hastening delivery.   It is hard to imagine that this magical conception of the finger print, which is an ingredient of indigenous folklore, should be credited to the discovery of Herschel.   There are, further, good reasons to presume that the marks on the finger bulbs were familiar to the Indian system of palmistry.   I recently had occasion to study an ancient Sanskrit treatise on painting, the Citralakshaṇa,[2] which is preserved in a Tibetan translation embodied in the Tanjur.   One chapter of this work is taken up with a detailed description of the physical qualities of the Cakravartin, the wheel-turning king, the hero and racial ideal who formed the principal object of ancient painting.   The majority of the marks of beauty attributed to him are derived from the rules of physiognomy, a system reaching back to remote times; some of these marks, by way of comparison of the Sanskrit with the old Persian terms, are traceable to the Aryan period when the Iranians and Indians still formed a united stock of peoples.   The interpretation of prominent physical qualities, as laid down by the physiognomists, led to artistic attempts of portrayal, and for this reason I was induced to study, in connection with the Citralakshaṇa, two Indian treatises on physiognomy contained likewise in the Tibetan Tanjur, with the result that the terminology of physiognomy and art theory are identical, and that the rules of the painter closely follow in the trail of the physiognomist and palmist.

---

[1] Bihār Peasant Life. Calcutta, 1885, p. 388.
[2] Edited and translated under the title Dokumente der Indischen Kunst, I, Leipzig, 1913.

It would lead too far away from our subject proper to enter into the manifold details of this quaint art, but the principal points relating to the fingers may be insisted upon. It is said in the Sāmudravyañjanāni, one of the works on physiognomy, that a woman, if the marks on her fingers are turned toward the right-hand side will obtain a son, but if turned toward the left, a daughter will be born.[1] The Indian painter paid minute attention to the hand, the fingers, and their lines. In the above-mentioned manual of painting, their measurements, inclusive of those of the ball of the thumb, are conscientiously given.[2] A peculiar term of Indian cheiromancy is *yava* (*lit.* a barleycorn), explained by Monnier Williams in his Sanskrit Dictionary as "a figure or mark on the hand resembling a barleycorn, a natural line across the thumb at the second joint compared to a grain of barley and supposed to indicate good fortune." In all probability, this term refers also to the marks on the finger tips, and there is further the Sanskrit word *angulīmudrā* (*lit.* finger seal) used in the sense of finger print and exactly corresponding to the Chinese term *chi yin* (likewise finger seal) of the same significance.[3]

An interesting case, though not directly bearing on our subject, may here be mentioned:

Hüan Tsang, the famous Chinese traveler to India, in the seventh century, relates a story in regard to the king of Takshaçilā in India who availed himself of his tooth impression stamped in red wax on official documents. In giving instructions to his son, the king said: "The affairs of a country are of serious importance; the feelings of men are contradictory; undertake nothing rashly, so as to endanger your authority; verify the orders sent you; my seal is the impression of my teeth; here in my mouth is my seal. There can be no mistake."[4] Only one analogy to this curious custom is known to me. In a charter of King Athelstan of Northumberland it is said:

> And for a certen truth
> I bite this wax with my gang-tooth.[5]

---

[1] Laufer, Dokumente, etc., p. 159.

[2] Ibid., pp. 163, 164.

[3] At the present time India is probably the country where the most extensive use of the finger-print system is made. It has been adopted since 1899 by the Director General of the Post Offices of India. On the forms of Indian Inland Money Orders, for example, it is printed: "Signature (in ink) of payee or thumb impression if payee is illiterate." In many other departments of government it has proved an efficient method of preventing perjury and personation. No objection can be raised on the ground of religion or caste, so there is no prejudice to be overcome in obtaining the finger print. The Government has been so fully convinced of the effectiveness of the new system, and of the certainty of the results it yields, that the Indian Legislature has passed a special act amending the law of evidence to the extent of declaring relevant the testimony of those who by study have become proficient in finger-print decipherment. In all registration offices, persons who, admitting execution, present documents for registration, are required to authenticate their identity by affixing the impression of their left thumb both on the document and in a register kept for the purpose. (Compare E. R. Henry, Classification and Uses of Finger Prints, London, 1905, pp. 6–9.)

[4] S. Beal, Buddhist Records of the Western World, Vol. 1, p. 140. St. Julien, Mémoires sur les contrées occidentales, Vol. 1, p. 156.

[5] *Folk-lore*, vol. 15, 1904, p. 342.

While it is likely that the people of ancient India were familiar with the striæ on the finger tips, there is, however, no evidence whatever that finger impressions were employed to establish the identity of a person. No mention of finger prints is made in the ancient Indian law books. The signature of an individual was a recognized institution of law and a requirement in all contracts. The debtor was obliged to sign his name at the close of the bond, and to add: "I, the son of such and such a one, agree to the above." Then came the witnesses signing their name and that of their father, with the remark: "I, so and so, am witness thereof." The scribe finally added: "The above has been written by me, so and so, the son of so and so, at the request of both parties." An illiterate debtor or witness was allowed to have a substitute write for him. A note of hand written by the debtor himself was also valid without the signatures of witnesses, provided there was no compulsion, fraud, bribery, or enmity connected with the operation. The cleverness of forgers is pointed out, and the necessity of comparison of handwritings and conscientious examination of documents are insisted on.[1]

Besides the documents pertaining to private law, there were public or royal deeds, among which those relating to foundations, grants of land to subjects as marks of royal favor, took a prominent place. They were written on copper plates or cotton cloth, and the royal seal (*mudrā*) was attached to them, a necessary act to legalize the document. The forgery of a deed was looked upon as a capital crime, in the same way as in China. The seals represented an animal like a boar or the mythical bird Garuḍa. It is thus shown by the legal practice in ancient India that there was no occasion in it for the use of finger prints, and it appears that the significance of the latter was recognized only in palmistry and magic.[2]

In recent times the finger-print system has been employed in China only in two cases, at the reception of foundlings in the foundling asylums and in the signing of contracts on the part of illiterate people. In regard to the former mode we owe valuable information to F. Hirth,[3] who has made a study of the regulations of Chinese benevolent institutions.[4] The foundling asylums established in all large cities receive orphan children, forsaken babies, or any others sent to them. These are placed by their relatives in a sliding drawer in the wall near the front gate and a bamboo drum is struck to notify the gatekeeper, who opens the drawer from the inside of the wall and

---

[1] Compare J. Jolly, Recht und Sitte, p. 113 (Grundriss der indo-arischen Philologie, Strassburg, 1896).

[2] Possibly, also the Malayan tribes, as shown by their notions of palmistry, may be acquainted with finger marks. W. W. Skeat (Malay Magic, London, 1900, p. 562) notes that a whorl of circular lines on the fingers is considered as the sign of a craftsman.

[3] *T'oung Pao*, vol. 7, 1896, p. 299.

[4] An interesting account of these is given also by W. Lockhart, The Medical Missionary in China, London, 1861, pp. 23–30, and recently by Yu-Yue Tsu, The Spirit of Chinese Philanthropy (Columbia University, 1912), who refers also to the finger marks (p. 61).

transfers the little one to the care of the matron. Every infant is subjected to a method by which its identity is permanently placed on record. Sex and age are entered on a register. If the age can not be made out—it may be inferred, for example, from the style of clothing varying from year to year—the time of the reception into the asylum according to year, month, day, and hour is noted. Then follows a description of bodily qualities, including remarks on the extremities, formation of the skull, crown of the head, birthmarks, and design on the finger tips, for later identification. Emphasis is laid on the latter, for each Chinese mother is familiar with the finger marks of her new born, and as there is a high degree of probability that a baby temporarily placed in the care of the asylum owing to distressed circumstances of the family will be claimed at a later time, this identification system is carefully kept up. The Chinese seem to be acquainted with the essential characteristics of finger marks. What in the technical language of our system is called "arches" and "whorls" is styled by them lo "snail," and our "loops" are designated _ki_ "sieve," "winnowing-basket." [1] The former are popularly looked upon as foreboding of luck.

Deeds of sale are sometimes signed with a finger print by the negotiating party. We reproduce (pl. 1) such a document after Th. T. Meadows [2] in preference to any other of recent date because this deed, executed and dated in 1839, furnishes actual evidence of the use of an individual finger impression in China before the system was developed in Europe. The transaction in question is the disposal of a plot of cultivated land for which a sum of 64 taels and 5 mace was paid. The receipt of the full value of this amount is acknowledged by the head of the family selling the land; in this case the mother née Ch'ên whose finger print is headed by the words "Impression of the finger of the mother née Ch'ên." It is evident that Mrs. Ch'ên was unable to write and affixed her finger print in lieu of her name. Sir Francis Galton [3] comments on this finger print in the words: "The impression, as it appears in the woodcut, is roundish in outline, and was therefore made by the tip and not the bulb of the finger. Its surface is somewhat mottled, but there is no trace of any ridges."

---

[1] A brief nomenclature pertaining to finger prints may here be given. The numbers in parentheses refer to Giles' Chinese-English Dictionary (2d edition). _Lo v'ên_ (No. 7291, _lit._ net-pattern), "the impress of a finger, hand, or foot, dipped in ink and appended as a signature to any kind of deed or other legal instrument." _Chi yin_ (No. 13282, _lit._ finger-seal), "seal on deeds, etc., made by dipping the finger or hand in ink and pressing it on paper." _Hua kung_ (No. 6752), "to sign one's deposition, usually by dipping the thumb in ink and making an impression of it on the paper." _Lien ki tou_ (No. 13133), "to verify the lines on a man's fingers, in connection with the impression on a deed, etc." Further, _chi mo_ (No. 8066), "finger-pattern" and _hua ya_ (_lit._ to paint, i. e., to ink and press down) are expressions in the sense of our signature; _hua chi_ (No. 1791), "to make a finger print, as a signature"; _chi jên_ (_ibid._), "to identify."

[2] Land Tenure in China (_Transactions of the China Branch of the Royal Asiatic Society_, Hongkong, 1848, p. 12).

[3] Finger Prints, London, 1892, p. 24.

In all contracts of civil law Chinese custom demands the auto-graphic signatures of the contracting parties, the middlemen, and the witnesses.[1] Also the writer of the bond is obliged to sign his name at the end with the title *tai pi* ("writing for another"). If the seller write the contract out in person, he should sign again at the end, with the addition *tse pi* ("self-written"). As the number of those able to write is very large, and as even those who have an imper-fect or no knowledge of writing are at least able to write their names, it will be seen that there is little occasion for the employment of finger prints in such contracts. Prof. Giles[2] states that title deeds and other legal instruments are still often found to bear, in addition to signatures, the finger prints of the parties concerned; sometimes, indeed, the imprint of the whole hand. This would indicate a sur-vival of the originally magical and ritualistic character of the custom.

From the fact that the signature has little or hardly any legal importance, it follows that the forgery of a signature does not fall under the provisions of the Penal Code. The Code of the Manchu dynasty provided only for the forging of imperial edicts and official seals with intent to defraud, and punished these as capital crimes.[3]

In plates 2 and 3 a Tibetan document written in the running hand is reproduced. It is a promissory note signed by the debtor with the impressions from the balls of both his thumbs. The Tibetans have apparently derived the practice from the neighboring Chinese; there is little probability, at least, that, to speak with Herschel, "a pas-senger of the *Mongolia*" may have carried the suggestion to Tibet. The language of the Tibetans proves that this procedure is an old affair with them, for a seal or stamp is called *t'e-mo*, which is derived from, or identical with, the word *t'e-bo*, "thumb." Sarat Chandra Das in his Tibetan-English Dictionary justly says that the word *t'e-mo* originally means the thumb or thumb impression. We may hence infer that the thumb print was the first mode of signature of the Tibetans, in vogue prior to the introduction of metal (brass, iron, or lead) seals which were named for the thumb print, as they were identical with the latter in the principle of utilization. In the related language of the Lepcha, which has preserved a more ancient condition, we find the same expression *t'e-tsu*, "seal," and even *t'e c'ung*, "small seal," meaning at the same time "little finger."[4]

---

[1] Numerous examples may be seen in P. Hoang, Notions techniques sur la propriété en Chine avec un choix d'actes et de documents officiels, Shanghai, 1897 (*Variétés sinologiques* No. 11).

[2] Adversaria Sinica, Shanghai, 1908, p. 184.

[3] G. Th. Staunton, Ta Tsing Leu Lee, being the Fundamental Laws * * * of the Penal Code of China, London, 1810, pp. 392, 396. E. Alabaster, Notes and Commentaries on Chinese Criminal Law, London, 1899, pp. 438, 439. B. H. Chamberlain (Things Japanese, 3d ed., 1898, p. 445) states: "It seems odd, considering the high esteem in which writing is held in Japan, that the signature should not occupy the same important place as in the West. The seal alone has legal force, the impression being made, not with sealing wax, but with vermilion ink."

[4] Mainwaring and Grünwedel, Dictionary of the Lepcha Language, Berlin, 1898, p. 155.

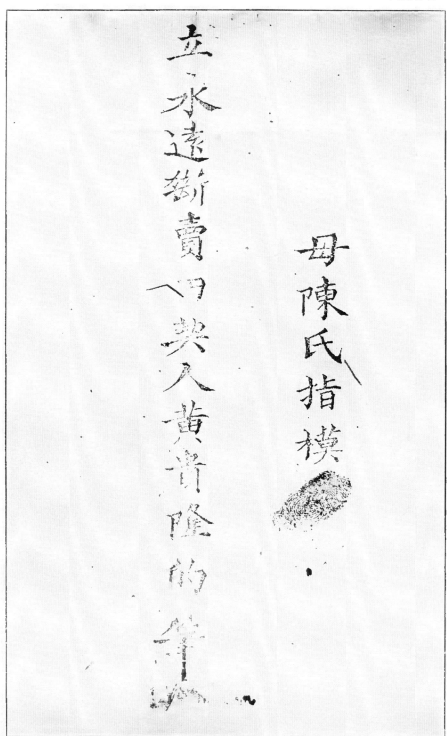

PORTION OF DEED OF SALE IN THE YEAR 1839 SIGNED WITH THE THUMB-PRINT
OF A WOMAN.

PLATE 2.

Smithsonian Report, 1912.—Laufer.

LEFT PART OF TIBETAN PROMISSORY NOTE SIGNED BY THE DEBTOR WITH THE IMPRINT OF HIS THUMBS. (SEE PLATE 3.)

PLATE 3.

RIGHT PART OF TIBETAN PROMISSORY NOTE SIGNED BY THE DEBTOR WITH THE IMPRINT OF HIS THUMBS. (SEE PLATE 2.)

Not many literary data are available with which to trace the history of the finger-print system in China. Indeed, it is striking that we do not find in any author a clear description of it and its application. The physicians, in their exposition of the anatomy of the human body, do not allude to it, and it is certain that it was not anatomical or medical studies which called it into existence. It formed part of the domain of folklore, but not of scholarly erudition. In a society where learning was so highly esteemed and writing was almost worshiped as a fetich there was little chance for the development of a process from which only the illiterate class could derive a benefit. An ingenious system of tallies and a highly organized system of official and private seals regulated by Government statutes took the function of verification. The personal signature never had any great importance in public or private transactions, and the style of handwriting as individually differentiated in China as among us would always allow of a perfectly safe identification. We have most successfully applied the finger-print system in two phases of our social life—in banking transactions and in the detection of criminals. These two institutions move on entirely different lines of organization in China, and for this reason finger prints never were a real necessity there. The Chinese banking system does not require any signature, and could accordingly introduce no substitute for it. A bank in China issues to its depositor a pass book of miniature size consisting of a long continuous sheet of paper folded in pages and held together by two stiff blue covers. The entire book may, therefore, be unfolded at once and exhibit the credits and debits at a glance. Every deposit is entered, with the date, by a clerk of the establishment, and should the depositor wish to draw a sum, he carries or sends his book to the bank, which, on payment of the amount, charges it against him by entry in the same book. There is no check system. If the customer would make payment to a third person, the procedure is the same. The draft system, which is highly developed in China, works well without a stroke of the brush being involved on the part of the person to whose credit the draft is issued. Mr. N. orders a draft from a Peking bank, payable in his name, at a bank in Si-ngan fu. The Peking house writes the document out on a rectangular paper bill containing the same matter on the right and left sides, one column of writing running exactly down the center. The document is then evenly divided into halves, the vertical column of characters being cut through in the middle. Mr. N. will receive the right half, while the left half will be forwarded in the mail by the Peking bank to Si-ngan fu. On arrival there Mr. N. will present his part of the document, which will be carefully checked off with the other half, and if both are found out on close examination to tally, the draft will be honored, no receipt and signature on the part of Mr. N. being required. The fact that both halves of the draft are in the hands of the Si-ngan fu banker is legal proof for the transaction having been closed. It is easy to see that this system is the natural offshoot of the ancient tallies in wood and metal. In regard to criminal persecution we must remember that crime had never assumed vast proportions in China, that detection and capture were comparatively easy, and that anything like a criminal science was not required for a patriarchal organization of government.[1] These are the reasons why the Chinese, though well acquainted with the character and significance of finger prints, did not develop them into a system; why they did not enter much into the speculations of their scholars, and why the records concerning them are brief and sparse.

The poet Su Shi (1036–1101) avails himself metaphorically of the expression "the whorls (snails) on the fingers" in the verse:

"Ngan, King of Ts'i, found on the bank of a river a fine stone veined like finger marks."[2]

During the Sung period (960–1278 A. D.) finger prints were taken in wax. This fact is reported by Wang Fu, the author of the *Po ku t'u lu*, the well-known catalogue of ancient bronzes first published in 1107 A. D. In chapter 6, p. 30, of this work, a bronze wine-cup of the Chou period is illustrated, on one side of which four large finger-shaped grooves appear, closely joined and looking like the fingers of a hand. The author explains the presence of these finger marks by saying that the ancients feared to drop such a vessel from their hands and therefore held it with a firm grip of their fingers in these grooves, "in order to indicate that they were careful to observe the rules of propriety." "At the present time," Wang Fu concludes, "finger marks are reproduced by means of wax, and are simply effected by pressing the fingers into wax."

Kia Kung-yen, an author of the T'ang period, who wrote about the year 650 A. D., makes a distinct allusion to finger impressions employed in his time for purposes of identification. He comments on the wooden tallies used in ancient times (before the invention of rag paper)—that is, a pair of wooden tablets on which the contract was inscribed. Each of the contracting parties received such a tablet, and notches were cut in the side of each tablet in identical places so that the two documents could be matched and easily verified.

---

[1] In Japan it is said the imprint of the left thumb (*bo-in* or *bo-han*) was formerly taken exclusively from criminals (H. Spörry, Das Stempelwesen in Japan, Zürich, 1901, p. 16; comp. *Globus*, vol. 81, 1902, p. 187). But from the way the matter is represented by this author it does not clearly follow that the desire of identification was the purpose of this method. A criminal, when placed in jail, was stripped of his clothing and money, and his thumb impression was taken, whereby he was deprived of his civil rights. During this term he was allowed to sign documents only with his thumb, also the record of his trial. A verdict, even a capital sentence, had formerly to be signed by the defendant with the thumb print. These cases indicate that the thumb print was looked upon in Japan as an inferior sort of signature; the criminal had lost his personality and name, and was therefore not allowed to use it as his signature. His thumb print, which took the place of it, was not intended to establish his identity.

[2] *P'ei wên yün fu*, Ch. 20 B, p. 50.

In explaining this ancient practice to his countrymen, Kia Kung-yen remarks: "The significance of these notches is the same as that of the finger prints (*hua chi*) of the present time." This comparison sufficiently shows that finger prints were utilized in the age of the T'ang dynasty (618–906), and not only this, but also that it was their purpose to establish the identity of a person. In the same manner, the author means to say, as the notches of the tallies served for the verification of a contract concluded between two persons, so the finger prints on two written contracts of the same tenor had the function of proving the identity of the contractors.[1]

The existence of the finger print system in the T'ang period (618–906) is confirmed by the contemporaneous account of the Arabic merchant Soleiman who made several voyages to India and China, and left an interesting series of notes on both countries written in 851 A. D. It has been translated by M. Reinaud (Relation des voyages faits par les Arabes et les Persans dans l'Inde et à la Chine, Paris, 1845) where it is said (Vol. I, p. 42): "The Chinese respect justice in their transactions and in judicial proceedings. When anybody lends a sum of money to another, he writes a bill to this effect. The debtor, on his part, drafts a bill and marks it with two of his fingers united, the middle finger and the index. The two bills are joined together and folded, some characters being written on the spot separating them; then, they are unfolded and the lender receives the bill by which the borrower acknowledges his debt." This bill was legally recognized and served to the creditor in the court as an instrument proving the validity of the debt. It will be recognized that the process described by our Arabic informant in the ninth century is identical with the modern system of bank drafts, as outlined above, except that the finger prints of the debtor were affixed to the document in the T'ang period.

In regard to the prevalence of the finger-print system in China during the T'ang period, K. Minakawa has furnished a valuable piece of information. Chūryō Katsurakawa, the Japanese antiquary (1754–1808), writes on the subject as follows:

According to the "Domestic Law" (Korei), to divorce the wife the husband must give her a document stating which of the seven reasons for divorce was assigned for the action. * * * All letters must be in the husband's handwriting, but in case he does not understand how to write he should sign with a finger print. An ancient commentary on this passage is: "In case a husband can not write, let him hire another man to write the document * * * and after the husband's name sign with his own index finger." Perhaps this is the first mention in Japanese literature of the finger-print method.

---

[1] Compare E. Chavannes, Les livres chinois avant l'invention du papier, p. 56. (Reprint from *Journal asiatique*, Paris, 1905.)

This "Domestic Law" forms a part of the "Laws of Taihō," enacted in 702 A. D. With some exceptions, the main points of these laws were borrowed from the Chinese "Laws of Yung-hui" (650–655 A. D.); so it appears, in the judgment of Minakata, that the Chinese of the seventh century had already acquired the finger-print method.

It is very likely that the Chinese code of the T'ang dynasty and the abundant Chinese law literature will yield more information on this question.

Some writers have supposed on merely speculative grounds a connection between finger prints and palmistry. Galton [1] remarks on this point:

The European practitioners of palmistry and cheiromancy do not seem to have paid particular attention to the ridges with which we are concerned. A correspondent of the American journal, Science, volume 8, page 166, states, however, that the Chinese class the striæ at the ends of the fingers into "pots" when arranged in a coil and into "hooks." They are also regarded by the cheiromantists in Japan.

K. Minakata (l. c., p. 200) makes the following statement:

That the Chinese have paid minute attention to the finger furrows is well attested by the classified illustrations given of them in the household Tá-tsáh-tsú—the "Great Miscellany" of magic and divination—with the end of foretelling the predestined and hence unchanging fortunes; and as the art of chiromancy is alluded to in a political essay written in the third century B. C. (Han-fei-tse, XVII), we have reason to suppose that the Chinese in such early times had already conceived, if not perceived, the "forever unchanging" furrows on the finger tips.

But close research of this subject does not bear out this alleged fact. The fact is that in the Chinese system of palmistry the lines on the bulbs of the fingers are not at all considered, and that Chinese palmistry is not based on any anatomical considerations of the hand but is merely a projection of astrological notions. We have an excellent investigation of this tedious and wearisome subject by G. Dumoutier,[2] further by Stewart Culin,[3] by H. Doré,[4] and finally by H. A. Giles.[5] Not one of these four authors makes any mention of the striæ on the finger tips, and I am myself unable to find anything to this effect in Chinese books on the subject. It is quite evident to me that Chinese finger prints do not trace their origin from the field of palmistry but are associated, as will be shown farther on, with another range of religious ideas. I do not doubt the antiquity of palmistry in China, though the date B. C. 3000, given in the last edition of the Encyclopædia Britannica on the authority of Giles, seems to be an exaggeration, but the conclusion of Minakata that for this reason the finger prints are equally old is unjus-

---

[1] Finger Prints, p. 26.

[2] Études d'ethnographie religieuse annamite in Actes du onzième congrès des orientalistes, Paris, 1898, pp. 313 et seq. The Annamite system there expounded is derived from the Chinese.

[3] Palmistry in China and Japan (Overland Monthly, 1894, pp. 476–480).

[4] Recherches sur les superstitions en Chine, Vol. II, Shanghai, 1912, pp. 223 et seq.

[5] Phrenology, Physiognomy, and Palmistry (Adversaria Sinica, Shanghai, 1908, pp. 178–184).

tified. We must remember, also, that no system of palmistry has been handed down to us from ancient times; we merely know the fact that the practice itself existed at an early date. The philosopher Wang Ch'ung, who wrote in 82 or 83 A. D., states in regard to palmisters that they examine the left palm, but neglect the right one, because the lines of the former are decisive, whereas diviners turn to the right side and neglect the left one, because the former are conclusive.[1]

The view of the independence of finger prints from palmistry is by no means contradicted by the following statement of A. H. Smith:[2]

The Chinese, like the gypsies and many other peoples, tell fortunes by the lines upon the inside of the fingers. The circular striæ upon the finger tips are called *tou*, a peck, while those which are curved, without forming a circle are styled *ki*, being supposed to resemble a dustpan. Hence the following saying: "One peck, poor; two pecks, rich; three pecks, four pecks, open a pawnshop; five pecks, be a go-between; six pecks, be a thief; seven pecks, meet calamities; eight pecks, eat chaff; nine pecks and one dustpan, no work to do—eat till you are old."

This is neither fortune telling nor palmistry, but harmless jocular play which merely goes to prove that the striæ on the finger bulbs are noticed by the people and made the object of slight reflections. The above saying belongs to a well-known category of folklore which may be described under the title "counting out."

We alluded above to the hand stamp and its fundamental difference from finger prints in that it is unsuitable for identification. Let us now enter more particularly into this subject.

W. G. Aston[3] has given three examples of the use of the hand stamp in the East. In the Chinese novel *Shui hu chuan* of the thirteenth century a writing of divorce is authenticated by the husband stamping on it the impress of his hand smeared with ink.[4] In Japan, deeds, notes of hand, certificates, and other documents to

---

[1] A Forke, Lun-hêng, Part II (Berlin, 1911), p. 275.—Many ideas of Chinese palmistry are directly borrowed from India. Prominent among these is the exaltation of long arms reaching down to the knees, which appears among the beauty marks of the Buddha and is in fact an ancient Aryan conception of the ruler (A. Grünwedel. Buddhist Art in India, p. 162; Laufer, Dokumente, pp. 166, 167). With the Indians as with the Persians, this is an old mark of noble birth (compare the name *Longimanus*, old Persian *Darghabāzu*, Sanskrit *Dīrghabahu*). In China we meet the notion that a man whose hand reaches below his knees will be among the bravest and worthiest of his generation, but one whose hand does not reach below his waist will ever be poor and lowly (Giles, l. c., p. 181). In regard to Liu Pei (162-223 A. D.), it is on record that his ears reached to his shoulders and his hands to his knees (Giles, Biographical Dictionary, p. 516).

[2] Proverbs and Common Sayings from the Chinese, Shanghai, 1902, p. 314.

[3] *Folk-lore*, vol. 17, 1907, p. 113.

[4] K. Minakata (l. c., p. 199) tries to make out finger prints occurring in this work, which seems to me an unwarranted statement. It is there plainly the question of hand impressions only. He states:

"In the novel *Shui hu chuan* examples are given of the use of finger prints, not only in divorce, but also in criminal cases. Thus the chapter narrating Lin Chung's divorce of his wife has this passage: 'Lin Chung, after his amanuensis had copied what he dictated, marked his sign character, and stamped his hand pattern.' And in another place, giving details of Wu Sung's capture of the two women, the murderers of his brother, we read: 'He called forth the two women; compelled them both to ink and stamp their fingers; then called forth the neighbors; made them write down the names and stamp (with fingers).'"

be used as proofs were formerly sealed in this way, a practice to which the word *tegata* (hand shape) still used of such papers remains to testify. Documents are in existence in which Mikados have authenticated their signatures by an impression of their hand in red ink.[1]

In the religious beliefs of the Tibetans impressions from the hands and feet of saints play an extensive rôle. These notions were apparently derived with Buddhism from India. In the Himalayan region of southern Tibet the pious believers are still shown foot imprints left by the famous mystic, ascetic, and poet, Milaraspa (1038–1122), and in an attractive book containing his legends and songs many accounts of this kind are given. By "Traces of the Snowshoes" is still designated a bowlder on which he performed a dance and left the traces of his feet and staff, and the fairies attending on the solitary recluse marked the rocks with their footprints.[2] In the life of the Lama Byams-c'en C'os-rje (1353–1434), who visited China at the invitation of the Emperor Yung-lo (1398–1419), of the Ming dynasty, it is narrated that when he was dwelling on the sacred Mount Wu-t'ai, in Shansi Province, he showed a miracle by kneading a solid, hard blue stone like soft clay and leaving on it an impression of his hand, which astounded all inhabitants of that region.[3] The fourth Dalai Lama, Yon-tan rgya-mts'o (1588–1615) produced on a stone the outlines of his foot.[4] In Tibet I myself had occasion to see, in the possession of a layman, an impression on silk of the hand of the Pan-c'en rin-po-c'e, the hierarch residing at Tashi-lhun-po. At least it was so ascribed to him; but the hand was almost twice as large as an ordinary human hand, and the vermilion color with which it was printed from a wooden block lent it a ghastly appearance. These talismans are sold to the faithful at goodly prices and secure for them the permanent blessings of the sacred hand of the pontifex.

---

[1] Red ink, as in many Chinese religious ceremonies, evidently is here a metaphorical substitute for blood, and the act of the Mikado retains its purely magical character. Il. Spörry (Das Stempelwesen in Japan, p. 18) remarks that the *tegata* is found on ancient documents usually in red, but also in black; it seems that they were chiefly employed on instruments of donations to temples, without having properly the sense and character of a signature. Sheets of white or red paper with the imprint of the left hand of the husband and the right hand of the wife are pasted over the doors of houses as charms against smallpox and other infectious diseases. Giles (Adversaria Sinica, No. 6, p. 184) narrates the following story: "A favorite concubine of the Emperor Ming Huang (713–756 A. D.) having several times dreamed that she was invited by some man to take wine with him on the sly, spoke about it to the Emperor. 'This is the work of some magician,' said his Majesty; 'next time you go, take care to leave behind you some record.' That very night she had the same dream; and accordingly she seized the opportunity of putting her hand on an ink slab and then pressing it on a screen. When she awaked, she described what had happened; and on a secret examination being made, the imprint of her hand was actually found in the Dawn-in-the-East Pavilion outside the palace. The magician, however, was nowhere to be seen." In regard to the same woman, Yang Kuei-fei, another anecdote is told to the effect that she once touched the petals of peonies with her fingers dipped into rouge, whereupon the coming year, after the flowers had been transplanted, red traces of her finger prints were visible on the opening blossoms (compare *P'ei wên yün fu*, ch 71, p. 19).

[2] Laufer, Aus den Geschichten und Liedern des Milaraspa, pp. 2, 16 (*Denkschriften Wiener Akademie*, 1902), and *Archiv für Religionswissenschaft*, vol. 4, 1901, pp. 26, 42.

[3] G. Huth, Geschichte des Buddhismus in der Mongolei, Vol. II, Strassburg, 1896, p. 196.

[4] Ibid., p. 245.

The use of the hand in sealing documents is by no means restricted to China and Japan. It occurred as well in western Asia. Malcolm,[1] in describing the conquests of Timur, states:

> The officers of the conqueror's army were appointed to the charge of the different provinces and cities which had been subdued, and on their commissions, instead of a seal, an impression of a red hand was stamped; a Tartar usage, that marked the manner in which the territories had been taken, as well as that in which it was intended they should be governed.

The symbolism of the hand is here clearly set forth; it was a political emblem of conquest and subjugation. W. Simpson[2] received, at Constantinople, the information that in early times when the Sultan had to ratify a treaty a sheep was killed, whereupon he put his hand into the blood, and pressed it on the document as his "hand and seal."[3]

The foot impressions attributed in India by popular belief to Buddha or Vishnu are well known. They occur, likewise, on the megalithic tombs of western Europe and in petroglyphs of upper Egypt. Likewise the numerous representations of hands in the European paleolithic caves[4] should be called to mind. Only vague guesses can be made as to their original meaning,[5] and they may be altogether different from the hand stamps much later in point of time. But these various examples of the occurrence of hand representations in different parts of the world should admonish us to exercise precaution in framing hasty conclusions as to hand impressions leading to finger prints in China. The former occur outside of China where no finger prints are in use, and do not pretend to serve for identification, nor can they answer this purpose. They have a purely religious significance; they may symbolize political power or subjugation or become the emblem of a cult.

K. Minakata, at the suggestion of his friend Teitarō Nakamura, believed that possibly the "finger stamp" was merely a simplified form of the "hand stamp." This view, in his opinion applies equally well to the case of the Chinese, for they still use the name "hand pattern" for the finger print.[6] This theory is untenable, if for no other reason than that the thumb print, as will be shown by actual archeological evidence, is very much older in China than the hand impression. The two Japanese, however, may have had a correct feeling in this matter which they were unable to express in words,

---

[1] History of Persia, Vol. I, p. 465.

[2] *Journal Royal Asiatic Society*, vol. 21, p. 309.

[3] The design of a hand is found also on Persian engraved gems of the Sassanian period (226–641 A. D.). See E. Thomas, Sassanian Mint Monograms and Gems (*Journal Royal Asiatic Society*, vol. 13, 1852, Pl. 3, No. 61).

[4] F. Regnault, Empreintes de mains humaines dans la grotte de Gargas (*Bulletins et Mémoires Société d'Anthropologie de Paris*, 1906, Vol. 1, pp. 331–2).

[5] Comp. e. g. G. Wilke, Südwesteuropäische Megalithkultur, Würzburg, 1912, p. 148.

[6] The latter statement is incorrect: the Chinese expression "hand pattern" only means what it implies, an impression taken from the palm, but never a finger print.

or to prove properly. If the finger print has not been evolved from the hand print, nor the latter from the former, there is a certain degree of inward relationship between the two; both are coexisting phenomena resting on a common psychological basis. In order to penetrate into the beginnings and original significance of finger prints, it is necessary to consider another subject, that of seals.

The antiquarian history of Chinese seals begins with the famous seal of the first Emperor Ts'in Shi (B. C. 246–210). This was carved from white jade obtained at Lan-t'ien in Shensi Province and is said to have contained the inscription "Having received the mandate of Heaven, I am in possession of longevity and eternal prosperity." It was, accordingly, the emblem of sovereignty conferred by Heaven on the Emperor.[1] The word *si*,[2] which had up to that date served for the designation of any seal, was henceforth reserved as the exclusive name of the imperial seal; in other words, a taboo was placed on it. Furthermore the character used in writing this word underwent a change: the symbol for "jade" (*yü*) entering into its composition, together with a phonetic element, was substituted for the previous symbol "earth" (*t'u*). The latter word denotes also clay, so that we are allowed to infer that prior to the time of the jade seal of Ts'in Shi seals were ordinarily made of clay.

The common name for these clay seals is *fêng ni*,[3] and they were utilized especially in sealing documents which were written at that time on slips of bamboo or wood. After the age of Emperor Wu (B. C. 156–87) of the Former Han dynasty they fell into disuse, but during his reign they were still employed, as attested by the biographies of the Gens. Chang K'ien and Su Wu. A. Stein[4] has discovered a large number of such tablets with clay seals attached to them in the ruins of Turkistan. A number of ancient clay seals having been discovered also on Chinese soil, particularly in the provinces of Shensi and Honan, they could not escape the attention of the native archeologists. One of these, Liu T'ie-yün, published at Shanghai in 1904 a small work in four volumes under the title *T'ie-yün ts'ang t'ao*, "Clay Pieces from the Collection of T'ie-yün." These volumes contain facsimiles of a number of clay seals as anciently employed for sealing official letters and packages.[5] The subject, however, is not investigated, and no identifications of the characters of old script with their modern forms are given. Their decipherment is difficult and remains a task for the future. A few such

---

[1] E. Chavannes (Les mémoires historiques de Se-ma Ts'ien, vol. 2, pp. 108–110) has recorded the various destinies of this now lost seal, according to the Chinese tradition, down to the T'ang dynasty.

[2] Giles, Chinese-English Dictionary, Nos. 4143, 4144.

[3] Literary references to them are scarce; some notes regarding them are gathered in the cyclopedia *Yen kien lei han*, ch. 205, p. 36.

[4] Ancient Khotan, Oxford, 1907, Vol. 1, p. 318.

[5] Laufer, Chinese Pottery, p. 287, and Chavannes in *Journal asiatique*, vol. 17, No. 1, 1911, p. 128.

clay seals secured by me at Si-ngan fu are likely to furnish an important contribution to the early history of the finger-print system.

The seal was considered in ancient China as a magical object suitable to combat or to dispel evil spirits, and the figures of tigers, tortoises, and monsters by which the metal seals were surmounted had the function of acting as charms. We read in Pao-p'o-tse [1] that in olden times people traveling in mountainous regions carried in their girdles a white seal 4 inches wide, covered with the design of the Yellow Spirit and 120 characters. This seal was impressed into clay at the place where they stopped for the night, each of the party made 100 steps into the four directions of the compass, with the effect that tigers and wolves did not dare approach. Jade boxes, and even the doors of the palaces, were sealed by means of clay seals to shut out the influence of devils. Numerous are the stories regarding Buddhist and Taoist priests performing miracles with the assistance of a magical seal.

On plate 4 six such clay seals from the collection of the Field Museum are illustrated. The most interesting of these is that shown in figure 2, consisting of a hard, gray baked clay, and displaying a thumb impression with the ridges in firm, clear, and perfect outlines, its greatest length and width being 2.5 cm. It is out of the question that this imprint is due to a mere accident caused by the handling of the clay piece, for in that case we should see only faint and imperfect traces of the finger marks, quite insufficient for the purpose of identification. This impression, however, is deep and sunk into the surface of the clay seal and beyond any doubt was effected with intentional energy and determination. Besides this technical proof there is the inward evidence of the presence of a seal bearing the name of the owner in an archaic form of characters on the opposite side. This seal, 1 cm. wide and 1.2 cm. long, countersunk 4 mm. below the surface, is exactly opposite the thumb mark, a fact clearly pointing to the intimate affiliation between the two. In reasoning the case out logically, there is no other significance possible than that the thumb print belongs to the owner of the seal who has his name on the obverse and his identification mark on the reverse, the latter evidently serving for the purpose of establishing the identity of the seal. This case, therefore, is somewhat analogous to the modern practice of affixing on title deeds the thumb print to the signature, the one being verified by the other. This unique specimen is the oldest document so far on record relating to the history of the finger-print system. I do not wish to enter here into a discussion of the exact period from which it comes down, whether the Chou

---

[1] Surname of the celebrated Taoist writer Ko Hung who died around 330 A. D., at the age of 81.

period or the Former Han dynasty is involved; this question is irrelevant; at all events it may be stated confidently that this object, like other clay seals, was made in the pre-Christian era. An examination of other pieces may reveal some of the religious ideas underlying the application of the thumb print. Many clay seals are freely fashioned by means of the finger and exhibit strange relations to these organs. The finger shape of the two seals in figures 6 and 7 on plate 4 is obvious. Our illustration shows the lower uninscribed sides, while the name is impressed by means of a wooden mold on the upper side. Examination of these two pieces brings out the fact that they were shaped from the upper portion of the small finger, and further from the back of the finger. The lower rounded portion of the object in figure 7 is evidently the nail of the small finger which was pressed against the wet clay lump; the seal has just the length of the first finger joint (2.6 cm. long), the clay mold follows the round shape of the finger, and the edges coiled up after baking. The lines of the skin, to become visible, were somewhat grossly enlarged in the impression. The clay seal in figure 6 (2.4 cm. long), I believe, was fashioned over the middle joint of the small finger of a male adult, the two joints at the upper and lower end of the seal being flattened out a little by pressure on the clay, and the lines of the epidermis being artificially inserted between them. The seal in figure 5, of red-burnt clay, with four characters on the opposite side (not illustrated), was likewise modeled from the bulb of the thumb by pressure of the left side against a lump of clay which has partially remained as a ridge adhering to the surface. The latter was smoothed by means of a flat stick so that no finger marks could survive. The groove in the lower part is accidental. Another square clay seal in our collection (No. 117032) has likewise a smoothed lower face, but a sharp mark from the thumb nail in it. These various processes suffice to show that the primary and essential point in these clay seals was a certain sympathetic relation to the fingers of the owner of the seal. Here we must call to mind that the seal in its origin was the outcome of magical ideas, and that, according to Chinese notions, it is the pledge for a person's good faith; indeed, the word *yin*, "seal," is explained by the word *sin*, "faith." [1] The man attesting a document sacrificed figuratively part of his body under his oath that the statements made by him were true, or that the promise of a certain obligation would be kept. The seal assumed the shape of a bodily member; indeed, it was immediately copied from it and imbued with the flesh and blood of the owner. It was under the sway of these notions of magic that the mysterious, unchangeable furrows on the finger bulbs came into prominence and received their importance. They not

---

[1] In the work *T'ie yün ts'ang t'ao*, p. 85b, above quoted, is illustrated a clay seal containing only this one character. The same book contains also a number of finger-shaped clay seals.

ANCIENT CHINESE CLAY SEALS. THE ONE IN FIG. 2 SHOWING THUMB-IMPRESSION ON THE REVERSE. THOSE IN FIGS. 6 AND 7 BEING MOLDED FROM THE SMALL FINGER.

INK-SKETCH BY KAO K'I-P'EI, EXECUTED BY MEANS OF THE FINGER-TIPS.

INK-SKETCH BY KAO K'I-P'EI, EXECUTED BY MEANS OF THE FINGER-TIPS.

INK-SKETCH BY YO YU-SUN, EXECUTED BY MEANS OF THE FINGER-TIPS.

only contributed to identify an individual unmistakably but also presented a tangible essence of the individuality and lent a spiritual or magical force to the written word.

Finally, I should like, in this connection to call attention to a peculiar method of painting practiced by the artists of China, in which the brush is altogether discarded and only the tips of the fingers are employed in applying the ink to the paper. This specialty is widely known in China under the name *chi hua*, which literally means "finger painting," and still evokes the highest admiration on the part of the Chinese public, being judged as far superior to brush painting. The first artist to have cultivated this peculiar style, according to Chinese traditions, was Chang Tsao, in the eighth or ninth century, of whom it is said that "he used a bald brush, or would smear color on the silk with his hand." [1] Under the Manchu dynasty, Kao K'i-pei, who lived at the end of the seventeenth and in the first part of the eighteenth century, was the best representative of this art. "His finger paintings were so cleverly done that they could scarcely be distinguished from work done with the brush; they were highly appreciated by his contemporaries," says Hirth. On plates 4 and 5 two ink sketches by this artist in the collection of the Field Museum are reproduced. Both are expressly stated in the accompanying legends written by the painter's own hand to have been executed with his fingers. The one representing two hawks fluttering around a tree trunk is dated 1685; the other presents the reminiscence of an instantaneous observation, a sort of flashlight picture of a huge sea fish stretching its head out of the waves for a few seconds and spurting forth a stream of water from its jaws. The large monochrome drawing shown on plate 6—cranes in a lotus pond by Yo Yu-sun—is likewise attested as being a finger sketch (*chi mo*), and the painter seems to prove that he really has his art at his fingers' ends. Hirth is inclined to regard this technique "rather a special sport than a serious branch of the art," and practiced "as a specialty or for occasional amusement." There was a time when I felt tempted to accept this view, and to look upon finger painting as an eccentric whim of the virtuosos of a decadent art who for lack of inner resources endeavored to burn incense to their personal vanity. But if Chang Tsao really was the father of this art, at a time when painting was at the culminating point of artistic development, such an argument can not be upheld. I am now rather disposed to believe that the origin of finger painting seems to be somehow linked with the practice of finger prints, and may have received its impetus from the latter. The relationship of the two terms is somewhat significant; *hua chi*, "to paint the finger," as we saw above

---

[1] H. A. Giles, An Introduction to the History of Chinese Pictorial Art, p. 61. F. Hirth, Scraps from a Collector's Note Book, p. 30.

in the passage quoted from Kia Kung-yen, is the phrase for "making a finger impression" in the T'ang period, and the same words reversed in their position, *chi hua*, mean "finger painting" or "painted with the finger." It seems to me that also in finger painting the idea of magic was prevalent at the outset, and that the artist, by the immediate bodily touch with the paper or silk, was enabled to instill part of his soul into his work. Eventually we might even go a step farther and make bold to say that finger painting, in general, is a most ancient and primitive method of drawing and painting, one practiced long before the invention in the third century B. C. of the writing brush of animal hair, and the older wooden stylus. The hand, with its versatile organs of fingers, was the earliest implement utilized by man, and the later artistic finger painting might well be explained as the inheritance of a primeval age revived under suggestions and impressions received from the finger-print system.

097

纪念 J.P. 摩根

OSTASIATISCHE ZEITSCHRIFT

THE FAR EAST · L'EXTRÊME ORIENT

BEITRÄGE ZUR KENNTNIS DER KULTUR UND KUNST DES FERNEN OSTENS

HERAUSGEGEBEN

VON

OTTO KÜMMEL UND WILLIAM COHN

1913/1914
ZWEITER JAHRGANG

OESTERHELD & CO. · VERLAG · BERLIN W. 15

AUS MUSEEN UND SAMMLUNGEN.

IN MEMORIAM J. PIERPONT MORGAN.

WITH A NOTICE OF CATALOGUE OF THE MORGAN COLLECTION OF CHINESE PORCELAINS[1].

J. Pierpont Morgan died in Rome on March 31, 1913, at the age of 76. In his quality as a patron of Chinese art, as the greatest collector of Chinese porcelain, he merits that a record of his aspirations in this line should be entered in a journal devoted to the art of the East. Mr. Morgan's restless activity as a collector has evoked world-wide comment, and his rather accumulative than discriminative tendencies were sometimes apt to expose him to criticism. In 1900 while serving the Metropolitan Museum of New York in the capacity of trustee, an honorary position which he had occupied since 1888 to be elected president in 1904, he purchased the Garland collection of Chinese porcelains for the Museum which had long sheltered it as a loan. From that time onward he steadily added to this collection which is now a part of the great treasures of that institution.

In 1904 Mr. Morgan published the first volume descriptive of his marvelons collection of ancient Chinese porcelains and containing 77 colored plates. Since the issue of this volume numerous important additions have been made to the collection, so that a continuation of the work seemed desirable, and a worthy successor to the former publication has now been presented in a no less magnificent and convenient volume embracing 81 colored plates of masterly execution. 1521 specimens are listed in the new catalogue, making with the 1115 of the previous volume a total of 2636. Though bearing the imprint 1911, the book was only distributed in the beginning of this year, much time having apparently been consumed in the completion of the sumptuous bindings.

Mr. Morgan has put all students of Chinese art under lasting obligation by placing them, so to speak, in joint partnership with the possession of his art-treasures, for the reproductions of these porcelains are so faithful in color and design including even the finest crackles and veins of the glaze, that they are real substitutes of the originals. Such a work which the student could take into his hands as a sound basis for his investigations has never before existed, and this unique enterprise denotes a new era in that it signals for the future the transition of this branch of knowledge from the point of view of a collector to that of a serious science. While our data of the history of porcelain are fairly sufficient, thoug many obscure points remain to be cleared up, it has been overlooked that porcelain can be studied from many other different points of view and may be subjected to tests which will eventually turn out to be fruitful to the work of science. To the field of esthetic psychology, e. g., it will offer a live and interesting source of inquiry. Only a few of the initiated seem heretofore to have understood that a Chinese ceramic product is more than the outcome of a material process, that it has also an inward significance, that it is suggestive of some emotional thought inspired by its form, its glaze, and the decorative motives displayed thereon. It elicits poetry, and poetry has been composed in its honor; it may be a trinket given to a dear friend, and the amicable relations of two persons may be symbolically expressed by the character of the ornaments. Moreover, each piece of pottery has an organic function in the surroundings where it is displayed; it forms an artistic unit with a group of other art-objects, it has a distinct relation to the milieu, to the room, the veranda, or the adjoining garden, it is linked in harmonious bonds with the colors of the sky, the

---

[1] Volume two in two parts. Privately printed by order of Mr. J. Pierpont Morgan. New York, 1911 (distributed January, 1913).

moon, or the surrounding vegetation, — a subject full of charm and grace wholly neglected by foreign students. The crude and shallow barbarism of our civilization could be no better illustrated than by the fact that we have foolishly deprived ourselves of the opportunity of grasping the Chinese soul in their ceramic masterpieces; no collector hunting in China for the choicest spoils has ever taken the trouble to inquire into their meaning, their artistic and social function. These treasure-seekers are content to give the name of a Chinese period which to them is no more than a name without substance, and may supply some collector's jargon phraseology in regard to the glaze, but the most precious information which could be obtained becomes for ever lost. What, after all, is the quality of the wisdom that a certain piece is determined to have been made, not in the K'ang-hi but in the K'ien-lung period? If we were in possession of a genealogical list of its former owners giving us their impressions and inward experiences relative to this work, we would enjoy a fine piece of psychological documentary evidence. I should distribute a fortune, if I had one, could I obtain this information on the vase figured on Plate CXVII of the present volume. Whoever can read Chinese thoughts, will at once recognize the poetic conception underlying this work of art. The ornamental waves laid around the foot, and an imposing dragon emerging from the depths symbolize the power of the sea, the sea-green celadon glaze being suggestive of the ocean, and in the bold outlines of its forceful shape the whole vessel stands with a feeling of eternity like the ocean itself. Thanks to such poetic inspirations, the Chinese have made of porcelain an art, and we can only hope to understand porcelain in its artistic merits by endeavoring to analyze the thoughts of the people which are manifest in these productions. Here is also the boundary barring for ever our own achievements in this field from those of the Chinese. We could imitate porcelain in its chemical composition and in its technical qualities, we quickly perceived its immense utility, and easily acquired a taste for it, but we could never reproduce the fine spirit of the Chinese masters; our attempts at utilizing the cold substance in the cause of figure-art are childisch absurdities when compared with their ideal creations. Only they were able to breathe life into the dead matter, and to imbue it with a warmth of feeling which coupled with their sense of color and form-beauty marks a triumph of spirit over matter.

Another significant problem would be an inquiry into the esthetic effects of porcelain, first as far as the Chinese are concerned, and secondly the outside world. The latter may be of still greater value to the psychologist, for while it is plausible that an art based on a national industry will meet with a ready response at home, it remains a singular fact that Chinese porcelain has made such a strong appeal to the world at large, that it has deeply impressed the peoples of Korea, Japan, India and Persia among whom ceramic art had already reached a high state of perfection, and that its triumphal procession has finally enthralled Europe and America, with such hypnotizing effects that even our large domestic output of the ware has never checked (and presumably never will do so) importation from the East. The supply can hardly keep pace with the ever-growing demand for ancient pieces, and the upward tendency of prices seems rather to enhance than to cool off the enthusiasm of collectors. The greatest record was achieved on July 9, 1912, in London when at the auction of the collection of John Edward Taylor a small quadrangular bottle, 48 cm high, of the K'ang-hi epoch was purchased by Duveen for the sum of £ 7,425, the highest price ever paid for a single piece of porcelain; in 1895, the same object had been acquired by its last owner for the small consideration of £ 336, which thus reached the twenty-twofold within seventeen years. Facts like this are symptoms of a deep appreciation and worthy of scientific speculation. The greater an art, the wider, as a rule, is the circle of its admirers; the more it is limited in its scope, the more restricted the community of adherents. There are poets and artists of importance only to the nation which has produced them; others are merely local or clannish heads. Is the universal reputation of Chinese craftsmanship in porcelain an index of its superior qualities, and if so, by what characteristics are they conditioned? What are the underlying causes of this world-wide admiration and the lasting effect of this art? The psychologist who would answer this problem will find the safest guide in Mr. Morgan's admirable publication where he may delve to his heart's delight among forms, color schemes and designs, and experiment on himself and others in subjecting their effects to a test of esthetic analysis.

There are veritable gems illustrated in this catalogue. There is the famous series of nine peach-bloom boxes (four on Plate CXXIII) the glazes of which display all stages in the gradual ripening of a peach from the initial green intermingled with slightly red flecks to the richest and deepest hues, — formerly in the possession of Marsden J. Perry of Providence. From his collection originates also the magnificent piece of Han pottery (Plate LXXIX) formerly described by the present writer; it is the first specimen of this pottery reproduced in colors, and all admirers of the ancient art of China will gratefully feast their eyes on these delicate shades of green, brown and iridescent gold. The ruby-glazed (alias *sang-de-boeuf*) pieces (Plates CXI, CXXXVII, CXLII, and CXLVII which is the frontispiece) belong to the finest of their class and are *hors de concours*. The specimen from the imperial collection in Plate CXXXI is a perfect wonder in shape and glaze ("moon-light"); the rose-colored vase in Plate CXXX is unique and of great scientific interest owing to the application of a silk textile pattern which is spread with great delicacy over the surface of the glaze.

"Mr. Morgan was a candid man", recently said an essayist in the *Nation* characterizing him as an art patron, "and it would be to affront his memory to claim for him a connoisseurship to which he never pretended when living. It was his weakness as a collector that he did not readily seek or win the confidence of critics and other amateurs, but depended too much on dealers". Criticizing pieces of the Morgan collection, therefore, does not signify a criticism of Mr. Morgan or his taste. There are also inferior pieces figured in this catalogue which do not come up to the high standard of the others. To these belong the two objects on the two last plates which are classified as K'ien-lung and K'ang-hi, respectively, yet are plainly modern; and whatever their date may be, they are ugly and should have no place in this collection. The decoration of the former piece is the childish production of a bungler who was not able to draw a straight line; as to the latter piece — no K'ang-hi potter would have been able to produce such cold, harsh and glaring colors, nor to conceive such inane and tasteless designs. Such specimens may be purchased for a few Mexicans in any new-porcelain shop of Shanghai, and they are not worth the cost of packing and transportation.

As admirable as are the plates of this volume, so is the text disappointing. 73 pages are taken up with a mere reprint of Dr. Bushell's popular article on porcelain formerly issued in the catalogue published by the Metropolitan Museum. The descriptions of the specimens move in the same enlightening manner to which sale and auction catalogues have accustomed us, and there is hardly a gleam of intelligence to be discovered. Nor is there any sensible division of the material nor any principle observed in the arrangement of the plates. Dr. Bushell, Mr. Laffan, the late editor of the *Sun*, and Mr. Th. B. Clarke have all contributed to the catalogue. It is a strange phenomenon that those men who have spent a lifetime in collecting and studying porcelain utterly lack the ability of presenting the subject intelligently and forcibly in a coherent essay. It is not catalogues that we are in need of, but intelligent and sympathetic interpretation evolved from serious research of Chinese thoughts. We are tired of being treated on every page to the same phraseology; it is not enough to say that a piece is made in such and such a shape, of such and such a glaze, and in such and such a period. This is all mere technical by-play. Above all we want to know what it means, for what idea it stands, what it symbolizes, what the Chinese say and think about it, and what function in their social and religious life it fulfills. In order to accomplish this task it is certainly indispensable to have a thorough knowledge of Chinese culture. To really understand Chinese porcelain, it is not sufficient to know porcelain alone, and nothing besides. But only such appreciative understanding will be fertile and give us something worth while. The "catalogues" of porcelain of the sanctioned style will eternally remain dry and dead matter.

In the descriptive text the date-mark Ch'êng-hua designating the period 1465—1487 of the Ming dynasty is frequently referred to as "fictitious". This attribute does not do justice to the facts. Mr. Laffan, in the introductory notes to the first volume had stated:

"Thousands upon thousands of pieces of this date-mark survive, but we have never seen a piece of porcelain bearing the Ch'êng-hua mark which was made in the reign of that monarch.

We have never seen a piece bearing it that was older than the beginning of the reign of K'ang-hi (1662—1722); but we have seen a vast number that were even more modern."

This is certainly very true, and it is probable that under the reign of Ch'êng-hua no porcelains bearing a date-mark were turned out. The Chinese themselves are well aware of this fact, but the Ming date-mark was added in the K'ang-hi and K'ien-lung periods not with the intention to deceive, but to bring out the idea that the piece in question was made in the style of the Ch'êng-hua period. ·It is therefore important for the history of decorative styles of porcelains to take due note of this interpretation. Chinese sentences, and especially those on seals and inscriptions, do not always mean what they seem to convey on the surface, and there is in many cases a hidden significance which only careful inquiry or regard for the special merits of the case may bring out. The Ch'êng-hua mark is therefore not truly fictitious but something real, and constitutes no greater crime than e. g. our California Port or our Swiss cheese made at Green Lake, Wisconsin.

If it is true that only the first word has been said in the matter of porcelain, and that we are still waiting for the magician who will revive the dead body and fill it with life and a new soul, we feel the more grateful to Mr. Morgan who has done so much to provide the student with authentic and choice material, and to enrich our esthetic enjoyment. It would be ungratefu to forget on this occasion that there are other collections of porcelain in this country which have their peculiar merits. Mr. Freer has specialized on the pottery of the Sung period and is easily without a rival in this department; while Mr. Carvalho owns the finest and most complete collection of blue and whites, not do mention many other collections.

Mr. Morgan was a man of impressive personality and comprehensive culture of mind for which he had doubtless laid the foundation in his student days at the University of Göttingen. His interests in subjects of art and science were wide and deep, and in his palatial library on Thirty-Sixth Street are hoarded untold treasures whetting the appetite of the orientalist. The first part of the Babylonian records in his possession has just been published by Mr. A. T. Clay. The present writer had the privilege of meeting Mr. Morgan twice in his sanctum. He was once summoned to make out for him a Tibetan manuscript in silver writing; on another occasion he procured for him a fine manuscript of the Tibetan Gesar Saga which had been especially written for the Pontifex of Tashilhunpo in the eighteenth century.   B. Laufer.

DARSTELLUNGEN VON EUROPÄERN IN EINEM CHINESISCHEN DORFTEMPEL.

Einige zwanzig Li innerhalb der Großen Mauer (nächster Paß ist das kleine Ma-tze-k'ou 馬子口) und noch unter den Türmen der doppelten Befestigungslinie längs des Si-yang-ho 西揚河 liegt an der Grenze der Provinzen Shansi und Chihli das Dörfchen Kuan-ti-miao 關帝廟. Der Tempel des Kriegsgottes, der ihm den Namen gab, wird von einem verfallenden Wachtturm und hohen alten Bäumen beschattet. Tritt man in den kleinen Hof hinein, so hat man das übliche Bild vor sich: rechts und links je eine kleine Halle — die Götterbilder sind daraus verschwunden, — geradeaus das Hauptgebäude mit der Lehmfigur Kuan-ti's. Dem Hauptgebäude schließt sich beiderseits je eine kleine Halle an. In der westlichen sitzt eine mehrarmige Gottheit, deren Namen niemand zu sagen weiß; die östliche beherbergt den Gott des Reichtums, den Ts'ai-shen 財神. Käschnachbildungen charakterisieren ihn hinreichend. Die Seitenwände dieses kleinen Raumes zeigen einen Bildschmuck, der unverkennbar buddhistischen Tempelbildern nachgebildet ist: links sehen wir den weißen Elefanten des Samantabhadra (P'u-hien p'u-sa), rechts den Löwen des Manjusri (Wen-shu p'u-sa). An Stelle der beiden Bodhisatvas aber stehen zwei Europäer.

Betrachten wir zunächst das linke Bild (Abb. 1). Den Mittelpunkt bildet ein weißer Elefant, reich geschirrt, mit hohem Sattel, von einem dienenden Knaben (t'ung-tze 童子) an einem Zaume geleitet. Ein zweiter Knabe, der hinter dem Sattel sitzt, hält auf diesem das „Schatzgefäß" (pao-pen 寶盆) fest, das ständig Edelsteine, Korallen und andere Kostbarkeiten hervorbringt. Ein dritter Knabe versucht, auf den Elefanten hinaufzuklettern, ein vierter treibt ihn mit langem Stabe an.

15